# INSTRUMENTATION FOR ENGINEERING MEASUREMENTS

**JAMES W. DALLY**

*IBM Corporation*

**WILLIAM F. RILEY**

*Iowa State University*

**KENNETH G. McCONNELL**

*Iowa State University*

**John Wiley & Sons, Inc.**

*New York* • *Chichester* • *Brisbane* • *Toronto* • *Singapore*

*Library of Congress Cataloging in Publication Data:*
Dally, James W.
    Instrumentation for engineering measurements.

    Includes bibliographical references and index.
    1. Engineering instruments.    I. Riley, William F.
(William Franklin), 1925–        II. McConnell,
Kenneth G.    III. Title.

TA165.D34      1983        681'.2        83-10452
ISBN 0-471-04548-9

Printed in the United States of America

10 9 8 7

# PREFACE

During the past decade, considerable progress was made in developing finite-element methods and other numerical techniques to predict the performance of vehicles, machines, and structures. The use of computer-aided design and computer-aided manufacturing (CAD/CAM) is growing at a very rapid rate. Many firms are reducing design and development times and increasing product quality and yield by using these modern technological methods that integrate the powers of the digital computer into the design and production processes. Engineering evaluations and tests are important components of the computer-aided design process. Carefully specified engineering tests should be performed during the design and development stages of all new products and processes.

Engineering evaluations and tests serve many purposes. Periodic sampling and testing of a product on a production line is an essential element of most quality control programs. Experiments are often used to verify mathematical or numerical models used in the design process. Prototypes are often tested thoroughly to establish their performance limits. Vehicles and systems are often subjected to the rigors of qualification testing to ensure their satisfactory performance over extended periods of time under very harsh environmental conditions. Continuous measurements of process variables, such as temperature, pressure, and flow rate, are used to optimize and/or control industrial processes. Evaluation, experimentation, and testing are essential to the continued development of sophisticated industrial products. The efficient and accurate measurement of quantities, such as voltage, current, strain, temperature, pressure, flow rate, and force, is of critical importance today and for the foreseeable future.

This textbook deals in considerable detail with many aspects of the instrumentation currently employed for engineering measurements and process control. The book was written for use in a first course in engineering experimentation or engineering measurements that may be part of an undergraduate program in agricultural, aerospace, chemical, civil, mechanical, or nuclear engineering. Such a course would normally follow an introductory course in electrical engineering; therefore, it is assumed that the student is familiar with the fundamentals of electricity and electronics.

The first four chapters of the book deal with instrumentation systems in general, experimental error, voltage measuring instruments, sensors for transducers, time, count, and frequency measurements, and signal conditioning circuits. This part of the text contains the basic core of material that is applicable to all types of measurements; therefore, it should be presented in as much detail as the course schedule permits. Experience indicates that the student's awareness of concepts, methods, and apparatus used in engineering measurements often lags his theoretical and numerical analysis capabilities.

Methods used to measure specific quantities are described in considerable detail in Chapters 5 through 9. The coverage includes strain, force, torque, pressure, displacement, velocity, acceleration, temperature, and flow rate. Limitations on the length of the text precluded coverage of thermal- and transport-property measurements, sound measurements, and nuclear radiation measurements. Statistical methods that are commonly employed in the analysis of experimental data are briefly discussed in Chapter 10.

Throughout the text, heavy emphasis is placed on electronic methods of measurement, since electronic systems provide accurate data that can be used for automatic data reduction or process control. Again because of limitations on the length of the book, this emphasis on electronic methods required omission of most mechanical instruments, such as dial gages, bourdon-tube pressure gages, manometers, and so on. We believe that most students can master the use of these simple devices without assistance.

The material presented in this text is sufficient for a two-semester or a two-quarter course. The first course could consist of a lecture/laboratory sequence covering Chapters 1 through 5, 8, and 10. The second course would then consist of a lecture/project sequence with material from Chapters 5 through 9 covered as required for the proper development of the projects. In spite of the very significant costs associated with a laboratory program in terms of capital equipment requirements and salaries for the faculty and support staff, exposure to the widely used transducers and recording instruments for measurement and process control represent an essential part of a modern engineering education.

We developed the material for this book in numerous classroom and laboratory situations over the past 20 years at Illinois Institute of Technology, IIT Research Institute, Iowa State University, and the University of Maryland. The mathematics employed in the treatment can easily be understood by junior and senior undergraduates. A great deal of effort was devoted to the selection and preparation of the illustrations used throughout the text. These illustrations, together with the exercises at the end of the chapters, complement the text and should aid appreciably in presenting the material to the students. We hope that students and instructors alike will find the presentation clear and enlightening and that many of the mysteries of the laboratory, with its numerous black boxes, will be clarified.

We thank the following reviewers whose valuable comments aided us in refining this material: Robert G. Leonard, Virginia Polytechnic Institute and State University; Byron Jones, Kansas State University; and John Ligon, Michigan Technological University.

**James W. Dally**
**William F. Riley**
**Kenneth G. McConnell**

# CONTENTS

# LIST OF SYMBOLS

| | |
|---|---|
| $a$ | Acceleration |
| $a$ | Radius |
| $a_x, a_y, a_z$ | Cartesian components of acceleration |
| $a_r, a_\theta, a_z$ | Polar components of acceleration |
| $a_R, a_\theta, a_\phi$ | Spherical components of acceleration |
| $A$ | Amplification factor |
| $A$ | Area |
| $B$ | Flux density |
| $B$ | Strength of a magnetic field |
| $c$ | Specific heat capacity |
| $c$ | Velocity of light in a vacuum |
| $c$ | Velocity of sound |
| $c_p$ | Specific heat at constant pressure |
| $c_v$ | Specific heat at constant volume |
| $C$ | Capacitance |
| $C$ | Discharge coefficient |
| $C$ | Viscous damping constant |
| $C_c$ | Cable capacitance |
| $C_C$ | Contraction coefficient |
| $C_e$ | Equivalent capacitance |
| $C_f$ | Feedback capacitance |
| $C_G$ | Galvanometer constant |
| $C_i$ | Input amplitude |
| $C_L$ | Leadwire capacitance |
| $C_o$ | Output amplitude |
| $C_p$ | Piezoelectric sensor capacitance |
| $C_t$ | Transducer capacitance |
| $C_v$ | Coefficient of variation |
| $C_V$ | Coefficient of velocity |
| $d$ | Damping ratio |
| $d$ | Deviation |
| $d$ | Displacement |
| $d^*$ | Full-scale displacement |
| $D$ | Damping coefficient |
| $D$ | Diameter |
| $D$ | Flexural rigidity of a plate |

| | |
|---|---|
| $e$ | Electron charge |
| $e$ | Junction potential per unit temperature |
| $E$ | Electromotive force, voltage, or potential |
| $E$ | Modulus of elasticity |
| $E'$ | Potential gradient |
| $E_i$ | Input voltage |
| $E_{ic}$ | Amplitude of a carrier signal |
| $E_{it}$ | Amplitude of a transducer signal |
| $E_m$ | Back electromotive force |
| $E_m$ | Voltage displayed by a meter |
| $E_o$ | Output voltage |
| $E_r$ | Reference voltage |
| $E_{sw}$ | An adjustable voltage |
| $E_x$ | Unknown voltage |
| $E^*$ | Full-scale voltage or range |
| $\mathscr{E}$ | Error |
| $\mathscr{E}_a$ | Accumulated error for a system |
| $\mathscr{E}_A$ | Amplifier error |
| $\mathscr{E}_R$ | Recorder error |
| $\mathscr{E}_{sc}$ | Signal-conditioner circuit error |
| $\mathscr{E}_T$ | Transducer error |
| $f$ | Frequency |
| $f_{bw}$ | Bandwidth |
| $f_r$ | Resonant frequency |
| $F$ | Force, applied load |
| $F_b$ | Bolt load |
| $g$ | Gravitational constant |
| $G$ | Gain |
| $G$ | Shear modulus of elasticity |
| $h$ | Convective heat-transfer coefficient |
| $h$ | Planck's constant |
| $h$ | Thickness |
| $I$ | Current |
| $I$ | Intensity of light |
| $I$ | Moment of inertia |
| $I'$ | Current density |
| $I_f$ | Feedback current |
| $I_g$ | Gage current |
| $I_G$ | Galvanometer current |
| $I_i$ | Input current |
| $I_i^*$ | Full-scale input current |
| $I_m$ | Meter current |
| $I_m^*$ | Full-scale meter current |
| $I_o$ | Output current |

| | |
|---|---|
| $I_s$ | Steady-state current |
| $I_{sh}$ | Shunt current |
| $I^*$ | Amplitude of a sinusoidal current input |
| $J$ | Polar moment of inertia |
| $k$ | Adiabatic exponent |
| $k$ | Boltzmann's constant |
| $k$ | Radius of gyration |
| $k$ | Spring constant or stiffness |
| $k$ | Transmission coefficient of a lens |
| $K$ | Dielectric constant |
| $K$ | Torsional spring constant |
| $K_t$ | Transverse sensitivity factor for a strain gage |
| $L$ | Inductance |
| $L$ | Length |
| $L_T$ | Transducer inductance |
| $l$ | Length |
| $l_g$ | Gage length of a strain gage |
| $\mathcal{L}$ | Loss factor |
| $m$ | Mass |
| $m_a$ | Accelerometer mass |
| $m_o$ | Object mass |
| $m_p$ | Plate mass |
| $m_t$ | Transducer mass |
| $M$ | Mach number |
| $M$ | Moment |
| $M_x, M_y, M_z$ | Cartesian components of a moment |
| $n$ | Index of refraction |
| $N$ | Number of charge carriers |
| $N$ | Number of cycles |
| $N$ | Number of turns |
| $N_{db}$ | Number of decibels |
| $p$ | Pressure |
| $p_d$ | Dynamic pressure |
| $p'_d$ | Measured dynamic pressure |
| $p_o$ | Static pressure |
| $p_s$ | Stagnation pressure |
| $p'_s$ | Measured stagnation pressure |
| $P$ | Force, applied load |
| $P$ | Power |
| $P_D$ | Power density |
| $P_g$ | Power dissipated by a strain gage |
| $P_T$ | Power dissipated by a transducer |
| $P_x, P_y, P_z$ | Cartesian components of a force |
| $q$ | Charge |

| | |
|---|---|
| $q$ | Rate of heat transfer |
| $q$ | Resistance ratio |
| $Q$ | Volume flow rate |
| $Q_i$ | Input quantity |
| $Q_o$ | Output quantity |
| $r, \theta, z$ | Polar coordinates |
| $r$ | Resistance ratio |
| $r, \theta, \phi$ | Spherical coordinates |
| $R$ | Radius |
| $R$ | Range |
| $R$ | Resistance, resistor |
| $R_A$ | amplifier resistance |
| $R_b$ | ballast resistance, ballast resistor |
| $R_B$ | Equivalent bridge resistance |
| $R_{ci}$ | Input resistance of a circuit |
| $R_{co}$ | Output resistance of a circuit |
| $R_e$ | Equivalent resistance |
| $R_e$ | Reynold's number |
| $R_f$ | Feedback resistance |
| $R_g$ | Gage resistance |
| $R_G$ | Galvanometer resistance |
| $R_L$ | Leadwire resistance |
| $R_m$ | Meter resistance |
| $R_M$ | Measuring-circuit resistance |
| $R_M$ | Recording-instrument resistance |
| $R_o$ | Reference resistance |
| $R_p$ | Parallel resistance, parallel resistor |
| $R_p$ | Piezoelectric-sensor resistance |
| $R_s$ | Series resistance, series resistor |
| $R_s$ | Source resistance |
| $R_S$ | Series resistance |
| $R_{sh}$ | Shunt resistance, shunt resistor |
| $R_{sr}$ | Series resistance, series resistor |
| $R_T$ | Transducer resistance |
| $R_x$ | External resistance, external resistor |
| $s$ | Displacement |
| $s$ | Distance |
| $s$ | Span |
| $S$ | Calibration constant |
| $S$ | Sensitivity |
| $s_a$ | Axial strain sensitivity of a strain gage |
| $S_A$ | Strain sensitivity of a material, alloy sensitivity |
| $S_c$ | Circuit sensitivity |
| $S_{cc}$ | Constant-current circuit sensitivity |

| | |
|---|---|
| $S_{cv}$ | Constant-voltage circuit sensitivity |
| $S_E$ | Voltage sensitivity |
| $S_f$ | Fatigue strength of a material |
| $S_g$ | Gage factor, strain sensitivity of a gage |
| $S_g^*$ | Corrected gage factor for a strain gage |
| $S_G$ | Galvanometer sensitivity |
| $S_I$ | Current sensitivity |
| $S_I$ | Instrument sensitivity |
| $S_N$ | Strouhal Number |
| $S_q$ | Charge sensitivity |
| $S_R$ | Reciprocal sensitivity |
| $S_R$ | Recorder sensitivity |
| $S_s$ | Shear strain sensitivity of a strain gage |
| $S_s$ | System sensitivity |
| $S_{sg}$ | Gage factor for a stress gage |
| $S_t$ | Transverse strain sensitivity of a strain gage |
| $S_x$ | Standard deviation |
| $S_{\bar{x}}$ | Standard error |
| $S_\tau$ | Torsional yield strength of a material |
| $t$ | Time |
| $t_r$ | Rise time |
| $T$ | Period |
| $T$ | Temperature |
| $T$ | Time constant |
| $T$ | Torque |
| $T_A$ | Absolute temperature |
| $T_e$ | Equivalent time constant |
| $T_n$ | Natural period of an oscillation |
| $T_o$ | Reference temperature |
| $T_s$ | Stagnation temperature |
| $u, v, w$ | Cartesian components of displacement |
| $u_r, u_\theta, u_z$ | Polar components of displacement |
| $u_x, u_y, u_z$ | Cartesian components of displacement |
| $U$ | Energy |
| $v$ | Velocity |
| $v_r, v_\theta, v_z$ | Polar components of a velocity |
| $v_x, v_y, v_z$ | Cartesian components of a velocity |
| $v_R, v_\theta, v_\phi$ | Spherical components of a velocity |
| $V$ | Fluid velocity |
| $V$ | Voltage |
| $V_o$ | Centerline velocity |
| $w$ | Specific weight of a solid |
| $w$ | Width |
| $W$ | Weight |

| | |
|---|---|
| $W_c$ | Calibration weight |
| $W_e$ | Equivalent weight |
| $W_x$ | External weight |
| $W_\lambda$ | Spectral radiation intensity |
| $x, y, z$ | Cartesian coordinates |
| $x$ | Position |
| $\bar{x}$ | Sample mean |
| $Z$ | Impedance |
| $Z_i$ | Input impedance |
| $Z_o$ | Output impedance |
| $Z_o$ | Zero offset |
| $\alpha$ | Angle of incidence |
| $\alpha, \beta, \gamma, \theta, \phi$ | Angles |
| $\alpha$ | Angular acceleration |
| $\alpha, \beta$ | Coefficient of thermal expansion |
| $\beta$ | Time constant for a temperature sensor |
| $\gamma$ | Shear strain |
| $\gamma$ | Specific weight of a fluid |
| $\gamma$ | Temperature coefficient of resistance |
| $\gamma$ | Temperature coefficient of resistivity |
| $\gamma_{r\theta}$ | Shear strain component in polar coordinates |
| $\left.\begin{array}{l}\gamma_{xy} = \gamma_{yx}\\ \gamma_{yz} = \gamma_{zy}\\ \gamma_{zx} = \gamma_{xz}\end{array}\right\}$ | Shear strain components in cartesian coordinates |
| $\delta$ | Displacement |
| $\varepsilon$ | Emissivity |
| $\varepsilon$ | Normal strain |
| $\varepsilon'$ | Apparent normal strain |
| $\varepsilon_a$ | Axial strain |
| $\varepsilon_c$ | Calibration strain |
| $\varepsilon_n$ | Normal strain component |
| $\varepsilon_r, \varepsilon_\theta, \varepsilon_z$ | Normal strain components in polar coordinates |
| $\varepsilon_t$ | Transverse strain |
| $\varepsilon_{xx}, \varepsilon_{yy}, \varepsilon_{zz}$ | Normal strain components in cartesian coordinates |
| $\varepsilon_1, \varepsilon_2, \varepsilon_3$ | Principal normal strains |
| $\eta$ | Nonlinear term |
| $\theta$ | Angle of rotation |
| $\theta$ | Angular deflection |
| $\theta$ | Angular displacement |
| $\theta_s$ | Steady-state rotation of a galvanometer |
| $\theta^*$ | Amplitude of a sinusoidal oscillation |
| $\lambda$ | Wavelength |
| $\mu$ | Absolute viscosity |
| $\mu$ | Arithmetic mean |

| | |
|---|---|
| $\mu$ | Mobility of charge carriers |
| $\nu$ | Poisson's ratio |
| $\pi$ | Piezoresistive proportionality constant |
| $\rho$ | Mass density |
| $\rho$ | Radius of curvature |
| $\rho$ | Resistivity coefficient |
| $\rho$ | Specific resistance |
| $\sigma$ | Normal stress |
| $\sigma$ | True standard deviation |
| $\sigma_n$ | Normal stress component |
| $\sigma_r, \sigma_\theta, \sigma_z$ | Normal stress components in polar coordinates |
| $\sigma_{xx}, \sigma_{yy}, \sigma_{zz}$ | Normal stress components in cartesian coordinates |
| $\sigma_1, \sigma_2, \sigma_3$ | Principal normal stresses |
| $\tau$ | Shear stress |
| $\tau$ | Time constant |
| $\tau_e$ | Effective time constant |
| $\tau_n$ | Shear stress component |
| $\tau_{r\theta}$ | Shear stress component in polar coordinates |
| $\left.\begin{array}{l}\tau_{xy} = \tau_{yx}\\ \tau_{yz} = \tau_{zy}\\ \tau_{zx} = \tau_{xz}\end{array}\right\}$ | Shear stress components in cartesian coordinates |
| $\phi$ | Phase shift |
| $\phi$ | Photon flux density |
| $\psi$ | Illumination |
| $\omega$ | Angular velocity |
| $\omega$ | Circular frequency |
| $\omega_n$ | Undamped natural frequency |
| $\omega_{nm}$ | Natural frequency of a measurement system |
| $\omega_{ns}$ | Structural resonance frequency |

# ONE

# APPLICATIONS OF ELECTRONIC INSTRUMENT SYSTEMS

## 1.1 INTRODUCTION

The primary objective of this textbook is to introduce electronic instrumentation systems in a manner sufficiently complete that the student will acquire an ability to make accurate and meaningful measurements of mechanical and thermal quantities. The mechanical quantities include strain, force, pressure, moment, torque, displacement, velocity, acceleration, flow velocity, mass flow rate, volume flow rate, frequency, and time. The thermal quantities include temperature and heat flux.

Most readers of this text will have a conceptual understanding of these quantities through exposure in previous mechanics or physics courses, such as statics, dynamics, strength of materials, or thermodynamics. The student's experience in actually measuring these quantities by conducting experiments, however, will usually be quite limited. It is an objective of this text to introduce methods commonly employed to make such measurements. Through this exposure to the experimental aspects of the problem, the student will improve his[1] understanding of many of the laws and/or formulas that were introduced in the analytically oriented courses. The individual will also become familiar with all elements of an electronic instrumentation system and will improve his ability to design effective experiments and measurement methods that can provide solutions to many practical engineering problems.

Emphasis in the text will be directed toward electronic instrumentation systems rather than mechanical systems. In most cases, electronic systems provide better data that more accurately and more completely characterize the design or process being experimentally evaluated. Also, the electronic system provides an electrical output signal that can be used for automatic data reduction or for control of the process. These advantages of the electronic measurement system over the mechanical measurement system have initiated and sustained the trend in instrumentation toward electronic methods.

## 1.2   THE ELECTRONIC INSTRUMENT SYSTEM

A complete electronic instrument system usually contains six subsystems or elements as indicated in Fig. 1.1.

The *transducer* is a device that converts a change in the mechanical or thermal quantity being measured into a change of an electrical quantity. For example, a strain transducer (gage) bonded to a specimen converts a change in strain $\Delta\varepsilon$ in the specimen to a change in electrical resistance $\Delta R$ in the gage. The change in resistance $\Delta R$ can then be converted to a change in voltage $\Delta E$, which can be measured accurately with relative ease. Since the voltage is proportional to the strain, the strain sensed by the transducer can be determined once the instrument system is properly calibrated.

The *power supply* provides the energy to drive the transducer. For instance, a differential transformer, which is a transducer used to measure displacement,

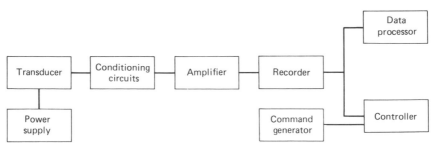

*Figure 1.1*   Block diagram representing an electronic instrumentation system.

---

[1] Throughout this book the impersonal "he" and "his" will be used for reasons of style and accepted English usage. It has been apparent for many years that women are active in all engineering disciplines. Pronouns encountered in the text should be mentally translated into "he and she" or "his and her" each time they appear.

requires an ac voltage supply to excite two coils that create a fluctuating magnetic field. Power supplies, such as constant dc voltage sources, constant dc current sources, and ac voltage sources, are selected to satisfy the requirements of the transducer being employed.

*Signal conditioners* are electronic circuits that convert, compensate, or manipulate the output from the transducer into a more usable electrical quantity. The Wheatstone bridge used with the strain transducer (gage) converts the change in resistance $\Delta R$ to a change in voltage $\Delta E$. Filters, compensators, modulators, demodulators, integrators, and differentiators are other examples of signal conditioning circuits in common usage in electronic instrument systems.

*Amplifiers* are required in the system when the voltage output from the transducer–signal conditioner combination is small. Output signals of a millivolt or less are common. Amplifiers with gains of 10 to 1000 are used to increase these signals to levels where they are compatible with the voltage-measuring devices used in the system.

*Recorders* are voltage-measuring devices that are used to display the measurement in a form that can be read and interpreted. Digital voltmeters are often used to measure static voltages. The display of a digital voltmeter is in the form of an array of easy to read illuminated numerals. Self-balancing potentiometers, oscillographs, oscilloscopes, and magnetic tape recorders are other examples of recording devices employed to display quasi-static and dynamic output signals.

*Data processors* are used to convert the output signals from the instrument system into data that can be easily interpreted by the engineer. Data processors are usually employed where large amounts of data are being collected and manual reduction of these data would be too time-consuming and costly. Suppose 50 transducers are installed on a development vehicle and the vehicle is operated for several hours on a test course to evaluate its performance. In an experiment of this type, the reduction of the data to graphs, charts, and tables that are sufficiently concise for engineering interpretation is a mammoth and time-consuming task. Data processors convert the analog input signal to digital form, which can be automatically processed on a digital computer in accordance with programmed instructions. The processed data are displayed as graphs and tables that illustrate the salient findings of the experimental program.

*Process controllers* are used to monitor and adjust mechanical and thermal quantities in a manufacturing process. The signal from the instrumentation system is compared to a command signal that reflects the required value of the quantity in the process. The process controller accepts both the command signal and the measured signal and forms the difference to give an error signal. This error signal is then used to automatically adjust the process. As a very simple example of automatic control, consider a time–temperature cycle in an industrial oven that is being used to cure plastic components. The temperature is measured and converted to a voltage output signal. The signal voltage is connected to the input terminals of a process controller, where it is compared to a command voltage. The command voltage is provided by a source that can be varied with time to give a voltage–time profile identical to the temperature–time profile required in the oven. The process controller forms the error signal by subtracting the measured voltage from the command voltage. The error signal is then used in an automatic control system to adjust the heat flow into the oven to maintain the required temperature–time profile within a specified error band.

Electronic instrument systems are used in three different areas of application, which include:

1. Engineering analysis of machine components, structures, and vehicles to ensure efficient and reliable performance.
2. Monitoring processes to provide on-line operating data pertaining to the process that allows an operator to make adjustments and thereby control the process.
3. Automatic process control to provide on-line operating data pertaining to the process that is used as feedback signals in closed-loop control systems to automatically control the process.

Each of these applications is described in the following sections.

## 1.3 ENGINEERING ANALYSIS

Engineering analyses are conducted to evaluate new or modified designs of a machine component, structure, or vehicle to ensure efficient and reliable performance when it is placed in operation. Two approaches can be followed in performing the engineering analysis: a theoretical approach or an experimental approach.

When the *theoretical approach* is used, an analytical model of the component is formulated and assumptions are made pertaining to the operating conditions, the loads imposed on the component, the properties of the material, and the mode of failure. Equations are then written that describe the behavior of the analytical model. These equations are solved by using either exact mathematical methods or, more frequently, by using finite element procedures. The results of the theoretical analysis provide the designer with an indication of the adequacy of the design and an estimate of the probable performance of the component or structure in service.

Uncertainties often exist pertaining to the validity of results from the finite-element model. Does the model accurately reflect all aspects of the prototype design? Do the assumed operating conditions properly cover the full range of loadings imposed on the component? Are the boundary conditions properly represented in the model? Have significant errors been introduced in the analysis through use of the numerical procedures?

With the *experimental approach*, a prototype of the component is fabricated and a test program is conducted to evaluate the performance of the component in service by making direct measurements of the important quantities that control the adequacy of the design. This approach eliminates two of the serious uncertainties in the theoretical approach. An analytical model is not required and the assumptions regarding operating conditions and material properties are not necessary. However, the experimental approach also has serious shortcomings. In comparison to the theoretical approach, it is extremely expensive. Also, uncertainties arise due to inevitable experimental error in the measurements. Finally, there is always a question whether the transducers were placed at the correct locations to record the quantities that actually control the adequacy of the design.

The preferred approach is a combination of the theoretical and experimental methods. The theoretical analysis should be conducted to ensure a thorough understanding of the problem. The significance of the results of the theoretical analysis should be completely evaluated and any shortcomings of the analysis should be clearly identified. An experimental program should be designed at this stage of the theoretical analysis to verify the analytical model and check the validity of assumptions pertaining to operating conditions and material properties.

The results of the theoretical analysis are extremely important in the design of the experimental program. The locations and orientations of the transducers can be specified more accurately and the number of measurements can be reduced appreciably. It is also possible to reduce the number of tests necessary to cover the full spectrum of operating conditions when theoretical results are available.

The results from the experimental program are then used to verify the analytical model and to check the validity of the assumptions and numerical procedures. If significant differences exist, the analytical model can be modified or the theoretical approach can be changed. When the theoretical approach is verified and confidence in the analysis is established, it is then possible to optimize the weight, strength, or cost of the component.

The combined theoretical/experimental approach to engineering analysis provides the most cost-effective method to ensure efficient and reliable performance of new or modified designs of mechanical or structural components.

## 1.4  PROCESS CONTROL

Electronic instrumentation systems are used in two types of process control: open-loop or monitoring control and closed-loop or automatic control.

*Open-loop control*, involving a process that is being monitored with several transducers, is illustrated in Fig. 1.2. The data from the transducers are displayed continuously on an instrument panel. An operator observes the quantities being monitored and, if necessary, makes adjustments to the process input parameters to maintain control of the process. The operator closes the loop in this type of process control. The accuracy and reliability of the data displayed on the instrument panel are extremely important as they provide the basis for the operator's decisions in adjusting the process. The operation of most freighters and tankers is done with open-loop control. An operator in the engine room monitors measurements of ship speed, engine speed, engine temperature, oil pressure, fuel consumption, etc., and manually makes the adjustments necessary to maintain the required speed.

A second type of process control, known as *automatic* or *closed-loop control*, is illustrated in Fig. 1.3. In the closed-loop control system, the operator has been eliminated. Instead, the signals from the electronic instrumentation system are compared to command signals that represent voltage–time relationships for important mechanical or thermal quantities associated with the process. The first controller measures the difference between the command signal and the transducer signal and develops an error or feedback signal. The feedback signal is then used in the second controller to drive devices that correct the process. As an example of closed-loop control, consider a hydraulically actuated positioning mechanism that moves an engine block, during machining, through a battery of drilling and tapping machines. The desired position of the engine block along a track, together with the time required at each position, are used by the command generator to establish a voltage–time trace that represents the required position of the block at any time. The actual position of the engine block is measured with a displacement transducer. The difference between the command signal and the measured displacement signal is used by the first controller to generate a feedback signal that is proportional to the adjustment needed to correct the position. The feedback signal is amplified and used to drive a servo-control valve in the second controller. The servo-control valve adjusts the flow of pressurized fluid to a hydraulic cylinder. The cylinder moves the engine block and zeros the feedback signal. The block is then correctly positioned for the machining operation.

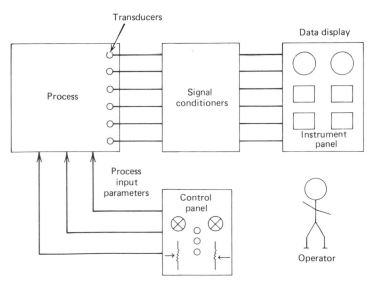

*Figure 1.2*  Schematic diagram of open-loop process control.

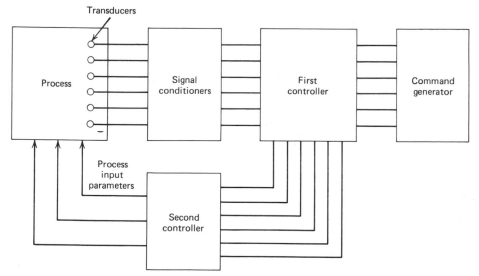

***Figure 1.3***  Schematic diagram of closed-loop process control.

## 1.5  EXPERIMENTAL ERROR

Error is the difference between the true value and the measured value of a quantity such as displacement, pressure, temperature, etc. The better electronic instrumentation systems are designed to limit the error, which is inevitable in any measurement, to a value that is acceptable in terms of the accuracies required in an engineering analysis or in the control of a process. Errors can occur due to the following causes:

1. Accumulation of accepted error in each element of the instrumentation system.
2. Improper functioning of any element in the system.
3. Effect of the transducer on the process.
4. Dual sensitivity of the transducer.
5. Other sources of error.

Each of these sources of error is described in terms of the general characteristics of the elements of the instrumentation system in the following subsections.

### Accumulation of Accepted Error

All elements of an instrumentation system have accuracy limits that are specified by the manufacturer. For instance, a recorder may have a specified accuracy of ±2 percent of full-scale values. The recorder can be expected to operate within these limits if it is properly maintained and periodically calibrated. Because of limits on accuracy, the recorder will introduce error in a measurement when it is placed in an instrumentation system; however, this error is acceptable or known provided the recorder is operating within specifications.

The specified accuracy limits should be clearly understood, since a recorder accurate to within ±2 percent can introduce larger errors than these limits seem to imply. Consider the input–output curve, shown in Fig. 1.4, that characterizes the recorder. The *deviation d* is defined as the product of the accuracy and the full-scale value of the response of the recorder. Lines drawn parallel to the true response of the recorder, but displaced by ±d, form the upper and lower bounds of the response of the instrument. The shaded area between these two bounds gives the region where the recorder (or any other element) is operating within the manufacturer's specifications. If an instrument is operated at one-half scale, the deviation *d* remains constant; however, the response (true value) is reduced by a factor of 2. Thus, the *error*, which is defined as the deviation divided by the true value, is doubled. This example indicates that errors of ±4 percent would be within specifications if the recorder is operated at half scale. Operation of instruments at less than full scale is sometimes convenient; however, any reduction in scale should be carefully considered, since the error increases rapidly as the percent of scale used is reduced. Instruments should not normally be used at less than $\frac{1}{3}$ to $\frac{1}{2}$ of full scale without carefully considering the effect of the errors involved in the measurement.

Since the usual instrumentation system contains several elements, and each element introduces error even when its operation is within specifications, error

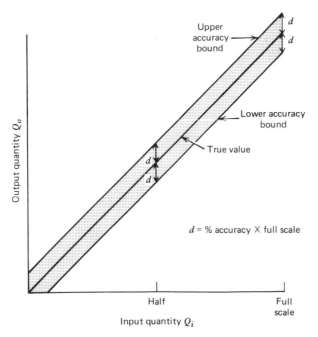

**Figure 1.4** Accuracy bounds for an instrument operating within specification.

accumulates. It is possible to estimate the accumulated error[2] $\mathscr{E}_a$ for the system as

$$\mathscr{E}_a = \sqrt{\mathscr{E}_T^2 + \mathscr{E}_{sc}^2 + \mathscr{E}_A^2 + \mathscr{E}_R^2} \qquad (1.1)$$

**where**  $\mathscr{E}_T$ is the transducer error.
$\mathscr{E}_{sc}$ is the signal conditioner error.
$\mathscr{E}_A$ is the amplifier error.
$\mathscr{E}_R$ is the recorder error.

It is evident from Eq. (1.1) that small but acceptable errors for each element can accumulate and beome unacceptably large for critical measurements where high accuracy is required.

## Improper Functioning of Instruments

If any element in the instrumentation system is not properly maintained or adjusted prior to use, calibration, zero offset, or range errors can occur. Before discussing these errors, consider the typical response curve for an instrument shown in Fig. 1.5. Here the output quantity $Q_o$ is measured as the input quantity $Q_i$ is varied. A significant portion of the response curve can be represented by a straight line that is fit to the data using a least-squares method (i.e., the instrument response is linear). The slope of the straight line is the calibration constant or sensitivity $S$ of the instrument. Thus,

$$S = \frac{\Delta Q_o}{\Delta Q_i} \qquad (1.2)$$

For a recorder, the sensitivity $S$ is given in units of displacement per volt. For a piezoelectric pressure gage, the sensitivity $S$ is given as the voltage output per unit of pressure.

If the response line does not pass through the origin, the deviation $d$ measured along the ordinate is called the *zero offset* $Z_o$. It is evident from Fig. 1.5 that

$$Q_o = SQ_i + Z_o \qquad (1.3)$$

Most instruments have a capability for adjusting the zero offset so that $Z_o$ can be set equal to zero. The relationship for the output quantity $Q_o$ then reduces to

$$Q_o = SQ_i \qquad (1.4)$$

---

[2] See Section 10.9 for a discussion of error propagation. Compare Eq. (1.1) with Eq. (10.50).

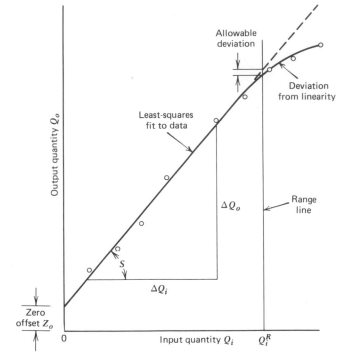

***Figure 1.5***   Input–output response curve for a typical instrument element.

For large values of the input quantity, the typical response curve frequently deviates from a straight line (linear relationship), as shown in the upper right portion of Fig. 1.5. When this deviation becomes excessive, say 1 or 2 percent, Eqs. (1.4) and (1.5) are no longer valid and the range of the instrument has been exceeded. If an allowable deviation is specified, a range line can be drawn on the response graph and the range of the instrument $Q_i^R$ can be established (see Fig. 1.5). The value $Q_i^R$ defines the upper limit of operation of the instrument. The lower limit of operation $Q_i^L$ is determined by excessive scale error (operation of the instrument at less than full scale). The difference between the upper limit of operation and the lower limit of operation defines the *span s* of the instrument. Thus,

$$s = Q_i^R - Q_i^L \tag{1.5}$$

Errors in the measurement of $Q_o$ will occur if the instrument is not properly calibrated and zeroed. Errors will also occur if the input $Q_i$ is greater than the range of the instrument $Q_i^R$. Illustrations of calibration error, zero-offset error, and range error are presented in Fig. 1.6.

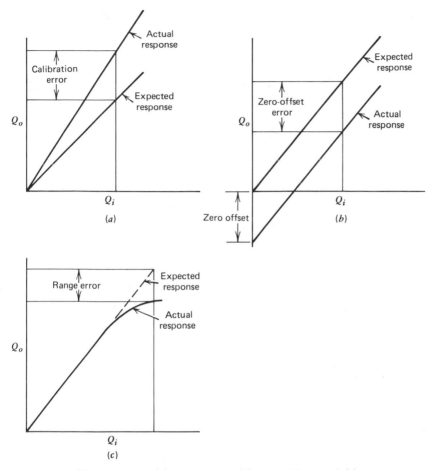

**Figure 1.6**  Illustration of ($a$) calibration, ($b$) zero-offset, and ($c$) range errors.

## Effect of the Transducer on the Process

The transducer must be selected and placed in the process in such a manner that it does not affect and/or change the process. If the installation of the transducer does affect the process, serious errors can result and the measurements may be misleading or meaningless. For most measurements, the size and weight of the transducer should be small relative to the size and weight of the component or process. Also, the transducer should require small forces or draw little energy from the process for its operation.

To illustrate the errors that can occur as a result of the presence of the transducer, consider an experiment designed to measure the frequency associated with the fundamental mode of vibration of a circular plate with clamped edges. The equation[3] governing the frequency of the first mode of vibration of a clamped circular plate with an additional concentrated mass at the center of the plate is

$$\omega = .\frac{\lambda^2}{a^2\sqrt{\rho/D}}$$ (1.6)

**where** $\omega$ is the circular frequency.
$\lambda$ is a constant that depends on the ratio of the concentrated mass to the plate mass.
$a$ is the radius of the plate.
$\rho$ is the mass density per unit area of the plate.
$D$ is the flexural rigidity of the plate. $D = Eh^3/12(1-v^2)$.
$E$ is the modulus of elasticity of the plate material.
$v$ is Poisson's ratio of the plate material.
$h$ is the thickness of the plate.

For the experiment under consideration, the value of the constant $\lambda^2$ depends upon the ratio of the concentrated mass of the accelerometer $m_a$ to the mass of the plate $m_p$. For $m_a/m_p$ equal to 0, 0.05, and 0.10, the constant $\lambda^2$ equals 10.214, 9.012, and 8.111, respectively. Thus, the error in this measurement of the first natural frequency, due to the mass of the accelerometer, will be

$$\left(\frac{10.214}{9.012} - 1\right)100 = 13.3\% \qquad \text{for } \frac{m_a}{m_p} = 0.05$$

and

$$\left(\frac{10.214}{8.111} - 1\right)100 = 25.9\% \qquad \text{for } \frac{m_a}{m_p} = 0.10$$

It is clear from this example that the mass of the transducer has a profound effect on the vibratory process and that significant errors may occur due to the presence of the transducer. To avoid excessive errors, the mass of the transducer in this case should not exceed 1 percent of the mass of the plate.

## Dual Sensitivity Errors

Transducers are usually designed to measure a single quantity such as pressure; however, they often exhibit some sensitivity to other quantities such as temperature or acceleration. If a transducer is employed to measure some quan-

---

[3] A. W. Leissa, "Vibration of Plates," NASA SP-160, 1969, p. 19.

tity, say pressure, in a process and if the temperature also changes as the measurement is made, error due to the dual sensitivity of the transducer will occur. The effect of dual sensitivity is illustrated in the input–output response graph of Fig. 1.7. As shown in this figure, two errors arise due to dual sensitivity when both quantities that affect the transducer or instrument are changing simultaneously during the time period of the measurement. First, a zero shift occurs due to the change in the secondary quantity. Second, a change in the sensitivity of the transducer occurs. These errors are illustrated in Fig. 1.7. Both can be significant in poorly designed transducers.

In some experiments, the secondary quantity changes as a function of time. In these cases, the zero offset and the sensitivity also vary as a function of time. The changing zero offset is referred to as *zero drift*. The varying sensitivity is termed *sensitivity drift*. It is very difficult to make accurate measurements under these conditions, since the continuous changes in the zero base and calibration constant of the instrument system preclude any possibility of making a single correction for the effect of the secondary quantity. A better approach is to carefully select the transducer so that its secondary sensitivity is negligible. Also, the remaining elements of the instrumentation system should be housed, if possible, in a temperature-controlled environment.

While the emphasis of this discussion has centered on the influence of dual sensitivity of the transducer, it should be recognized that all elements in the

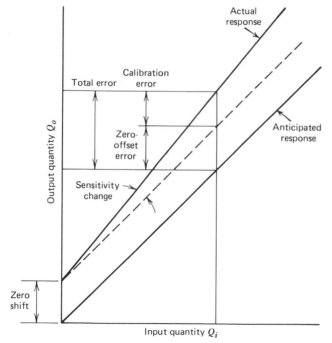

*Figure 1.7*  Change in response of an instrument due to dual sensitivity.

instrumentation system exhibit dual sensitivity. This dual sensitivity of the other elements becomes particularly important if the study is of long duration (several days or weeks). Time then becomes the secondary parameter and the stability characteristics of the signal conditioner, power supply, amplifiers, and recorder control the accuracy of the measurements. Since zero drift will occur in most instruments, particularly amplifiers, provision must be made in any long-duration experiment to periodically check and reestablish the zero base (rezero) or correct for the zero shift.

## Other Sources of Error

Other important sources of error include lead-wire effects, electronic noise, and the human operator.

The effects of lead wires, used to connect the transducer to the instrumentation system, can be significant if the transducers contain resistive sensing elements. Lead wires, which are long and of small gage, exhibit a resistance that is not negligible relative to the transducer resistance. The added resistance of the lead wires can change the sensitivity or calibration constant of the transducer. The lead wires can also produce erroneous signals due to temperature-induced resistance changes in the wires. When long lead wires are placed in the arms of a Wheatstone bridge that is being used for strain measurements, the accuracy of the measurements can be easily compromised.

Electronic noise usually results from spurious signals that are picked up by the lead wires. When lead wires are positioned in close proximity to electrical devices, such as motors or lights, the fluctuating magnetic fields in the vicinity of the devices generate small voltages in the adjoining lead wires that superimpose on the measurement signal. Since the measurement signal is usually small, the error produced by lead-wire noise can be significant. The noise picked up by the lead wires can be minimized with proper shielding, which isolates the leads from the effects of the fluctuating magnetic fields. In certain measurements, where the measurement signal is very small, noise from a properly shielded lead-wire installation may still be objectionable. In these cases, notched filters that block passage of a narrow band of frequencies can be employed to eliminate most of the noise, since it usually exhibits a power-line frequency of 60 Hz.

Another source of error is due to the operator. The operator must properly record the sensitivity $S$ of each element in the instrumentation system and he must accurately zero each element. Finally, the output that is displayed on the recorder must be read. Reading errors of 1 or 2 percent due to parallax and tracewidth are common.

## 1.6 MINIMIZING EXPERIMENTAL ERROR

In the preceding section, the many sources of experimental error were identified so that the reader would become aware of the difficulties commonly encountered in conducting experiments. Measurement systems designed to yield accuracies of 0.1 or even 1 percent are usually unrealistic when the cost of the system and the time required to make the measurements are considered. Accuracies of 2 to 5 percent can usually be achieved at reasonable cost; however, procedures must be followed that minimize error at each step of the experiment. A single mistake can easily degrade the system beyond acceptable limits of error. In the worst case, the mistake will degrade the system to the point where the data are misleading or meaningless. Accepted procedures for minimizing error in a measurement system are:

1.  Carefully select the transducer. Pay particular attention to its size, weight, and energy requirements to ensure that it does not affect the variable being measured.
2.  Check the accuracy of each element in the instrumentation system. Compute the accumulated "accepted" error.
3.  Calibrate each instrument in the system to verify that it is operating within specifications.
4.  Examine the process and the environment in which the instrumentation system must operate. Pay particular attention to temperature variations and the time required for the measurement. Estimate the errors that will be produced by dual sensitivity effects of all elements in the instrumentation system.
5.  Connect the system together with properly shielded and terminated lead wires. Use wiring procedures that minimize lead-wire errors. Estimate the errors that may be introduced by the lead wires.
6.  Check the system for electronic noise. If necessary, reroute the lead wires and/or insert suitable filters to minimize the noise.
7.  Perform a system calibration by measuring the variable in a known process, as illustrated in Fig. 1.8.
8.  Estimate the total error in the system due to all known sources.

This procedure does not insure a perfect measurement since some error is always inherent in any experimental determination of unknown quantities; however, it does provide a systematic approach to minimizing error and to estimating the error involved in the measurement.

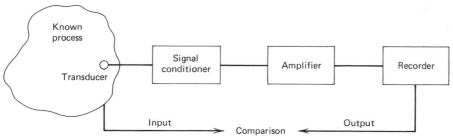

*Figure 1.8*  System calibration.

## 1.7 SUMMARY

An electronic instrumentation system usually contains a transducer, a power supply, a signal conditioner, an amplifier, and a recorder. Such a system is used to experimentally determine unknown quantities, such as force, pressure, displacement, temperature, etc. The data or output from an instrumentation system may also be used for an engineering analysis of machine components or structures, for process monitoring, for open-loop process control, or for closed-loop automatic control.

Error will always occur in an experimental determination of an unknown quantity. The error, which accumulates, may be due to many causes, such as summation of accepted error, improper functioning of an instrument, transducer interaction with the process, or dual sensitivity of the transducer. Measurement systems that yield accuracies of 0.1 to 1 percent are usually too costly for most applications. Accuracies of 2 to 5 percent are more realistic and can be achieved if careful procedures are employed in the design and installation of the instrumentation system. It is imperative that meticulous attention to detail be observed so that errors can be kept within acceptable bounds.

## REFERENCES

1.  Ambrosius, E. E., R. D. Fellows, and A. D. Brickman: *Mechanical Measurement and Instrumentation*, Ronald Press, New York, 1966.

2.  Bartholomew, D.: *Electrical Measurements and Instrumentation*, Allyn & Bacon, Boston, 1963.

3.  Beckwith, T. G., W. L. Buck, and R. D. Marangoni: *Mechanical Measurements*, 3rd ed., Addison-Wesley, Reading, Massachusetts, 1982.

4.  Bendat, J. S., and A. G. Piersol: *Measurement and Analysis of Random Data*, Wiley, New York, 1966.

5.  Benedict, R. P.: *Fundamentals of Temperature, Pressure, and Flow Measurements*, 2nd ed., Wiley, New York, 1977.

6.  Cerni, R. H., and L. E. Foster: *Instrumentation for Engineering Measurement*, Wiley, New York, 1962.

7.  Considine, D. M., (Ed.): *Process Instruments and Controls Handbook*, McGraw-Hill, New York, 1957.

8.  Cook, N. H., and E. Rabinowicz: *Physical Measurement and Analysis*, Addison-Wesley, Reading, Massachusetts, 1963.

9.  Dally, J. W., and W. F. Riley: *Experimental Stress Analysis*, 2nd ed., McGraw-Hill, New York, 1978.

10. Doebelin, E. O.: *Measurement Systems: Application and Design*, 3rd ed., *McGraw-Hill*, New York, 1983.

11. Dove, R. C., and P. H. Adams: *Experimental Stress Analysis and Motion Measurement,* Merrill, Columbus, Ohio, 1964.

12. Frank, E.: *Electrical Measurement Analysis,* McGraw-Hill, New York, 1959.

13. Holman, J. P.: *Experimental Methods for Engineers*, 3rd ed., McGraw-Hill, New York, 1978.

14. Kallen, H. P. (Ed.): *Handbook of Instrumentation and Controls*, McGraw-Hill, New York, 1961.

15. Keast, D. N.: *Measurements in Mechanical Dynamics*, McGraw-Hill, New York, 1967.

16. Lion, K. S.: *Instrumentation in Scientific Research,* McGraw-Hill, New York, 1959.

17. Malmstadt, H. V., and C. G. Enke: *Digital Electronics for Scientists*, Benjamin, New York, 1969.

18. Neubert, H. K. P.: *Instrument Transducers*, Oxford University Press, Fair Lawn, New Jersey, 1963.

19. Schenck, H.: *Theories of Engineering Experimentation*, 2nd ed., McGraw-Hill, New York, 1968.

20. Stein, P. K.: *Measurement Engineering*, Stein Engineering Services, Phoenix, Arizona, 1964.

21. Sweeney, R. J.: *Measurement Techniques in Mechanical Engineering*, Wiley, New York, 1953.

22. Tuve, G. L., and L. C. Domholdt: *Engineering Experimentation*, McGraw-Hill, New York, 1966.

23. Wilson, E. B., Jr.: *An Introduction to Scientific Research*, McGraw-Hill, New York, 1952.

## EXERCISES

1.1 List the elements that are used in an instrumentation system.

1.2 Give an example of an electronic instrumentation system employed in engineering analysis.

1.3 Give an example of an electronic instrumentation system employed in process monitoring.

1.4 Give an example of an electronic instrumentation system employed in process control.

1.5 Why is it frequently necessary to conduct an experimental program in conjunction with an analytical engineering analysis?

1.6 Why is a combined analytical/experimental approach preferred for an engineering analysis?

**1.7** Explain the difference between open-loop and closed-loop control.

**1.8** List several sources of error that must be considered in the design of an instrumentation system.

**1.9** A recorder is specified accurate to $\pm 2$ percent of full scale and full scale is set at 50 mV. Determine the deviation that can be anticipated. Compute the probable percent error when the instrument is used at $\frac{3}{4}$, $\frac{1}{2}$, $\frac{1}{4}$, and $\frac{1}{8}$ scale. State the conclusion that can be drawn from the results of your computation.

**1.10** An instrumentation system that is composed of a transducer, power supply, signal conditioner, amplifier, and recorder will exhibit what accumulated error $\mathscr{E}$ if the accuracies of the individual elements are:

|  | Case 1 | Case 2 | Case 3 | Case 4 |
| --- | --- | --- | --- | --- |
| Transducer | 0.01 | 0.01 | 0.02 | 0.05 |
| Power supply | 0.01 | 0.01 | 0.02 | 0.01 |
| Signal conditioner | 0.02 | 0.01 | 0.05 | 0.01 |
| Amplifier | 0.02 | 0.01 | 0.02 | 0.01 |
| Recorder | 0.03 | 0.01 | 0.02 | 0.01 |

**1.11** Define range and span of an instrument.

**1.12** Determine the error produced by a zero offset $Z_o$ if it is not taken into account in determining the output quantity $Q_o$.

**1.13** Determine the error produced if an instrument sensitivity is $S_1$ instead of the anticipated sensitivity $S$.

**1.14** Determine the error produced if an instrument sensitivity is $S_1$ instead of the anticipated sensitivity $S$ and if a zero offset $Z_o$ is not taken into account in determining the output quantity $Q_o$.

**1.15** Give an example of a transducer that produces error because of its influence on the quantity being measured.

**1.16** Give an example of an instrument with dual sensitivity and explain how it may produce unanticipated error in a measurement.

**1.17** An amplifier in an instrumentation system exhibits a zero drift of 1 percent of full scale per hour. Determine the error if the measurement of $Q_o$ is taken 2.4 hours after the initial zero was established and if the amplifier is operated at one-half of full scale.

**1.18** A pressure transducer exhibits a temperature sensitivity of 0.1 units per degree Celsius and a pressure sensitivity of 2.5 units per megapascal. If the temperature changes 20°C during a measurement of a pressure of 120 MPa, determine the error due to the dual sensitivity of the transducer.

**1.19** The sensitivity of an electrical resistance strain gage is defined as

$$S = \frac{\Delta R/R}{\varepsilon}$$

**Where** $\Delta R$ is the resistance change of the gage due to an applied strain $\varepsilon$.

$R$ is the resistance of the gage.

If the sensitivity $S = 2.0$ for a gage with a resistance of 120 $\Omega$, compute the sensitivity if the gage is connected to the instrument system with lead wires having a total resistance of 12 $\Omega$.

**1.20** Determine the apparent strain indicated by the strain gage lead-wire system described in Exercise 1.19 if the lead wires are subjected to a temperature change of 16°C after the initial zero is established for the system. Note that the lead wires change resistance with temperature according to:

$$\Delta R = R \, \gamma \, \Delta T$$

where $\gamma$ is the temperature coefficient of resistance (0.0039/°C for copper).

**1.21** Describe a suitable transducer for measuring pressure in a shock tube.

**1.22** Place a weight limit on a transducer used to determine the natural frequency of a clamped circular plate fabricated from aluminum and having a diameter of 250 mm and a thickness of 1 mm.

**1.23** Describe calibration procedures for:
(a) A power supply
(b) A pressure transducer
(c) A Wheatstone bridge
(d) An amplifier
(e) A voltmeter

**1.24** Describe a calibration procedure to check the entire instrumentation system if the quantity being measured is:
(a) Strain
(b) Pressure
(c) Temperature
(d) Displacement
(e) Acceleration

**1.25** How is it possible to reduce noise in an electronic measurement system?

# TWO
# VOLTAGE RECORDING INSTRUMENTS

## 2.1 INTRODUCTION

Voltage recording instruments are used in an electronic measurement system to display an output voltage $E_o$ proportional to the quantity being measured $Q_i$ in a form that may easily be read by an operator. If the quantity $Q_i$ is constant with respect to time (static), voltmeters or ammeters are usually employed. Self-balancing potentiometers or a digital voltmeter with an output printer may be used if the voltage varies slowly with time (quasi-static). Quantities that vary rapidly with respect to time (dynamic signals) must be displayed with oscillographs, oscilloscopes, or magnetic tape or disk recorders that can respond to the dynamic input.

In the design of an instrumentation system, it is important that the correct voltage recorder be selected so that the output displayed is first accurate and second in a form that can be easily interpreted and processed. The general characteristics of voltage recorders must be understood before the best recording instrument can be specified for an instrumentation system.

## 2.2 GENERAL CHARACTERISTICS OF RECORDING INSTRUMENTS

The general characteristics that describe the behavior of a recording instrument are input impedance, sensitivity, range, zero drift, and frequency response. Each of these characteristics is described in the following subsections.

### Input Impedance

*Input impedance Z* controls the energy removed from the system by the recording instrument in order to display the input voltage. Consider a simple dc voltmeter used to measure the voltage $E$ of a source. The power loss $P$ through the meter is given by

$$P = E^2/R_m \qquad (2.1)$$

where $R_m$, the resistance of the meter, is the input impedance.[1] It is evident from Eq. (2.1) that the ideal voltmeter should have an input impedance $Z = R_m$ that approaches infinity to reduce the energy removed from the system to zero. Unfortunately, it is impossible to achieve extremely high input impedances in recording instruments at reasonable cost; therefore, some small power loss through the meter must be accepted.

To determine the error produced by a finite input impedance, it is necessary to consider the interaction between two adjacent elements in the instrumentation system. Consider, for example, the Wheatstone bridge–voltage recorder combination shown in Fig. 2.1a. The Wheatstone bridge converts the resistance change $\Delta R_1$ to a voltage $E_i$ with a source resistance $R_s$. By applying Thevenin's theorem to the bridge and the recording instrument, the equivalent circuit shown in Fig. 2.1b is obtained. The Wheatstone bridge is replaced by a voltage generator with a potential $E_i$ and a source resistance $R_s = R_1$. A current $I$ flows in the loop and the recorder input resistance $R_m$ acts as a load on the source in series with $R_s$. The voltage displayed by the recorder $E_m$ is the $IR_m$ drop across the resistor $R_m$. Thus,

$$E_m = IR_m \qquad (2.2)$$

---

[1] In general, the input impedance $Z$ for elements in series is given by

$$Z = \sqrt{R^2 + \left(L\omega - \frac{1}{C\omega}\right)^2}$$

**where**   $R$ is the resistance.
             $C$ is the capacitance.
             $L$ is the inductance.
             $\omega$ is the circular frequency of the input signal.

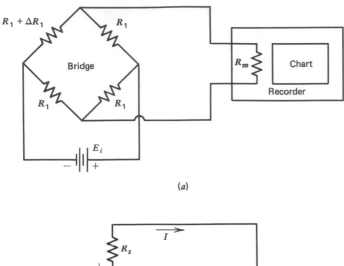

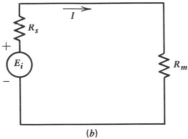

*Figure 2.1* (*a*) Combination of a Wheatstone bridge and voltage recorder. (*b*) Equivalent circuit by Thevenin's theorem.

Since

$$I = \frac{E_i}{R_s + R_m} \qquad (2.3)$$

it is evident that

$$E_m = \frac{E_i}{1 + (R_s/R_m)} \qquad (2.4)$$

Inspection of Eq. (2.4) shows that the meter indication $E_m$ will be less than the source potential $E_i$. The error $\mathscr{E}$ is

$$\mathscr{E} = \frac{E_i - E_m}{E_i} = \frac{R_s/R_m}{1 + (R_s/R_m)} \qquad (2.5)$$

The load error as a function of the ratio of source impedance to recorder impedance is shown in Fig. 2.2. Examination of Fig. 2.2 shows that a ratio $R_s/R_m <$

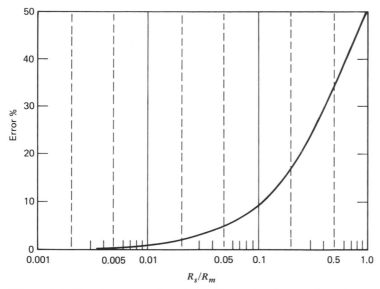

**Figure 2.2** Load error as a function of the ratio of source impedance to recorder impedance.

0.01 gives a load error of less than 1 percent. The rule that the input impedance should be 100 times the source impedance is based on Eq. (2.5) and limits load error to less than 1 percent.

## Sensitivity

The sensitivity $S$ of a voltage recording instrument is given by Eq. (1.2) as

$$S = \frac{d}{E} \tag{2.6}$$

**where** $d$ is the displacement of the pointer or pen.
$E$ is the voltage being measured.

Sensitivity of the recorder is important when measurements of small voltages are to be made. High sensitivity is required to give a sufficiently large pen displacement $d$ for accurate readout.

From Eq. (2.6) it is clear that the voltage $E$ is determined by measuring $d$ and dividing by $S$. Since division is more difficult than multiplication, most manufacturers of recording instruments define a voltage sensitivity $S_R$ as

$$E = dS_R \tag{2.7}$$

where $S_R = 1/S$ is expressed in terms of volts per division of displacement.

High sensitivity $S$ or low reciprocal sensitivity $S_R$ is usually achieved with amplifiers that are incorporated in the recorder. The amplification factor of the amplifier is varied to provide a recorder with several different sensitivities to accommodate a wide range of input voltages.

## Range

The range, which represents the maximum voltage that can be recorded, is determined from Eq. (2.6) as

$$E^* = \frac{d^*}{S} = d^* S_R \tag{2.8}$$

**where** $E^*$ is the maximum voltage or range.
$d^*$ is the width of the chart (fixed for a given instrument).

The form of Eq. (2.8) shows the trade-off that must be made between range and sensitivity . When the sensitivity $S$ is high, the range $E^*$ will be low; and conversely, if the range is high, the sensitivity will be low. A voltage amplifier with a variable amplification factor extends the applicability of a recorder by matching appropriate sensitivity with the input voltage.

## Zero Drift

Most voltage recorders have provisions for adjusting the zero offset so that the pen (pointer) displacement or print out is zero when the input voltage is zero. The position of the zero on the chart may change with time, however, due to instabilities in one or more of the circuits in the recorder. Zero drift is usually due to circuit changes in the amplifier that occur with temperature fluctuations, variations in line voltage, and time.

Zero drift is specified for most recording instruments and can be minimized by using a regulated line voltage, by turning instruments on for a suitable time period before recording, and by controlling the temperature of the room in which the instrument is housed. If measurements are to be made over a long period of time, provisions should be made to periodically check to determine the zero position, thereby accounting for the drift.

## Frequency Response

If the voltage being recorded is dynamic, the recorder should reproduce the transient input without amplitude or time distortion. The ability of a recorder to respond to dynamic signals is determined by its frequency response, which is based on the recorder's steady-state response to a sinusoidal input

$$E_i = C_i \sin \omega t \tag{2.9}$$

The output $E_o$ of the recorder is of the form

$$E_o = C_o \sin (\omega t + \phi) \qquad (2.10)$$

The amplitude of the output represented by $C_o$ may be different from the input amplitude $C_i$. A phase shift $\phi$ may also occur. Both amplitude ratio $C_o/C_i$ and phase shift $\phi$ change as the circular frequency $\omega$ of the input is varied. Curves such as those given in Fig. 2.3 for $C_o/C_i$ and $\phi$ as a function of $\omega$ define the frequency response of a recording system. A more complete discussion of the frequency response of second-order systems is given in Section 2.5.

Frequently, specifications for recorders give the amplitude ratio $C_o/C_i$ in terms of decibels. It should be noted that

$$N_{\mathrm{dB}} = 20 \log_{10}(C_o/C_i) \qquad (2.11)$$

**where**    $N_{\mathrm{dB}}$ is the number of decibels.
$C_o$ and $C_i$ are voltages.

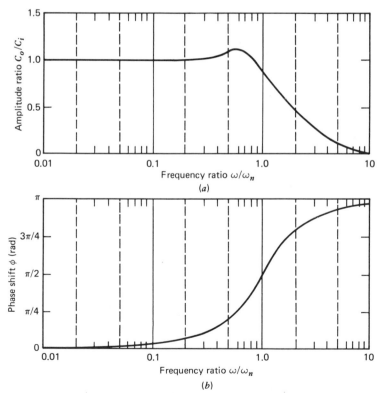

*Figure 2.3*  Response of a recorder to harmonic excitation. (*a*) Amplitude as a function of frequency. (*b*) Phase shift as a function of frequency.

Several values from Eq. (2.11) which are often useful are presented in Table 2.1.

Reference to Table 2.1 shows that significant errors result in recording dynamic signals even for relatively small $N_{dB}$. For instance, a recorder specification that indicates that the frequency response is within $\pm 3dB$ from direct current to 100 Hz implies an error of $+41$ percent (1.413) for $+3dB$ and $-30$ percent (0.708) for $-3dB$ over the range of frequencies specified.

## 2.3 STATIC VOLTMETERS

There are four types of voltmeters in general use today for measurement of static phenomena: the analog voltmeter, the amplified analog voltmeter, the potentiometer, and the digital voltmeter. All but the digital voltmeter use the D'Arsonval galvanometer to indicate the voltage.

### The D'Arsonval Galvanometer

The *D'Arsonval galvanometer* illustrated in Fig. 2.4 is the basic device used in detecting and measuring dc current. The galvanometer design incorporates a coil of wire that is supported in a magnetic field with either jeweled bearings or torsion springs. When a current $I$ flows in the coil, it rotates in the magnetic field until restrained by springs that are a part of the suspension system.

The torque $T_1$ developed by the current flow in the coil is

$$T_1 = NBlDI \tag{2.12}$$

**where** $N$ is the number of turns in the coil.
$B$ is the flux density of the magnetic field.
$l$ is the axial length of the field.
$D$ is the mean coil diameter.

**TABLE 2.1** Conversion of Voltage Ratio $C_o/C_i$ to $N_{dB}$

| $C_o/C_i$ | $N_{dB}$ | $C_o/C_i$ | $N_{dB}$ |
|-----------|----------|-----------|----------|
| 1 | 0 | 1 | 0 |
| 1.01 | 0.086 | 0.99 | $-0.087$ |
| 1.02 | 0.172 | 0.98 | $-0.175$ |
| 1.05 | 0.424 | 0.95 | $-0.446$ |
| 1.10 | 0.827 | 0.90 | $-0.915$ |
| 1.20 | 1.583 | 0.80 | $-1.938$ |
| 1.50 | 3.522 | 0.707 | $-3.012$ |
| 2.00 | 6.020 | 0.500 | $-6.021$ |

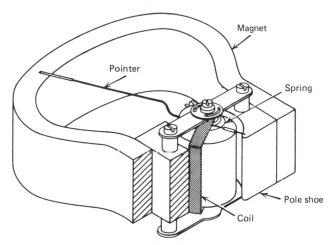

*Figure 2.4*   Construction details of a D'Arsonval galvanometer.

The torque $T_2$ developed by the restraining springs is

$$T_2 = K\theta \tag{2.13}$$

**where**   $\theta$   is the angle of rotation of the coil.
$K$   is the spring constant.

The equilibrium condition $T_1 - T_2 = 0$ leads to

$$\theta = \frac{NBlD}{K}I = SI \tag{2.14}$$

where $S = ((NBlD)/K)$ is the sensitivity or calibration constant for the galvanometer.

The sensitivity of the galvanometer can be changed by varying the parameters $N$, $B$, $l$, $D$, or $K$; however, for economic reasons, galvanometers are designed to measure small currents. A typical D'Arsonval galvanometer will exhibit a full-scale deflection at 20 μA with a coil resistance of 30 Ω.

## Ammeter

An *ammeter* consists of a D'Arsonval galvanometer with a shunt resistance as shown in Fig. 2.5. The input current $I_i$ divides with $I_m$ passing through the meter and $I_{sh}$ passing through the shunt. It is clear that

$$I_m = \frac{I_i}{1 + (R_m/R_{sh})} \tag{2.15}$$

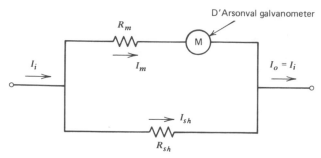

*Figure 2.5*   A D'Arsonval galvanometer being used as an ammeter.

A D'Arsonval galvanometer with a full-scale current capability of 1 mA and a resistance $R_m$ of 50 $\Omega$ can be used to measure any current greater than 1 mA by properly selecting the shunt resistance from the expression

$$R_{sh} = \frac{I_m^*}{I_i^* - I_m^*} R_m \qquad (2.16)$$

where $I_i^*$ and $I_m^*$ are full-scale input and meter currents, respectively. For example, the 1-mA–50-$\Omega$ galvanometer can be used to measure 5 A (full scale) if $R_{sh} = 0.01$ $\Omega$.

One difficulty encountered in using an ammeter of this type is the need to cut or rearrange the wires in the circuit so that the ammeter may be placed in the path of the current flow. To alleviate this problem, a clamp-like probe is clipped over the wire in which the current is to be determined. The probe contains a magnetic core that concentrates the magnetic field (which is proportional to the current in the wire) around the wire. A Hall-effect transducer, mounted in an air gap in the magnetic core, produces an output voltage that is directly proportional to the magnetic field and, thus, to the current in the wire. A clip-on-type ammeter, that can measure from 1 to 400 A over a frequency range from dc to 1 KHz on conductors up to $1\frac{1}{4}$ in. (32 mm) in diameter is shown in Fig. 2.6.

## DC Voltmeters

A D'Arsonval galvanometer is converted to a dc voltmeter by using a series resistor as shown in Fig. 2.7. The appropriate series resistor $R_{sr}$ is determined from the expression

$$R_{sr} = \frac{E^*}{I_m^*} - R_m \qquad (2.17)$$

**where**   $E^*$ is the full-scale voltage.
   $I_m^*$ is the full-scale current for the meter.

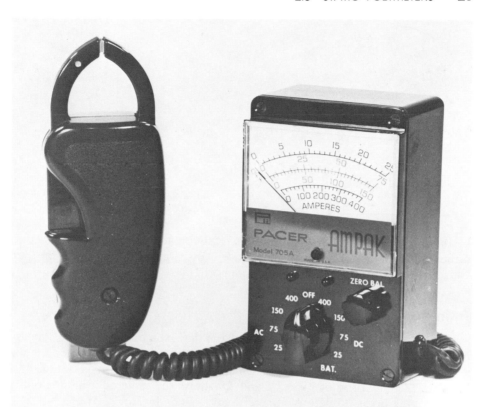

**Figure 2.6** An analog AC/DC ammeter with clamp-on probe. (Courtesy of Pacer Industries, Inc., Chippewa Falls, Wisconsin.)

For example, a 20-μA–30-Ω galvanometer is converted to a 100-mV voltmeter with a series resistor $R_{sr} = 4970$ Ω.

A dc analog multimeter is illustrated in Fig. 2.8 where several different series resistors are used with a single meter to give a multirange instrument. Typical *multimeters* have approximately nine ranges with full-scale readings from 100 mV to 1000 V. Analog multimeters are low-cost instruments with accuracies of ±2 to 3 percent of full scale. When operated on the lower voltage ranges,

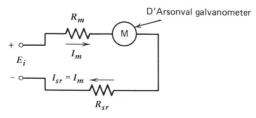

**Figure 2.7** A D'Arsonval galvanometer being used as a voltmeter.

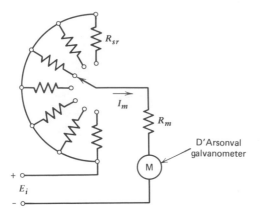

**Figure 2.8** Circuit diagram for a multi-range dc voltmeter.

the input impedance of the instrument is relatively low and loading errors due to the voltmeter can occur. The impedance is often given in terms of ohms per volt full scale with 20,000 Ω/V being common.

## Voltmeter Loading Errors

Whenever a voltmeter draws current from the voltage source during a measurement, an error will result due to voltmeter load. To illustrate voltmeter load, consider the equivalent circuit in Fig. 2.9 that represents the voltage source and the meter. Comparison of Figs. 2.1 and 2.9 shows the same equivalent circuit; therefore, the error $\mathscr{E}$ due to voltmeter load is given by Eq. (2.5) as

$$\mathscr{E} = \frac{R_s/(R_m + R_{sr})}{1 + R_s/(R_m + R_{sr})}$$

where $R_s$ is the output impedance of the voltage source and $(R_m + R_{sr})$ replaces $R_m$ in Eq. (2.5). Since $R_m \ll R_{sr}$

$$\mathscr{E} = \frac{R_s/R_{sr}}{1 + (R_s/R_{sr})} = \frac{1}{1 + (R_{sr}/R_s)} \qquad (2.18)$$

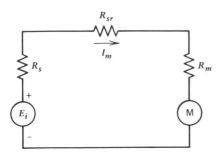

**Figure 2.9** Equivalent circuit for a voltage source and a voltmeter.

Reference to Fig. 2.2 shows that a ratio $R_{sr}/R_s > 100$ is required to reduce voltmeter loading errors to less than 1 percent. Since $R_{sr}$ decreases as the full-scale range is reduced, the input impedance is low when the multimeter is used on sensitive scales and the voltmeter readings must be adjusted for loading errors.

## Amplified Voltmeters

Difficulties in measuring very small voltages while maintaining a high input impedance can be resolved by using a high-gain amplifier in conjunction with the D'Arsonval meter. A schematic of this circuit is shown in Fig. 2.10 where an amplifier is used between the voltage source and the meter. The voltage output from the amplifier is

$$E_o = GE_i \tag{2.19}$$

where $G$ is the gain of the amplifier.

Since $R_{sr} >> R_m$, the effect of the amplifier is to permit an increase in $R_{sr}$ by a factor equal to the gain $G$. Note that Eq. (2.17) becomes

$$R_{sr} = G\frac{E^*}{I_m^*} \tag{2.20}$$

Thus, the input impedance can be increased by a factor $G$ while maintaining the sensitivity of the meter.

The amplifier also permits the meter to be used to measure very small voltages, since

$$E_i = \frac{E_m^*}{G} \tag{2.21}$$

Thus, the meter sensitivity can be increased by a factor $G$ while maintaining the input impedance of the meter.

An example of a multimeter with high amplification is the Keithley Model 148, which has 18 ranges from 10-nV full scale to 100-mV full scale. The input

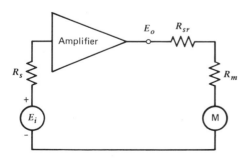

*Figure 2.10* Equivalent circuit for a voltage source and an amplified voltmeter.

impedance varies from 1 kΩ to 1 MΩ as the range is increased from 10 nV to 100 mV.

## Potentiometric Voltmeters

*Potentiometers* are null-balance instruments in which an unknown voltage $E_x$ is compared to a precision reference voltage $E_r$. The basic potentiometer, shown in Fig. 2.11, contains a reference source that energizes a slide-wire resistor of length $l$. As the wiper is moved along the slide-wire, an adjustable voltage $E_{sw}$ is obtained, which is given by

$$E_{sw} = \frac{x}{l}E_r$$

where $x$ is the wiper position along the slide-wire resistor.

If $E_{sw} \neq E_x$, the galvanometer deflects to indicate this out-of-balance. The slide-wire wiper is then adjusted until the galvanometer returns to the null position and

$$E_x = E_{sw} = \frac{x}{l}E_r \tag{2.22}$$

With the potentiometer balanced, no current flows; therefore, there is no voltage drop across the instrument. The slide-wires used in these instruments exhibit a uniform resistance (linear variation with wiper position); therefore, the scale reading $x$ provides an accurate means of reading the voltage without a meter movement and an associated load error. Since the galvanometer is used only to indicate null or zero voltage, it can be very sensitive yet inexpensive. Extremely accurate readings of voltage can be made with potentiometric voltmeters.

For example, the Leeds Northrup Model 7555 potentiometer covers the

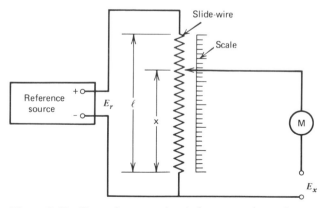

*Figure 2.11*   Potentiometer circuit for measuring voltage.

range from $-0.5$ $\mu$V to $+1.611050$ V with six decimal precision. Accuracy of $\pm 0.005$ percent of full scale or $\pm 0.1$ $\mu$V can be achieved with this relatively low-cost instrument.

## Digital Voltmeters

Digital voltmeters (DVM) offer many advantages over the analog-type meter, such as speed in reading, increased accuracy, better resolution, and the capability of automatic operation. Digital voltmeters display the measurement with lighted numerals, as shown in Fig. 2.12, rather than as a pointer deflection on a continuous scale as with analog meters. Digital multimeters are available to read current, resistance, and ac and dc voltages. The DVM may be used together with a multiplexer and a digital printer to provide simple but reliable automatic data logging systems.

The range of a DVM is determined by the number of full digits in the display. For example, a four-digit DVM can record a count of 9999. If the full scale of the DVM is set at 1 V, the count of 9999 provided by the four digits would register a reading of 0.9999 V. Some DVMs are equipped with partial digits to extend the range. The partial digit can only display the numbers 0 and 1; nevertheless, it is very useful since it permits readings in the overrange region beyond full scale. As an example, consider use of a four-digit DVM for measuring 10.123 V. Since only four digits are available, the meter set on the 10-V scale would read 10.12 V. The last digit (3) would be truncated and lost. If a $4\frac{1}{2}$-digit DVM is employed for the same measurement, the extra partial digit permits 100 percent overranging and a maximum count of 19999. With the $4\frac{1}{2}$-digit meter, the voltage 10.123 would be accurately displayed.

***Figure 2.12*** A $5\frac{1}{2}$-digit digital multimeter. (Courtesy of Keithley Instruments, Inc.)

Overranging may be expressed as a percentage of full scale. For instance, a four-digit DVM with 100 percent overrange displays a maximum reading of 19999. Similarly, with a 20 percent overrange, the maximum display is 11999. In some instances, the overrange capability of the DVM is expressed in terms of the specified range. The four-digit DVM with 100 percent overrange, which has a maximum display of 19999, could be specified with full-scale ranges of 2, 20, 200 V, etc., and with no overrange specification.

Resolution of a DVM is the ratio of the maximum number of counts that can be displayed to the least number of counts. For the four-digit DVM, resolution is 10,000 to 1 or 0.01 percent. Usually, the overrange capability is not considered in determining resolution.

The sensitivity of a DVM is the smallest increment of voltage that can be detected and is determined by multiplying the lowest full-scale range by the resolution. Therefore, a four-digit DVM with a 100-mV lowest full-scale range has a sensitivity of $0.0001 \times 100$ mV = 0.01 mV.

Accuracy of a DVM is usually expressed as $\pm x$ percent of the reading $\pm N$ digits.

The simplified signal flow diagram for a digital multimeter, shown in Fig. 2.13, illustrates the overall features of integrating-type digital meters. The input to the multimeter may be an ac voltage, a dc voltage, a current, or a resistance;

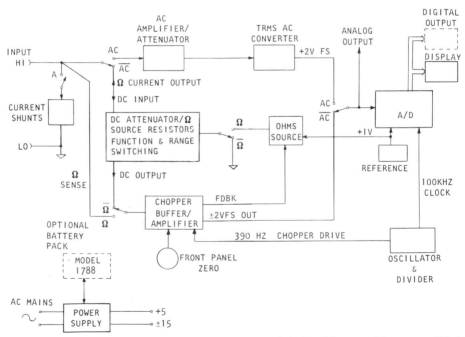

***Figure 2.13***  Simplified signal flow diagram for a digital multimeter. (Courtesy of Keithley Instruments, Inc.)

however, in all cases, the input is ultimately amplified or attenuated and converted to a dc voltage with variable gain amplifiers such as the ac amplifier/attenuator and the chopper buffer/amplifier shown in Fig. 2.13. The gain of these amplifiers is automatically adjusted by control logic so that the voltage applied to the analog-to-digital (A/D) converter is within specifications so as to avoid an overload condition.

The A/D converter changes the dc voltage input to a proportional clock count by using the dual-slope integration technique illustrated in Fig. 2.14. There are three different operations in the dual-slope integration technique for A/D conversion. First, during auto zero, the potential at the integrator output is zeroed for a fixed time, say 100 ms. Second, the dc input is integrated with respect to time for a fixed period, say 100 ms. The output of the integrator is a linear ramp with respect to time, as shown in Fig. 2.14. At the end of the run up, the dc input voltage is disconnected from the integrator and the third operation, run down, is initiated. Run-down time may vary from zero to, say, 200 ms and will depend on the charge developed on the integrating capacitor during run up. Since the discharge rate is fixed during run down, the larger the charge on the integrating capacitor, the longer the discharge time. Since both run up and run down produce slopes on the voltage–time trace, this conversion method from voltage to time is called dual-slope integration.

A counter is started at the beginning of run down and operates until the output voltage from the integrator crosses zero. The accumulated time is proportional to the dc voltage applied to the integrator. This time count is then displayed as the voltage. Polarity, range, and function information are provided by the controller and are also displayed.

The characteristics of a digital voltmeter may be altered by changing the

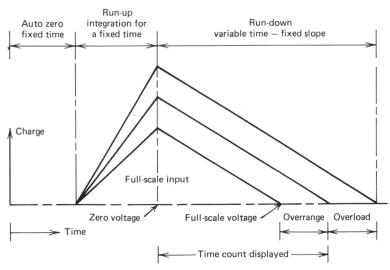

*Figure 2.14*  Dual-slope integration technique for A/D conversion.

number of digits, the time interval for integrator run up, and the frequency of the clock. A typical bench-type DVM with $3\frac{1}{2}$ digits and a maximum count of 1999 is designed with five different ranges: $\pm 199.9$ mV, $\pm 1.999$ V, $\pm 19.99$ V, $\pm 199.9$ V, and $\pm 1999$ V. The highest sensitivity is 100 $\mu$V on the 200-mV range. Accuracy is $\pm 0.1$ percent of the reading plus two digits. The clock frequency is 200 kHz and the integration time is 100 ms. The reading rate varies from 2.4 to 4.7 readings per second, depending on the input.

System DVMs are more complex than bench-type DVMs, since the former are provided with the digital control needed to interface with other components of an automatic data processing system. A typical data processing system consists of a scanner for switching input voltages into the DVM, a data log consisting of a random access memory (RAM), where the output is stored, and special-purpose calculators/computers that take the stored data and perform the required reduction, manipulation, and analysis of data according to programed instructions. System DVMs are higher performance devices than bench DVMs. The number of digits is usually increased to $6\frac{1}{2}$, which gives a resolution greater than 1 part in $10^6$. Clock frequencies are increased to 20 MHz to give a reading rate of 20 readings per second while maintaining an integration time of 50 ms. Microprocessors are added to control the voltage measurement and to control the interface with associated data processing instruments.

## 2.4 QUASI-STATIC VOLTMETERS

There are two different approaches used to record *quasi-static voltages*, which are voltages that change relatively slowly with respect to time. The first approach utilizes a servo-balanced potentiometric circuit where the null condition is achieved automatically with servomotors that are driven by an amplified error signal. The second approach utilizes a digital voltmeter and an associated printer or data storage instrument. With the servo-balanced recorder, a pen is driven over a paper chart to provide a continuous record of the voltage fluctuations with time. With DVM data logging, a number of voltage readings are made over the period of voltage fluctuation, say 10 readings per second, so that sufficient discrete points are available to establish the voltage–time trace.

### Strip-Chart Recorders

A strip-chart recorder utilizes a servomotor-driven null-balance potentiometric circuit similar to the one illustrated in Fig. 2.15. The input signal from the transducer is amplified and used as a command signal for a servo amplifier. The signal from the servo amplifier drives a servomotor that positions a wiper along a slide-wire resistor. Since the slide-wire resistor is across a reference voltage, the wiper picks up a feedback voltage from the slide wire that is proportional to the position of the wiper along the wire. The servo amplifier receives

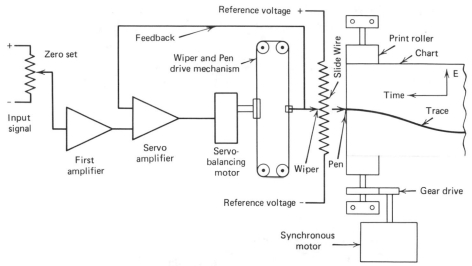

***Figure 2.15*** Components of a strip-chart recorder.

this feedback voltage and compares it to the command signal. The output from the servo amplifier is proportional to the difference between the two signals. When the servomotor has adjusted the wiper so that the difference or error signal from the servo amplifier is zero, the system is in balance and the wiper position provides an indication of the input voltage.

A permanent record of the input voltage is obtained by connecting a pen to the wiper. As the wiper is positioned along the slide-wire resistor, the voltage is recorded as an ink trace on a roll of chart paper. The chart paper is moved at a constant velocity in a direction perpendicular to the wiper motion by a print roller that is driven by a clock motor and a suitable gear train. Distance along the length of the chart is then proportional to time.

A typical example of a commercial strip-chart recorder is illustrated in Fig. 2.16. Chartwidths are from 5 to 10 in. (120 to 250 mm), and the *response time* (the time required for the servomotor to move the pen across the width of the chart) is typically 0.5 s. The sensitivity of strip-chart recorders can be varied by attenuating the output from the first amplifier. Typical sensitivities range from 5 mV to 100 V for the full width of the chart paper. Chart speeds can be varied by changing the speed of the print roller. Speeds from 1 in./hr to 8 in./ min (25 mm/hr to 250 mm/min) are common. The input impedance will depend on the details of the design and may be potentiometric (i.e., no current flow at null balance) or about 1 MΩ, which corresponds to the input impedance of the first amplifier. Accuracy of ±0.2 percent is typical.

Strip-chart recorders can be modified to permit intermittent multisignal recording instead of continuous single-signal recording. With this modification, the input from several transducers (say, thermocouples) are switched in sequence into the servo system. Balance is achieved for each input in 2 to 5 s, and then

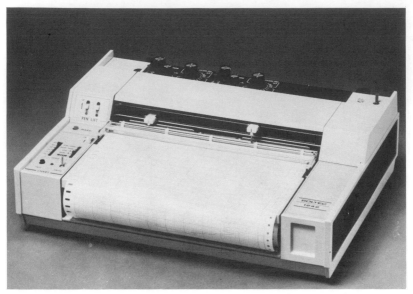

***Figure 2.16*** A strip-chart recorder. (Courtesy of Soltec, Corporation.)

the corresponding voltage is printed as a single point on the chart with a rotary printing mechanism instead of a pen. Each point on the chart is printed with a channel number so that the outputs from individual transducers can be quickly identified. Up to 24 input signals can be accommodated on a single strip chart; thus, this type of recorder provides an extremely low-cost method of accurately recording several voltages that are changing slowly with time.

## *X-Y* Recorders

The *x-y* recorder is another type of instrument that utilizes servo-driven motors and null-balance potentiometric circuits. The operation of the *x-y* recorder is similar to the strip-chart recorder, except that the *x-y* recorder simultaneously records two voltages along orthogonal axes (usually referred to as the *x* and *y* axes) as indicated in Fig. 2.17. The *x-y* recorder uses sheets of graph paper (either $8\frac{1}{2}$ by 11 in. or 11 by 17 in.) for recording purposes instead of a strip chart.

The sensitivities of an *x-y* recorder will depend on the design of the input amplifiers. Some models have plug-in modules that permit the characteristics of the recorder to be changed. A typical recorder has an attenuated input amplifier with about 10 different sensitivities ranging from 0.5 mV/in. to 10 V/in. in conventional models or 0.025 mV/mm to 0.5 V/mm in models with metric calibration. The input impedance of both amplifiers is usually about 1 MΩ. The *deadband,* that small zone about the balance point where friction inhibits exact zeroing of the error signal, is about 0.1 percent of full scale. Accuracy, which includes deadband error, is typically 0.2 percent.

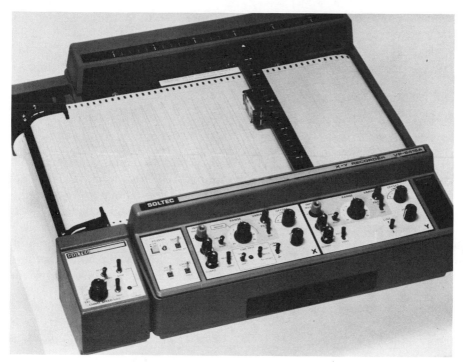

**Figure 2.17** An *x-y* recorder displaying one voltage along the *x* axis and another along the *y* axis. (Courtesy of Soltec, Corporation.)

The dynamic recording characteristics of an *x-y* recorder depend upon the amplitude of the quantity being measured, the acceleration capabilities of the servodrives, and the slewing speed. The *slewing speed* is the maximum velocity of the pen when it is being driven by both servodrives. Slewing speeds of 20 in./ s (500 mm/s), with peak accelerations of 1000 in./s² (25 m/s²) of an individual servodrive, are common. Typical dynamic performance is represented in Fig. 2.18, where the linear region of operation is defined in terms of the amplitude and frequency of the input signals.

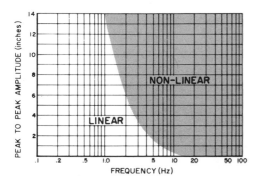

**Figure 2.18** Dynamic frequency response of a typical *x-y* recorder.

Many x-y recorders are equipped with a time base so that the recorder can also be used to record the variable y as a function of time t instead of the variable x. In this mode of operation, the x-y recorder (acting as a y-t recorder) is similar to a very slow oscillograph. The time base provides the input signal for the x-axis servo system. Sweep speeds of 0.5 to 100 s/in. (0.025 to 5 s/mm) are available in eight calibrated ranges in a typical model.

## Data Loggers

A basic data logging system consists of a scanner, a digital voltmeter, and a recorder. Such a system can be employed to record the output from a large number of transducers (as many as 1000) at a rate of approximately 15 readings per second, or it can be used to continuously monitor a single channel. Since data loggers are relatively fast (10 to 20 readings per second), a system controller is needed to direct the scanner to each new channel, to control the integration time for the DVM, and to transfer the output from the DVM to the recorder. A block diagram of a typical multichannel data logging system is shown in Fig. 2.19.

The system controller is a microprocessor that uses two separate busses—

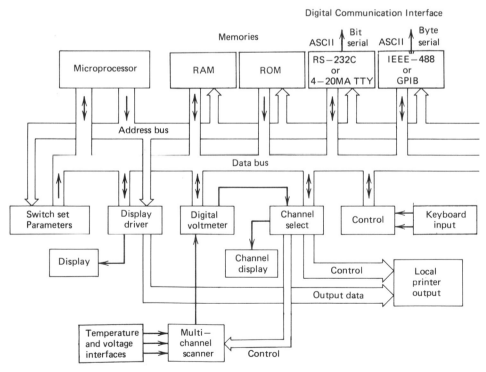

*Figure 2.19* Block diagram for a typical data logging system.

one for data transfer and the other for memory addressing. The software, which directs the operation of the controller, is stored in a read-only memory (ROM) and a random-access memory (RAM). The system operating programs are permanently stored in the ROMs, which are programed during manufacture of the instrument. The operator uses a keyboard to enter individual channel parameters and other program routines. The input is stored in the RAM.

The scanner contains a bank of switches (usually three pole) that serve to switch the two leads and the shield from the input cable to the integrating digital voltmeter. In most cases, high-speed (1000 channels per second) solid-state switching devices (J field-effect transistors) are employed. The scanner operation is directed by the system controller, and several modes of operation are possible that include single-channel recording, single scan of all channels, continuous scan, and periodic scan. In the single-channel mode, a preselected channel is continuously monitored at the reading rate of the system. In the single-scan mode, the scanner makes a single sequential sweep through a preselected group of channels. The continuous-scan mode is identical to the single-scan mode, except that the system automatically resets and recycles on completion of the previous scan. The periodic scan is simply a single scan that is initiated at preselected time intervals such as 1, 5, 15, 30, or 60 min. The scanner also provides a visual display of the channel number and a code signal to the controller to identify the transducer being monitored.

The transducer signal is switched through the scanner to a high-quality integrating digital voltmeter that serves as an A/D (analog-to-digital) converter. The coverage of digital voltmeters beginning on page 33 describes the operation of the DVM. Data loggers often incorporate $4\frac{1}{2}$ digit DVMs capable of counting between $+19999$ and $-19999$. With this type of DVM, the decimal is positioned automatically, depending upon the range that has been selected. For example, on the 20-mV range, a count of 09400 is automatically interpreted and stored as 9.400 mV. With this type of DVM, resolution is $\pm 1$ $\mu$V and accuracy is typically $\pm 0.5$ percent of full scale. The time required for the DVM to measure the voltage is about 0.05s. This required reading period establishes the speed of the system at about 20 readings per second. Higher speeds are possible, but resolution and accuracy of the readout are sacrificed. For example, a rate of 125 channels per second is possible with a $3\frac{1}{2}$ digit DVM that provides 100-$\mu$V resolution.

The output from a data logger is displayed with a digital panel meter that indicates the voltage units, polarity, and channel number. A permanent record is usually made with a line printer that records the output data and identification on a paper tape. However, the output of most data logging systems can be recorded with numerous other devices, such as magnetic tapes, auxiliary printers, or disk memories. One of the principal advantages of a data logging system is the capability for processing the data in real time with an on-line computer.

A modern data logging system is shown in Fig. 2.20.

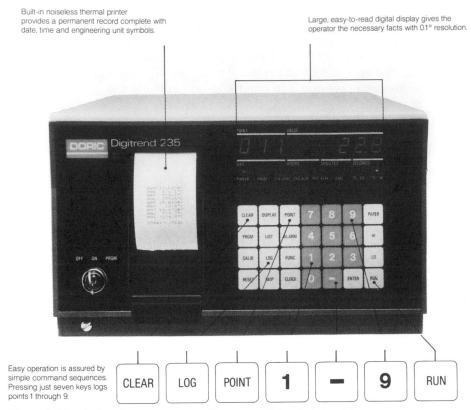

Built-in noiseless thermal printer provides a permanent record complete with date, time and engineering unit symbols.

Large, easy-to-read digital display gives the operator the necessary facts with 0.1° resolution.

Easy operation is assured by simple command sequences. Pressing just seven keys logs points 1 through 9.

CLEAR LOG POINT 1 — 9 RUN

*Figure 2.20* A data logging system. (Courtesy of Doric Scientific Division of Emerson Electric Company.)

## Data Acquisition Systems

Data acquisition systems are similar to data logging systems in that they accept input from a large number of transducers and automatically process the data. There are two principal differences between data logging and data acquisition systems. First, data acquisition systems are much faster (sample rates up to 20,000 per second), and second, the computer architecture and software available is different. The higher sampling rates are achieved by replacing the integrating DVM with high-speed, successive-approximation, analog-to-digital converters. These converters utilize sample and hold amplifiers which sample and hold the data while the A-to-D converter is performing a previous conversion. Systems available today are based on 16-bit microprocessors and smart terminals (terminals with auxiliary memory and control features) which permit a significant amount of on-site processing. Also, software is available which provides the operator with user friendly menus that guide and verify the setup sequence. Many programs are also available for data analysis. A typical example of a modern data acquisition system is shown in Fig. 2.21.

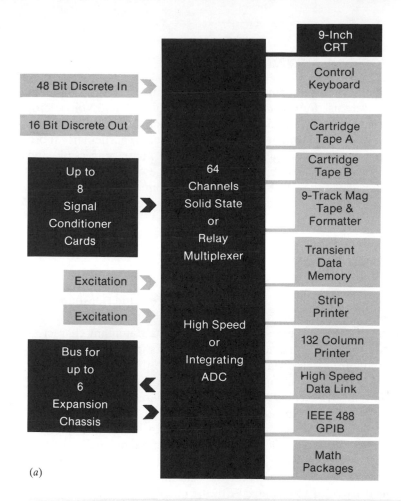

| | | 9-Inch CRT |
| 48 Bit Discrete In | | Control Keyboard |
| 16 Bit Discrete Out | | Cartridge Tape A |
| | 64 Channels Solid State or Relay Multiplexer | Cartridge Tape B |
| Up to 8 Signal Conditioner Cards | | 9-Track Mag Tape & Formatter |
| Excitation | | Transient Data Memory |
| Excitation | | Strip Printer |
| | High Speed or Integrating ADC | 132 Column Printer |
| Bus for up to 6 Expansion Chassis | | High Speed Data Link |
| | | IEEE 488 GPIB |
| | | Math Packages |

(a)

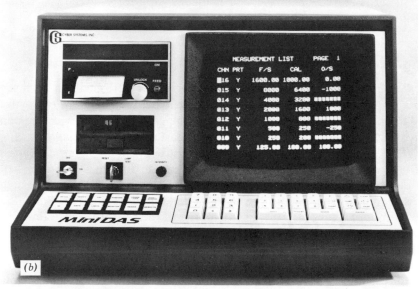

(b)

***Figure 2.21*** Microprocessor-based data acquisition system. (Courtesy of Cyber Systems, Inc.) (*a*) Input and output from a data acquisition system. (*b*) Data acquisition system showing CRT, printer, key pad, and cartridge.

Significant progress can be anticipated in the near future in the development of data acquisition systems. Future models will incorporate the 32-bit microprocessor with 64k memory chips and thus provide a microcomputer with substantial power at the test site. Future software developments, particularly those related to graphics, will permit complete data acquisition, data processing, and graphical display of results in essentially real time at the test location. These advances will significantly enhance the value of developmental testing in prototype evaluations and will greatly reduce the time required in a product development cycle.

## 2.5   DYNAMIC VOLTMETERS

Measuring transient phenomena, where the signal from the transducer is a rapidly changing function of time, is the most difficult and most expensive measurement in experimental work. Frequency response is the dominant characteristic required of the recording instrument in dynamic measurements and usually accuracy and economy are sacrificed in order to improve the response capabilities. Four totally different instruments are used to record transient voltages; namely, the oscillograph, the oscilloscope, the digital oscilloscope, and the magnetic tape recorder. The oscillograph utilizes a galvanometer, which incorporates a highly refined D'Arsonval movement, to drive a pen or a light beam over a moving strip of chart paper. The oscilloscope utilizes a focused beam of electrons to produce a voltage–time trace on a phosphor screen. The digital oscilloscope utilizes an analog-to-digital converter to digitize the incoming signal, which is then stored in a memory prior to display. With the magnetic tape recorder, the dynamic signal is stored on high-speed magnetic tape for later playback and display on either an oscillograph or an oscilloscope.

### Oscillograph Recorders

*Oscillograph recorders* employ galvanometers to convert the dynamic input signal to a displacement on a moving strip of chart paper. There are two types of oscillographs: the *direct-writing type*, where the galvanometer drives the pen or hot stylus used to write on a moving strip of chart paper, and the *light-writing type*, where the galvanometer drives a mirror that deflects the light beam used to write on a moving strip of photosensitive paper. The frequency responses of the two different types of oscillographs differ markedly. The relatively high inertia associated with the pen or hot stylus of the direct-writing type limits the frequency response to about 150 Hz. The inertia of the mirror is much lower in a light-writing type, and a frequency response as high as 13 kHz has been achieved.

Both types of recorders can be used to record low-frequency signals; however, the direct-writing oscillograph is usually preferred over the light-writing type, since the records on chart paper with either the pen or hot stylus are less expensive, more permanent, and of higher quality than comparable recordings made on photosensitive paper. A modern, four-channel, pen-type, direct-writing oscillograph is shown in Fig. 2.22.

The light-writing oscillograph is often used as the recording instrument for dynamic signals that contain frequency components between zero and 13 kHz. A schematic diagram illustrating the operating principle of the light-writing oscillograph is shown in Fig. 2.23a. A typical oscillograph utilizes several galvanometers that are mounted in a row of holes in magnetic blocks. A mirror, mounted on the moving member of the galvanometer, reflects a focused beam of light onto a moving strip of photosensitive paper. The angular rotation of the mirror produces a deflection of the light beam, which is amplified optically to provide a trace on the photosensitive paper. The deflection of the trace from a null position is proportional to the dynamic input voltage. The speed of the strip of photosensitive paper is controlled by a motor and gear train. The paper speed can be adjusted to give a specified time scale on the abscissa of the record (i.e., along the length of the strip of paper).

The galvanometer for an oscillograph is a highly refined version of the D'Arsonval movement described in Section 2.3. The essential components of this type of galvanometer (see Fig. 2.23b) include a filament suspension system,

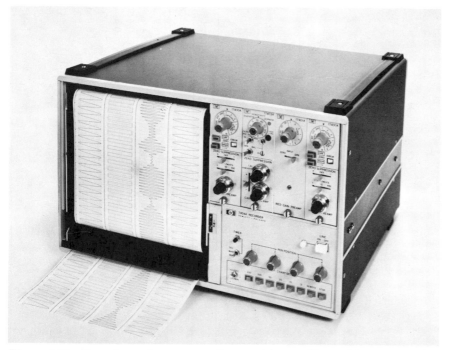

*Figure 2.22*   Four-channel, pen-type, direct-writing, oscillographic recorder. (Courtesy of Hewlett-Packard.)

**TABLE 2.2** Galvanometer Characteristics (Metric)

| Type No. | Nominal Undamped Natural Frequency (Hz) | Flat (±5%) Frequency Response (Hz) | Required External Damping Resistance (Ω) | Nominal Coil Resistance (Ω) | Sensitivity in 30-cm Optical Arm of all Honeywell Oscillographs | | | | | | Maximum Safe Current (mA) | Maximum P-P Deflection, w/ ±2% Linear (cm) | Balance Standard (cm/g) | Precision (cm/g) |
| | | | | | Current (±5%) | | Galvanometer Voltage | | Circuit Voltage | | | | | |
| | | | | | $I_g$ (μA/cm) | $\frac{1}{I_g}$ (cm/μA) | $V_g$ (mV/cm) | $\frac{1}{V_g}$ (cm/mV) | $V_c$ (mV/cm) | $\frac{1}{V_c}$ (cm/mV) | | | | |
|---|---|---|---|---|---|---|---|---|---|---|---|---|---|---|
| | | | | | Electromagnetic-Damped Types | | | | | | | | | |
| M18-5000 | 18 | 0–12 | 4200 | 220.0 | 0.169 | 5.920 | 0.037 | 37.2 | 0.749 | 1.33 | 5 | 20 | | NA |
| M24-350A | 24 | 0–15 | 350 | 135.0 | 0.59 | 1.700 | 0.080 | 12.4 | 0.288 | 3.48 | 5 | 20 | 0.124 | NA |
| M40-350A | 40 | 0–24 | 350 | 66.0 | 1.62 | 0.617 | 0.107 | 9.4 | 0.673 | 1.49 | 15 | 20 | 0.089 | 0.045 |
| M100-350 | 100 | 0–60 | 350 | 75.5 | 2.48 | 0.401 | 0.187 | 5.33 | 1.06 | 0.95 | 10 | 20 | 0.056 | 0.028 |
| M200-350 | 200 | 0–180 | 350 | 62.0 | 10.1 | 0.100 | 0.626 | 1.61 | 4.13 | 0.24 | 10 | 20 | 0.041 | 0.020 |
| M400-350 | 400 | 0–360 | 350 | 125.0 | 30.4 | 0.033 | 3.80 | 0.264 | 14.4 | 0.068 | 15 | 20 | 0.041 | 0.020 |
| M600-350 | 600 | 0–540 | 350 | 320.0 | 51.2 | 0.020 | 16.4 | 0.061 | 34.2 | 0.028 | 15 | 20 | 0.041 | 0.020 |
| M40-120A | 40 | 0–24 | 120 | 30.0 | 3.15 | 0.317 | 0.095 | 10.6 | 0.472 | 2.12 | 15 | 20 | 0.089 | 0.045 |
| M100-120A | 100 | 0–60 | 120 | 52.2 | 3.94 | 0.254 | 0.206 | 4.8 | 0.671 | 1.48 | 15 | 20 | 0.040 | 0.020 |
| M200-120 | 200 | 0–120 | 120 | 62.0 | 10.1 | 0.100 | 0.626 | 1.61 | 1.83 | 0.548 | 15 | 20 | 0.040 | 0.020 |
| M400-120 | 400 | 0–240 | 120 | 125.0 | 30.4 | 0.033 | 3.80 | 2.64 | 7.54 | 0.134 | 10 | 20 | 0.040 | 0.020 |

**Fluid-Damped Types**

| | Hz | Hz | a | Ω | mA/cm | cm/mA | V/cm | cm/V | mA | cm | cm/g |
|---|---|---|---|---|---|---|---|---|---|---|---|
| M1000 | 100 | 0–600 | | 39.0 | 1.04 | 0.965 | 0.040 | 2.50 | 100 | 20 | 0.020 |
| M1650 | 1,650 | 0–1,000 | | 26.8 | 3.66 | 0.272 | 0.098 | 1.02 | 100 | 20 | 0.020 |
| M3300 | 3,300 | 0–2,000 | | 32.0 | 7.88 | 0.127 | 0.252 | 0.396 | 100 | 15 | 0.020 |
| M3300T | b | b | | 32.0 | 7.88 | 0.127 | 0.252 | 0.396 | 100 | 15 | 0.020 |
| M5000 | 5,000 | 0–3,000 | | 39.5 | 12.3 | 0.080 | 0.485 | 0.206 | 100 | 9 | 0.020 |
| M8000 | 8,000 | 0–4,800 | | 35.0 | 15.7 | 0.063 | 0.552 | 0.182 | 100 | 5 | 0.020 |
| M10000 | 10,000 | 0–6,000 | | 35.0 | 15.7 | 0.069 | 0.552 | 0.182 | 100 | 5 | 0.020 |
| M13000 | 13,000 | 0–13,000 | | 81.6 | 32.3 | 0.030 | 2.61 | 0.038 | 55 | 2.5 | 0.020 |

a No requirement with fluid-damped galvanometers.
b No specification.

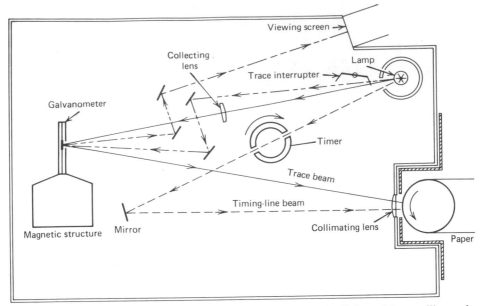

*Figure 2.23(a)* Schematic illustration of the components of a light-writing oscillograph.

a rotating coil, a mirror, and a pair of stationary pole pieces. The design features of the components are varied to change the dynamic characteristics (sensitivity and frequency response) of different galvanometers. The characteristics of a line of commercially available galvanometers are presented in Table 2.2. A light-writing oscillograph is shown in Fig. 2.23c.

### Transient Response of Galvanometers

Since the galvanometer is a highly refined version of a D'Arsonval movement, its transient response can be studied by considering the equation of motion for the coil. Thus

$$T_1 - T_2 - T_3 = J\frac{d^2\theta}{dt^2} \tag{2.23}$$

where     $T_1$ is the applied torque (Eq. 2.12).
            $T_2$ is the opposing torque due to the restraining spring (Eq. 2.13).
            $T_3$ is the opposing torque due to damping.

$$T_3 = D_1 \frac{d\theta}{dt} \tag{2.24}$$

      $D_1$ is the damping coefficient.
      $J$   is the moment of inertia of the rotating mass.

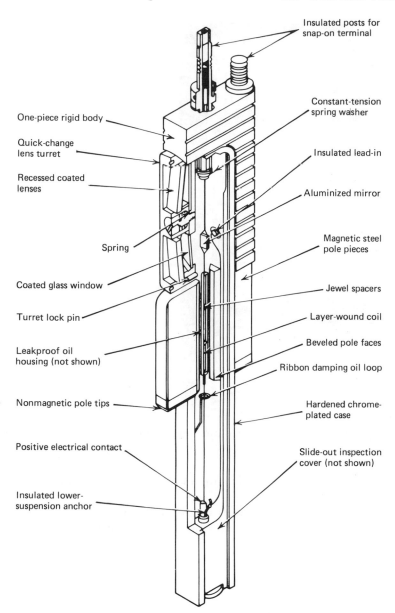

*Figure 2.23(b)* Construction details of a galvanometer.

Substituting Eqs. (2.12), (2.13), (2.14), and (2.24) into Eq. (2.23) yields:

$$J\frac{d^2\theta}{dt^2} + D_1\frac{d\theta}{dt} + K\theta = SKI \qquad (2.25)$$

**Figure 2.23(c)** A light-writing oscillograph. (Courtesy of Consolidated Electrodynamics Corporation.)

where $I$ is the instantaneous current in the coil and $\theta$ is small enough that the approximation $\cos \theta \approx 1$ is valid.

The dynamic response of a galvanometer can be determined by considering the transient condition associated with a step-voltage input, which is applied as indicated in Fig. 2.24. For a step-voltage input, the instantaneous current $I$ in the coil is given by

$$I = \frac{V - E_m}{R_s + R_G} \tag{2.26}$$

**where**   $V$  is the applied signal voltage (constant).

           $R_s$  is the resistance of the source.

           $R_G$  is the resistance of the galvanometer coil.

           $E_m$  is the back electromotive force induced as the coil rotates in the magnetic field. Thus

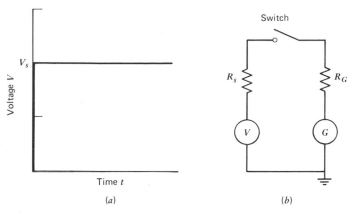

**Figure 2.24**  Circuit for applying a step pulse of voltage to a galvanometer. (*a*) Step-function input. (*b*) Galvanometer circuit.

$$E_m = NBlD \frac{d\theta}{dt} = SK \frac{d\theta}{dt} \tag{2.27}$$

Substituting Eq. (2.27) into Eq. (2.26) and solving for the instantaneous current $I$, gives

$$I = I_s - \frac{SK}{R_s + R_G} \frac{d\theta}{dt} \tag{2.28}$$

where $I_s$ is the steady-state current $I_s = V/(R_s + R_G)$. Substituting Eq. (2.28) into Eq. (2.25) and simplifying gives

$$J \frac{d^2\theta}{dt^2} + D_2 \frac{d\theta}{dt} + K\theta = SKI_s \tag{2.29}$$

where the damping coefficient $D_2$ is a combination of fluid and electromagnetic damping, which can be expressed as

$$D_2 = D_1 + \frac{(SK)^2}{R_s + R_G} \tag{2.30}$$

It is important to note from Eq. (2.30) that the damping coefficient $D_2$ can be varied by changing the resistance of the source $R_s$; thus, the dynamic response of a galvanometer can be adjusted.

Equation (2.29) describes the angular movement of the pen arm or mirror of a galvanometer with respect to time. Since this is a second-order differential equation, the galvanometer is referred to as a second-order instrument. Second-order systems are important in many aspects of measurements and in shock and vibration studies in particular; therefore, the solution of Eq. (2.29) will be thoroughly explored. Once it is recognized that Eq. (2.29) describes a damped second-order system, it is convenient to note that the undamped natural frequency $\omega_n$ is given by

$$\omega_n = \sqrt{\frac{K}{J}} \tag{2.31}$$

Also, the damping ratio $d$ can be defined as

$$d = \frac{D_2}{2\sqrt{KJ}} \tag{2.32}$$

By using Eqs. (2.31) and (2.32), Eq. (2.29) can be rewritten as

$$\frac{1}{\omega_n^2}\frac{d^2\theta}{dt^2} + \frac{2d}{\omega_n}\frac{d\theta}{dt} + \theta = SI_s \tag{2.33}$$

The auxiliary equation associated with the complementary solution of Eq. (2.33) is

$$\frac{1}{\omega_n^2}\lambda^2 + \frac{2d}{\omega_n}\lambda + 1 = 0 \tag{2.34}$$

which exhibits roots of

$$\lambda = \omega_n(-d \pm \sqrt{d^2 - 1}) \tag{2.35}$$

Inspection of Eq. (2.35) shows that three different solutions of Eq. (2.33) must be considered. These solutions are

*Case 1.*   Overdamped when $d > 1$.

*Case 2.*   Critically damped when $d = 1$.

*Case 3.*   Underdamped when $d < 1$.

The solutions of Eq. (2.33) for these three cases are as follows.

*Case 1:*   Overdamped ($d > 1$).

$$\frac{\theta}{\theta_s} = 1 + \frac{d - \sqrt{d^2 - 1}}{2\sqrt{d^2 - 1}}e^{-(d + \sqrt{d^2 - 1})\omega_n t}$$

$$-\frac{d + \sqrt{d^2 - 1}}{2\sqrt{d^2 - 1}}e^{-(d - \sqrt{d^2 - 1})\omega_n t} \tag{2.36}$$

where $\theta_s = SI_s$ is the steady-state deflection of the galvanometer.

In the overdamped case, the response of the galvanometer is sluggish, as indicated by the response curve of Fig. 2.25. The time required for the galvanometer to reach the steady-state position where $\theta/\theta_s = 1$ is quite long. In most applications the overdamped condition is avoided, since the response time is prohibitively long.

*Case 2:* Critically damped ($d = 1$).

$$\frac{\theta}{\theta_s} = 1 - (1 + \omega_n t)e^{-\omega_n t} \tag{2.37}$$

When the galvanometer is critically damped, the response of the galvanometer is improved with respect to the overdamped condition. The rotation $\theta$ approaches the steady-state rotation $\theta_s$, but never exceeds this value. The response curve for a critically damped galvanometer is also shown in Fig. 2.25.

*Case 3:* Underdamped ($d < 1$).

$$\frac{\theta}{\theta_s} = 1 - e^{-d\omega_n t}\left[\frac{d}{\sqrt{1 - d^2}}\sin(\sqrt{1 - d^2}\,\omega_n t) + \cos(\sqrt{1 - d^2}\,\omega_n t)\right] \tag{2.38}$$

Underdamped galvanometers respond quickly to a transient signal, initially overshoot the steady-state rotation, and then oscillate with decaying amplitude about the steady-state rotation $\theta_s$ for some time. The response curve for an underdamped galvanometer is also shown in Fig. 2.25.

The amount of overshoot in the underdamped case depends upon the damping ratio used with the galvanometer. The damping ratio that reduces the re-

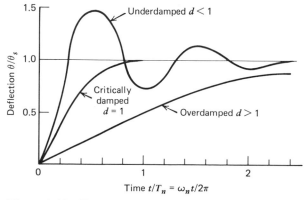

**Figure 2.25** Response curves for underdamped, critically damped, and overdamped galvanometers for a step-pulse input.

sponse time of the galvanometer to a minimum is dependent on the overshoot that is permitted. For example, if the accuracy required in a transient measurement is ±5 percent, a damping ratio is selected such that the overshoot gives $(\theta/\theta_s)_{max} = 1.05$. This example, which is illustrated in Fig. 2.26, shows that the response curve for the underdamped galvanometer is tangent to the upper accuracy limit at point $A$. The response time is defined as the time when the response curve first intersects the lower accuracy limit (see point $B$ of Fig. 2.26). For this example, the response time $t$ equals $0.454T_n$, where $T_n$, the natural period of oscillation, is given by the expression

$$T_n = \frac{2\pi}{\omega_n} = 2\pi\sqrt{\frac{J}{K}} \tag{2.39}$$

It should be noted that the response time for the critically damped galvanometer (point $C$ in Fig. 2.26) is considerably longer ($t = 0.754T_n$) than the response time for the underdamped case.

In most applications, galvanometers are employed in an underdamped condition such that the first overshoot and the subsequent oscillations about $\theta_s$ do not exceed the bandwidth imposed by the error bounds. The damping ratio required is given by the expression

$$d = \sqrt{\frac{ln^2\mathscr{E}}{\pi^2 + ln^2\mathscr{E}}} \tag{2.40}$$

where $\mathscr{E}$ is the error associated with the accuracy limits.

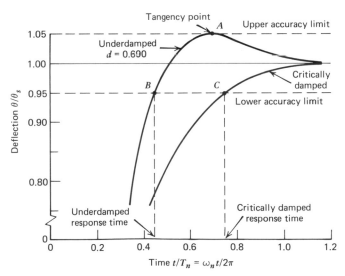

**Figure 2.26** Accuracy limits superimposed on the response curves of a galvanometer.

Use of Eq. (2.40) ensures that the overshoot at the peak value of $\theta/\theta_s$ will be tangent to the upper accuracy limit and that the response time is minimum for this level of accuracy. Minimum response time as a function of percent accuracy is shown in Fig. 2.27. Superimposed on this curve at selected points is the damping ratio required to obtain the level of accuracy. Since galvanometers are electromagnetically damped, the damping coefficient $D_2$ and thus the damping ratio can be varied by adjusting the source resistance $R_s$.

## Response of a Galvanometer to a Periodic Signal

When a galvanometer is used to monitor a periodic signal such as a sinusoidal voltage, the initial transient response discussed in the previous subsection is not a major consideration. Instead, it is more important to determine whether the galvanometer is precisely following the input signal. In some cases, the galvanometer will distort the amplitude of the sinusoidal signal and will introduce a nonlinear phase shift.

The response of a galvanometer to a periodic input $I_i$ of the form

$$I_i = I_* \sin \omega t \tag{2.41}$$

can be studied by considering the equation of motion for the galvanometer (Eq. 2.33), which has been modified to account for the periodic input. Thus

$$\frac{1}{\omega_n^2} \frac{d^2\theta}{dt^2} + \frac{2d}{\omega_n} \frac{d\theta}{dt} + \theta = \theta_* \sin \omega t \tag{2.42}$$

where $\theta_* = SI_*$ is the amplitude of the sinusoidal oscillation.

The particular solution of Eq. (2.42) is required since it describes the steady-state response of the galvanometer to a periodic input. The particular solution

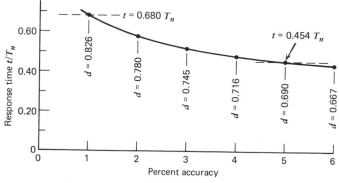

*Figure 2.27*  Response time as a function of accuracy for a step input.

of Eq. (2.42) is

$$\frac{\theta}{\theta_*} = \frac{1}{[1 - (\omega/\omega_n)^2]^2 + 4d^2(\omega/\omega_n)^2}$$

$$\{[1 - (\omega/\omega_n)^2]\sin \omega t - 2d(\omega/\omega_n)\cos \omega t\} \quad (2.43)$$

The results in Eq. (2.43) can be converted to a more useful form by letting

$$\sin(\omega t - \phi) = \frac{[1 - (\omega/\omega_n)^2]\sin \omega t - 2d(\omega/\omega_n)\cos \omega t}{\sqrt{[1 - (\omega/\omega_n)^2]^2 + 4d^2(\omega/\omega_n)^2}} \quad (2.44)$$

where $\phi$ is a phase angle given by

$$\phi = \arctan \frac{2d(\omega/\omega_n)}{1 - (\omega/\omega_n)^2} \quad (2.45)$$

Substituting Eq. (2.44) into Eq. (2.43) yields

$$\frac{\theta}{\theta_*} = a \sin(\omega t - \phi) \quad (2.46)$$

where $a$ is the nondimensional amplitude of the galvanometer response, which can be expressed as

$$a = \frac{1}{\sqrt{[1 - (\omega/\omega_n)^2]^2 + 4d^2(\omega/\omega_n)^2}} \quad (2.47)$$

The frequency response of a galvanometer is described by Eq. (2.47). The amplitude $a = \theta/\theta_*$ is a function of the frequency ratio $(\omega/\omega_n)$ and the damping ratio $d$. The results of Eq. (2.47) are used in Fig. 2.28 to plot the amplitude $a$ as a function of $\omega/\omega_n$ for different damping ratios. The frequency response of the galvanometer depends upon the damping ratios and upon the accuracy required. In Fig. 2.29, an error band of $\pm 5$ percent has been superimposed on a response curve ($d = 0.59$) for a large range of frequencies. Inspection of the response curve relative to the error bands shows that $d = 0.59$ optimizes the frequency response of the galvanometer for these particular accuracy limits. As illustrated in Fig. 2.29, the response curve stays within the error band for the range of frequencies $0 < \omega/\omega_n < 0.87$ and is tangent to the upper accuracy limit at $\omega/\omega_n = 0.60$. As the allowable error band is decreased, the range of the frequency response of the galvanometer also decreases, as indicated in Table 2.3.

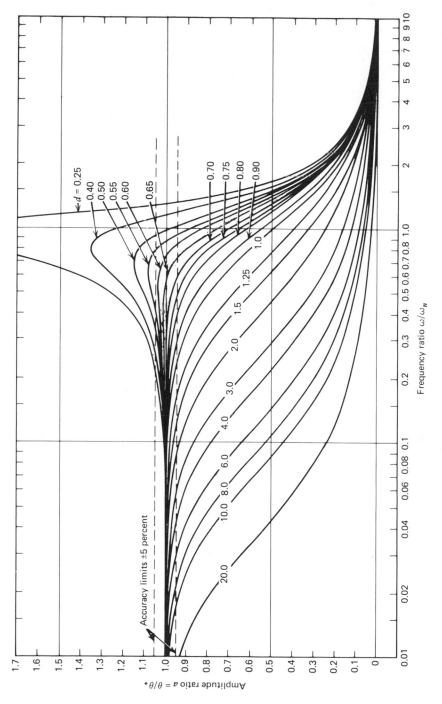

*Figure 2.28* Amplitude ratio as a function of frequency ratio.

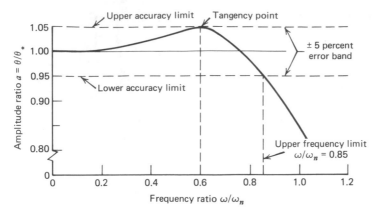

*Figure 2.29*  Response curve for $d = 0.59$.

Accurate recording of a dynamic signal also depends upon the phase angle $\phi$, which is given by Eq. (2.45). If the phase angle is a constant, it simply represents a shift on the time scale of the oscillograph record; therefore, it is not important when measuring a signal with a constant frequency $\omega$. However, if the galvanometer is used to measure a transient pulse where the input $I_i$ is composed of several frequencies

$$I_i = \sum_{k=1}^{N} A_k \sin k\omega t \qquad (2.48)$$

then the phase angle must be a linear function of $\omega$ to prevent distortion in the recording.

To illustrate the requirement that $\phi$ must be a linear function of $\omega$, consider the response of a galvanometer to an input of the form given by Eq. (2.48). For this example, the output may be written as

$$\theta = a_1 \sin(\omega t - \phi_1) + a_2 \sin(2\omega t - \phi_2) + a_3 \sin(3\omega t - \phi_3) + \cdots \qquad (a)$$

**TABLE 2.3**  Frequency Response, Accuracy, and Optimum Damping for Galvanometers Responding to a Periodic Input

| Allowable Error (%) | Optimum Damping $d$ | Frequency Response $(\omega/\omega_n)_{max}$ |
|---|---|---|
| $\pm 10$ | 0.540 | 1.028 |
| $\pm 5$ | 0.589 | 0.870 |
| $\pm 2$ | 0.634 | 0.692 |
| $\pm 1$ | 0.655 | 0.585 |

If the phase angle $\phi$ is a linear function of the frequency, then

$$\phi_1 = c\omega, \qquad \phi_2 = 2c\omega, \qquad \phi_3 = 3c\omega, \cdots \qquad \text{(b)}$$

where $c$ is a proportionality constant associated with the linear phase shift. Substituting Eq. (b) into Eq. (a) gives

$$\theta = a_1 \sin \omega(t - c) + a_2 \sin 2\,\omega(t - c) + a_3 \sin 3\omega(t - c) + \cdots \qquad \text{(c)}$$

Examination of Eq. (c) indicates that the time shift $c$ in the recording of each harmonic is the same; therefore, the composite record of all of the harmonics will be shifted in time by the amount $c = \phi_1/\omega$. No distortion will occur, since the time shift for all of the terms is the same.

Examination of the relationship for the phase angle $\phi$, Eq. (2.45), shows that $\phi$ is not, in general, linear with respect to $\omega/\omega_n$. This nonlinearity is evident in Fig. 2.30, where $\phi$ is shown as a function of $\omega/\omega_n$ and the damping ratio $d$. Only in the special case when $d = 0.64$ is the phase angle linear with respect to $\omega/\omega_n$ over the range $0 < (\omega/\omega_n) < 1$. It is fortunate that the damping ratios normally associated with extended frequency response (see Table 2.3) also provide a linear or near linear phase angle with respect to frequency. This fact means that the damping ratio for extended frequency recording and for distortion-free recording are approximately the same.

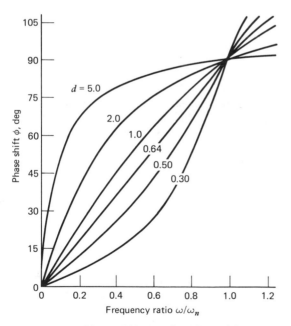

**Figure 2.30**  Phase shift as a function of frequency ratio.

In more modern high-frequency oscillographs, the galvanometers and associated optical components illustrated in Fig. 2.23 have been replaced with a fiber-optic cathode-ray tube. The face plate of this special-purpose tube consists of approximately $10 \times 10^6$ glass fibers which have been fused together to form a 0.2 by 8-in. (5 by 200-mm) rectangular area. The fibers transmit light from the phosphor coating on the inside surface of the cathode-ray tube to the photosensitive direct-print paper which is driven past the face plate of the tube. This arrangement provides excellent trace resolution (sharp traces) at fast writing speeds.

Each input channel in this type of oscillograph contains a voltage to time converter which produces a pulse having a duration proportional to the input voltage. These pulses activate the beam of the cathode-ray tube which continuously sweeps at a fixed frequency of 50 kHz. A dotted-trace representation is avoided at high frequencies by using a memory circuit to produce a continuous trace.

This major advance in oscillographic recording, which replaces the galvanometers and associated optical components with a special-purpose fiber-optic cathode-ray tube, eliminates the need for impedance matching and current amplifiers. Also, most of the concerns related to sensitivity and frequency response are eliminated. The new systems can be used to record up to 18 channels of data on an 8-in. (200-mm)-wide strip of photosensitive paper at speeds up to 120 in. per second (3000 mm/s). Frequency response is flat from dc to 5 kHz, and with a high-gain differential amplifier, sensitivities as low as 1 mV per division can be attained. A photograph of a modern fiber-optic cathode-ray oscillograph is shown in Fig. 2.31.

## Oscilloscopes

The cathode-ray-tube oscilloscope is a voltage measuring instrument that is capable of recording extremely high-frequency signals. The cathode-ray tube (CRT), which is the most important component in an oscilloscope, is illustrated in Fig. 2.32. The CRT is an evacuated tube in which electrons are produced, controlled, and used to provide a voltage–time record of a transient signal. The electrons are produced by heating a cathode. Then, the electrons are collected, accelerated, and focused onto the face of the tube with a grid and a series of hollow anodes. The impinging stream of electrons forms a bright point of light on a fluorescent screen at the inside face of the tube. Voltages are applied to horizontal and vertical deflection plates in the CRT (see Fig. 2.32) to deflect the stream of electrons and thus move the point of light over the face of the tube. It is this ability to deflect the stream of electrons that enables the CRT to act as a dynamic voltmeter with essentially an inertialess indicating system.

An oscilloscope can be used to record a signal $y$ as a function of time, or it can be used to simultaneously record two unknown signals $x$ and $y$. A block diagram of an oscilloscope, presented in Fig. 2.33, shows the inputs and the

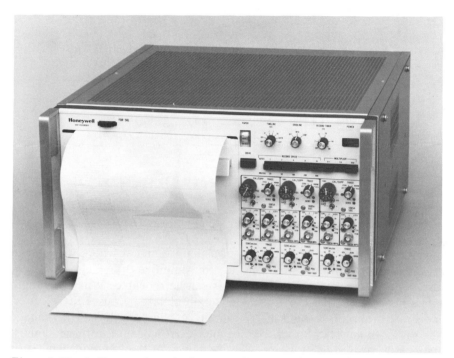

***Figure 2.31***  A fiber optic cathode-ray oscillograph. (Courtesy of Honeywell Test Instruments Division, Denver, Colorado.)

connections to the deflection plates in the CRT. The *y* and the *x* inputs are connected to the vertical and horizontal deflection plates through amplifiers. Since the sensitivity of the CRT is relatively low (approximately 100 V are required on the deflection plates to deflect the beam of electrons 1 in. (25 mm) on the face of the tube), high gain amplifiers are used to increase the voltage of the input signal.

When the oscilloscope is used as a *y-t* recorder, the input to the horizontal amplifier is switched to a sawtooth generator. The sawtooth generator produces

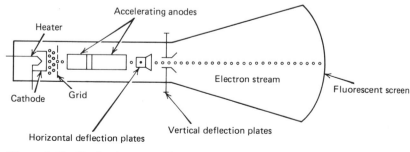

***Figure 2.32***  Basic elements of a cathode-ray tube.

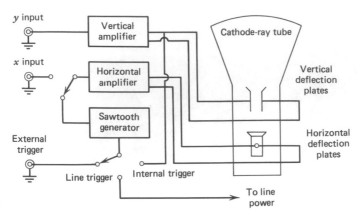

*Figure 2.33* Block diagram for a typical oscilloscope.

a voltage–time output, having the form of a ramp function where the voltage increases uniformly with time from zero to a maximum and then almost instantaneously returns to zero so that the process can be repeated. When this ramp function is imposed on the horizontal deflection plates, it causes the electron beam to sweep from left to right across the face of the tube. When the voltage from the sawtooth generator goes to zero, the electron beam is returned almost instantaneously to its starting point. The frequency of the sawtooth generator can be varied to give different sweep times. Typical sweep rates can be varied from 10 ns/div to 5 s/div in calibrated steps in a 1-2-5-10 sequence. Since the face of the CRT is divided into 10 divisions in the $x$ or $t$ direction, observation times associated with a single-sweep range from 100 ns to 50 s.

Since the observation time can be relatively short, the horizontal sweep must be synchronized with the event to ensure that a recording of the signal from the transducer is made at the correct time. Three different triggering modes are used to synchronize the oscilloscope with the event: the external trigger, the line trigger, and the internal trigger. Trigger signals from any one of these three sources activate the sawtooth generator and initiate the horizontal sweep. The *external trigger* requires an independent triggering pulse from an external source usually associated with the dynamic event being measured. A sharp front pulse of about 2 to 5 V is recommended for the input to the external trigger.

The *line trigger* utilizes the signal from the power line to activate the sawtooth generator. Since the line-trigger signal is repetitive at 60 Hz, the horizontal sweep triggers 60 times each second; therefore, the trace on the CRT appears continuous. The line trigger is quite useful when the oscilloscope is used to measure periodic waveforms that exhibit a fundamental frequency of 60 Hz.

The *internal trigger* makes use of the $y$ input signal to activate the sweep. The level of the trigger signal required to initiate the sawtooth generator can be adjusted to very low levels; therefore, only a small region of the record is lost in measuring a transient pulse. If the $y$ input signal is repetitive, the frequency of the sawtooth generator can be adjusted to be nearly equal to some multiple of the frequency of the input signal. The sawtooth generator is then synchronized in both frequency and phase with the input signal, and the trace appears stationary on the CRT screen.

The trace on the screen of the CRT is produced when the electron beam impinges on a phosphor coating on the inside face of the tube. Of the total energy of the beam, 90 percent is converted to heat and 10 percent to light. The light produced by fluorescence of the phosphor has a degree of persistence that enhances the visual observation of the trace. For instance, the phosphor identified as P31 produces a yellowish-green trace that requires 38 μs to decay to 10 percent of its original intensity. Permanent records of the traces can be made with special-purpose oscilloscope cameras (see Fig. 2.34) that attach directly to the mainframe of the oscilloscope.

Many cathode-ray tubes are of the storage type and continue to display the trace after the input signal ceases. The period of retention varies from a few seconds to several hours, depending upon the type of phosphor used in fabricating the screen of the CRT. Most storage tubes utilize a bistable phosphor that permits the tube to be used in both the storage mode and the conventional (nonstorage) mode. The writing speed of early model storage oscilloscopes, operating in the storage mode, was not as high as the writing speed in the conventional mode. Modern models, however, have been improved and writing rates now permit analysis of single-shot events without use of a camera. The advantages of trace storage are numerous. Storage permits easy and accurate evaluation of slowly changing events that would appear as slowly moving dots on the conventional CRT. Storage is useful also in observing rapidly changing nonrepetitive signals that would flash across the screen too quickly to be evaluated on a conventional CRT. Operation in the storage mode also permits one to carefully select the data to be recorded photographically. Unwanted displays can be erased, and the expense of photographing is avoided. Finally, many storage oscilloscopes have split-screen viewing, which allows each half of the screen (top and bottom) to be used independently for stored trace displays. With the split screen, a reference trace can be stored on one-half of the screen and the other half can be used to display the unknown trace. In this manner, comparisons can be made quickly and accurately.

The amplifier used in an oscilloscope is quite important since it controls the sensitivity, bandwidth, rise time, input impedance, and the number of channels that can be recorded with the instrument. Because of the importance of the amplifier to the operational characteristics of the oscilloscope, many instruments are designed to accept plug-in-type amplifiers (see Fig. 2.34). This arrangement permits the amplifier to be changed quickly and easily to alter the characteristics of the oscilloscope.

*Bandwidth* is defined as the frequency range over which signals are recorded with less than a 3-dB loss compared to midband performance. Since modern amplifiers perform very well at low frequencies (down to dc), bandwidth refers

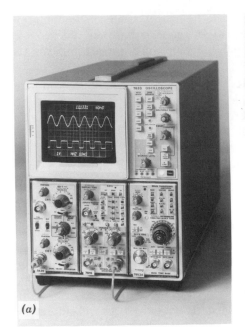

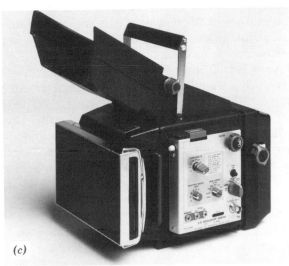

*Figure 2.34* A modern cathode-ray oscilloscope with auxiliary equipment. (Courtesy of Tektronix, Inc.) (*a*) Modern cathode-ray oscilloscope with variable persistance storage. (*b*) Oscilloscope with plug-in amplifiers and cart. (*c*) Oscilloscope camera.

to the highest frequency that can be recorded with an error less than 3 dB (30 percent). Bandwidth and rise time are related such that

$$f_{bw}t_r = 0.35 \qquad\qquad (a)$$

**where**   $f_{bw}$ is the bandwidth expressed in megahertz.

$t_r$   is the rise time expressed in nanoseconds.

Since good practice dictates utilization of a vertical amplifier capable of responding five times as fast as an applied step signal, Eq. (a) is modified in practice to

$$f_{bw}t_r = 1.70 \tag{2.49}$$

It is apparent from Eq. (2.49) that an amplifier–oscilloscope combination with a bandwidth of 80 MHz is capable of recording signals with a rise time of approximately 21 ns.

Sensitivity[2] refers to the voltage input needed to produce a prescribed deflection of the electron beam. The sensitivity is usually given in terms of millivolts per division (mV/div). Sensitivity of a typical amplifier ranges from 5 mV/div to 10 V/div in calibrated steps arranged in a 1-2-5 sequence. Higher sensitivity can be achieved; however, a reduction in bandwidth is required. Increasing bandwidth increases the noise pick-up, decreases sensitivity, and increases the cost of the amplifier. For general-purpose mechanical measurements, rise times shorter than 100 ns are seldom needed, therefore, low-performance amplifiers (500-kHz bandwidth) are usually adequate. For very-high-speed electronic or optic measurements, amplifiers with 18-GHz bandwidth may be required and are available. With respect to sensitivity–bandwidth trade-off, it is advisable to specify only the minimum bandwidth required and to work with the higher sensitivity and lower noise amplifiers. The input impedance of most amplifiers used with oscilloscopes is 1 MΩ paralleled by 20 to 50 pF of capacitance.

The input amplifier also controls the number of traces displayed on the oscilloscope screen. In some models an electronic switch is housed in the amplifier that alternately connects two input signals to the vertical deflection system in the CRT. The principal advantages of using this feature to produce a dual-trace oscilloscope are lower cost and better comparison capabilities. Both of these advantages are due to the fact that only one horizontal amplifier and one set of deflection plates is used in making both traces. However, high-speed transient events are difficult to record in this manner since a significant variation might occur on one channel while the beam is tracing on the other channel. Since the electronic switch operates at a frequency of approximately 250 kHz, dynamic events with frequencies between 25 and 50 kHz ($\frac{1}{10}$ to $\frac{1}{5}$ of the switching frequency) can be recorded. Whenever two nonrecurrent signals of very short duration must be recorded together, dual-beam oscilloscopes are employed. A dual-beam oscilloscope has independent deflection plates within the CRT for each beam and employs independent horizontal and vertical amplifiers for each beam. The dual-beam system is superior to the dual-trace system, since it can display two signals separately and simultaneously; however, it is more costly.

---

[2] Oscilloscope manufacturers use reciprocal sensitivity $S_R$ to describe the deflection–voltage relationship (see Eq. 2.7).

## Digital Oscilloscopes

The *digital oscilloscope* is identical to the conventional oscilloscope except for the manipulation of the input signal prior to its display on the CRT and for the permanent storage capabilities of the instrument. With a digital oscilloscope, the input signal is converted to digital form, stored in a buffer memory, and then transferred to a mainframe memory prior to display. A microprocessor controls the storage, transfer, and display of the data. A photograph of a modern digital oscilloscope is presented in Fig. 2.35.

Since the input data are stored in addition to being displayed, operation of the digital oscilloscope differs from the operation of the conventional oscilloscope. The display on the CRT of the digital oscilloscope is a series of points produced by the electron beam at locations controlled by the data in storage.

Operation begins when a trigger signal is received. First, the input signal is converted into digital form by an analog-to-digital converter that measures the signal at preselected intervals. The converted signal is in binary form with either 8- or 12-bit resolution. With 12-bit resolution, the voltage range is 4096 increments (from $-2048$ at the bottom of the screen to $+2047$ at the top of the screen). Resolution with 12 bits is 0.025 percent; with 8 bits, it is 0.4 percent. The analog-to-digital converter measures the signal periodically at preselected sampling intervals (termed sweep speeds). Sampling intervals can be varied from 500 ns per point to 200 s per point in calibrated steps arranged in a 1-2-5-10 sequence. When a measurement is completed, the data are stored in a buffer memory at an address that is proportional to the time at which the data were taken. When all measurements have been made (usually 1024, 2048, or 4096 measurements), the sweep ends (the buffer memory is full) and the data are transferred from the buffer memory to the mainframe memory. The microprocessor monitors the mainframe memory and produces the display of the

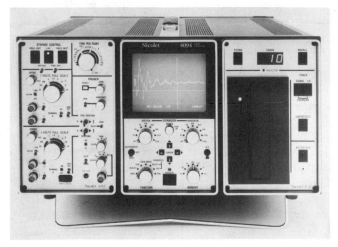

*Figure 2.35*   A digital oscilloscope with auxiliary-disk memory. (Courtesy of Nicolet Instrument Corp.)

voltage–time trace on the CRT. The resolution of the trace is excellent since a 12-bit digital-to-analog converter gives 4096 vertical data locations on the screen of the CRT.

The fact that the input signal has been stored in the mainframe or on an auxiliary-disk memory offers many advantages for data display or data processing. The data are displayed on the CRT in a repetitive manner so that traces of one-shot, transient events appear stationary. The trace can also be manipulated by expanding either the horizontal or vertical scales or both. This expansion feature permits a small region of the record to be enlarged and examined in detail, as illustrated in Fig. 2.36. Readout of the data from the trace is also much easier and more accurate with digital oscilloscopes. A pair of marker lines (one vertical and the other horizontal) can be positioned anywhere on the screen. The procedure is to position the vertical line at a time on the trace when a reading of the voltage is needed. The horizontal marker (or cross hair) automatically positions itself on the trace. The coordinates of the cross-hair intersection with the trace are presented as a numerical display on the screen, as illustrated in Fig. 2.37.

The auxiliary storage is a magnetic disk with significant storage capacity. In operation, data are transferred from the mainframe memory to the disk for storage or from the disk to the mainframe memory for display.

There are several advantages of auxiliary-disk storage of data. The data can be stored permanently on inexpensive disks and quickly recalled for display and analysis. Signals occurring in sequence can be automatically recorded and stored. The disk memory increases the mainframe memory size by a factor of 8, so that an uninterrupted record 32 s long may be stored with 1-ms time resolution. The data bus is compatible with computers and calculators; therefore, data from the oscilloscope can be transmitted directly to the computer. Once the data are processed, they can be transmitted back to the disk storage unit and then to the mainframe storage for display of the externally processed results.

Digital oscilloscopes are relatively new; early models were introduced only in 1972. Initially, performance of digital oscilloscopes was limited due to the relatively low bandwidth capability (10 kHz or less). Recent improvements in digital electronics (particularly high-speed analog-to-digital converters) and the introduction of the microprocessor have greatly enhanced the speed of conversion and storage of the data. Today, plug-in units with 8-bit resolution are available with a rise-time capability of 50 ns; with 12-bit resolution, the rise time is 500 ns. Except for very high-speed transient signals, where rise times are less than 50 ns, the digital oscilloscope is superior in every respect to the conventional oscilloscope. As costs for the digital oscilloscope decrease, they should become more widely used than the conventional oscilloscope for mechanical measurements.

## Magnetic Tape Recorders

Magnetic tape recorders are used to store dynamic signals when the frequency components in the signals range from dc to approximately 40 kHz. The

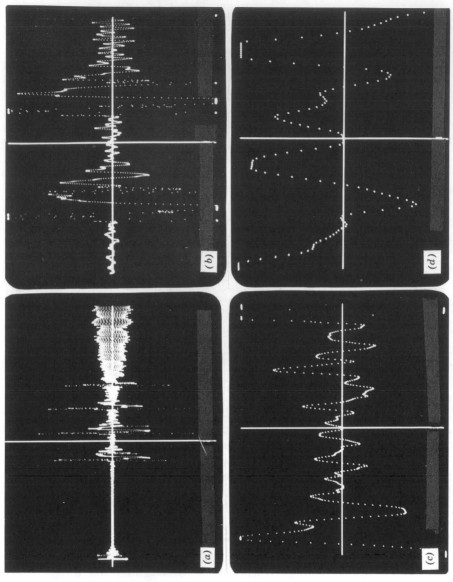

*Figure 2.36* Expansion of the display on the screen of a digital oscilloscope. (*a*) Unmagnified. (*b*) Both axes expanded by a factor of 4. (*c*) Both axes expanded by a factor of 16. (*d*) Both axes expanded by a factor of 64. (Courtesy of Nicolet Instrument Corp.)

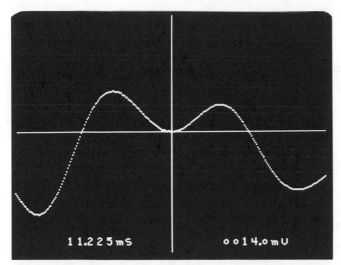

**Figure 2.37**  Numeric display of data on the screen of a digital oscilloscope. (Courtesy of Nicolet Instrument Corp.)

recording is accomplished by applying a magnetizing field to a magnetic film coating on a Mylar tape. The recording process is illustrated in Fig. 2.38. The magnetic flux in the record head fluctuates due to variations in the input signal and a magnetic record of these variations is permanently imposed on the coating. The data are retrieved by moving the tape under the reproduce head where the variations in the magnetic field stored on the tape induce a voltage in the windings of the reproduce head that produces the output signal.

In a magnetic tape recorder, the $\frac{1}{2}$-in. (12.7-mm)- or 1-in. (25.4-mm)-wide tape is driven at a constant speed by a servo-type dc capstan motor over either the record or reproduce heads, as shown in Fig. 2.39. The speed of the capstan motor is monitored with a photocell and tone wheel. The frequency of the signal from the photocell is compared with the frequency from a crystal oscillator to

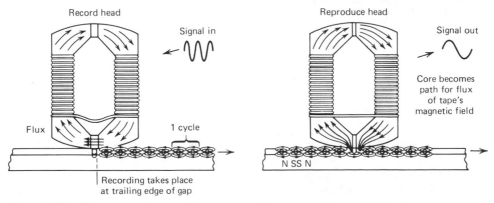

**Figure 2.38**  The magnetic recording process. (Courtesy of Ampex Corp.)

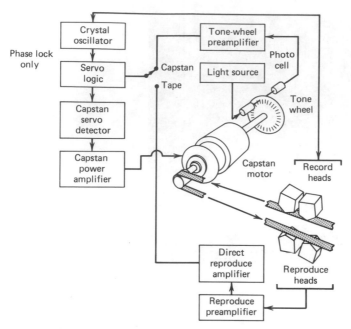

*Figure 2.39* Constant-speed tape drive. (Courtesy of Honeywell Test Instruments Division, Denver, Colorado.)

produce a feedback signal that is used in a closed-loop servo system to maintain a constant tape speed. Tape speeds have been standardized at $\frac{15}{16}$, $1\frac{7}{8}$, $3\frac{3}{4}$, $7\frac{1}{2}$, 15, 30, 60, and 120 in./s (23.8, 47.6, 95.3, 191, 381, 762, 1524, and 3048 mm/s). Tape with a thin base ($\frac{1}{2}$ mil) is available on $10\frac{1}{2}$-in. (267-mm) reels to provide 7200 ft (2200 m) of tape, which is equivalent to over 24 h of recording time at a tape speed of $\frac{15}{16}$ in./s (23.8 mm/s).

Multichannel recorders employ four stacked-head assemblies that are precisely positioned on a single base plate to ensure alignment of the tape. Two of the heads are for recording and the other two are for reproducing as indicated in Fig. 2.39. As the tape passes the first head assembly, odd-channel data are recorded; as it passes the second head assembly, even-channel data are recorded. This recording procedure minimizes interchannel cross talk by maximizing the spacing between individual heads in the stacked-head assembly. Use of two recording-head stacks permits seven channels to be recorded on $\frac{1}{2}$-in. (12.7-mm) tape or 14 channels on 1-in. (25.4-mm) tape.

There are three different types of recording in common usage: direct or AM, FM, and digital. The characteristics of each type of recording are shown in Fig. 2.40. The direct or AM (amplitude modulation) recording method is the most commonly used, since it is simple, low cost, and suitable for most audio (speech and music) recordings. The signal to be recorded is amplified, mixed with a high-frequency bias, and then used to drive the record head. In playback or reproduction, the tape is driven under the reproduce head at the same speed that was used in recording. The output of the head is proportional to the frequency of the recorded signal. The output from the reproduce head is then fed into a reproduce amplifier that must have a frequency response that is the inverse of the frequency response of the reproduce head in order to obtain a flat frequency response for the system.

There are two very serious disadvantages to AM-type recording. First, the lowest frequency that can be recorded is about 50 Hz; therefore, dc or slowly varying signals cannot be stored. Second, imperfections in the coating on the tape can produce significant reductions in signal levels for short periods, which can cause serious errors in the recording of transient signals. This type of error can be tolerated in speech or music recording, but not in data recording where precision is critical. Because of these two limitations, AM or direct recording is used only on one track of a multitrack recorder for voice commentary relative to the event being recorded (identification and/or experimental description).

The method used most frequently to record data is the FM (frequency modulation) method, since it overcomes both of the limitations of AM recording.

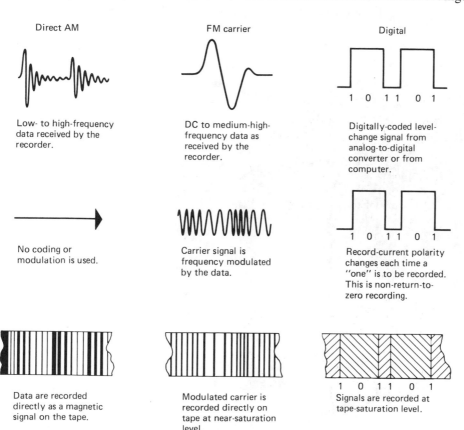

*Figure 2.40* Features of direct AM, FM carrier, and digital methods for magnetic recording.

With FM recording, the input signal is used to drive a voltage-controlled oscillator (VCO). The VCO outputs its center frequency with zero input voltage. A positive dc input signal produces an increase in the frequency of the carrier signal issued from the VCO, while an ac input signal produces carrier frequencies on both sides of the center frequency. Thus, with frequency modulation (FM) the voltage–time data are recorded on the tape in the frequency domain. Low-frequency or dc input signals can be recorded and amplitude instabilities due to imperfections in the coating on the tape do not markedly affect the output.

The signal is retrieved from the magnetic tape in a playback process that includes filtering and demodulation. The signal from the reproduce head is filtered to remove the carrier frequency and then demodulated to give the output signal. This output signal is displayed on some other type of recorder, such as an oscillograph or an oscilloscope. The magnetic tape system is used only to record and store data. Display of the data can be enhanced, since the recording can be reproduced at a different tape speed than that used for recording. This procedure alters the time base; therefore, voltage–time traces can be stretched or compressed. Data stored on magnetic tape can also be processed automatically

*Figure 2.41*  A modern magnetic tape recorder. (Courtesy of Honeywell Test Instruments Division, Denver, Colorado.)

without visual display. For FM records, the output signal can be fed into an analog-to-digital converter. The digitized data can then be processed on a computer according to programmed instructions. A photograph of a modern, portable, magnetic tape recorder is shown in Fig. 2.41.

Digital recording involves storing two-level data (0 and 1) and is accomplished by magnetizing the tape to saturation in either one of the two possible directions. In one type of digital recording (return to zero), the positive state of saturation represents the binary digit 1 and the negative state of saturation represents the digit 0. Data are recorded as a series of pulses that represent the decimal number (expressed in binary code) of the voltage input averaged over a sampling interval. Bits that express the number are recorded simultaneously in parallel across the width of the tape, with each bit on a separate track.

While digital recording is sensitive to tape dropouts, thus requiring high-quality tape and tape transports that ensure excellent head-to-tape contact, other aspects of digital recording are easier than those associated with FM recording. The output is not strongly dependent on tape speed and the record and reproduce amplifiers are simple and therefore inexpensive. Also, since the output is in digital form it can be processed directly on a computer.

The primary disadvantage of digital recording has been the need to digitize the input data prior to recording. This disadvantage is currently being overcome, since significant improvements are being made in high-speed analog-to-digital converters. It is possible that developments in high-speed A-to-D converters will permit digital recorders to replace FM recorders in the near future for dynamic recording of long-term events.

## 2.6  SUMMARY

A voltage recording instrument, the final component in a measuring system, is used to convert a voltage representing the unknown quantity $Q_i$ into a display for visual readout or into a digital code that is suitable for automatic data processing. The instruments in use today (1983) range from simple analog voltmeters to complex digital oscilloscopes with auxiliary magnetic storage. While all five of the general characteristics of a recording instrument (input impedance, sensitivity, range, zero drift, and frequency response) are important, the single characteristic that dominates selection of a recorder is frequency response.

Static measurements, where frequency response is not important and the unknown is represented by a single number that is independent of time, can be made quickly and accurately with relatively inexpensive voltmeters. When the phenomena being studied begin to vary with time, the recording instrument becomes more complex, less accurate, and more expensive. The major difficulty experienced in measuring unknown parameters associated with time-dependent phenomena is the need to display the data with respect to time.

Display of the data for quasi-static measurements, where the unknown is varying with a frequency of less than a few hertz, can be accomplished with a servo-driven potentiometer recorder. The servo-driven potentiometer provides an accurate and inexpensive means of measuring and displaying a voltage; the time display of the voltage is presented on a strip of chart paper that is driven by a simple clock motor. For very slowly varying quantities (say one cycle every few minutes), switching can be used with a single potentiometer recorder to handle multiple inputs. However, for many studies, an independent channel is needed for each unknown quantity being measured. This type of requirement greatly increases the complexity and cost of the measurements.

As the frequency of the unknown quantity increases to about 10 kHz, recording instruments with adequate frequency response must be used in the measurement system. Oscillographs with galvanometer driven pens or hot stylus are used for frequencies between 0 and 150 Hz. For frequencies between 0 and 10 kHz, light-writing oscillographs with galvanometer driven mirrors or fiber-optic cathode-ray tubes are used. Magnetic tape recorders are useful for frequencies up to about 50 kHz. Signals with frequencies above 50 kHz must be recorded with either digital or conventional oscilloscopes. As the frequency increases, it is evident that the recording instrument becomes more sophisticated, more difficult to operate, less accurate, and more expensive. The data obtained are usually in the form of voltage–time records or traces that require considerable time for analysis.

Recent advances in digital electronics are resulting in new instruments that offer significant advantages in making measurements at both ends of the frequency spectrum. Digital voltmeters and data acquisition systems are easy to use and offer a relatively low-cost method for acquiring and processing large amounts of low-frequency data. The digital oscilloscope has the capacity to store high-speed transient signals that can be easily displayed, compared, and processed externally on a computer. Perhaps in the near future a digital recorder will be developed for the middle range of frequencies (10 Hz to 10 kHz) that will offer advantages over the galvanometer type of oscillographic recorder.

---

## EXERCISES

**2.1**   List the general characteristics of a recording instrument.

**2.2**   Determine the power loss in a voltmeter with an input impedance of 20,000 $\Omega$ if it is used to measure voltages of 1 mV, 5 mV, 10 mV, 50 mV, 100 mV, 500 mV, 1 V, 5 V, 10 V, 50 V, and 100 V.

**2.3**   Prepare a graph showing power loss in a voltmeter as a function of voltage with input impedance as a parameter. (Will the use of semilog paper simplify this task?)

**2.4**   Determine the errors if the voltmeter in Exercise 2.2 is used to measure voltages from sources with output impedances of 0.1 $\Omega$, 0.5 $\Omega$, 1 $\Omega$, 5 $\Omega$, 10 $\Omega$, 50 $\Omega$, 100 $\Omega$, 500 $\Omega$, and 1000 $\Omega$.

**2.5**  If a voltmeter load error of 2 percent is acceptable, what limit must be placed on the resistance ratio $R_s/R_m$?

**2.6**  An oscilloscope being used to measure a voltage–time function exhibits a deflection of 3.6 divisions. The sensitivity of the oscilloscope is 0.2 div/V. Determine the reciprocal sensitivity and the voltage represented by the deflection.

**2.7**  The oscilloscope described in Exercise 2.6 has eight divisions in the vertical direction on the face of the tube. For the sensitivity given in Exercise 2.6, determine the range. If the sensitivity is increased to 0.5 div/V, determine the new range. Can both sensitivity and range be increased simultaneously?

**2.8**  An amplifier that is being used in an instrumentation system to measure a voltage of 9 mV over a period of two weeks exhibits a drift of 0.1 mV/h. Determine the error that may result from zero drift.

**2.9**  Specifications for a recorder indicate that it is down 2 dB at 100 Hz. Determine the error if the recorder is used to measure a signal with a frequency of 100 Hz.

**2.10**  Tests with a recorder at frequencies of 10, 20, 40, 60, 80, and 100 Hz provided the following output-to-input ratios $C_o/C_i$: 1.01, 1.03, 1.05, 1.00, 0.93, and 0.80. Determine the amplitude ratio in terms of decibels at each frequency.

**2.11**  The sensitivity of a galvanometer is listed by the manufacturer as 20-μA full scale. The full-scale rotation of the pointer is 60 degrees. Determine the sensitivity $S$ of the galvanometer by using the definition of sensitivity given in Eq. (2.14).

**2.12**  A galvanometer with a 40-Ω coil is rated at 10-mA full scale. Determine the required shunt resistance if it is to be used to measure a 20-A current.

**2.13**  Determine the series resistors needed to convert a 50-μA full-scale galvanometer to a multimeter with full-scale voltages of 10 mV, 30 mV, 50 mV, 100 mV, 300 mV, 500 mV, 1 V, 3 V, 5 V, and 10 V. The coil of the galvanometer has a resistance of 35 Ω.

**2.14**  Determine the loading error for the multimeter of Exercise 2.13 if it is used to measure voltage from a source with a 1000-Ω output resistance. Determine the error for each of the 10 scales. Why does this error differ and for what scale is it the largest?

**2.15**  Outline the advantages of the amplified voltmeter when compared to the conventional voltmeter.

**2.16**  What is the most significant advantage of a null-balance instrument?

**2.17**  A potentiometer with a slide-wire resistor 10 in. (254 mm) long is balanced with the wiper at the 4.5-in. (114-mm) position. If the reference voltage is 2 mV, determine the input voltage.

**2.18** Define resolution for a digital voltmeter (DVM).

**2.19** Determine the resolution for three-, four-, five-, and six-digit digital voltmeters.

**2.20** Describe how a single-channel strip-chart recorder can be converted so that it can record data from several different sources.

**2.21** Define the terms deadband and slewing speed as they apply to an $x$-$y$ recorder.

**2.22** List the advantages of a data logging system over a strip-chart recorder for the following types of recording:

(a) Single-channel                  (b) Multiple-channel

**2.23** List the disadvantages of a data logging system over a strip-chart recorder for the following types of recording:

(a) Single-channel                  (b) Multiple-channel

**2.24** Verify Eq. (2.36).

**2.25** Verify Eq. (2.37).

**2.26** Verify Eq. (2.38).

**2.27** For an error band of $\pm 5$ percent, show that the response time of a critically damped galvanometer to a step-function input is $t/T_n = 0.754$.

**2.28** Verify Eq. (2.39).

**2.29** Determine the degree of damping $d$ to be specified for a galvanometer measuring step-type transients if the specified error band is $\pm 1$ percent, $\pm 2$ percent, $\pm 5$ percent, and $\pm 10$ percent.

**2.30** Beginning with Eq. (2.43), verify Eqs. (2.44) and (2.45).

**2.31** Verify the results presented in Table 2.3.

**2.32** A current represented by

$$I = 10 \sin 400t + 2 \sin 800t + \sin 1200t$$

is recorded by a galvanometer, with $d = 0.55$ and $\omega_n = 1200$ rad/s. Determine the output. Compare the input and output signals and determine amplitude and time distortion of the recorded pulse.

**2.33** Describe measurements that would be made with each of the following trigger modes:

(a) Internal
(b) Line
(c) External

**2.34** Outline the differences between conventional and storage cathode-ray tubes.

**2.35** Define bandwidth and describe its importance in measuring transient signals.

**2.36** Prepare a graph showing rise time as a function of bandwidth for use with oscilloscopes.

**2.37** Outline the differences between dual-trace and dual-beam oscilloscopes.

**2.38** Describe measurements where a dual-trace oscilloscope would be adequate.

**2.39** Describe measurements where a dual-beam oscilloscope would be necessary.

**2.40** What are the essential differences between digital and conventional oscilloscopes?

**2.41** Tabulate the counts associated with signal conversion in binary form for 4, 6, 8, 10, and 12 bits.

**2.42** How many data locations are possible on a CRT screen if an eight-bit analog-to-digital converter is used with an eight-line data bus?

**2.43** List the advantages of a digital oscilloscope when compared to a conventional oscilloscope.

**2.44** List the disadvantages of a digital oscilloscope when compared to a conventional oscilloscope.

**2.45** What is the role of auxiliary storage in applying digital oscilloscopes to transient measurements?

**2.46** If the sweep rate is as shown below, determine the observation period with a digital oscilloscope having eight data lines:

    (a)   500 ns/point           (c)   10 ms/point
    (b)   2 μs/point             (d)   200 s/point

**2.47** If the sweep rate is as shown below, determine the observation period with a digital oscilloscope having 12 data lines:

    (a)   500 ns/point           (c)   10 ms/point
    (b)   2 μs/point             (d)   200 s/point

**2.48** Discuss the advantages and disadvantages of the direct or AM method of recording.

**2.49** Discuss the advantages and disadvantages of the FM method of recording.

**2.50** Discuss the advantages and disadvantages of the digital method of recording.

# THREE

# SENSORS FOR
# TRANSDUCERS

## 3.1 INTRODUCTION

*Transducers* are electromechanical devices that convert a change in a mechanical quantity such as displacement or force into a change in an electrical quantity that can be monitored as a voltage after signal processing. A wide variety of transducers are available for use in measuring mechanical quantities. Transducer characteristics include range, sensitivity, linearity, and operating temperature limits. Transducer characteristics are determined primarily by the sensor that is incorporated into the transducer to produce the electrical output. For example, a set of strain gages on a tension link provides a transducer that produces a resistance change $\Delta R/R$ that is proportional to the load applied to the tension link. The strain gages serve as the sensor in this force transducer and play a dominant role in establishing the characteristics of the transducer.

Many different sensors are utilized in transducer design, such as potentiometers, differential transformers, strain gages, capacitor sensors, piezoelectric elements, piezoresistive crystals, thermistors, etc. The important features of sensors that are commonly used either directly for measurements or as a part of a transducer are described in this chapter.

## 3.2 POTENTIOMETERS

The simplest form of potentiometer is the slide-wire resistor shown schematically in Fig. 3.1. The sensor consists of a length $l$ of resistance wire attached across a voltage source $E_i$. A wiper moves along the length of the wire. The relationship between the output voltage $E_o$ and the position $x$ of the wiper can be expressed as

$$E_o = \frac{x}{l} E_i \qquad \text{or} \qquad x = \frac{E_o}{E_i} l \qquad (3.1)$$

Thus, the slide-wire potentiometer can be used to measure a displacement $x$. Straight-wire resistors are not feasible for most applications, since the resistance of a short length of wire is low and low resistance imposes excessive power requirements on the voltage source.

High-resistance wire-wound potentiometers are obtained by winding the resistance wire around an insulating core, as shown in Fig. 3.2. The unit illustrated in Fig. 3.2a is used for linear displacement measurements. Cylindrically shaped potentiometers similar to the one illustrated in Fig. 3.2b are used for angular displacement measurements. The resistance of wire-wound potentiometers ranges between 10 and $10^6$ $\Omega$, depending upon the diameter of the wire used to wind the coil, the length of the coil, and the material used for the coil.

The resistance of the wire-wound potentiometer increases in a stepwise manner as the wiper moves from one turn to the adjacent turn. This step change in resistance limits the resolution of the potentiometer to $L/n$, where $n$ is the number of turns in the length $L$ of the coil. Resolutions from 0.05 to 1 percent are common, with the lower limit obtained by using many turns of very small diameter wire.

The range of the potentiometer is controlled by the active length $L$ of the coil. Linear potentiometers are available in many lengths up to about 1 m. The range of the angular displacement potentiometer can be extended by arranging

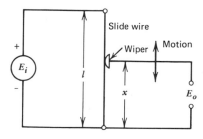

**Figure 3.1**   Slide-wire resistance potentiometer.

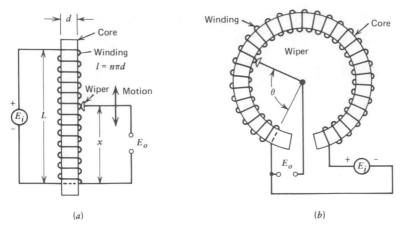

(a)                                    (b)

**Figure 3.2** (a) Wire-wrapped resistance potentiometer for longitudinal displacements. (b) Wire-wrapped resistance potentiometer for angular displacements.

the coil in the form of a helix. Helical potentiometers are commercially available with as many as 20 turns; therefore, angular displacements as large as 7200 degrees can be measured quite easily.

In recent years, potentiometers have been introduced that utilize a film of conductive plastic instead of a wire-wound coil. The film resistance on the in-

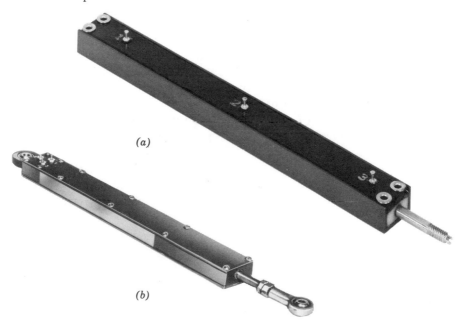

(a)

(b)

**Figure 3.3** (a) Precision wirewound and (b) precision conductive-plastic linear potentiometers. (Courtesy of Maurey Instrument Corp.)

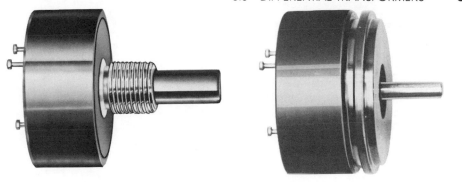

***Figure 3.4***  Precision conductive-plastic angular potentiometers. (Courtesy of Maurey Instrument Corp.)

sulating substrate exhibits essentially infinite resolution together with lower noise and longer life. A resistance of 50 to 100 $\Omega$/mm can be obtained with the conductive plastics. Potentiometers with a resolution of 0.001 mm are commercially available. Photographs of linear and angular potentiometers are shown in Figs. 3.3 and 3.4, respectively.

The dynamic response of the linear potentiometer is limited by the mass of the wiper. The response of the circular potentiometer is limited by the inertia of the shaft and wiper assembly.

Electronic noise often occurs as the brush on the wiper moves from one turn to the next. Much of the noise can be eliminated by ensuring that the coil is clean and free of oxide films and by applying a light lubricating film to the coil. Under ideal conditions, the life of a wire-wound potentiometer exceeds 1 million cycles; the life of a conductive-plastic potentiometer exceeds 10 million cycles.

Potentiometers are used primarily to measure relatively large displacements, that is, 10 mm or more for linear motion and 15 degrees or more for angular motion. Potentiometers are relatively inexpensive and accurate; however, their main advantage is simplicity of operation, since only a voltage source and a simple voltmeter comprise the instrumentation system. Their primary disadvantage is limited frequency response which precludes their use for dynamic measurements.

## 3.3  DIFFERENTIAL TRANSFORMERS

*Differential transformers*, based on a variable-inductance principle, are also used to measure displacement. The most popular variable-inductance transducer for linear displacement measurements is the linear variable differential transformer (LVDT). The LVDT illustrated in Fig. 3.5*a* consists of three symmetrically spaced coils wound onto an insulated bobbin. A magnetic core, which moves through the bobbin without contact, provides a path for magnetic flux

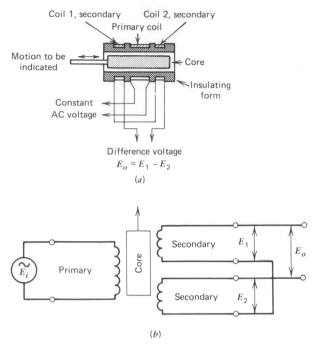

**Figure 3.5**   (*a*) Sectional view of a linear variable differential transformer (LVDT). (*b*) Schematic diagram of the LVDT circuit.

linkage between coils. The position of the magnetic core controls the mutual inductance between the center or primary coil and the two outer or secondary coils.

When an ac carrier excitation is applied to the primary coil, voltages are induced in the two secondary coils that are wired in a series-opposing circuit, as shown in Fig. 3.5*b*. When the core is centered between the two secondary coils, the voltages induced in the secondary coils are equal but out of phase by 180 degrees. With the series-opposing circuit, the voltages in the two coils cancel and the output voltage is zero. When the core is moved from the center position, an imbalance in mutual inductance between the primary and secondary coils occurs and an output voltage develops. The output voltage is a linear function of core position, as shown in Fig. 3.6, as long as the motion of the core is within the operating range of the LVDT. The direction of motion can be determined from the phase of the output voltage.

The frequency of the voltage applied to the primary winding can range from 50 to 25,000 Hz. If the LVDT is to be used to measure dynamic displacements, the carrier frequency should be 10 times greater than the highest frequency component in the dynamic signal. Highest sensitivities are attained with excitation frequencies between 1 and 5kHz. The input voltage ranges from 5 to 15 V. The power required is usually less than 1 W. Sensitivities of different LVDTs vary from 0.02 to 0.2 V/mm of displacement per volt of excitation applied to

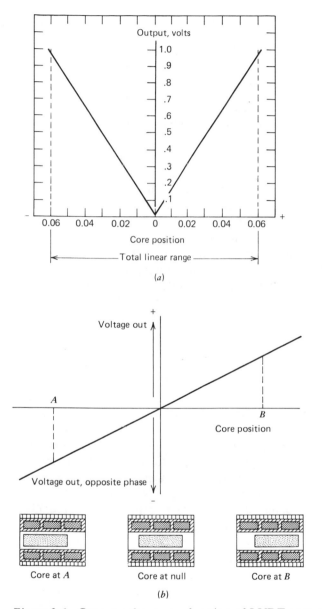

**Figure 3.6** Output voltage as a function of LVDT core position. (*a*) Magnitude of the output voltage. (*b*) Phase-referenced output voltage.

the primary coil. At rated excitation voltages, sensitivities vary from 0.16 to 2.6 V/mm of displacement. The higher sensitivities are associated with short-stroke LVDTs, with an operating range of ±2 mm; the lower sensitivities are for long-stroke LVDTs, with a range of ±150 mm.

Since the LVDT is a passive sensor requiring ac excitation at a voltage and

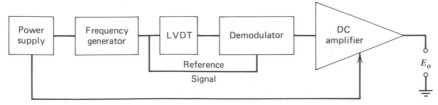

***Figure 3.7***   Block diagram of a signal conditioning circuit for an LVDT.

frequency not commonly available, signal conditioning circuits must be employed in its operation. A typical signal conditioner (see Fig. 3.7 for a block diagram) provides a power supply, a frequency generator to drive the LVDT, and a demodulator to convert the ac output signal from the LVDT to a dc output voltage. Finally, a dc amplifier is incorporated in the signal conditioner to provide a higher output voltage than can be obtained directly from the LVDT.

During the past decade, solid-state electronic devices have been developed that permit production of miniature signal conditioning circuits that can be packaged within the cover of an LVDT. The result is a small self-contained sensor known as a direct current differential transformer (DCDT). A DCDT operates from a battery or a regulated power supply and provides an amplified output signal. The output impedance of a DCDT is relatively low (about 100 $\Omega$).

The LVDT and the DCDT are both used to measure linear displacement; however, an analogous device known as a rotary variable differential transformer (RVDT) has been developed to measure angular displacements. As shown in Fig. 3.8, the RVDT consists of two primary coils and two secondary coils wound symmetrically on a large-diameter insulated bobbin. A cardioid-shaped rotor, fabricated from a magnetic material, is mounted on a shaft that extends through the bobbin and serves as the core. As shaft rotation turns the core, the mutual inductance between the primary and secondary windings varies and the output voltage-versus-rotation response curve shown in Fig. 3.9 is produced.

Although the RVDT is capable of a complete rotation (360 degrees), the range of linear operation is limited to ±40 degrees. The linearity of a typical RVDT having a range of ±40 degrees is about 0.5 percent of the range. Reducing the operating range improves the linearity, and an RVDT operating within a range of ±5 degrees exhibits a linearity of about 0.1 percent of this range.

The LVDT, DCDT, and RVDT have many advantages as sensors for measuring displacement. There is no contact between the core and the coils; therefore, friction is eliminated, thereby giving infinite resolution and no hysteresis. Noncontact also ensures that life will be very long with no significant deterioration of performance over this period.[1] The small core mass and freedom from friction give the sensor some capability for dynamic measurements. Finally, the sensors are not damaged by overtravel; therefore, they can be employed as feedback

---

[1] Mean time between failures for a typical DCDT is 33,000 h.

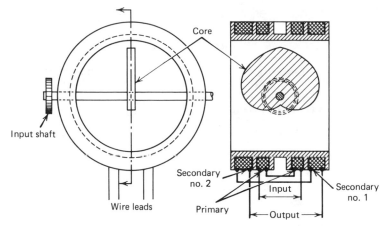

***Figure 3.8*** Simplified cross section of a rotary variable differential transformer (RVDT).

transducers in servo-controlled systems where overtravel occasionally may occur. Typical performance characteristics for LVDTs, DCDTs, and RVDTs are listed in Tables 3.1, 3.2, and 3.3, respectively.

## 3.4 RESISTANCE-TYPE STRAIN GAGES

*Electrical resistance strain gages* are thin metal-foil grids (see Fig. 3.10) that can be adhesively bonded to the surface of a component or structure. When the component or structure is loaded, strains develop and are transmitted to the foil grid. The resistance of the foil grid changes in proportion to the load-induced strain. The strain sensitivity of metals (copper and iron) was first observed by Lord Kelvin in 1856. The effect can be explained by the following simple analysis.

The resistance $R$ of a uniform metallic conductor can be expressed as

$$R = \frac{\rho L}{A} \tag{3.2}$$

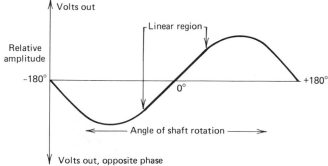

***Figure 3.9*** Output signal as a function of angular displacement for a typical RVDT.

**TABLE 3.1**   Performance Characteristics of Linear Variable Differential Transformers (LVDTs)
(*Courtesy of Schaevitz Engineering*)

| Model Number | Nominal Linear Range (in.) | Linearity ± Percent | | | | Sensitivity ((mV/V)/0.001 in.) | Impedance (Ω) | |
| --- | --- | --- | --- | --- | --- | --- | --- | --- |
| | | Percent of Full Range | | | | | | |
| | | 50 | 100 | 125 | 150 | | Primary | Secondary |
| 050 HR | ± 0.050 | 0.10 | 0.25 | 0.25 | 0.50 | 6.3 | 430 | 4000 |
| 100 HR | ± 0.100 | 0.10 | 0.25 | 0.25 | 0.50 | 4.5 | 1070 | 5000 |
| 200 HR | ± 0.200 | 0.10 | 0.25 | 0.25 | 0.50 | 2.5 | 1150 | 4000 |
| 300 HR | ± 0.300 | 0.10 | 0.25 | 0.35 | 0.50 | 1.4 | 1100 | 2700 |
| 400 HR | ± 0.400 | 0.15 | 0.25 | 0.35 | 0.60 | 0.90 | 1700 | 3000 |
| 500 HR | ± 0.500 | 0.15 | 0.25 | 0.35 | 0.75 | 0.73 | 460 | 375 |
| 1000 HR | ± 1.000 | 0.25 | 0.25 | 1.00 | 1.30[a] | 0.39 | 460 | 320 |
| 2000 HR | ± 2.000 | 0.25 | 0.25 | 0.50[a] | 1.00[a] | 0.24 | 330 | 330 |
| 3000 HR | ± 3.000 | 0.15 | 0.25 | 0.50[a] | 1.00[a] | 0.27 | 115 | 375 |
| 4000 HR | ± 4.000 | 0.15 | 0.25 | 0.50[a] | 1.00[a] | 0.22 | 275 | 550 |
| 5000 HR | ± 5.000 | 0.15 | 0.25 | 1.00[a] | — | 0.15 | 310 | 400 |
| 10000 HR | ±10.000 | 0.15 | 0.25 | 1.00[a] | — | 0.08 | 550 | 750 |

[a] Requires reduced core length.

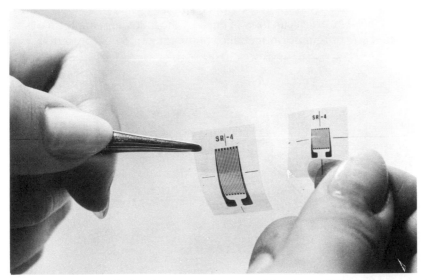

*Figure 3.10*   Electrical resistance strain gages. (Courtesy of BLH Electronics.)

**where**  $\rho$ is the specific resistance of the metal.
   $L$ is the length of the conductor.
   $A$ is the cross-sectional area of the conductor.

Differentiating Eq. (3.2) and dividing by the resistance $R$ gives

$$\frac{dR}{R} = \frac{d\rho}{\rho} + \frac{dL}{L} - \frac{dA}{A} \qquad (a)$$

The term $dA$ represents the change in cross-sectional area of the conductor resulting from the applied load. For the case of a uniaxial tensile stress state, recall that

$$\varepsilon_a = \frac{dL}{L} \qquad (b)$$

$$\varepsilon_t = - \nu\varepsilon_a = -\nu \frac{dL}{L}$$

**where**  $\varepsilon_a$ is the axial strain in the conductor.
   $\varepsilon_t$ is the transverse strain in the conductor.
   $\nu$ is Poisson's ratio of the metal used for the conductor.

**TABLE 3.2**  Performance Characteristics of Direct-Current Differential Transformers (DCDTs)
(*Courtesy of Schaevitz Engineering*)

| Model Number | Nominal Linear Range (in.) | Scale Factor (V/in.) | Response −3 dB (Hz) |
|---|---|---|---|
| 050 DC-D | ± 0.050 | 200 | 500 |
| 100 DC-D | ± 0.100 | 100 | 500 |
| 200 DC-D | ± 0.200 | 50 | 500 |
| 500 DC-D | ± 0.500 | 20 | 500 |
| 1000 DC-D | ± 1.000 | 10 | 200 |
| 2000 DC-D | ± 2.000 | 5.0 | 200 |
| 3000 DC-D | ± 3.000 | 3.3 | 200 |
| 5000 DC-D | ± 5.000 | 2.0 | 200 |
| 10000 DC-D | ±10.000 | 1.0 | 200 |

If the diameter of the conductor before application of the axial strain is $d_o$, the diameter of the conductor after it is strained $d_f$ is given by

$$d_f = d_o\left(1 - \nu\frac{dL}{L}\right) \qquad (c)$$

**TABLE 3.3** Performance Characteristics of Rotary Variable Differential
Transformers (RVDTs)
(*Courtesy of Schaevitz Engineering*)

| Model Number | Linearity ± Percent | | | Sensitivity ((mV/V)/degree) | Impedance (Ω) | |
|---|---|---|---|---|---|---|
| | ±30° | ±40° | ±60° | | Primary | Secondary |
| (@ 2.5 kHz) | | | | | | |
| R30A | 0.25 | 0.5 | 1.5 | 2.3 | 125 | 500 |
| R36A | 0.5 | 1.0 | 3.0 | 1.1 | 750 | 2000 |
| (@ 10 kHz) | | | | | | |
| R30A | 2.5 | 0.5 | 1.5 | 2.9 | 370 | 1300 |
| R36A | 0.5 | 1.0 | 3.0 | 1.7 | 2500 | 5400 |

From Eq. (c) it is clear that

$$\frac{dA}{A} = -2v\frac{dL}{L} + v^2\left(\frac{dL}{L}\right)^2 \approx -2v\frac{dL}{L} \tag{d}$$

Substituting Eq. (d) into Eq. (a) and simplifying yields

$$\frac{dR}{R} = \frac{d\rho}{\rho} + \frac{dL}{L}(1 + 2v) \tag{3.3}$$

which can be written as

$$S_A = \frac{dR/R}{\varepsilon_a} = \frac{d\rho/\rho}{\varepsilon_a} + (1 + 2v) \tag{3.4}$$

The quantity $S_A$ is defined as the sensitivity of the metal or alloy used for the conductor.

It is evident from Eq. (3.4) that the strain sensitivity of a metal or alloy is due to two factors; namely, the changes in dimensions of the conductor as expressed by the term $(1 + 2v)$ and the change in specific resistance as represented by the term $(d\rho/\rho)/\varepsilon$. Experimental studies show that the sensitivity $S_A$ ranges between 2 and 4 for most alloys used in strain-gage fabrication. Since the quantity $(1 + 2v)$ is approximately 1.6 for most of these materials, the contribution due to the change in specific resistance with strain varies from 0.4 to 2.4. The change in specific resistance is due to variations in the number of free electrons and their increased mobility with applied strain.

A list of the alloys commonly employed in commercial strain gages together with their sensitivities is presented in Table 3.4. The most commonly used strain gages are fabricated from the copper–nickel alloy known as Advance or Constantan. The response curve for this alloy ($\Delta R/R$ as a function of strain) is shown in Fig. 3.11. This alloy is widely used because its response is linear over a wide range of strain, it has a high specific resistance, and it has excellent thermal stability.

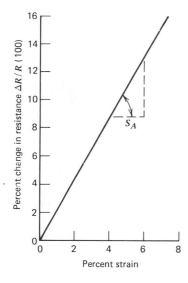

**Figure 3.11**   Change of resistance $\Delta R/R$ as a function of strain for an Advance alloy.

Most resistance strain gages are of the metal-foil type, where the grid configuration is formed by a photoetching process. Since the process is very versatile, a wide variety of gage sizes and grid shapes can be produced. Typical examples are shown in Fig. 3.12. The shortest gage available is 0.20 mm; the longest is 102 mm. Standard gage resistances are 120 and 350 $\Omega$; however, special-purpose gages with resistances of 500 and 1000 $\Omega$ are available.

The etched metal-film grids are very fragile and easy to distort, wrinkle, or tear. For this reason, the metal grid is bonded to a thin plastic film that serves as a backing or carrier before photoetching. The carrier film, shown in Fig. 3.10, also provides electrical insulation between the gage and the component after the gage is mounted.

A strain gage exhibits a resistance change $\Delta R/R$ that is related to the strain $\varepsilon$ in the direction of the grid lines by the expression

$$\frac{\Delta R}{R} = S_g \varepsilon \tag{3.5}$$

**TABLE 3.4**   Strain Sensitivity $S_A$ for Common Strain-Gage Alloys

| Material | Composition (%) | $S_A$ |
|---|---|---|
| Advance or Constantan | 45 Ni, 55 Cu | 2.1 |
| Nichrome V | 80 Ni, 20 Cr | 2.1 |
| Isoelastic | 36 Ni, 8 Cr, 0.5 Mo, 55.5 Fe | 3.6 |
| Karma | 74 Ni, 20 Cr, 3 Al, 3 Fe | 2.0 |
| Armour D | 70 Fe, 20 Cr, 10 Al | 2.0 |
| Platinum-Tungsten | 92 Pt, 8 W | 4.0 |

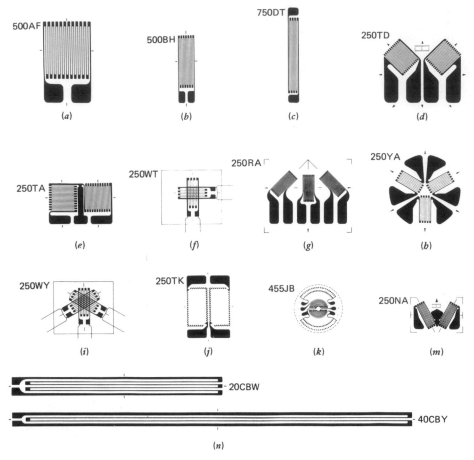

***Figure 3.12*** Configurations of metal-foil resistance strain gages. (Courtesy of Micro-Measurements.)

*(a) Single-element gage.*
*(b) Single-element gage.*
*(c) Single-element gage.*
*(d) Two-element rosette.*
*(e) Two-element rosette.*
*(f) Two-element stacked rosette.*
*(g) Three-element rosette.*
*(h) Three-element rosette.*
*(i) Three-element stacked rosette.*
*(j) Torque gage.*
*(k) Diaphragm gage.*
*(m) Stress gage.*
*(n) Single-element gage for use on concrete.*

where $S_g$ is the gage factor or calibration constant for the gage. The gage factor $S_g$ is always less than the sensitivity of the metallic alloy $S_A$ because the grid configuration of the gage is less responsive to strain than a straight uniform conductor.

The output $\Delta R/R$ of a strain gage is usually converted to a voltage signal with a Wheatstone bridge, as illustrated in Fig. 3.13. If a single gage is used in one arm of the Wheatstone bridge and equal but fixed resistors are used in the other three arms, the output voltage is

$$E_o = \frac{E_i}{4}(\Delta R_g/R_g) \qquad (3.6)$$

Substituting Eq. (3.5) into Eq. (3.6) gives

$$E_o = \frac{1}{4}E_iS_g\varepsilon \qquad (3.7)$$

The input voltage is controlled by the gage size (the power it can dissipate) and the initial resistance of the gage. As a result, the output voltage $E_o$ usually ranges between 1 and 10 $\mu$V/microunit of strain ($\mu$m/m or $\mu$in./in.).

## 3.5 CAPACITANCE SENSORS

The *capacitance sensor*, illustrated in Fig. 3.14, consists of two metal plates separated by an air gap. The capacitance $C$ between terminals is given by the expression

$$C = \frac{kKA}{h} \qquad (3.8)$$

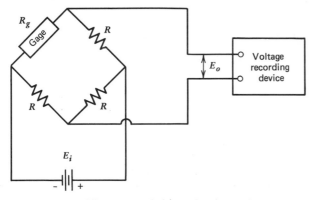

**Figure 3.13** Wheatstone bridge circuit used to convert resistance change $\Delta R/R$ of a strain gage to an output voltage $E_o$.

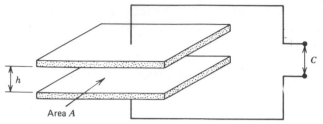

***Figure 3.14*** Flat-plate capacitance sensor.

**where** $C$ is the capacitance in picofarads (pF).
$K$ is the dielectric constant for the medium between the plates.
$A$ is the overlapping area for the two plates.
$h$ is the thickness of the gap between the two plates.
$k$ is a proportionality constant ($k = 0.225$ for dimensions in inches, and $k = 0.00885$ for dimensions in millimeters).

Capacitance sensors are used to measure displacement in one of two different ways; namely, by changing the plate separation $h$ or by changing the overlapping area $A$.

If the plate separation is changed by an amount $\Delta h$, then the capacitance can be expressed as

$$C + \Delta C = \frac{kKA}{h + \Delta h} \tag{a}$$

which, after substituting Eq. (3.8) and simplifying, yields

$$\Delta C = -\frac{kKA}{h + \Delta h}\left(\frac{\Delta h}{h}\right) \tag{b}$$

Equation (b) indicates that the response of this type of capacitance sensor is nonlinear because of the presence of the $\Delta h$ term in the denominator.

The sensitivity $S$ of the capacitance sensor is defined as

$$S = \frac{\Delta C}{\Delta h} = -\frac{kKA}{h(h + \Delta h)} \tag{c}$$

Unfortunately, the sensitivity $S$ of the capacitance sensor is also a function of $\Delta h$. This fact severely limits the useful range of the sensor.

For those cases where the change in spacing is a very small fraction of the original spacing ($\Delta h << h$), the response and sensitivity of the capacitance sensor can be approximated by

$$\frac{\Delta C}{C} = -\frac{\Delta h}{h} \tag{3.9}$$

$$S = -\frac{kKA}{h^2} \tag{3.10}$$

The error $\mathscr{E}$ associated with the assumption of linearity implied by the use of Eqs. (3.9) and (3.10) can be determined from the expression

$$\mathscr{E} = \left(\frac{\Delta h}{h + \Delta h}\right)100 \tag{3.11}$$

A capacitance sensor having the overlapping area change is illustrated in Fig. 3.15. In this case, the capacitance can be written as

$$C = \frac{kKlw}{h} \tag{3.12}$$

where $l$ and $w$ are the length and width of the overlapping area of the capacitor plates. As the movable plate displaces an amount $\Delta l$ relative to the fixed plate, the capacitance changes such that

$$C + \Delta C = \frac{kKw}{h}(l + \Delta l) \tag{d}$$

Substituting Eq. (3.12) into Eq. (d) and simplifying gives

$$\frac{\Delta C}{C} = \frac{\Delta l}{l} \tag{3.13}$$

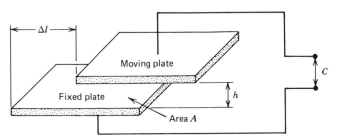

**Figure 3.15**   Capacitance sensor with changing area.

It is evident from Eq. (3.13) that the response of this type of capacitance sensor is linear; therefore, the range of the sensor is not limited by linearity restrictions. The sensitivity $S$ of this type of capacitance sensor is

$$S = \frac{\Delta C}{\Delta l} = \frac{kKw}{h} \tag{3.14}$$

The sensitivity of capacitance-type sensors is inherently low. For example, a sensor with $w = 10$ mm and $h = 0.2$ mm has a sensitivity $S = 0.4425$ pF/mm. Theoretically, the senstivity can be increased without limit by decreasing the airgap $h$; however, there are practical electrical and mechanical limits that preclude high sensitivities.

One of the primary advantages of the capacitance transducer is that the forces involved in moving one plate relative to the other are extremely small. A second advantage is stability. The sensitivity of the sensor is not influenced by pressure or temperature of the environment.

Many types of circuits can be used to measure the change in capacitance associated with these sensors. The circuit shown in Fig. 3.16 is capable of resolving changes as small as 0.00001 pF. The system consists of a capacitive potentiometer circuit driven with an ac carrier (3.4 kHz). The output from this circuit goes to a charge amplifier that is used to accommodate the high output impedance of the circuit and to reduce the detrimental effects of the cable between the sensor and the charge amplifier. The signal is amplified prior to demodulation and a dc voltmeter is used for the readout.

## 3.6 EDDY CURRENT SENSORS

An *eddy current sensor* measures distance between the sensor and an electrically conducting surface, as illustrated in Fig. 3.17. Sensor operation is based on eddy currents that are induced at the conducting surface as magnetic flux

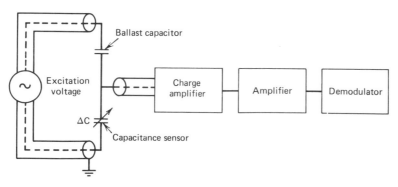

*Figure 3.16* Circuit for measuring small changes in capacitance.

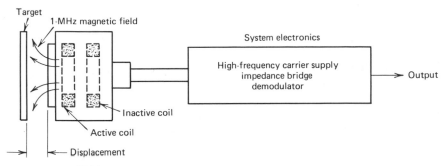

***Figure 3.17***  Eddy current sensor.

lines from the sensor pass into the conducting material being monitored. The magnetic flux lines are generated by the active coil in the sensor, which is driven at a very high frequency (1 MHz). The eddy currents produced at the surface of the conducting material are a function of distance between the active coil and the surface. The eddy currents increase as the distance decreases.

Changes in the eddy currents are sensed with an impedance (inductance) bridge. Two coils in the sensor are used for two arms of the bridge. The other two arms are housed in the associated electronic package illustrated in Fig. 3.17. The first coil in the sensor (active coil) is wired into the active arm of the bridge, which changes inductance with target movement. The second coil is wired into an opposing arm of the bridge, where it serves as a compensating coil to balance and cancel most of the effects of temperature change. The output from the impedance bridge is demodulated and becomes the analog signal, which is linearly proportional to distance between the sensor and the target.

The sensitivity of the sensor is dependent upon the target material with higher sensitivity associated with higher conductivity materials. The output for a number of materials is shown as a function of specific resistivity in Fig. 3.18. For aluminum targets, the sensitivity is typically 100 mV/mil (4 V/mm). Thus, it is apparent that eddy current sensors are high-output devices.

For nonconducting or poorly conducting materials, it is possible to bond a thin film of aluminum foil to the surface of the target at the location of the sensor to improve the conductivity. Since the penetration of the eddy currents into the material is minimal, the thickness of the foil can be as small as 0.7 mil (ordinary kitchen aluminum foil).

The effect of temperature on the output of the eddy current sensor is small. The sensing head with dual coils is temperature compensated; however, a small signal can be produced by temperature changes in the target material since the resistivity of the target material is a function of temperature. For instance, if the temperature of an aluminum target is increased by 500°F, its resistivity increases from 0.03 to 0.06 $\mu\Omega \cdot$ m. Figure 3.19 shows that the bridge output is reduced by about 2 percent for this change in resistivity, which is equivalent to a temperature sensitivity of 0.004 percent/°F.

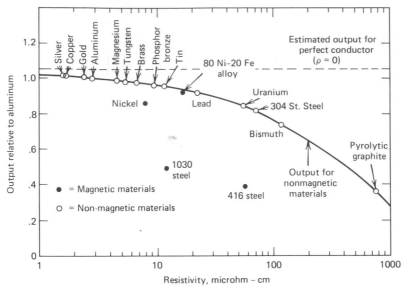

**Figure 3.18**    Relative output from an eddy current sensor as a function of resistivity of the target material.

The range of the sensor is controlled by the diameters of the coils with the larger sensors exhibiting the larger ranges. A typical range to diameter ratio is 0.25. Linearity is typically better than ±0.5 percent and resolution is better than 0.05 percent of full scale. The frequency response is typically 20 kHz, although small-diameter coils can be used to increase this response to 50 kHz.

The fact that eddy current sensors do not require contact for measuring displacement is quite important. As a result of this feature, they are often used in transducer systems for automatic control of dimensions in fabrication processes. They are also applied extensively to determine thicknesses of organic coatings that are nonconducting. A modern portable eddy-current tester is shown in Fig. 3.19.

## 3.7  PIEZOELECTRIC SENSORS

A piezoelectric material is, as its name implies, a material that produces an electric charge when subjected to a force or pressure. Piezoelectric materials, such as single-crystal quartz or polycrystalline barium titanate, contain molecules with asymmetrical charge distributions. When pressure is applied, the crystal deforms and there is a relative displacement of the positive and negative charges within the crystal. This displacement of internal charges produces external charges of opposite sign on two surfaces of the crystal. If these surfaces are coated with metallic electrodes, as illustrated in Fig. 3.20, the charge $q$ that develops can be

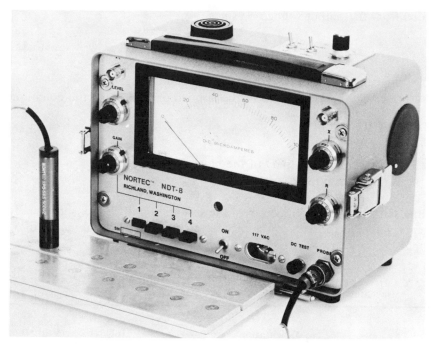

**Figure 3.19** A portable eddy-current tester. (Courtesy of Nortec Corp.)

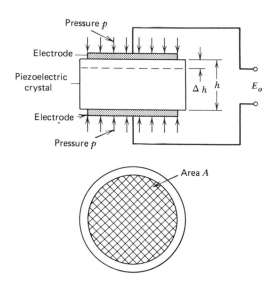

**Figure 3.20** Piezoelectric crystal deforming under the action of applied pressure.

determined from the output voltage $E_o$ since

$$q = E_o C \qquad (3.15)$$

where $C$ is the capacitance of the piezoelectric crystal.

The surface charge $q$ is related to the applied pressure $p$ by the equation

$$q = S_q A p \qquad (3.16)$$

**where**    $S_q$ is the charge sensitivity of the piezoelectric crystal.
$A$ is the area of the electrode.

The charge sensitivity $S_q$ is a function of the orientation of the sensor (usually a cylinder) relative to the axes of the piezoelectric crystal. Typical values of $S_q$ for common piezoelectric materials are given in Table 3.5.

**TABLE 3.5**    Typical Charge and Voltage Sensitivities $S_q$ and $S_E$ of Piezoelectric Materials

| Material | Orientation | $S_q$ (pC/N) | $S_E$ (V·m/N) |
|---|---|---|---|
| Quartz SiO$_2$ Single crystal | X-cut length longitudinal | 2.2 | 0.055 |
|  | X-cut thickness longtudinal | −2.0 | −0.05 |
|  | Y-cut thickness shear | 4.4 | 0.11 |
| Barium titanate BaTiO$_3$ Ceramic, poled polycrystalline | Parallel to polarization | 130 | 0.011 |
|  | Perpendicular to polarization | −56 | − 0.004 |

The output voltage $E_o$ developed by the piezoelectric sensor is obtained by substituting Eqs. (3.8) and (3.16) into Eq. (3.15). Thus,

$$E_o = \left(\frac{S_q}{kK}\right) hp \qquad (3.17)$$

The voltage sensitivity $S_E$ of the sensor can be expressed as

$$S_E = \frac{S_q}{kK} \qquad (3.18)$$

The output voltage of the sensor is then

$$E_o = S_E h p \tag{3.19}$$

Again, the voltage sensitivity $S_E$ of the sensor is a function of its orientation relative to the axes of the crystal. Typical values of $S_E$ are also presented in Table 3.5.

Most piezoelectric transducers are fabricated from single-crystal quartz, since it is the most stable of the piezoelectric materials. Mechanically and electrically it is nearly loss free. Its modulus of elasticity is 86 GPa; its resistivity is about $10^{12}$ $\Omega \cdot$ m; and its dielectric constant is 40.6 pF/m. It exhibits excellent high-temperature properties and can be operated up to 550°C. The charge sensitivity of quartz is low when compared to barium titanate; however, with high-gain charge amplifiers available for processing the output signal, the lower sensitivity is not a serious disadvantage.

Barium titanate is a polycrystalline material that can be polarized by applying a high voltage to the electrodes while the material is at a temperature above the curie point (125°C). The electric field aligns the ferroelectric domains in the barium titanate and it becomes piezoelectric. If the polarization voltage is maintained while the material is cooled well below the Curie point, the piezoelectric characteristics are permanent and stable after a short aging period.

The mechanical stability of barium titanate is excellent; it exhibits high mechanical strength and has a high modulus of elasticity (120 GPa). It is more economical than quartz and can be fabricated in a wide variety of sizes and shapes. While its application in transducers is second to quartz, it is frequently used in ultrasonics as a source. In this application, a voltage is applied to the electrodes and the barium titanate deforms and delivers energy to the work piece or test specimen.

Most transducers exhibit a relatively low output impedance (from about 100 to 1000 $\Omega$). When piezoelectric crystals are used as the sensing elements in transducers, the output impedance is extremely high. The output impedance of a small cylinder of quartz depends upon the frequency $\omega$ associated with the applied pressure. Since the sensor acts like a capacitor, the output impedance is given by

$$Z = \frac{1}{\omega C} \tag{3.20}$$

Thus, the impedance ranges from infinity for static applications of pressure to about 10 k$\Omega$ for very high-frequency applications (100 kHz). With this high output impedance, care must be exercised in monitoring the output voltage; otherwise, serious errors can occur.

A circuit diagram of a measuring system with a piezoelectric sensor is shown in Fig. 3.21. The piezoelectric sensor acts as a charge generator. In addition to the charge generator, the sensor is represented by a capacitor $C_p$ (from 10 to 1000 pF) and a leakage resistor $R_p$ (about $10^{14}$ $\Omega$). The capacitance of the lead wires $C_L$ must also be considered, since even relatively short lead wires have a capacitance larger than the sensor. The amplifier is either a cathode follower or a charge amplifier with sufficient impedance to isolate the piezoelectric sensor. If a pressure is applied to the sensor and maintained for a long period of time, a charge $q$ and an output voltage $E_o$ is developed by the piezoelectric material; however, the charge $q$ leaks off by way of a small current flow through $R_p$ and the amplifier resistance $R_A$. The time available for readout of the signal depends upon the effective time constant $\tau_e$ of the circuit which is given by

$$\tau_e = R_e C_e = \frac{R_p R_A}{R_p + R_A} (C_p + C_L + C_A) \tag{3.21}$$

**where**   $R_e$ is the equivalent resistance of the circuit.
           $C_e$ is the equivalent capacitance of the circuit.

Time constants ranging from 1000 to 100,000 s can be achieved with quartz sensors and commercially available charge amplifiers. These time constants are sufficient to permit measurement of quantities that vary slowly with time or measurement of static quantities for short periods of time. Problems associated with output voltage measurements diminish as the frequency of the mechanical input increases.

    The inherent dynamic response of the piezoelectric sensor is very high, since the resonant frequency of the small cylindrical piezoelectric element is very large. The resonant frequency of the transducer depends upon the mechanical design of the transducer as well as the mass and stiffness of the sensor. For this reason, specification of frequency response will be deferred to later sections on force,

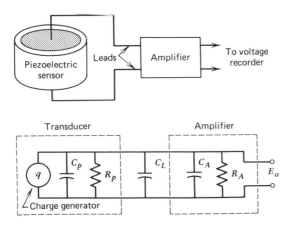

***Figure 3.21***   Schematic diagram of a measuring system with a piezoelectric sensor.

pressure, and acceleration measurements. It should be noted here, however, that one of the primary advantages of the piezoelectric sensor is its very high-frequency response.

## 3.8  PIEZORESISTIVE SENSORS

*Piezoresistive sensors*, as the name implies, are materials that exhibit a change in resistance when subjected to a pressure. The development of piezo-resistive materials was an outgrowth of research on semiconductors by Bell Telephone Laboratories in the early 1950s that eventually led to the transistor.

Piezoresistive sensors are fabricated from semiconductive materials—usu-ally silicon containing boron as the trace impurity for the P-type material and arsenic as the trace impurity for the N-type material. The resistivity of the semiconducting materials can be expressed as

$$\rho = \frac{1}{eN\mu} \qquad (3.22)$$

**where**  $e$  is the electron charge, which depends on the type of impurity.

$N$ is the number of charge carriers, which depends on the concentration of the impurity.

$\mu$ is the mobility of the charge carriers, which depends upon strain and its direction relative to the crystal axes.

Equation (3.22) shows that the resistivity of the semiconductor can be adjusted to any specified value by controlling the concentration of the trace impurity. The impurity concentrations commonly employed range from $10^{16}$ to $10^{20}$ atoms/cm$^3$, which permits a wide variation in the initial resistivity. The resistivity for P-type silicon with a concentration of $10^{20}$ atoms/cm$^3$ is 500 $\mu\Omega \cdot$ m, which is about 30,000 times higher than the resistivity of copper. This very high resistivity facilitates the design of miniaturized sensors.

Equation (3.22) also indicates that the resistivity changes when the piezo-resistive sensor is subjected to either stress or strain. This change of resisitivity is known as the piezoresistive effect and can be expressed by the equation

$$\rho_{ij} = \delta_{ij}\rho + \pi_{ijkl}\tau_{kl} \qquad (3.23)$$

**where**  the subscripts $i$, $j$, $k$, and $l$ range from 1 to 3.

$\pi_{ijkl}$ is a fourth rank piezoresistivity tensor.

$\tau_{kl}$ is the stress tensor.

$\delta_{ij}$ is the Kroneker delta.

Fortunately, silicon is a cubic crystal; therefore, the 36 piezoresistive coefficients reduce to 3 and Eq. (3.23) can be expressed as

$$\rho_{11} = \rho[1 + \pi_{11}\sigma_{11} + \pi_{12}(\sigma_{22} + \sigma_{33})]$$

$$\rho_{22} = \rho[1 + \pi_{11}\sigma_{22} + \pi_{12}(\sigma_{33} + \sigma_{11})]$$

$$\rho_{33} = \rho[1 + \pi_{11}\sigma_{33} + \pi_{12}(\sigma_{11} + \sigma_{22})]$$

$$\rho_{12} = \rho\pi_{44}\tau_{12}$$

$$\rho_{23} = \rho\pi_{44}\tau_{23}$$

$$\rho_{31} = \rho\pi_{44}\tau_{31} \tag{3.24}$$

where the subscripts 1, 2, and 3 identify the axes of the crystal.

These equations indicate that the piezoresistive crystal, when subjected to a state of stress, becomes electrically anisotropic. The resistivity depends upon both direction and the stresses in each direction. Because of this electrical anisotropy, Ohm's law must be written as

$$E'_i = \rho_{ij}I'_j \tag{3.25}$$

**where**    $E'$ is the potential gradient.
       $I'$ is the current density.

Substituting Eqs. (3.24) into Eq. (3.25) gives

$$\frac{E'_1}{\rho} = I'_1[1 + \pi_{11}\sigma_{11} + \pi_{12}(\sigma_{22} + \sigma_{33})]$$

$$+ \pi_{44}(I'_2\tau_{12} + I'_3\tau_{31})$$

$$\frac{E'_2}{\rho} = I'_2[1 + \pi_{11}\sigma_{22} + \pi_{12}(\sigma_{33} + \sigma_{11})]$$

$$+ \pi_{44}(I'_3\tau_{23} + I'_1\tau_{12})$$

$$\frac{E'_3}{\rho} = I'_3[1 + \pi_{11}\sigma_{33} + \pi_{12}(\sigma_{11} + \sigma_{22})]$$

$$+ \pi_{44}(I'_1\tau_{31} + I'_2\tau_{23}) \tag{3.26}$$

These results show that the voltage drop across a sensor will be dependent upon the current density $I'$, the state of stress $\tau$, and the three piezoresistive coefficients. The piezoresistive coefficients can be adjusted by controlling the concentration of the impurity, and by optimizing the direction of the axis of the sensor with respect to the crystal axes. As a consequence, the sensitivity of a typical sensor is quite high (for example, a piezoresistive strain gage exhibits a gage factor of 100, while a conventional metal-foil strain gage exhibits a gage factor of 2).

The high sensitivity and high resistivity of piezoresistive sensors have fostered the design of miniaturized transducers that respond to small mechanical inputs with high electrical outputs. Examples of a miniature pressure transducer and a miniature accelerometer are shown in Fig. 3.22.

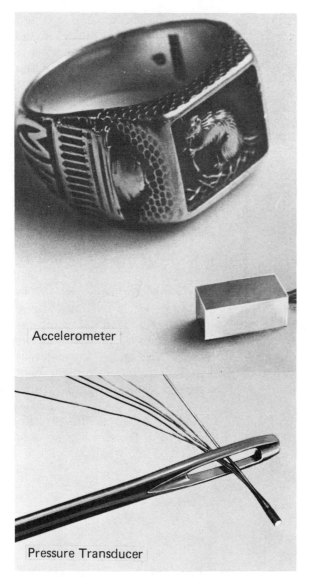

Accelerometer

Pressure Transducer

*Figure 3.22*   Miniaturized transducers that use piezoresistive sensing elements. (Courtesy of Kulite Semiconductor Products.)

## 3.9  PHOTOELECTRIC SENSORS

In certain applications where contact cannot be made with the test specimen, a photoelectric sensor can be used to monitor changes in light intensity, which can be related to the quantity being measured. Three different types of photoelectric detectors (photocells) are used to convert a radiation input to a voltage output. These include photoemissive cells, photoconductive cells, and photovoltaic cells.

The *photoemissive cell*, illustrated in Fig. 3.23, contains a cathode $C$ and an anode $A$ mounted in a vacuum tube. Incident radiation impinging on the cathode material frees electrons that flow to anode $A$ to produce an electric current. The photoelectric current $I$ is proportional to the illumination $\psi$ imposed on the cathode. Thus

$$I = S\psi \tag{3.27}$$

where $S$ is the sensitivity of the photoelectric cell. The sensitivity $S$ depends primarily upon the photoemissive material deposited on the cathode surface. *Photoemissive materials* are usually compounds of the alkali metals. One common photoemissive material consists of silver, oxygen, and cesium. The cathode is plated with a layer of silver that is oxidized and covered with a layer of cesium. The sensitivity $S$ is strongly dependent upon the wavelength $\lambda$ of the radiation. For this reason, photoemissive sensors should be employed with monochromatic light; otherwise, a careful calibration must be performed to account for the variation of sensitivity with wavelength.

*Photoconductive cells* are fabricated from semiconductor materials, such as cadmium sulfide (CdS) or cadmium selenide (CdSe), which exhibit a strong photoconductive response. The electrical resistivity of these materials decreases when they are exposed to light. A typical circuit diagram for a photoconductive cell is shown in Fig. 3.24. Photoconduction in these semiconducting materials is due to the absorption of incident photons that excite electrons to the level of the conduction band. These electrons are then free to move through the crystal lattice to produce a current $I_o$.

When the photoconductor is maintained in a dark environment, its resistance is high and only a small dark current is produced. If the sensor is exposed

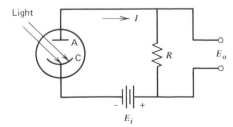

**Figure 3.23**  Photoemissive-type photocell and associated circuit.

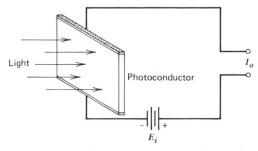

***Figure 3.24*** Photoconductor type photocell and associated circuit.

to light, the resistance decreases significantly (the ratio of maximum to minimum resistance ranges from 100 to 10,000 in common commercial sensors); therefore, the output current $I_o$ can be quite large. The sensitivity of photoconductive sensors, which depend upon the power dissipation capability of the cell and the applied voltage $E_i$, varies from about 0.00001 to 10 A/lm (amperes per lumen).

Photoconductors respond to radiation ranging from long thermal radiation through the infrared, visible, and ultraviolet regions of the electromagnetic spectrum. The sensitivity $S$, which is not constant with wavelength, drops sharply at the longer wavelengths; consequently, photoconductive cells exhibit the same disadvantage as photoemissive cells when they are exposed to radiation that undergoes a change in wavelength (high-temperature radiation).

In photoconductive cells, the photocurrent requires some time to develop after the excitation is applied and some time to decay after the excitation is removed. The rise and fall times for commercially available photoconductors range from $10^{-6}$ to 10 s. Therefore, in dynamic applications, frequency response of the photoconductor being used must be given careful consideration.

Photovoltaic cells have improved significantly in the past decade with the advent of transistor technology. The *photovoltaic cells* in common use today are P-N-type diffused-silicon guard-ring photodiodes (see Fig. 3.25). Operating features of these devices include a wide spectral range, fast response time, high sensitivity, excellent linearity, low noise, and simplicity of circuit design.

When the active area of a photodiode is illuminated and a connection is made between the P- and N-diffused regions, a current flows during the period of illumination. This phenomenon is the well-known *photovoltaic effect*, which is the operating mechanism for solar cells. For sensor usage, an external bias (voltage) $E_b$ is applied in the reverse direction between the P and N regions and the photodiode becomes a photoconductor. In the photoconductive mode, the photodiode acts similarly to a current generator and, unlike a photovoltaic cell, will deliver a constant current into any load $R_L$ at a fixed level of illumination. This current is composed of photocurrent and dark (reverse leakage) current. The dark current remains constant, provided the external bias and ambient temperature are fixed. The photocurrent varies linearly with illumination inten-

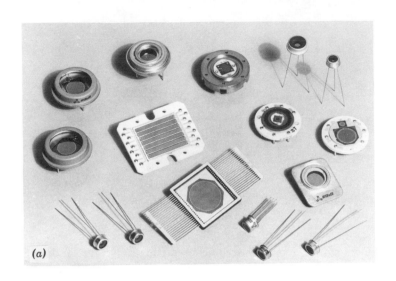

(a)

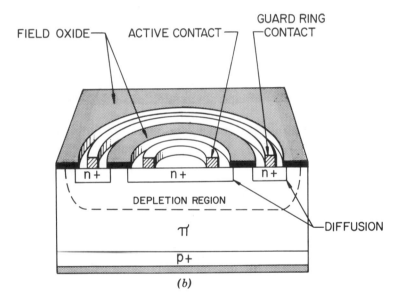

FIELD OXIDE — ACTIVE CONTACT — GUARD RING CONTACT

n+    n+    n+

DEPLETION REGION

π

p+

DIFFUSION

(b)

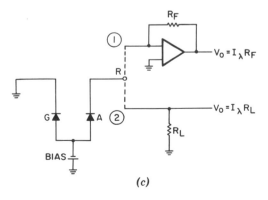

$R_F$

$V_O = I_\lambda R_F$

$R$

$G$    $A$

$V_O = I_\lambda R_L$

$R_L$

BIAS

(c)

sity. An advantage of operating in the photoconductive mode is that the sensitivity is higher than that for a photovoltaic cell of the same size.

The recommended circuit for use with a photodiode is shown in Fig. 3.25c. Connection 1 is used for best noise performance in applications which do not require wide bandwidth; otherwise, connection 2 is used.

The frequency response of a photodiode is excellent (a bandwidth of 45 MHz and a rise time of approximately 5 ns). Linearity of $\pm 1$ percent can be achieved over a range of seven decades.

## 3.10   RESISTANCE TEMPERATURE DETECTORS (RTDs)

The change in resistance of metals with temperature provides the basis for a family of temperature measuring sensors known as resistance temperature detectors (RTDs). The sensor is simply a conductor fabricated either as a wire-wound coil or as a film or foil grid. The change in resistance of the conductor with temperature is given by the expression

$$\frac{\Delta R}{R_o} = \gamma_1(T - T_o) + \gamma_2(T - T_o)^2 + \cdots + \gamma_n(T - T_o)^n \qquad (3.28)$$

**where**   $T_o$ is a reference temperature.

$R_o$ is the resistance at temperature $T_o$.

$\gamma_1, \gamma_2, \ldots, \gamma_n$ are temperature coefficients of resistance.

Resistance temperature detectors are often used in ovens and furnaces where inexpensive but accurate and stable temperature measurements and controls are required.

Platinum is widely used for sensor fabrication since it is the most stable of all the metals, is the least sensitive to contamination, and is capable of operating over a very wide range of temperatures (4°K to 1064°C). Platinum also provides an extremely reproducible output; therefore, it has been selected for use in interpolating instruments for the measurement of the International Practical Temperature Scale (1968) over the range from 13.81°K to 630.74°C.

The performance of platinum RTDs depends strongly on the design and construction of the package containing the sensor. The most precise sensors are fabricated with a minimum amount of support; therefore, they are fragile and will often fail if subjected to rough handling, shock, or vibration. Most sensors used in trandsucers for industrial applications have platinum coils supported on ceramic or glass tubes. These fully supported sensors are quite rugged and will withstand shock levels up to 100 g's. Unfortunately, the range and accuracy of such sensors are limited to some degree by the influence of the constraining

---

*Figure 3.25*   (p. 106) Diffused-silicon photodiode with guard-ring construction. (Courtesy of EG&G Electro-Optics.) (*a*) Selection of Diffused Silicon Photodiodes. (*b*) Construction details. (*c*) Recommended circuit.

package. More information on the performance characteristics of packaged RTDs and the circuits used for their application is presented in Chapter Eight.

The sensitivity $S_R$ of a platinum RTD is relatively high ($S_R = 0.39$ Ω/°C at 0°C); however, the sensitivity varies with temperature, as indicated by Eq. (3.28), and $S_R$ decreases to 0.378, 0.367, 0.355, 0.344, and 0.332 Ω/°C at temperatures of 100, 200, 300, 400, and 500°C, respectively.

The dynamic response of an RTD depends almost entirely on construction details. For large coils mounted on heavy ceramic cores and sheathed in stainless steel tubes, the response time may be several seconds or more. For film or foil elements mounted on ceramic or polyimide substrates, the response time can be less than 0.1 s.

## 3.11 THERMISTORS

A second type of temperature-measuring sensor based on the fact that the resistance of a material may change with temperature is known as a *thermistor*. Thermistors differ from resistance temperature detectors (RTDs) in that they are fabricated from semiconducting materials instead of metals. The semiconducting materials, which include oxides of copper, cobalt, manganese, nickel, and titanium, exhibit very large changes in resistance with temperature. As a result, thermistors can be fabricated in the form of extremely small beads as shown in Fig. 3.26.

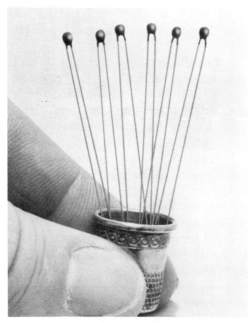

*Figure 3.26*   Miniature bead-type thermistors. (Courtesy of Dale Electronics, Inc.)

Resistance change with temperature can be expressed by an equation of the form

$$\ln \rho = A_0 + \frac{A_1}{T} + \frac{A_2}{T^2} + \cdots + \frac{A_n}{T^n} \qquad (3.29)$$

**where**   $\rho$ is the specific resistance of the material.
   $A_1, A_2, \ldots, A_n$ are material constants.
   $T$ is the absolute temperature.

The temperature–resistance relationship as expressed by Eq. (3.29) is usually approximated by retaining only the first two terms. The simplified equation is then expressed as

$$\ln \rho = A_0 + \frac{\beta}{T} \qquad (3.30)$$

Use of Eq. (3.30) is convenient and acceptable when the temperature range is small and the higher-order terms of Eq. (3.29) are negligible.

Thermistors have many advantages over other temperature sensors and are widely used in industry. They can be small (0.005-in. diameter) and, consequently, permit point sensing and rapid response to temperature change. Their high resistance minimizes lead-wire problems. Their output is more than 10 times that of a resistance temperature detector (RTD) as shown in Fig. 3.27. Finally, thermistors are very rugged, which permits use in those industrial environments where shock and vibrations occur. The disadvantages of thermistors include nonlinear output with temperature, as indicated by Eqs. (3.29) and (3.30), and limited range, unless the output is processed in accordance with Eq. (3.29). Significant advances have been made in thermistor technology over the past decade, and it is now possible to obtain stable, reproducible, interchangeable thermistors that are accurate to 0.5 percent over a specified temperature range.

## 3.12 THERMOCOUPLES

When two dissimilar materials are brought into contact, a potential develops as a result of an effect known as the *Seebeck effect*. A thermocouple is a temperature-measuring device whose operation depends upon the Seebeck effect. Many pairs of metals can be used for thermocouples. Thermoelectric sensitivities ($\mu V/°C$) for a number of different materials in combination with platinum are listed in Table 3.6.

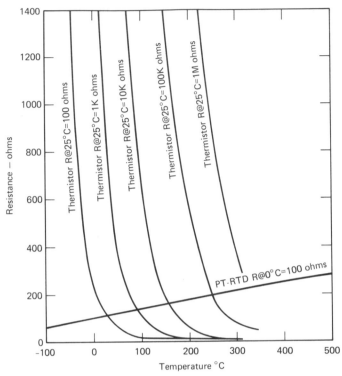

*Figure 3.27* Resistance-temperature characteristics of thermistors and RTDs.

The data of Table 3.6 can be used to determine the sensitivity of any junction by noting, for example, that

$$S_{\text{chromel/alumel}} = S_{\text{chromel/platinum}} - S_{\text{alumel/platinum}}$$

$$= +25.8 - (-13.6) = 39.4 \ \mu V/^\circ C$$

**TABLE 3.6** Thermoelectric Sensitivities for Different Materials Junctioned with Platinum

| Material | Sensitivity ($\mu V/^\circ C$) | Material | Sensitivity ($\mu V/^\circ C$) |
|---|---|---|---|
| Constantan | $-35$ | Copper | $+\ \ 6.5$ |
| Nickel | $-15$ | Gold | $+\ \ 6.5$ |
| Alumel | $-13.6$ | Tungsten | $+\ \ 7.5$ |
| Carbon | $+\ \ 3$ | Iron | $+\ 18.5$ |
| Aluminum | $+\ \ 3.5$ | Chromel | $+\ 25.8$ |
| Silver | $+\ \ 6.5$ | Silicon | $+440$ |

Commonly employed thermocouple material combinations include iron/constantan, chromel/alumel, chromel/constantan, copper/constantan, and platinum/platinum–rhodium.

The output voltage from a thermocouple junction is measured by connecting the junction into a circuit as shown in Fig. 3.28. The output voltage $E_o$ from such a circuit is related to temperature by an expression of the form

$$E_o = S_{A/B}(T_1 - T_2) \qquad (3.31)$$

**where**   $S_{A/B}$ is the sensitivity of material combination $A$ and $B$.
  $T_1$   is the temperature at junction 1.
  $T_2$   is the temperature at junction 2.

In practice, junction $J_2$ is a reference junction that is maintained at a carefully controlled reference temperature $T_2$. The junction $J_1$ is placed in contact with the body at the point where a temperature is to be measured. When a meter is inserted into the thermocouple circuit, junctions $J_3$ and $J_4$ are created. If the input terminals of the meter are at the same temperature $(T_3 = T_4)$, these added junctions $(J_3$ and $J_4)$ do not affect the output voltage $E_o$.

The use of thermocouples in temperature measurement is covered in much more detail in Chapter Eight. As a sensor, the thermocouple can be made quite small (wire diameter as small as 0.0005 in. is available); therefore, the response time is rapid (milliseconds) and essentially point measurements are possible. The thermocouple can cover a very wide range of temperatures; however, the output is nonlinear, and calibration charts are required to convert output voltage $E_o$ to temperature. In addition to nonlinear output, thermocouple sensors suffer the disadvantage of very low signal output and the need for a very carefully controlled reference temperature.

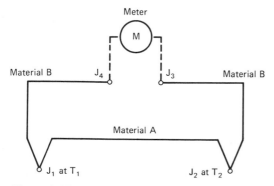

*Figure 3.28*  A typical thermocouple circuit.

## 3.13 SENSORS FOR TIME AND FREQUENCY MEASUREMENTS

*Time* can be defined as the interval between two events. Measurements of this interval are made by making comparisons with some reproducible event, such as the time required for the earth to orbit the sun or the time required for the earth to rotate on its axis. *Ephemeris time* is based on astronomical measurements of the time required for the earth to orbit the sun. *Sidereal time* is earth rotation time measured with respect to a distant star. *Solar time* is earth rotation time measured with respect to the sun. Sidereal time is used primarily in astronomical laboratories. Solar time is used for navigation on earth and for daily living purposes.

The fundamental unit of time in both the English and International Systems of units is the second. Prior to 1956, the second was defined as 1/86,400 of a mean solar day (the average period of revolution of the earth on its axis). This definition was satisfactory for most engineering work; however, the earth's rotation is somewhat irregular and is slowing at a rate of approximately 0.001 s per century, therefore, a more precise definition was needed for some exact scientific work. In 1956, the International Committee on Weights and Measures redefined the second as 1/31,556,925.9747 of the time required for the earth to orbit the sun in the year 1900. This definition had the desired high degree of exactness required for some scientific work; however, it suffers from the limitation that direct comparison of a time interval with the standard requires astronomical observations extending over a period of several years. In the late 1950s, atomic research revealed that certain atomic transitions can be measured with excellent repeatability. As a result, the Thirteenth General Conference on Weights and Measures, which was held in Paris in 1967, redefined the second as "the duration of 9,192,631,770 periods of the radiation corresponding to the transition between the two hyperfine levels of the fundamental state of the atom of Cesium-133." The estimated accuracy of this standard is two parts in $10^9$. Currently, standards laboratories throughout the world have cesium-beam oscillators that agree in frequency to within a few parts in $10^{11}$.

Precise time and frequency standards are available in the United States through National Bureau of Standards radio transmissions from Fort Collins, Colorado. Station WWVB is a low-frequency station that transmits at 60 kHz. Station WWVL is a very-low-frequency station that transmits at 20 kHz. Frequencies, as transmitted, are accurate to one part in $10^8$. Time intervals are accurate to within a microsecond.

Any device that is used to indicate passage of time is referred to as a *clock*. Reproducible events commonly used as sensing mechanisms for clocks include the swing of a pendulum, oscillation of a torsional pendulum, oscillation of a spiral spring and balance wheel, vibration of a tuning fork, and oscillation of a piezoelectric crystal. The time required for one of these devices to complete one cycle of motion is known as the period. The frequency of the motion is the number of cycles occurring in a given unit of time. Thus, the frequency $f$ is the reciprocal of the period $T$.

## Physical Pendulum

A rigid object oscillating about a fixed horizontal axis under the action of its own weight is known as a *physical pendulum*. As an example, consider the body shown in Fig. 3.29 which is constrained to oscillate in a vertical plane about the horizontal axis that passes through point $P$. The center of mass $B$ of the body is located a distance $b$ from point $P$. If the body is displaced from its position of equilibrium, the weight $W = mg$ acting through the center of mass together with the support reaction $R$ at $P$ provides a restoring couple $C$ that tends to return the body to its equilibrium position once it is released in the displaced position. Summation of moments about the fixed axis of rotation at $P$, with $\theta$ positive in the counterclockwise direction, yields

$$C = -Wb \sin \theta = -mgb \sin \theta = I \frac{d^2\theta}{dt^2} \qquad (3.32)$$

or

$$\frac{d^2\theta}{dt^2} = \alpha = -\frac{mgb \sin \theta}{I} \qquad (3.33)$$

**where**   $I$  is the moment of inertia of the body about the axis of rotation.
$m$ is the mass of the body.
$b$ is the distance from the center of mass to the axis of rotation.
$g$ is the local acceleration of gravity.
$\theta$ is the angular displacement of the body from its equilibrium position.
$\alpha$ is the angular acceleration of the body about the axis of rotation.

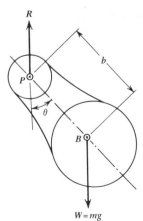

$W = mg$          *Figure 3.29*   Physical pendulum.

If the angular displacement $\theta$ is small (less than 10 degrees), $\sin \theta \approx \theta$ and Eq. (3.33) becomes

$$\alpha = \frac{d^2\theta}{dt^2} = -\frac{mgb\theta}{I} \tag{3.34}$$

Equation (3.34) indicates that the motion associated with a physical pendulum is angular simple harmonic motion if the angular displacements are small. For this motion, the period $T$ and the frequency $f$ can be expressed as

$$T = \frac{1}{f} = 2\pi\sqrt{\frac{I}{mgb}} \tag{3.35}$$

### Simple Pendulum

Pendulums used in timing devices often consist of a concentrated mass (bob) on the end of a cord or a slender rod. This form, known as a *simple pendulum*, can be idealized as a point mass on the end of a weightless rod of length $L$. Thus, $I = mL^2$ and $b = L$. Equation (3.35) then reduces to

$$T = \frac{1}{f} = 2\pi\sqrt{\frac{mL^2}{mgL}} = 2\pi\sqrt{\frac{L}{g}} \tag{3.36}$$

### Torsional Pendulum

A *torsional pendulum* consists of a disk or other body with a large amount of inertia supported by a torsionally flexible rod, as shown in Fig. 3.30. If the disk is given an angular displacement $\theta$ from its position of equilibrium, a restoring couple $C = K\theta$ develops (depends on the size and elastic properties of the supporting rod) that tends to return the disk to its equilibrium position once

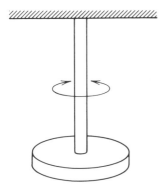

**Figure 3.30**   Torsional pendulum.

it is released in the displaced position. Summation of moments about the axis of rotation of the disk yields

$$C = -K\theta = I\frac{d^2\theta}{dt^2} \tag{3.37}$$

or

$$\frac{d^2\theta}{dt^2} = \alpha = -\frac{K\theta}{I} \tag{3.38}$$

**where**   $I$ is the moment of inertia of the disk about the axis of rotation.
$K$ is the torsional spring constant of the supporting rod.
$\theta$ is the angular displacement of the disk from its equilibrium position.
$\alpha$ is the angular acceleration of the disk about the axis of rotation.

Equation (3.38) indicates that the motion associated with the torsional pendulum is angular simple harmonic motion. The magnitude of the angle $\theta$ is restricted only by the requirement that the stress level in the supporting rod remain below the proportional limit for the material used in its fabrication. Also, gravitation plays no role in the motion of the torsional pendulum. The period $T$ and the frequency $f$ of the torsional pendulum are

$$T = \frac{1}{f} = 2\pi\sqrt{\frac{I}{K}} \tag{3.39}$$

In the previous discussions of motion associated with the different types of pendulums the effects of friction were neglected. This is never true; therefore, means must be provided to introduce energy into the systems. Weights and coil springs are commonly used in pendulum clocks to provide the driving forces. Simple ratchet and pawl mechanisms together with gear trains convert the vibrational motions of the pendulums into the rotary motion of the hands of the clock that provide the time indications.

## Tuning Fork

A simple mechanical device consisting of two prongs and a handle, as shown in Fig. 3.31$a$, is known as a tuning fork. The frequency of vibration of a specific tuning fork depends on the exact geometry of the instrument; however, an approximate value can be obtained from the exact solution for a cantilever beam (see Fig. 3.31*b*), which can be expressed as

$$f = 0.5596 \sqrt{\frac{EI}{\rho L^4 A}}$$

$$= 0.5596 \sqrt{\frac{Ek^2}{\rho L^4}} \tag{3.40}$$

**where**   $E$ is the modulus of elasticity (Young's modulus) of the material used in the fabrication of the beam.

$\rho$ is the density of the material.

$L$ is the length of the beam.

$I$ is the moment of inertia of the cross-sectional area of the beam about a centroidal axis perpendicular to the direction of vibration.

$A$ is the cross-sectional area of the beam.

$k$ is the radius of gyration of the cross section $k = \sqrt{I/A}$. For a rectangular cross section, $k = t/\sqrt{12}$.

$t$ is the thickness of the beam in the direction of the vibrations.

Thus, for a tuning fork with a rectangular cross section, the frequency depends directly on the thickness of the prongs in the direction of the vibrations and inversely on the square of the length.

Tuning forks are widely used as standards of frequency in musical applications. In electrically driven forms, tuning forks are used to control electric circuits that require stable and accurate frequencies. Tuning forks have been constructed for frequencies ranging from 20 to 20,000 Hz. In the Accutron type of watch or clock, a precision tuning fork, driven with energy from a small battery, replaces the balance wheel and mainspring of a conventional watch or clock. The tuning fork vibrates at 360 Hz. The vibrations of the tuning fork ratchet a fine-toothed index wheel that drives the gear train used to turn the hands.

*(a)*

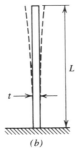

*(b)*

**Figure 3.31**   Tuning fork and cantilever beam.

## Electromagnetic Oscillator

An electrical circuit consisting of an inductor $L$ and a capacitor $C$, as shown in Fig. 3.32, exhibits an oscillatory electrical behavior that is similar to the oscillatory mechanical behavior exhibited by the physical or torsional pendulum. If the capacitor $C$ in Fig. 3.32 is given an initial charge by connecting a charging battery across its terminals, a fixed amount of energy is stored in the electric field of the capacitor when the battery is removed. At any instant of time after the switch $S$ is closed, this total energy in the system will be divided between the electric field of the capacitor and the magnetic field of the inductor. Thus, the total energy $U$ can be expressed as

$$U = \frac{1}{2}\frac{q^2}{C} + \frac{1}{2}LI^2 \tag{3.41}$$

For an ideal circuit with zero resistance (equivalent to no friction in the mechanical systems), this energy must remain constant; therefore, $dU/dt = 0$ and Eq. (3.41) yields

$$\frac{dU}{dt} = \frac{q}{C}\frac{dq}{dt} + LI\frac{dI}{dt} = 0 \tag{3.42}$$

Recall, however, that

$$I = \frac{dq}{dt}$$

Therefore, from Eq. (3.42)

$$\frac{d^2q}{dt^2} + \frac{q}{LC} = 0 \tag{3.43}$$

Equation (3.43) describes the behavior of the ideal $LC$ circuit. Since this equation is similar in form to Eqs. (3.33) and (3.38), the period and frequency of oscillation can be expressed as

$$T = \frac{1}{f} = 2\pi\sqrt{LC} \tag{3.44}$$

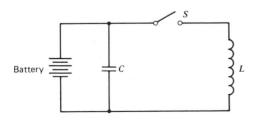

Figure 3.32   Electromagnetic oscillator.

Once started, such *LC* oscillations would continue indefinitely if no resistance was present to remove energy from the system. Sustained electromagnetic oscillations are maintained by supplying enough energy from an outside source, automatically and periodically, to compensate for the energy lost from the system.

## Piezoelectric Crystal Oscillators

As discussed in Section 3.7, a piezoelectric crystal has the ability to develop a difference in electric potential between two of its faces when it is deformed by being subjected to a force or pressure. Conversely, application of a voltage across the two faces produces a mechanical deformation of the crystal. These properties have proved to be useful in the development of dynamic force and pressure transducers (see Chapters Six and Seven), accelerometers (see Chapter Seven), and crystal oscillators for use in electronic counters (see Chapter Four).

Since a piezoelectric material such as quartz or barium titanate is an elastic material, it has a natural frequency of vibration that depends on its size and shape. For example, as a bar, it may exhibit either longitudinal or flexural vibrations. Similarly, as a plate, it may exhibit either longitudinal, flexural, or thickness vibrations. Frequencies from about 5 kHz to about 10 MHz can be obtained by using these different vibration modes.

When a piezoelectric crystal is introduced into an oscillating electric circuit having nearly the same frequency as the natural frequency of the crystal, the crystal vibrates at its natural frequency and the frequency of oscillation of the circuit becomes the same as the natural frequency of the crystal. The long-term frequency stability obtained by using the crystal to control the frequency of the circuit is about one part in $10^8$. In a quartz clock, for example, the accumulated error in time after one year is about 0.1 s.

## 3.14  SUMMARY

A wide variety of basic sensors have been described in this chapter. These sensors are sometimes used directly to measure an unknown quantity (such as use of a thermocouple to measure temperature); however, in many other instances the sensors are used as the critical element in a transducer (such as use of a piezoelectric crystal in a force gage or accelerometer). Important characteristics of each sensor that must be considered in the selection process include:

1. Size—with smaller being better because of enhanced dynamic response.
2. Range—with extended range being preferred so as to increase the latitude of operation.
3. Sensitivity—with the advantage to higher output devices that require less amplification.

4. Accuracy—with the advantage to devices exhibiting errors of 1 percent or less after considering zero shift, linearity, and hysteresis.
5. Frequency Response—with the advantage to wide-response sensors that permit application in both static and dynamic loading situations.
6. Stability—with very low drift in output over extended periods of time and with changes in temperature and humidity preferred.
7. Temperature Limits—with the ability to operate from cryogenic to elevated temperatures considered advantageous.
8. Economy—with reasonable costs preferred.
9. Ease of Application—with reliability and simplicity always a significant advantage.

---

## EXERCISES

**3.1** Briefly describe the difference between a transducer and a sensor.

**3.2** A slide-wire potentiometer having a length of 200 mm is fabricated by winding wire having a diameter of 0.25 mm around a cylindrical insulating core. Determine the resolution limit of this potentiometer.

**3.3** If the potentiometer of Exercise 3.2 has a resistance of 1000 $\Omega$ and can dissipate 4 W of power, determine the voltage required to maximize the sensitivity. What voltage change corresponds to the resolution limit?

**3.4** A 10-turn potentiometer with a calibrated dial (100 divisions/turn) is used as a balance resistor in a Wheatstone bridge. If the potentiometer has a resistance of 10 k$\Omega$ and a resolution of 0.1 percent, what is the minimum incremental change in resistance $\Delta R$ that can be read from the calibrated dial?

**3.5** Why are potentiometers limited to static or quasi-static applications?

**3.6** List several advantages of the conductive-film type of potentiometer.

**3.7** A new elevator must be tested to determine its performance characteristics. Design a displacement transducer that utilizes a 20-turn potentiometer to monitor the position of the elevator over its 50-m range of travel.

**3.8** Compare the potentiometer and LVDT as displacement sensors with regard to the following characteristics: range, accuracy, resolution, frequency response, reliability, complexity, cost.

**3.9** List the basic elements of the electronic circuit associated with an LVDT.

**3.10** Prepare a sketch of the output signal as a function of time for an LVDT with its core located in a fixed off-center position if:

    (a) The demodulator is functioning
    (b) The demodulator is removed from the circuit

**3.11**   Prepare a sketch of the output signal as a funciton of time for an LVDT with its core moving at constant velocity from one end of the LVDT through the center to the other end if:

(a)   The demodulator is functioning
(b)   The demodulator is removed from the circuit

**3.12**   Describe the basic differences between an LVDT and a DCDT.

**3.13**   Design a 50-mm strain extensometer to be used for a simple tension test of mild steel. If the strain extensometer is to be used only in the elastic region and to detect the onset of yielding, specify the maximum range. What is the advantage of limiting the range?

**3.14**   Compare the cylindrical potentiometer (helipotentiometer) and the RVDT as sensors for measuring angular displacement.

**3.15**   What two factors are responsible for the resistance change $dR/R$ in an electrical resistance strain gage? Which is the most important for gages fabricated from constantan?

**3.16**   What function does the thin plastic film serve for an electrical resistance strain gage?

**3.17**   A 120-$\Omega$ strain gage with a gage factor $S_g = 2.05$ is used as a sensor for a strain of 800 $\mu$m/m. Determine $\Delta R$ and $\Delta R/R$.

**3.18**   For the situation described in Exercise 3.17, determine the output voltage $E_o$ from an initially balanced Wheatstone bridge if $E_i = 9$ V.

**3.19**   If the strain on the gage of Exercises 3.17 and 3.18 is reduced until $E_o = 2.0$ mV, determine the new strain.

**3.20**   A short-range displacement transducer utilizes a cantilever beam as the mechanical element and a strain gage as the sensor. Derive an expression for the displacement $\delta$ of the end of the beam in terms of the output voltage $E_o$.

**3.21**   Consider the flat-plate capacitance sensor shown in Fig. 3.14. If $h$ is initially 1 mm, determine the maximum change in spacing $\Delta h$ that can be tolerated if the linearity of the sensor is to be rated at

(a)   0.1 percent
(b)   1.0 percent
(c)   2.0 percent

**3.22**   Determine the sensitivity of an overlap-type capacitance sensor if air is the dielectric medium, $w = 4$ mm, and $h = 0.5$ mm. If one plate moves a distance $\Delta l = 2$ mm, what is the change in capacitance $\Delta C$?

**3.23**   An eddy current sensor is calibrated for use on a 304 stainless steel target material. The gage is then used to monitor the displacement of a specimen fabricated from aluminum. Is an error produced? If so, estimate the magnitude of the error.

**3.24** Determine the charge $q$ developed when a piezoelectric crystal having $A = 20$ mm$^2$ and $h = 10$ mm is subjected to a pressure $p = 10$ MPa if the crystal is

(a)  X-cut, length-longitudinal quartz
(b)  Parallel to polarization barium titanate

**3.25** Determine the output voltages for the piezoelectric crystals described in Exercise 3.24.

**3.26** Compare the use of quartz and barium titanate as materials for

(a)  Piezoelectric sensors
(b)  Ultrasonic signal sources

**3.27** If the equivalent circuit (see Fig. 3.21) for a measuring system incorporating a piezoelectric crystal consists of the following: $R_p = 10$ T$\Omega$, $R_A = 10$ M$\Omega$, $C_p = 30$ pF, $C_L = 10$ pF, and $C_A = 20$ pF, determine the effective time constant $\tau_e$ for the circuit. If the error must be limited to 5 percent, determine the time available for measurement of the magnitude of a step pulse of unit magnitude.

**3.28** Compare the characteristics of a piezoresistive sensor with those of a piezoelectric sensor.

**3.29** What advantages does the piezoresistive sensor have over the common (metal) electrical resistance strain gage? What are some of the disadvantages?

**3.30** What are some of the advantages of a photodiode (a photovoltaic device operated with a reverse bias) over a photoemissive device or a photoconductive device?

**3.31** A large plate 25 in. wide by 60 in. long is supported as a physical pendulum that is constrained to oscillate in a vertical plane by drilling a hole 3 in. from the narrow edge along the centerline of the plate and inserting a circular rod to serve as a pivot. The weight of the plate is 100 lb. What is the natural frequency for small oscillations?

**3.32** A clock is to have a pendulum consisting of a thin wire with a weight at its end. If the weight of the wire is negligible, what weight must be placed at the end of the wire and what length of wire must be used if the pendulum is to complete one cycle in 1 s?

**3.33** A simple pendulum has a period of 1.95 s at a point on the surface of the earth where $g = 32.17$ ft/s$^2$. What is the value of $g$ at another point on the surface of the earth where the period is 1.97 s?

**3.34** A pendulum clock keeps correct time at a location where $g = 32.17$ ft/s$^2$ but loses 10 s/day at a higher altitude. Find the value of $g$ at the new location.

**3.35**   The moment of inertia of the propeller on a large ship is 2000 lb · s² · in. The propeller shaft is 5 in. in diameter, is 40 ft long, and is made of steel having a shear modulus of $12(10^6)$ psi. A very large flywheel is present on the engine end of the propeller shaft. Determine the natural frequency of vibration of this system.

**3.36**   An oscillating $LC$ circuit has an inductance $L = 20$ mH and a capacitance $C = 2.0$ μF. What is the frequency of oscillation of this circuit

# FOUR

# SIGNAL CONDITIONING CIRCUITS

## 4.1 INTRODUCTION

An instrumentation system, as noted in Chapter One, contains many elements that are used either to supply power to the transducer or to condition the output from the transducer so that it can be displayed by a voltage measuring instrument. Such elements are common to instrumentation systems designed to measure acceleration, displacement, flow, force, strain, etc.; therefore, each type of element will be discussed in this chapter independent of its application in a particular measuring system.

A wide variety of signal conditioning circuits are available today; thus, a complete coverage of the subject is beyond the scope of this text. However, it will be possible to cover the general characteristics of those circuits that are frequently encountered in engineering measurements.

## 4.2 POWER SUPPLIES

With few exceptions, transducers are driven (provided the energy required for their operation) with either a constant-voltage or a constant-current power supply. The simplest and least expensive constant-voltage power supply is the common battery that can provide a reasonably constant voltage with large current flow for short periods of time. The difficulty experienced with batteries is that the voltage decays with time under load; therefore, they must be replaced or recharged periodically.

The problem of voltage decay can be easily solved by using a simple regulating circuit. A circuit containing Zener diodes will maintain the voltage output at a constant value for long periods of time. Regulated battery-type power supplies, which can be recharged, are often superior to much more expensive and complex power supplies that convert an ac line voltage to a dc output voltage, since problems of noise and ripple are eliminated.

The use of general-purpose power supplies that convert an ac line voltage (either 110 V or 220 V) to a lower dc output voltage (often variable) is quite common. A block diagram for a simple dc power supply, which is capable of delivering nearly constant voltage, is shown in Fig. 4.1. This power supply uses a full-bridge diode rectifier to convert the ac line voltage to a dc output voltage and a filter to reduce the ripple. The ripple and regulation are further improved by incorporating a voltage regulator between the filter and the output. Performance characteristics of a typical high-performance power supply that is capable of providing either a dc voltage from 0 to 40 V or a current from 0 to 3 A is described in the following paragraphs to indicate the important features of a power supply.

The load effect (formerly known as the load regulation), which is the voltage

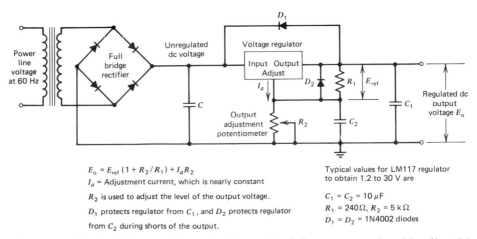

$E_o = E_{ref} (1 + R_2/R_1) + I_a R_2$

$I_a$ = Adjustment current, which is nearly constant

$R_2$ is used to adjust the level of the output voltage.

$D_1$ protects regulator from $C_1$, and $D_2$ protects regulator from $C_2$ during shorts of the output.

Typical values for LM117 regulator to obtain 1.2 to 30 V are

$C_1 = C_2 = 10 \, \mu F$
$R_1 = 240\Omega$, $R_2 = 5 \, k\Omega$
$D_1 = D_2 = $ 1N4002 diodes

***Figure 4.1*** Typical elements for a simple regulated dc power supply with adjustable output voltage.

drop from an initial setting as the current is increased from zero to the maximum rated value, is 0.01 percent plus 200 μV when the unit is operated as a constant-voltage source. When operated as a constant-current source, the current increases 0.02 percent plus 500 μA as the voltage is increased from zero to its maximum rated value.

The source effect (formerly known as the line regulation), which is the change in output for a change in line voltage (between 104 and 127 V for 110-V units), is 0.01 percent plus 200 μV for the voltage and 0.02 percent plus 500 μA for the current. The ripple and noise, which is a small ac signal superimposed on the dc output, is 10 mV peak to peak and 3 mA rms.

The temperature effect coefficient, which is the change in output voltage or current per degree Celsius following a warm-up period of 30 min, is 0.01 percent plus 200 μV for the voltage and 0.01 percent plus 1 mA for the current.

The drift stability, which is the change in output under constant load over an 8-h period following a 30-min warm-up period, is 0.03 percent plus 500 μV and 0.03 percent plus 3 mA for the voltage and current, respectively.

The output impedance of the power supply can be represented by a resistor and an inductor in series. Low output impedances are usual for voltage supplies; 2 mΩ and 1 μH can be considered typical.

## 4.3  THE POTENTIOMETER CIRCUIT (Constant Voltage)

The potentiometer circuit, which is often employed with resistance-type transducers to convert the transducer output $\Delta R/R$ to a voltage signal $\Delta E$, is shown in Fig. 4.2. With fixed-value resistors in the circuit, the open-circuit output voltage $E_o$ can be expressed as

$$E_o = \frac{R_1}{R_1 + R_2}E_i = \frac{1}{1 + r}E_i \qquad (4.1)$$

**where**  $E_i$ is the input voltage.
      $r$  is the resistance ratio $R_2/R_1$.

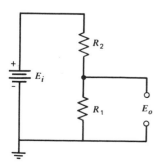

*Figure 4.2*   The constant-voltage potentiometer circuit.

If the resistors $R_1$ and $R_2$ are varied by $\Delta R_1$ and $\Delta R_2$, the change $\Delta E_o$ in the output voltage can be determined from Eq. (4.1) as

$$E_o + \Delta E_o = \frac{R_1 + \Delta R_1}{R_1 + \Delta R_1 + R_2 + \Delta R_2} E_i \tag{a}$$

Solving for $\Delta E_o$ gives

$$\Delta E_o = \left[ \frac{R_1 + \Delta R_1}{R_1 + \Delta R_1 + R_2 + \Delta R_2} - \frac{R_1}{R_1 + R_2} \right] E_i \tag{b}$$

This equation can be reduced and expressed in a more useful form by introducing the resistance ratio $r$. Thus

$$\Delta E_o = \frac{\dfrac{r}{(1 + r)^2} \left( \dfrac{\Delta R_1}{R_1} - \dfrac{\Delta R_2}{R_2} \right) E_i}{1 + \dfrac{1}{1 + r} \left( \dfrac{\Delta R_1}{R_1} + r \dfrac{\Delta R_2}{R_2} \right)} \tag{4.2}$$

Equation (4.2) indicates that the change in output voltage $\Delta E_o$ for the potentiometer circuit is a nonlinear function of the inputs $\Delta R_1/R_1$ and $\Delta R_2/R_2$. The nonlinear effects associated with the circuit can be expressed as a nonlinear term $\eta$, where

$$\eta = 1 - \frac{1}{1 + \dfrac{1}{1 + r} \left( \dfrac{\Delta R_1}{R_1} + r \dfrac{\Delta R_2}{R_2} \right)} \tag{4.3}$$

Equation (4.2) then becomes

$$\Delta E_o = \frac{r}{(1 + r)^2} \left( \frac{\Delta R_1}{R_1} - \frac{\Delta R_2}{R_2} \right) (1 - \eta) E_i \tag{4.4}$$

The nonlinear effects of the potentiometer circuit can be evaluated by considering a situation that is typical of many applications ($r = 9$ and $\Delta R_2 = 0$). For this simplified case the nonlinear term $\eta$ can be expressed as

$$\eta = 1 - \frac{1}{1 + \left( 0.1 \dfrac{\Delta R_1}{R_1} \right)}$$

$$= \left( 0.1 \frac{\Delta R_1}{R_1} \right) - \left( 0.1 \frac{\Delta R_1}{R_1} \right)^2 + \left( 0.1 \frac{\Delta R_1}{R_1} \right)^3 + \cdots \tag{4.5}$$

Values from Eq. (4.5) are plotted in Fig. 4.3. Note that linearity within 1 percent can be obtained if $\Delta R_1/R_1 < 0.1$.

The range of the potentiometer circuit is defined as the maximum $\Delta R_1/R_1$ that can be recorded without exceeding some specified value of the nonlinear term (usually 1 or 2 percent). In the special case with $r = 9$ and $\Delta R_2 = 0$, the range is 0.101 for linearity within 1 percent and 0.204 for linearity within 2 percent.

The sensitivity of the potentiometer circuit is defined for a case where $\Delta R_2 = 0$ as

$$S_c = \frac{\Delta E_o}{\dfrac{\Delta R_1}{R_1}} = \frac{r}{(1 + r)^2}E_i \tag{4.6}$$

Equation (4.6) indicates that the sensitivity can be increased without limit simply by increasing the input voltage $E_i$; however, all transducers have limited power-dissipation capabilities that restrict the input voltage. The power $P_T$ dissipated by a transducer in a potentiometer circuit is given by the expression

$$P_T = \frac{E_T^2}{R_T} \tag{4.7}$$

**where**   $E_T$ is the voltage across the transducer.
$R_T$ is the transducer resistance.

From Eq. (4.1)

$$E_T = \frac{E_i}{1 + r} \tag{4.8}$$

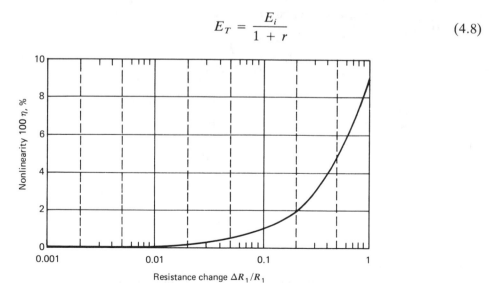

*Figure 4.3*   Nonlinear term $\eta$ as a function of resistance change $\Delta R_1/R_1$ for a constant-voltage potentiometer circuit with $r = 9$ and $\Delta R_2 = 0$.

The upper limit of the voltage that can be applied to the potentiometer circuit as obtained from Eqs. (4.7) and (4.8) is

$$E_{i_{\max}} = (1 + r)\sqrt{P_T R_T} \tag{4.9}$$

A realistic expression for the sensitivity of the constant-voltage potentiometer circuit $S_{cv}$ is obtained by substituting Eq. (4.9) into Eq. (4.6). Thus

$$S_{cv} = \frac{r}{1 + r}\sqrt{P_T R_T} \qquad \text{with } \Delta R_2 = 0 \tag{4.10}$$

It is clear from Eq. (4.10) that maximum sensitivity is achieved with large $r$, with a high-resistance transducer, and with a transducer capable of dissipating a large amount of power. In practice, sensitivity is usually limited by voltage requirements. For $r > 9$, the higher voltages required cannot be justified by the small additional gains in sensitivity.

The preceding equations for the potentiometer circuit have been based on the assumption that the input impedance of the voltage recording instrument is infinite (open-circuit voltage) and that no power is required to measure $E_o + \Delta E_o$. In practice, recording instruments have a finite resistance, and some power is drawn from the circuit.

The effect of input impedance of the recording instrument on the quantity being measured $\Delta E_o$ can be determined by considering the circuit shown in Fig. 4.4a, which incorporates a transducer with a resistance $R_T$ in position $R_1$ and a fixed or ballast resistor $R_b$ in position $R_2$. Also, the recording instrument is shown with a resistance $R_M$. This circuit can be reduced to an equivalent circuit where the parallel resistances $R_T$ and $R_M$ are replaced by a single resistance $R_e$, having the value

$$R_e = \frac{R_T R_M}{R_T + R_M} \tag{a}$$

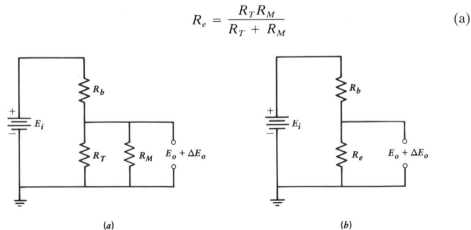

(a)  (b)

**Figure 4.4** Constant-voltage potentiometer circuit with a recording instrument. (a) Resistive load associated with the recording instrument. (b) The equivalent circuit.

The change in output voltage from this circuit with $\Delta R_b = 0$ and $\eta \approx 0$ is given by Eq. (4.4) as

$$\Delta E_o]_{R_M} = \frac{r}{(1 + r)^2} \frac{\Delta R_e}{R_e} E_i$$

$$= \frac{R_b R_e}{(R_e + R_b)^2} \frac{\Delta R_e}{R_e} E_i \tag{b}$$

Equation (b) can be expressed in terms of $R_M$ and $R_T$ by substituting Eq. (a) into Eq. (b). Thus

$$\Delta E_o]_{R_M} = \frac{R_b R_T}{\left[ R_T + R_b \left( \dfrac{R_T}{R_M} \right) + R_b \right]^2} \frac{\Delta R_T}{R_T} E_i \tag{4.11}$$

A similar expression for the open-circuit voltage ($R_M = \infty$) is

$$\Delta E_o]_{R_M \to \infty} = \frac{R_b R_T}{(R_T + R_b)^2} \frac{\Delta R_T}{R_T} E_i \tag{4.12}$$

Equation (4.11) can be expressed in terms of Eq. (4.12) and a loss factor $\mathcal{L}$ as

$$\Delta E_o]_{R_M} = \Delta E_o]_{R_M \to \infty} (1 - \mathcal{L}) \tag{4.13}$$

where $\mathcal{L}$ is the loss in output due to the presence of $R_M$. It can be shown that

$$\mathcal{L} = 2r \frac{R_T}{R_M} \frac{\left[ 1 + r \left( 1 + \dfrac{1}{2} \dfrac{R_T}{R_M} \right) \right]}{\left[ 1 + r \left( 1 + \dfrac{R_T}{R_M} \right) \right]^2} \tag{4.14}$$

The loss factor $\mathcal{L}$ depends upon the resistance ratios $R_T/R_M$ and $r = R_b/R_T$ as shown in Fig. 4.5. It is evident that the loss factor is very small for $R_T/R_M < 0.005$; therefore, load effects can be neglected in this range, irrespective of the magnitude of $r$. Load effects become more significant for $R_T/R_M > 0.01$ and exceed practical limits (say 2 percent) when $R_T/R_M > 0.0111$ for $r = 10$ and when $R_T/R_M > 0.0203$ for $r = 1$.

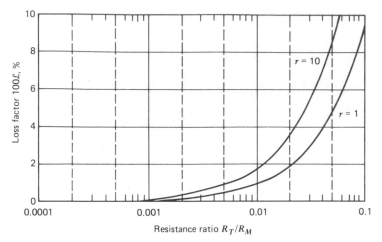

**Figure 4.5** Loss factor $\mathscr{L}$ as a function of resistance ratio $R_T/R_M$ for a potentiometer circuit loaded with a voltage measuring instrument.

## 4.4  THE POTENTIOMETER CIRCUIT (Constant Current)

The potentiometer circuit described in Section 4.3, which was driven with a constant-voltage power supply, exhibited a nonlinear output voltage $\Delta E_o$ when the input $\Delta R/R$ exceeded certain limits. In many applications, this nonlinear behavior limits the usefulness of the circuit; therefore, means are sought to extend the linear range of operation.

Constant-current power supplies with sufficient regulation for instrumentation systems have been made possible by recent advances in solid-state electronics. The constant-current power supply automatically adjusts the output voltage with a changing resistive load to maintain the current at a constant value.

A potentiometer circuit with a constant-current power supply is shown schematically in Fig. 4.6a. The open-circuit output voltage $E_o$ (measured with a very-high-impedance recording instrument so that loading errors are negligible) is

$$E_o = IR_1 \tag{4.15}$$

When the resistances $R_1$ and $R_2$ are changed by the amounts $\Delta R_1$ and $\Delta R_2$, the output voltage becomes

$$E_o + \Delta E_o = I(R_1 + \Delta R_1) \tag{a}$$

From Eqs. (4.15) and (a),

$$\Delta E_o = I\Delta R_1 = IR_1 \frac{\Delta R_1}{R_1} \tag{4.16}$$

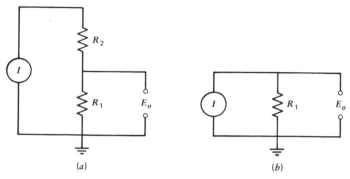

**Figure 4.6** Constant-current potentiometer circuits. (*a*) Two-element circuit. (*b*) Single-element circuit.

Equation (4.16) indicates that neither $R_2$ nor $\Delta R_2$ influences the output of the constant-current potentiometer circuit; therefore, it is possible to eliminate $R_2$ and use the simple circuit shown in Fig. 4.6*b*. It should also be observed that the change in output voltage $\Delta E_o$ is a linear function of the input $\Delta R_1/R_1$, regardless of the magnitude of $\Delta R_1$. This linear behavior extends the usefulness of the potentiometer circuit for many applications.

The circuit sensitivity $S_{cc}$ for the constant-current potentiometer circuit is

$$S_{cc} = \frac{\Delta E_o}{\dfrac{\Delta R_1}{R_1}} = IR_1 \tag{4.17}$$

If the constant-current source is adjustable so that the current $I$ can be increased to the power-dissipation limit of the transducer, then

$$I = \sqrt{\frac{P_T}{R_1}} \tag{b}$$

Substituting Eq. (b) into Eq. (4.17) yields

$$S_{cc} = \sqrt{P_T R_T} \tag{4.18}$$

Equations (4.10) and (4.18) indicate that the sensitivity of the potentiometer circuit is improved by a factor of $(1 + r)/r$ by using the constant-current source.

## 4.5 THE WHEATSTONE BRIDGE (Constant Voltage)

The Wheatstone bridge (see Fig. 4.7) is a second type of circuit that is commonly used to convert a change in resistance to an output voltage. The output voltage $E_o$ of the bridge shown in Fig. 4.7 can be determined by treating

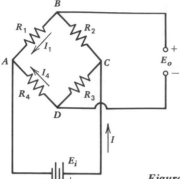

**Figure 4.7**   The constant-voltage Wheatstone bridge circuit.

the top and bottom parts of the bridge as individual voltage dividers. Thus

$$E_{AB} = \frac{R_1}{R_1 + R_2} E_i \tag{a}$$

$$E_{AD} = \frac{R_4}{R_3 + R_4} E_i \tag{b}$$

The output voltage $E_o$ of the bridge is

$$E_o = E_{BD} = E_{AB} - E_{AD} \tag{c}$$

Substituting Eqs. (a) and (b) into (c) yields

$$E_o = \frac{R_1 R_3 - R_2 R_4}{(R_1 + R_2)(R_3 + R_4)} E_i \tag{4.19}$$

Equation (4.19) indicates that the initial output voltage will vanish ($E_o = 0$) if

$$R_1 R_3 = R_2 R_4 \tag{4.20}$$

When Eq. (4.20) is satisfied, the bridge is said to be *balanced*. The ability to balance the bridge (set $E_o = 0$) represents a significant advantage, since it is much easier to measure small values of $\Delta E_o$ from a zero voltage base than from a base $E_o$, which may be as much as 1000 times greater than $\Delta E_o$.

With an initially balanced bridge, an output voltage $\Delta E_o$ develops when resistances $R_1$, $R_2$, $R_3$, and $R_4$ are varied by amounts $\Delta R_1$, $\Delta R_2$, $\Delta R_3$, and $\Delta R_4$, respectively. From Eq. (4.19), with these new values of resistance

$$\Delta E_o = \frac{(R_1 + \Delta R_1)(R_3 + \Delta R_3) - (R_2 + \Delta R_2)(R_4 + \Delta R_4)}{(R_1 + \Delta R_1 + R_2 + \Delta R_2)(R_3 + \Delta R_3 + R_4 + \Delta R_4)} E_i \tag{d}$$

Expanding, neglecting higher-order terms, and substituting Eq. (4.20) yields

$$\Delta E_o = \frac{R_1 R_2}{(R_1 + R_2)^2} \left( \frac{\Delta R_1}{R_1} - \frac{\Delta R_2}{R_2} + \frac{\Delta R_3}{R_3} - \frac{\Delta R_4}{R_4} \right) E_i \qquad (4.21)$$

Another form of the equation for the output voltage is obtained by substituting $r = R_2/R_1$ in Eq. (4.21). Thus

$$\Delta E_o = \frac{r}{(1 + r)^2} \left( \frac{\Delta R_1}{R_1} - \frac{\Delta R_2}{R_2} + \frac{\Delta R_3}{R_3} - \frac{\Delta R_4}{R_4} \right) E_i \qquad (4.22)$$

Equations (4.21) and (4.22) indicate that the output voltage from the bridge is a linear function of the resistance changes. This apparent linearity results from the fact that the higher-order terms in Eq. (d) were neglected. If the higher-order terms are retained, the output voltage $\Delta E_o$ is a nonlinear function of the $\Delta R/R$'s, which can be expressed as

$$\Delta E_o = \frac{r}{(1 + r)^2} \left( \frac{\Delta R_1}{R_1} - \frac{\Delta R_2}{R_2} + \frac{\Delta R_3}{R_3} - \frac{\Delta R_4}{R_4} \right) (1 - \eta) E_i \qquad (4.23)$$

where

$$\eta = \frac{1}{1 + \dfrac{r + 1}{\dfrac{\Delta R_1}{R_1} + \dfrac{\Delta R_4}{R_4} + r \left( \dfrac{\Delta R_2}{R_2} + \dfrac{\Delta R_3}{R_3} \right)}} \qquad (4.24)$$

In a widely used form of the bridge, $R_1 = R_2 = R_3 = R_4$. In this case, Eq. (4.24) reduces to

$$\eta = \frac{\displaystyle\sum_{i=1}^{4} \frac{\Delta R_i}{R_i}}{\displaystyle\sum_{i=1}^{4} \frac{\Delta R_i}{R_i} + 2} \qquad (4.25)$$

The error in percent ($100\eta$) due to the nonlinear effect is shown as a function of $\Delta R_1/R_1$ in Fig. 4.8 for a bridge with one active transducer in arm $R_1$ and fixed-value resistors in the other three arms. From these results it is clear that $\Delta R_1/R_1$ must be less than 0.02 if the error due to the nonlinear effect is not to exceed 1 percent. While this may appear quite restrictive, the Wheatstone bridge is usually employed with transducers that exhibit very small changes in $\Delta R/R$.

The sensitivity of a Wheatstone bridge with a constant-voltage power supply

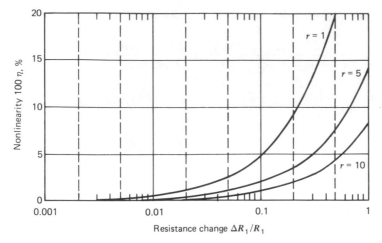

**Figure 4.8**   Nonlinear term $\eta$ as a function of resistance change $\Delta R_1/R_1$ for a constant-voltage Wheatstone bridge circuit with one active gage.

and a single active arm is determined from Eq. (4.22) as

$$S_{cv} = \frac{\Delta E_o}{\dfrac{\Delta R_1}{R_1}} = \frac{r}{(1 + r)^2}E_i \tag{4.26}$$

Again it is clear that increasing $E_i$ produces an increase in sensitivity; however, the power $P_T$ that can be dissipated by the transducer limits the bridge voltage $E_i$ to

$$E_i = I_T(R_1 + R_2) = I_T R_T(1 + r) = (1 + r)\sqrt{P_T R_T} \tag{4.27}$$

Substituting Eq. (4.27) into Eq. (4.26) gives

$$S_{cv} = \frac{r}{1 + r}\sqrt{P_T R_T} \tag{4.28}$$

Equation (4.28) indicates that the circuit sensitivity of the constant-voltage Wheatstone bridge is due to two factors—a circuit efficiency $r/(1 + r)$ and the characteristics of the transducer as indicated by $P_T$ and $R_T$. Increasing $r$ increases circuit efficiency; however, $r$ should not be so high as to require unusually large supply voltages. For example, a 500-$\Omega$ transducer capable of dissipating 0.2 W in a bridge with $r = 4$ (80 percent circuit efficiency) will require a supply voltage $E_i = 50$ V.

The selection of a transducer with a high resistance and a high power dissipating capability is much more effective in maximizing circuit sensitivity than increasing the circuit efficiency beyond 80 or 90 percent. The product $P_T R_T$ for commercially available transducers can range from about 1 W · Ω to 1000 W · Ω; therefore, more latitude exists for increasing circuit sensitivity $S_{cv}$ by transducer selection than by increasing circuit efficiency.

Circuit sensitivity $S_{cv}$ can also be increased, as indicated by Eq. (4.22), by using multiple transducers (one in each arm of the bridge). In most cases, however, the cost of the additional transducers is not warranted. Instead, it is usually more economical to use a high-gain differential amplifier to increase the output signal $\Delta E_o$ from the Wheatstone bridge.

Load effects in a Wheatstone bridge are usually negligible if a high-impedance voltage measuring instrument (such as a DVM for static signals or an oscilloscope for dynamic signals) is used with the bridge. The output impedance $R_B$ of the bridge can be determined by using Thevenin's theorem. Thus

$$R_B = \frac{R_1 R_2}{R_1 + R_2} + \frac{R_3 R_4}{R_3 + R_4} \tag{4.29}$$

In most bridge arrangements, $R_B$ rarely exceeds $10^4$ Ω. Since the input impedance $R_M$ of most modern voltage recording devices is at least $10^6$ Ω, the ratio $R_B/R_M < 0.01$; therefore, loading errors are usually very small. Furthermore, circuit calibration (discussed in Section 5.6) will automatically include any bridge-loading effects.

## 4.6 THE WHEATSTONE BRIDGE (Constant Current)

Use of a constant-current power supply with the potentiometer circuit improved the circuit sensitivity and eliminated nonlinear effects. The effects of using a constant-current power supply with the Wheatstone bridge can be determined by considering the circuit shown in Fig. 4.9. The current $I$ delivered to the bridge by the power supply divides at point $A$ into currents $I_1$ and $I_2$ where

$$I = I_1 + I_2 \tag{a}$$

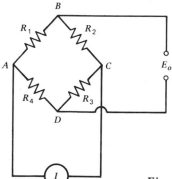

Figure 4.9   The constant-current Wheatstone bridge circuit.

The voltage drop across resistance $R_1$ is

$$E_{AB} = I_1 R_1 \tag{b}$$

Similarly, the voltage drop across resistance $R_4$ is

$$E_{AD} = I_2 R_4 \tag{c}$$

Thus, the output voltage $E_o$ from the bridge is

$$E_o = E_{BD} = E_{AB} - E_{AD} = I_1 R_1 - I_2 R_4 \tag{4.30}$$

From Eq. (4.30) it is clear that the bridge will be balanced ($E_o = 0$) if

$$I_1 R_1 = I_2 R_4 \tag{d}$$

This balance equation is not in a useful form, since the currents $I_1$ and $I_2$ are unknowns. The magnitudes of these currents can be determined by observing that the voltage $E_{AC}$ can be expressed in terms of $I_1$ and $I_2$ as

$$E_{AC} = I_1(R_1 + R_2) = I_2(R_3 + R_4) \tag{e}$$

From Eqs. (a), (d), and (e),

$$I_1 = \frac{R_3 + R_4}{R_1 + R_2 + R_3 + R_4} I$$

$$I_2 = \frac{R_1 + R_2}{R_1 + R_2 + R_3 + R_4} I \tag{f}$$

Substituting Eqs. (f) into Eq. (4.30) gives

$$E_o = \frac{I}{R_1 + R_2 + R_3 + R_4} (R_1 R_3 - R_2 R_4) \tag{4.31}$$

Thus, the balance requirement for the constant-current Wheatstone bridge is

$$R_1 R_3 = R_2 R_4 \tag{4.32}$$

This is the same condition as that required for balance of the constant-voltage Wheatstone bridge.

The open-circuit output voltage $\Delta E_o$ from an initially balanced bridge ($E_o = 0$) due to resistance changes $\Delta R_1$, $\Delta R_2$, $\Delta R_3$, and $\Delta R_4$ is given by Eq. (4.31) as

$$
\begin{aligned}
\Delta E_o &= \frac{I}{\Sigma R + \Sigma \Delta R} [(R_1 + \Delta R_1)(R_3 + \Delta R_3) \\
&\quad - (R_2 + \Delta R_2)(R_4 + \Delta R_4)] \\
&= \frac{IR_1 R_3}{\Sigma R + \Sigma \Delta R} \left[ \frac{\Delta R_1}{R_1} - \frac{\Delta R_2}{R_2} + \frac{\Delta R_3}{R_3} \right. \\
&\quad \left. - \frac{\Delta R_4}{R_4} + \frac{\Delta R_1}{R_1}\frac{\Delta R_3}{R_3} - \frac{\Delta R_2}{R_2}\frac{\Delta R_4}{R_4} \right]
\end{aligned}
\tag{4.33}
$$

**where**   $\Sigma R = R_1 + R_2 + R_3 + R_4$.

$\Sigma \Delta R = \Delta R_1 + \Delta R_2 + \Delta R_3 + \Delta R_4$.

Equation (4.33) shows that the constant-current Wheatstone bridge exhibits a nonlinear output voltage $\Delta E_o$. The nonlinearity is due to the $\Sigma \Delta R$ term in the denominator and to the two second-order terms within the bracketed quantity. For a typical application with a transducer in arm $R_1$ and fixed-value resistors in the other three arms of the bridge such that

$$
R_1 = R_4 = R_T, \qquad R_2 = R_3 = rR_T, \qquad \Delta R_2 = \Delta R_3 = \Delta R_4 = 0 \quad \text{(g)}
$$

Eq. (4.33) reduces to

$$
\Delta E_o = \frac{IR_T r}{2(1 + r) + \left(\dfrac{\Delta R_T}{R_T}\right)} \frac{\Delta R_T}{R_T}
\tag{4.34}
$$

which can also be expressed as

$$
\Delta E_o = \frac{IR_T r}{2(1 + r)} (1 - \eta) \frac{\Delta R_T}{R_T}
\tag{4.35}
$$

where

$$
\eta = \frac{\dfrac{\Delta R_T}{R_T}}{2(1 + r) + \dfrac{\Delta R_T}{R_T}}
\tag{4.36}
$$

It is clear from Eq. (4.36) that the nonlinear effect can be reduced by increasing $r$. Percent error ($100\eta$) as a function of $\Delta R_T/R_T$ and $r$ are shown in Fig. 4.10. A comparison of the errors illustrated in Figs. 4.8 and 4.10 clearly shows the advantage of the constant-current power supply in extending the range of the Wheatstone bridge circuit.

The circuit sensitivity $S_{cc}$ as obtained from Eq. (4.35) is

$$S_{cc} = \frac{\Delta E_o}{\dfrac{\Delta R_T}{R_T}} = \frac{IR_Tr}{2(1 + r)} \tag{4.37}$$

For the example being considered, the bridge is symmetric; therefore, the current $I_T = I/2$. The power dissipated by the transducer is

$$P_T = I_T^2 R_T = \frac{1}{4}I^2 R_T \tag{h}$$

Substituting Eq. (h) into Eq. (4.37) yields

$$S_{cc} = \frac{r}{1 + r}\sqrt{P_T R_T} \tag{4.38}$$

Equations (4.28) and (4.38) show that the circuit sensitivity is the same for constant-voltage and constant-current Wheatstone bridges.

The principal advantage of a Wheatstone bridge over a potentiometer circuit is related to the fact that the Wheatstone bridge can be initially balanced to

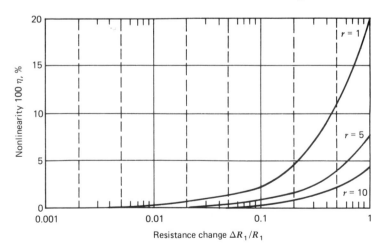

**Figure 4.10**   Nonlinear term $\eta$ as a function of resistance change $\Delta R_1/R_1$ for a constant-current Wheatstone bridge circuit with one active gage.

produce a zero output voltage ($E_o = 0$). A second advantage is realized when the bridge is used in a null-balance mode (see Section 5.5). This capability eliminates the need for a precise voltage measuring instrument and, therefore, offers the advantage of high accuracy at relatively low cost.

## 4.7 AMPLIFIERS

An amplifier is one of the most important components in an instrumentation system. It is used in nearly every system to increase low-level signals from a transducer to a level sufficient for recording with a voltage measuring instrument. An amplifier is represented in schematic diagrams of instrumentation systems by the triangular symbol shown in Fig. 4.11. The voltage input to the amplifier is $E_i$; the voltage output is $E_o$. The ratio $E_o/E_i$ is the gain $G$ of the amplifier. As the input voltage is increased, the output voltage increases in the linear range of the amplifier according to the relationship

$$E_o = GE_i \tag{4.39}$$

The linear range of an amplifier is finite since the output voltage is limited by the supply voltage and the characteristics of the amplifier components. A typical input–output graph for an amplifier is shown in Fig. 4.12. If the amplifier is driven beyond the linear range (overdriven) serious errors can result if the gain $G$ is treated as a constant.

If the gain from a single amplifier is not sufficient, two or more amplifiers can be series connected (cascaded), as shown in Fig. 4.13. Such an amplifier system has an output voltage $E_o$ given by the expression

$$E_o = G^3 \left( \frac{Z_i}{Z_i + Z_1} \right) \left( \frac{Z_i}{Z_i + Z_o} \right)^2 \left( \frac{Z_2}{Z_o + Z_2} \right) E_i \tag{4.40}$$

**where**   $Z_i$ is the amplifier input impedance.
                $Z_o$ is the amplifier output impedance.
                $Z_1$ is the internal impedance of the source.
                $Z_2$ is the input impedance of the voltage recorder.

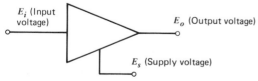

$E_i$ (Input voltage)

$E_o$ (Output voltage)

$E_s$ (Supply voltage)

*Figure 4.11*  Symbol for an amplifier.

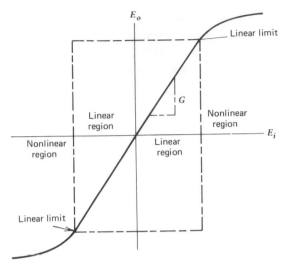

Figure 4.12    A typical voltage-input/voltage-output curve for an amplifier.

In properly designed voltage amplifiers $Z_i >> Z_1$ and $Z_i >> Z_o$; therefore, Eq. (4.40) reduces to

$$E_o = G^3 \left( \frac{Z_2}{Z_o + Z_2} \right) E_i \qquad\qquad (a)$$

The term $Z_2/(Z_o + Z_2)$ in Eq. (a) represents the voltage attenuation due to the current required to drive the voltage recorder. This topic was discussed in Section 4.3. By maintaining $Z_2 >> Z_o$ (using a recorder with a high input impedance), this attenuation term approaches unity and Eq. (a) becomes

$$E_o = G^3 E_i \qquad\qquad (4.41)$$

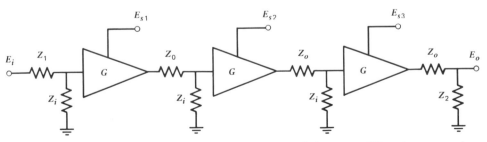

Figure 4.13    A high-gain amplifer system consisting of three amplifiers in a cascade arrangement.

With proper selection of $Z_o$, $Z_i$, $Z_1$, and $Z_2$, the overall gain of a cascaded amplifier system equals the product of the gains of the individual stages.

Frequency response of an amplifier must also be given careful consideration during design of an instrumentation system. The gain of an amplifier is a function of the frequency of the input voltage; therefore, there will always be some high frequency at which the gain of the amplifier will be less than its value at the lower frequencies. This frequency effect is similar to inertia effects in a mechanical system. A finite time (transit time) is required for current entering the input terminal of an amplifier to pass through all of the components and reach the output terminal. Also, time is required for the output voltage to develop, since some capacitance will always be present in the recording instrument.

The frequency response of an amplifier–recorder system can be illustrated in two different ways. First, the output voltage can be plotted as a function of time for a step input as shown in Fig. 4.14a. The rise in output voltage for this

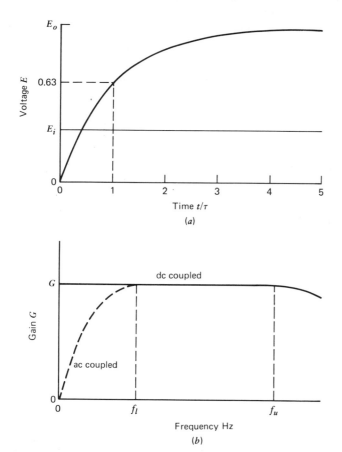

*Figure 4.14*  Frequency response of an amplifier-recorder system. (*a*) Amplifier response to a step-input voltage. (*b*) Gain as a function of frequency of the input voltage.

representation can be approximated by an exponential function of the form

$$E_o = G(1 - e^{-t/\tau})E_i \tag{4.42}$$

where $\tau$ is the time constant for the amplifier.

The second method of illustrating frequency effects utilizes a graph showing gain plotted as a function of frequency as shown in Fig. 4.14b. The output of the amplifier is flat between the lower and upper frequency limits $f_l$ and $f_u$. Thus, a dynamic signal with all frequency components within the band between $f_l$ and $f_u$ will be amplified with a constant gain. Amplifiers can be designed with coupling circuits that maintain a constant gain down to zero frequency. These amplifiers are known as *dc* or *dc-coupled amplifiers*. In some cases, a capacitor is placed in series with the input to the amplifier to block the dc components of the input signal. These *ac-coupled amplifiers* exhibit a gain $G = 0$ when the frequency of the input signal drops to zero.

Two very simple single-stage amplifiers, which utilize field-effect transistors, are illustrated in Fig. 4.15. The dc-coupled amplifier, shown in Fig. 4.15a, exhibits a gain of 15, while the ac-coupled amplifier, shown in Fig. 4.15b, exhibits a gain of 8. The field-effect transistor, with its high input resistance, low noise,

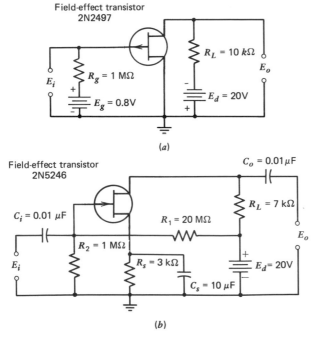

**Figure 4.15**  Single-stage amplifiers incorporating field-effect transistors (from data by Brophy). (*a*) DC-coupled amplifier. (*b*) AC-coupled amplifier.

and high stability, is used extensively in amplifiers designed to accommodate low-level signals from a wide variety of transducers.

The amplifiers illustrated in Fig. 4.15 can be classified as *single-ended amplifiers* since both the input and output voltages are referenced to ground. Single-ended amplifiers can be used only when the output from the signal conditioning circuit is referenced to ground, as is the case for the potentiometer circuit described in Section 4.3. Since the output from a Wheatstone bridge is not referenced to ground, single-ended amplifiers cannot be used, so *differential amplifiers* must be employed. In a differential amplifier (see Fig. 4.16b), two separate voltages, each referenced to ground, are connected to the input. The output is single-ended and referenced to ground. The output voltage from the differential amplifier is

$$E_o = G(E_{i1} - E_{i2}) \tag{4.43}$$

As a result of the form of Eq. (4.43), the differential amplifier rejects *common-mode signals* (those voltages that are identical on both inputs). Common-mode signals include spurious pickup (noise), temperature-induced drift, and power supply ripple. The ability of the differential amplifier to essentially eliminate these undesirable components of the input signal is an extremely important feature. The differential amplifier can be used with all signal conditioning circuits.

Differential amplifiers of excellent quality are readily available today thanks to recent developments in solid-state electronics. Typical specifications for a high-quality differential amplifier are

Frequency response: DC to 1-MHz bandwidth

Gain: from 1 to $10^4$ in calibrated steps in a 1, 2, 5, sequence.

Input impedance: $R = 1$ MΩ and $C = 47$ pF

Common-mode rejection ratios: 100,000 to 1

DC stability: 10 μV over a 1-h period

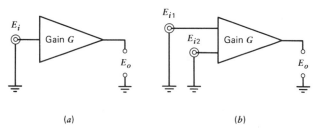

(a)     (b)

**Figure 4.16** Single-ended and differential amplifiers. (a) Single-ended input and output. (b) Double-ended input and single-ended output.

## 4.8 OPERATIONAL AMPLIFIERS

An *operational amplifier* (op-amp) is a complete amplifier circuit (an integrated circuit where components such as transistors, diodes, resistors, etc., have been miniaturized into a single element) that can be employed in a number of different ways by adding a small number of external passive components, such as resistors or capacitors. Operational amplifiers have an extremely high gain ($G = 10^5$ is a typical value), which can be considered infinite for the purpose of analysis and design of circuits containing the op-amp. The input impedance (typically $R = 4$ M$\Omega$ and $C = 8$ pF) is so high that circuit loading usually is

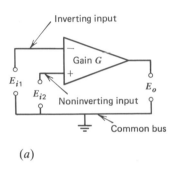

*(a)*

*(b)*

*(c)*

*Figure 4.17* Schematic circuit diagram and photographs of operational amplifiers. (*a*) An operational amplifier circuit. (*b*) An operational amplifier. (Courtesy of Teledyne Philbrick.) (*c*) A selection of operational amplifiers. (Courtesy of Burr-Brown Research Corp.)

not a consideration. Output resistance (of the order of 100 $\Omega$) is sufficiently low to be considered negligible in most applications.

Figure 4.17 shows the symbols used to represent the internal op-amp circuit in schematic diagrams and the physical size of a typical op-amp. The two input terminals are identified as the *inverting* $(-)$ *terminal* and the *noninverting* $(+)$ *terminal*. The output voltage $E_o$ of an op-amp is given by the expression

$$E_o = G(E_{i2} - E_{i1}) \tag{4.44}$$

It is evident from Eq. (4.44) that the op-amp is a differential amplifier; however, it is not used as a conventional differential amplifier because of its high gain and poor stability. The op-amp can be used effectively, however, as a part of a larger circuit (with more accurate and more stable passive elements) for many applications. Several applications of the op-amp, including inverting amplifiers, voltage followers, summing amplifiers, integrating amplifiers, and differentiating amplifiers, will be discussed in subsequent subsections.

### Inverting Amplifier

An *inverting amplifier* with single-ended input and output can be built with an op-amp and resistors, as shown in Fig. 4.18. In this circuit, the input voltage $E_i$ is applied to the negative terminal of the op-amp through an input resistor $R_1$. The positive terminal of the op-amp is connected to the common ground bus. The output voltage $E_o$ is fed back to the negative terminal of the op-amp through a feedback resistor $R_f$.

The gain of the inverting amplifier can be determined by considering the sum of the currents at point $A$ in Fig. 4.18. Thus

$$I_1 + I_f = I_a \tag{a}$$

If $E_a$ is the voltage drop across the op-amp,

$$I_1 = \frac{E_i - E_a}{R_1}$$

$$I_f = \frac{E_o - E_a}{R_f}$$

$$I_a = \frac{E_a}{R_a} \tag{b}$$

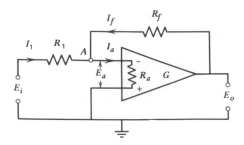

**Figure 4.18** An inverting amplifier with single-ended input and output.

The voltage drop across the op-amp $E_a$ is related to the output voltage $E_o$ by the gain relation. Therefore,

$$E_a = -\frac{E_o}{G} \tag{c}$$

From Eqs. (a), (b), and (c),

$$\frac{E_o}{E_i} = -\frac{R_f}{R_1}\left[\frac{1}{1 + \dfrac{1}{G}\left(1 + \dfrac{R_f}{R_1} + \dfrac{R_f}{R_a}\right)}\right] \tag{4.45}$$

As an example, consider a typical op-amp with a gain of 106 dB ($2 \times 10^5$) and $R_a = 4\ \text{M}\Omega$, with $R_1 = 0.1\ \text{M}\Omega$ and $R_f = 1\ \text{M}\Omega$. Substituting these values into Eq. (4.45) yields

$$\frac{E_o}{E_i} = -10\left(\frac{1}{1 + 5.6 \times 10^{-5}}\right) = -9.999 \approx -10$$

Thus, it is obvious that the op-amp gain can be neglected without introducing appreciable error (0.1 percent in this example), and the gain of the circuit $G_c$ can be accurately approximated by

$$G_c = \frac{E_o}{E_i} \approx -\frac{R_f}{R_1} \tag{4.46}$$

Op-amps can also be used to design noninverting amplifiers and differential amplifiers in addition to the inverting amplifiers. The circuits for each of these amplifiers are shown in Fig. 4.19. The governing equations for each of these circuits are as follows:

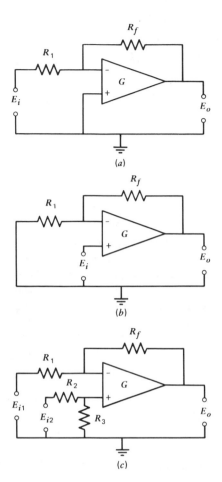

**Figure 4.19** Instrument amplifiers that use operational amplifiers as the central circuit element. (*a*) Inverting amplifier; (*b*) noninverting amplifier; (*c*) differential amplifier.

**Inverting Amplifier:**
$$G_c = \frac{E_o}{E_i} = -\frac{R_f}{R_1}\left[\frac{1}{1 + \dfrac{1}{G}\left(1 + \dfrac{R_f}{R_1} + \dfrac{R_f}{R_a}\right)}\right] \qquad (4.45)$$

$$G_c \approx -\frac{R_f}{R_1} \qquad (4.46)$$

**Noninverting Amplifier:** $\;G_c = \dfrac{E_o}{E_i} = \dfrac{G}{1 + \dfrac{GR_1}{R_1 + R_f}} \qquad (4.47)$

$$G_c \approx 1 + \frac{R_f}{R_1} \qquad (4.48)$$

$$\text{Differential Amplifier:} \quad E_o \approx \frac{R_3}{R_2}\left(\frac{1 + \dfrac{R_f}{R_1}}{1 + \dfrac{R_3}{R_2}}\right)E_{i2} - \left(\frac{R_f}{R_1}\right)E_{i1} \tag{4.49}$$

If $R_f/R_1 = R_3/R_2$:

$$G_c \approx \frac{E_o}{E_{i2} - E_{i1}} \approx \frac{R_f}{R_1} \tag{4.50}$$

The circuits shown in Fig. 4.19 have been simplified to illustrate the concept of developing an amplifier with a gain $G_c$ that is essentially independent of the op-amp gain $G$. In practice, these circuits must be modified to account for zero-offset voltages since, ideally, the output voltage $E_o$ of the amplifier should be zero when the inputs $(+)$ and $(-)$ of the op-amp are connected to the common bus (i.e., grounded). In practice, this does not occur automatically, since the op-amps exhibit a zero-offset voltage; therefore, it is necessary to add a biasing circuit to the amplifier that can be adjusted to restore the output voltage to zero, otherwise, serious measurement errors can occur. Since the magnitude of the offset voltage changes (drifts) as a result of temperature, time, and power-supply voltage variations, it is advisable to adjust the bias circuit periodically to restore the zero output conditions.

A biasing circuit for an inverting amplifier with single-ended input and output is shown in Fig. 4.20. Common values of resistances $R_2$, $R_3$, and $R_4$ are $R_3 = R_1$, $R_2 = 10 \ \Omega$ and $R_4 = 25$ k$\Omega$. A voltage $E_1 = \pm 15$ V is often used, since the zero-offset voltage of the op-amp can be either positive or negative. The magnitude of the bias voltage that must be applied to the op-amp seldom exceeds a few millivolts.

The frequency response of instrument amplifiers constructed with op-amps depends upon the frequency response of the op-amp and the feedback fraction. Since the gain of an op-amp depends on frequency (decreases with increasing frequency), the gain of the circuit also decreases with increasing frequency. The frequency responses of op-amps vary appreciably (depending upon design characteristics); however, a frequency response of 10 kHz or above is common. This is sufficient frequency response for most mechanical measurements.

## The Voltage Follower

An op-amp can also be used to construct an instrument with a very high input impedance for use with transducers that incorporate piezoelectric sensors. This high-impedance circuit, shown in Fig. 4.21, is known as a *voltage follower* and has a circuit gain of unity $(G_c = 1)$. The purpose of the voltage follower is

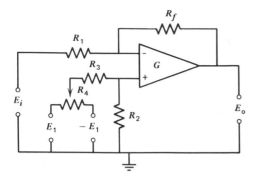

***Figure 4.20***   Biasing circuit for a single-ended input and output amplifier.

to serve as an insulator between the transducer and the voltage recording instrument. The voltage follower is also known as a unity-gain buffer amplifier.

The gain $G_c$ of the voltage follower can be determined from Eq. (4.44). Thus,

$$E_o = G(E_{i2} - E_{i1}) = G(E_i - E_o) \tag{a}$$

Solving Eq. (a) for the circuit gain $G_c$ gives

$$G_c = \frac{E_o}{E_i} = \frac{G}{1 + G} \tag{4.51}$$

When the gain $G$ of the op-amp is very large, the gain $G_c$ of the circuit becomes unity.

The input resistance of the voltage follower circuit is given by Ohm's law as

$$R_{ci} = \frac{E_i}{I_a} \tag{b}$$

The input current $I_a$ can be expressed in terms of the input and output voltages of the op-amp and the input resistance of the op-amp as

$$I_a = \frac{E_i - E_o}{R_i} \tag{c}$$

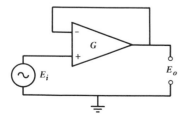

***Figure 4.21***   A high-impedance voltage follower circuit.

Combining Eqs. (a), (b), and (c) gives

$$R_{ci} = \frac{E_i R_i}{E_i - E_o} = \frac{E_i R_i}{E_i - \dfrac{GE_i}{1 + G}} = (1 + G)R_i \qquad (4.52)$$

Since both $G$ and $R_i$ are very large for op-amps (i.e., $G = 10^5$ and $R_i = 1$ to 10 M$\Omega$), the input impedance of the voltage follower circuit is of the order of $10^{11}$ to $10^{12}$ $\Omega$. This input impedance is sufficient to minimize any drain of charge from a piezoelectric transducer during a readout period of short duration.

The output resistance $R_{co}$ of the voltage follower circuit is extremely low and is given by the expression

$$R_{co} = \frac{R_o}{1 + G} \qquad (4.53)$$

where $R_o$ is the output resistance of the op-amp.

### Summing Amplifiers

In some data analysis applications, signals from two or more transducers must be added to obtain an output signal that is proportional to the sum of the input signals. This can be accomplished with the op-amp circuit, known as a *summing amplifier*, shown in Fig. 4.22.

Operation of the summing amplifier can be established by considering current flow at point $A$ of Fig. 4.22, which can be expressed as

$$I_1 + I_2 + I_3 + I_f = 0 \qquad (a)$$

Applying Ohm's law to Eq. (a) yields

$$\frac{E_{i1}}{R_1} + \frac{E_{i2}}{R_2} + \frac{E_{i3}}{R_3} + \frac{E_o}{R_f} = 0 \qquad (b)$$

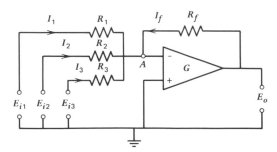

*Figure 4.22*   A summing amplifier incorporating an op-amp circuit.

Solving Eq. (b) for $E_o$ gives

$$E_o = -R_f \left( \frac{E_{i1}}{R_1} + \frac{E_{i2}}{R_2} + \frac{E_{i3}}{R_3} \right) \tag{4.54}$$

Equation (4.54) indicates that the input signals $E_{i1}$, $E_{i2}$, and $E_{i3}$ are scaled by ratios $R_f/R_1$, $R_f/R_2$, and $R_f/R_3$, respectively, and then summed. If $R_1 = R_2 = R_3 = R_f$, the inputs sum without scaling and Eq. (4.54) reduces to

$$E_o = -(E_{i1} + E_{i2} + E_{i3}) \tag{4.55}$$

If the gain $G$ and the input impedance $R_i$ are finite, the analysis of the circuit illustrated in Fig. 4.22 is more involved; however, it can be shown that

$$E_o = - \frac{\dfrac{E_{i1}}{R_1} + \dfrac{E_{i2}}{R_2} + \dfrac{E_{i3}}{R_3}}{\dfrac{1}{R_f} + \dfrac{1}{G} \left( \dfrac{1}{R_1} + \dfrac{1}{R_2} + \dfrac{1}{R_3} + \dfrac{1}{R_f} + \dfrac{1}{R_i} \right)} \tag{4.56}$$

Equations (4.54) and (4.56) indicate that the term

$$\frac{1}{G} \left( \frac{1}{R_1} + \frac{1}{R_2} + \frac{1}{R_3} + \frac{1}{R_f} + \frac{1}{R_i} \right)$$

represents an error in the scaling and summing operation. The magnitude of the error is small if $G$ and $R_i$ are large.

The circuit shown in Fig. 4.22 can be modified to produce an adding-subtracting amplifier if the positive terminal of the op-amp is used. The adding-subtracting amplifier circuit is shown in the figure associated with Exercise 4.28.

## Integrating Amplifiers

An *integrating amplifier* utilizes a capacitor in place of the feedback resistor of Fig. 4.22 as shown in Fig. 4.23. An expression for the output voltage from the integrating amplifier can be established by following the procedure used for

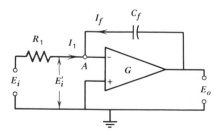

*Figure 4.23* An integrating amplifier incorporating an op-amp circuit.

the summing amplifier. Thus, by considering current flow at point $A$ of Fig. 4.23,

$$I_1 + I_f = 0 \tag{a}$$

If Ohm's law is applied, Eq. (a) can be written as

$$\frac{E_i - E_i'}{R_1} + I_f = 0 \tag{b}$$

where

$$E_i' = -\frac{E_o}{G} \tag{c}$$

The voltage $E_i' \rightarrow 0$ when the gain $G$ is large and Eq. (b) becomes

$$\frac{E_i}{R_1} + I_f = 0 \tag{d}$$

The charge $q$ on the capacitor is given by the expression

$$q = \int_0^t I_f \, dt = C_f E_o \tag{e}$$

Substituting Eq. (d) into Eq. (e) and solving for the output voltage $E_o$ yields

$$E_o = -\frac{1}{R_1 C_f} \int_0^t E_i \, dt \tag{4.57}$$

It is clear from Eq. (4.57) that the output voltage $E_o$ from the circuit of Fig. 4.23 is the integral of the input voltage $E_i$ with respect to time multiplied by the constant $-1/R_1 C_f$.

## Differentiating Amplifier

The *differentiating amplifier* is similar to the integrating amplifier except that the positions of the resistor and capacitor of Fig. 4.23 are interchanged, as shown in Fig. 4.24. An expression for the output voltage $E_o$ of the differentiating amplifier can be developed by following the procedure outlined for the integrating amplifier. The results are

$$E_o = -R_f C_1 \frac{dE_i}{dt} \tag{4.58}$$

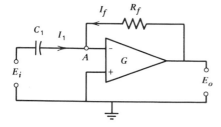

**Figure 4.24**  A differentiating amplifier incorporating an op-amp circuit.

Considerable care must be exercised to minimize noise when the differentiating amplifier is used, since noise is simply superimposed on the input voltage and differentiated in such a way that it contributes to the output voltage and produces error. The effects of high-frequency noise can be suppressed by placing a capacitor across resistance $R_f$; however, the presence of this capacitor affects the differentiating process, and Eq. (4.58) must be modified to account for its effects.

## 4.9  FILTERS

In many instrumentation applications, the signal from the transducer is combined with noise or some other undesirable voltage. These parasitic voltages can often be eliminated with a filter that is designed to attenuate the undesirable signals, but transmit the transducer signal without significant attenuation or distortion. Filtering of the signal is possible if the frequencies of the parasitic and transducer signals are different. Some of the different filters that are employed in signal conditioning include: the *RL* filter, the *RC* high-pass filter, the *RC* low-pass filter, and the Wein-bridge notched filter. Schematic diagrams of these filters are shown in Fig. 4.25.

### The *RL* Filter

A schematic diagram of the resistance–inductance (*RL*) filter is shown in Fig. 4.25a. The performance characteristics of this filter can be established by using the Kirchhoff voltage law, which can be expressed as

$$L \frac{dI}{dt} + RI = E_i \tag{a}$$

For a current $I$ that is sinusoidal and can be expressed as

$$I = I_o \sin \omega t \tag{b}$$

where $\omega$ is the circular frequency of the input signal, Eq. (a) becomes

$$E_i = I_o (R \sin \omega t + \omega L \cos \omega t) \tag{c}$$

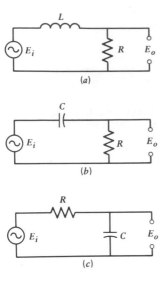

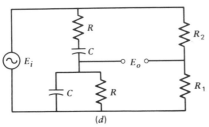

**Figure 4.25**  Filters commonly used for signal conditioning. (*a*) The *RL* filter; (*b*) the high-pass *RC* filter; (*c*) the low-pass *RC* filter; (*d*) the Wein-bridge notched filter.

Equation (c) can be expressed in a more convenient form by introducing a phase angle $\phi$, which is defined as

$$\phi = \tan^{-1}\frac{\omega L}{R} \qquad\qquad (d)$$

Thus,

$$E_i = I_o\sqrt{R^2 + (\omega L)^2}\ \sin(\omega t + \phi) \qquad\qquad (e)$$

The output voltage $E_o$ is the voltage drop across the resistance $R$; therefore, from Eq. (b)

$$E_o = IR = I_o R\ \sin \omega t \qquad\qquad (f)$$

The ratio of the amplitudes of the output and input voltages $E_o/E_i$ is obtained from Eqs. (e) and (f) as

$$\frac{E_o}{E_i} = \frac{1}{\sqrt{1 + \left(\frac{\omega L}{R}\right)^2}} \tag{4.59}$$

Equation (4.59) indicates that $E_o/E_i \to 1$ as $\omega L/R \to 0$. Thus, the $RL$ filter transmits the low-frequency components ($\omega \to 0$) of an input signal without significant attenuation. The amount by which a filter attenuates a particular frequency is given by the response curve or transfer function for the filter. A response curve showing $E_o/E_i$ as a function of $\omega L/R$ for the $RL$ filter is presented in Fig. 4.26a. This curve shows that the transfer function decreases from 0.995 for $\omega L/R = 0.10$ to 0.01 for $\omega L/R = 100$. Both ends of this response curve are important. The low-frequency end, which is presented in expanded form in Fig. 4.26b, is important since it controls the attenuation of the transducer signal. Note that a 2 percent attenuation of the transducer signal occurs when $\omega_t L/R = 0.203$ ($\omega_t$ is the circular frequency of the transducer signal). To avoid errors greater than 2 percent, $L$ and $R$ must be selected when designing the filter such that $\omega_t L/R \leq 0.203$.

The high-frequency response of the filter is also important since it controls the attenuation of the parasitic or noise signal. A reduction of 90 percent in the noise signal can be achieved if $\omega_p L/R = 10$ ($\omega_p$ is the circular frequency of the parasitic signal). It is not always possible to simultaneously limit the attenuation of the transducer signal to 2 percent while reducing the parasitic voltages by 90 percent, since this requires that $\omega_p/\omega_t \geq 20$. If $\omega_p/\omega_t < 20$, it will be necessary to accept a higher ratio of parasitic signal.

## High-Pass *RC* Filter

A simple yet effective high-pass resistance–capacitance ($RC$) filter is illustrated in Fig. 4.25b. The behavior of this filter in response to a sinusoidal input voltage of the form

$$E_i = E_a \sin \omega t \tag{a}$$

can be determined by summing the voltage drops around the loop of Fig. 4.25b. Thus,

$$E_i - \frac{q}{C} - RI = 0 \tag{b}$$

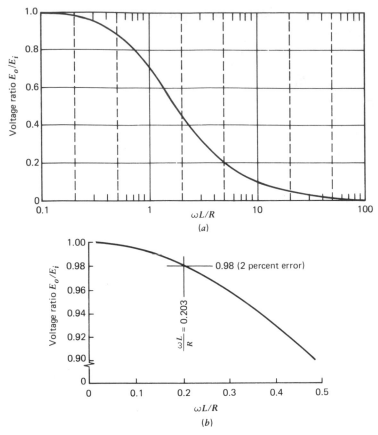

**Figure 4.26** Response curve for a low-pass $RL$ filter. (*a*) Complete response curve. (*b*) Low-frequency end of response curve.

where $q$ is the charge on the capacitor. Current $I$ and charge $q$ are related by the expression

$$I = \frac{dq}{dt} \qquad (c)$$

Equation (b) can be expressed in a more useful form by differentiating with respect to time and substituting Eqs. (a) and (c). Thus,

$$RC \frac{dI}{dt} + I = \omega C E_a \cos \omega t \qquad (d)$$

Solving Eq. (d) for the instantaneous current $I$ yields

$$I = \frac{\omega C E_a}{\sqrt{1 + (\omega RC)^2}} \sin(\omega t + \phi)$$

$$= \frac{E_a}{\sqrt{R^2 + \left(\dfrac{1}{\omega C}\right)^2}} \sin(\omega t + \phi) \tag{e}$$

where the phase angle $\phi$ is given by the expression

$$\phi = \tan^{-1} \frac{1}{\omega RC} \tag{f}$$

The output voltage $E_o$ is the voltage drop across the resistance $R$; therefore, from Eq. (e)

$$E_o = IR = \frac{\omega RC E_a}{\sqrt{1 + (\omega RC)^2}} \sin(\omega t + \phi)$$

$$= \frac{R E_a}{\sqrt{R^2 + \left(\dfrac{1}{\omega C}\right)^2}} \sin(\omega t + \phi) \tag{g}$$

The ratio of the amplitudes of the output and input voltages $E_o/E_i$ is obtained from Eqs. (a) and (g) as

$$\frac{E_o}{E_i} = \frac{\omega RC}{\sqrt{1 + (\omega RC)^2}} = \frac{1}{\sqrt{1 + \left(\dfrac{1}{\omega RC}\right)^2}} \tag{4.60}$$

Equation (4.60) indicates that $E_o/E_i \rightarrow 1$ as the frequency becomes large; thus, this filter is known as a high-pass filter. The response curve for a high-pass $RC$ filter is shown in Fig. 4.27. At zero frequency (dc), the voltage ratio $E_o/E_i$ vanishes, which indicates that the filter completely blocks any dc component of the output voltage. This dc blocking capability of the high-pass $RC$ filter can be used to great advantage when a low-amplitude transducer signal is superimposed on a large dc output voltage (see, for example, the potentiometer circuit of Section 4.3). Since the $RC$ filter eliminates the dc voltage, the low-magnitude dynamic transducer signal can be amplified to produce a satisfactory display.

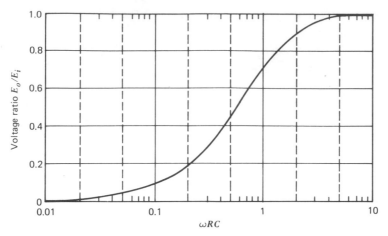

*Figure 4.27* Response curve for a high-pass *RC* filter.

When a high-pass *RC* filter is used, care must be exercised to ensure that the transducer signal is transmitted through the filter without significant attenuation. Figure 4.27 indicates that the attenuation will be less than 2 percent if $\omega RC \geq 5$.

## Low-Pass *RC* Filter

A low-pass *RC* filter is produced by interchanging the positions of the resistor and capacitor of the high-pass *RC* filter shown in Fig. 4.25*b*. This modified *RC* circuit, shown in Fig. 4.25*c*, has transmission characteristics that are opposite to those of the high-pass *RC* filter; namely, it transmits low-frequency signals and attenuates high-frequency signals.

The output voltage $E_o$ for the circuit shown in Fig. 4.25*c* is measured across the capacitor; therefore,

$$E_o = \frac{q}{C} \tag{a}$$

The charge $q$ can be determined by using Eq. (e) from the high-pass *RC* filter section for the current $I$. Thus,

$$q = \int I \, dt = \frac{E_a}{\sqrt{R^2 + \left(\dfrac{1}{\omega C}\right)^2}} \int \sin(\omega t + \phi) \, dt \tag{b}$$

and

$$E_o = \frac{E_a \cos(\omega t + \phi)}{\omega C \sqrt{R^2 + \left(\dfrac{1}{\omega C}\right)^2}} \qquad \text{(c)}$$

The ratio of the amplitudes of the output and input voltages for this filter is

$$\frac{E_o}{E_i} = \frac{1}{\sqrt{1 + (\omega RC)^2}} \qquad (4.61)$$

Equation (4.61) has the same form as Eq. (4.59), except that the term $\omega RC$ appears instead of $\omega L/R$; therefore, the modified $RC$ circuit is also a low-pass filter. The response curve for this filter, shown in Fig. 4.28, indicates that a transducer signal can be transmitted with an attenuation of less than 2 percent if $\omega RC < 0.203$.

## Wein-Bridge Notched Filter

The *Wein-bridge filter*, shown in Fig. 4.25d, is a notched filter that produces high attenuation at a selected filter frequency $\omega_f$. The voltage ratio for this filter can be expressed as

$$\frac{E_o}{E_i} = \left[ \frac{1}{3 + i\left(\dfrac{\omega}{\omega_f} - \dfrac{\omega_f}{\omega}\right)} - \frac{1}{1 + \dfrac{R_2}{R_1}} \right] \qquad (4.62)$$

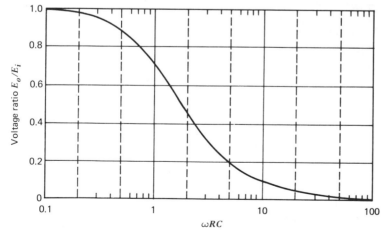

*Figure 4.28* Response curve for a low-pass $RC$ filter.

where $i$ is the imaginary number $\sqrt{-1}$.

The critical frequency $\omega_f$ for this filter is

$$\omega_f = \frac{1}{RC}$$

If the Wein-bridge filter is tuned to the frequency of a parasitic signal ($\omega_f = \omega_p$), the reactive term in Eq. (4.62) vanishes and the voltage ratio equation reduces to

$$\frac{E_o}{E_i} = \left[ \frac{1}{3} - \frac{R_1}{R_1 + R_2} \right] \qquad \text{for } \omega_p = \omega_f \qquad (4.63)$$

For the special case where $R_2 = 2R_1$, the voltage ratio $E_o/E_i = 0$, which indicates that the noise or parasitic signal is completely eliminated at the critical frequency of the filter. This high attenuation at a selected filter frequency can be used very effectively to eliminate 60-Hz noise that is difficult to handle by other means; however, this benefit is not gained without a penalty. The Wein-bridge filter will always attenuate the transducer signal; therefore, a correction must be introduced to compensate for the attenuation. The response curve for the filter, shown in Fig. 4.29, indicates that the magnitude of the correction will, in general, be large and will be a function of frequency. For this reason, the Wein-bridge filter should be used only when the transducer signal is a pure sinusoid with a known frequency $\omega$.

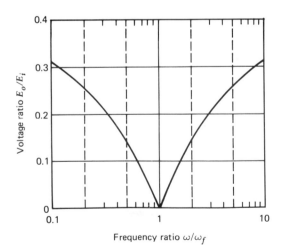

**Figure 4.29**  Response curve for a Wein-bridge filter.

## 4.10 AMPLITUDE MODULATION AND DEMODULATION

*Amplitude modulation* is a signal conditioning process in which the signal from a transducer is multiplied by a carrier signal of constant frequency and amplitude. The carrier signal can have any periodic form, such as a sinusoid, square wave, sawtooth, or triangle. The transducer signal can be sinusoidal, transient, or random. The only requirement for mixing carrier and transducer signals is that the frequency $\omega_c$ of the carrier signal must be much higher than the frequency $\omega_t$ of the transducer signal. Reasons for using amplitude modulation in instrumentation systems include better stability of ac carricr amplifiers and lower power requirements associated with transmission of high-frequency signals over long distances.

The significant aspects of data transmission with amplitude modulation can be illustrated by considering a case where both the carrier and transducer signals are sinusoidal. The output voltage $E_o$ is then given by the expression

$$E_o = (E_{it} \sin \omega_t t)(E_{ic} \sin \omega_c t)$$

$$= \frac{E_{it} E_{ic}}{2} \left[ \cos(\omega_c - \omega_t)t - \cos(\omega_c + \omega_t)t \right] \tag{4.64}$$

**where** $E_{it}$ is the amplitude of the transducer signal.
$E_{ic}$ is the amplitude of the carrier signal.

An amplitude-modulated output signal $E_o$ is illustrated in Fig. 4.30.

Equation (4.64) indicates that the output signal is being transmitted at two discrete frequencies $(\omega_c - \omega_t)$ and $(\omega_c + \omega_t)$. The amplitude associated with each frequency is the same. This data transmission at the higher frequencies permits use of high-pass filters to eliminate noise signals that usually occur at much lower frequencies. For example, consider use of a carrier frequency of 4000 Hz with a transducer signal frequency of 50 Hz. Normally any 60-Hz noise would be difficult to eliminate because of the small difference between the frequencies of the transducer signal and the noise. However, with amplitude modulation, the data in this example are transmitted at frequencies of 3950 and 4050 Hz; therefore, any 60-Hz noise superimposed onto the transmission signal can be easily eliminated with a high-pass filter. Thus, a 50-Hz signal can be successfully transmitted over lines with 60-Hz noise; in fact, even a 60-Hz signal can be transmitted in the presence of 60-Hz noise.

While amplitude modulation offers several advantages in data transmission (stability, low power, and noise suppression), the output signal is not suitable for display and interpretation until the transducer signal is separated from the carrier signal. The process of separating the transducer signal from the carrier signal is known as *demodulation*. The demodulation process is illustrated in block diagram form in Fig. 4.31a.

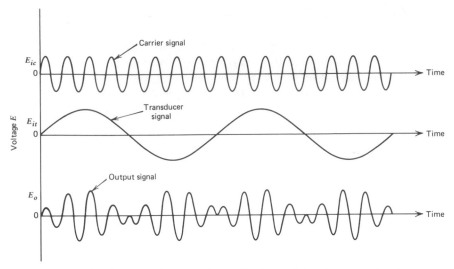

***Figure 4.30*** Carrier, transducer, and amplitude-modulated output signals.

The first step in the demodulation process involves rectifying the signal. This is usually accomplished with a full-wave, phase-sensitive rectifier. The output from this type of rectifier is a series of half-sine waves with amplitude and sense corresponding to the output signal from the transducer. The frequency spectrum associated with a rectified signal is shown in Fig. 4.31b. The frequency spectrum contains the transducer signal frequency as a single line at $\omega_t$ and four other lines for the carrier frequencies. There are many other carrier lines at higher frequencies; however, these have been omitted since they can be easily eliminated. The transducer signal is separated with a low-pass filter that transmits $\omega_t$ and severely attenuates the frequencies $2\omega_c \pm \omega_t$, $4\omega_c \pm \omega_t$, etc. If the carrier frequency is ten times the highest transducer signal frequency, the rejection of the carrier frequency can be accomplished with a minimum of distortion of the transducer signal.

## 4.11 A/D AND D/A CONVERTERS

The output from a transducer is usually an analog signal that varies linearly and continuously with the quantity being measured. With the larger and more complex instrumentation systems, it is often useful to convert this analog signal to a digital signal so that it can be analyzed or processed with a computer or microprocessor. The device employed to perform this function is an analog-to-digital (A/D) converter. A schematic illustration of the analog-to-digital conversion process is shown in Fig. 4.32.

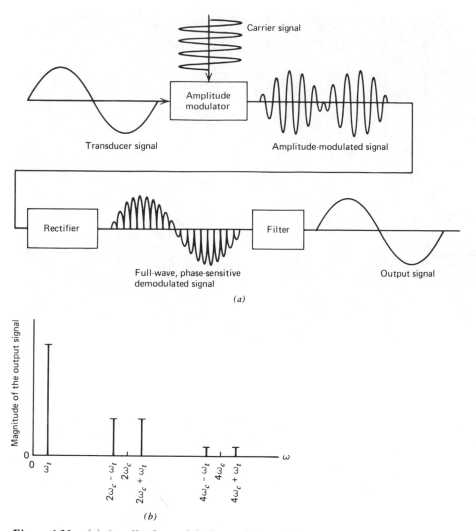

**Figure 4.31** (a) Amplitude modulation and demodulation. (b) Frequency spectrum for the output from a full-wave, phase-sensitive rectifier.

## A/D Converters

Many types of A/D converters are produced and marketed today; however, two types dominate the market: the shift-programmed, successive-approximation A/D converter (for use in high-speed applications) and the dual-slope, integrating A/D converter (for use in lower speed, low-cost applications).

The principle of operation of the successive-approximation A/D converter is illustrated in block-diagram form in Fig. 4.33. This A/D converter operates

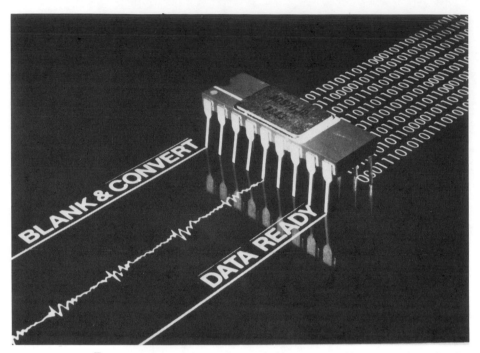

*Figure 4.32* The analog-to-digital (A/D) conversion process. (Courtesy of Analog Devices.)

by successively comparing the analog input voltage with programmed fractions of a reference voltage (which is larger than any analog input voltage expected from the instrumentation system). The reference voltage $E_r$ is divided into $2^N$ parts, where $N$ is the converter resolution expressed in terms of binary bits (for a 10-bit converter, the reference voltage is divided into $2^{10} = 1024$ parts) and each one of the parts ($E_r/2^N$, $2E_r/2^N$, $4E_r/2^N$, $\cdots$, $E_r/2$) is assigned to a different bit position.

As each comparison is made, the output of the comparitor is monitored to determine whether the analog signal is greater or less than the fraction of the reference voltage assigned to the bit position. The comparison process proceeds rapidly through the $N$-bit positions until the least significant bit has been identified in the comparitor. The digital word in the digital-to-analog (D/A) converter at this instant is the digital representation of the analog signal in binary code (see Fig. 4.34). At this point, an end-of-conversion signal is issued that indicates that the conversion is complete and that the digital signal can be recorded or observed with an LED display. In a continuous mode of operation, the next clock pulse would clear and reset the output register, enable the busy signal (an indicator that shows the converter is working), and begin the next conversion.

The speed of the successive-approximation A/D converters is amazing. State-of-the-art models exhibit a conversion time of 2.5 μs with 10-bit resolution. Since

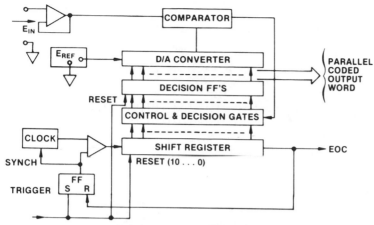

**Figure 4.33**  Block diagram for a shift-programed, successive-approximation A/D converter. (Courtesy of Analogic Corporation.)

the error involved is $\pm\frac{1}{2}$ of the least significant bit, accuracies of $\pm 0.05$ percent can be achieved with this very-high-speed A/D converter.

The principle of operation of the dual-slope integrating A/D converter is illustrated in Fig. 4.35. The conversion from analog to digital is accomplished by first charging a capacitor integrator with the analog input voltage $E_i$ for a fixed period of time $t_i$. The time interval $t_i$ is established from considerations of the converter's clock rate and the counter capacity. Next, the counter is reset and an internal reference voltage $E_r$ is substituted for the input signal. The reference voltage $E_r$ causes the capacitor integrator to discharge to zero over a

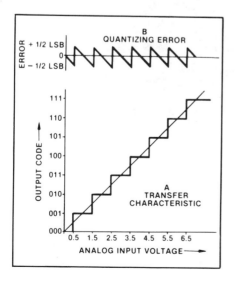

**Figure 4.34**  Digital code representation of an analog ramp function. (Courtesy of Analogic Corporation.)

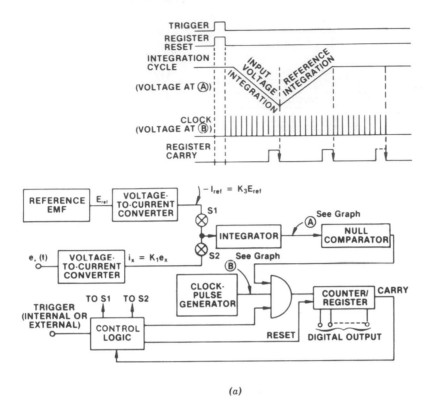

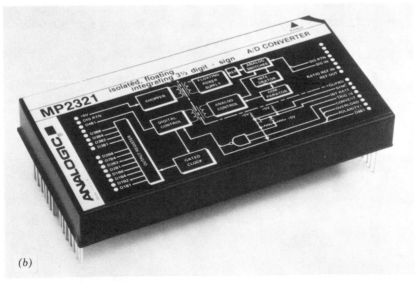

*(a)*

*(b)*

*Figure 4.35* A dual-slope, integrating A/D converter. (Courtesy of Analogic Corp.) (*a*) Block diagram. (*b*) Photograph of an A/D converter.

time $t_r$ that is related to the input voltage $E_i$, the reference voltage $E_r$, and the initial charging time $t_i$ by the expression

$$t_r = \frac{E_i t_i}{E_r}$$

Since $E_r$ and $t_i$ are constants,

$$E_i = \frac{E_r}{t_i} t_r = C t_r \tag{4.65}$$

The constant $C$ can be scaled so that the counter reading of $t_r$ gives a direct reading of $E_i$.

The integrating converter is preferred over the successive-approximation converter, except in those applications that require high speed. The integrating converter has the advantages of true averaging of the analog input signal during conversion, autozeroing, lower cost for a given resolution, higher accuracy, and linear stability. A high-performance integrating A/D converter exhibits a conversion time of 4 ms with 17-bit resolution. With 17-bit resolution, the full-scale analog signal can be divided into $2^{17} = 131,072$ parts and if the error is $\pm 1$ least significant bit ($E_r/2^{17}$), an accuracy of 0.000763 percent is achieved. An additional significant feature associated with the dual-slope method is that 60-cycle noise is averaged out when the integration period is 1/60th of a second.

## D/A Converters

*Digital-to-analog (D/A) converters* are electronic devices that accept digital signals at their input terminals and generate an analog output voltage corresponding to the digital input word. Digital-to-analog converters are often employed in digital control systems between a computer or microprocessor and the process control element. For example, parameters that control a process are usually monitored with transducers. In general, signals from a transducer are passed through an A/D converter so that they can be used as input to a control program stored in a computer or microprocessor. After executing the program for each input of transducer data, the computer issues digital signals that are to be used to adjust control elements (servovalves, heaters, motors, etc.) in the process. These control elements are usually analog devices that will not respond to digital commands; therefore, a D/A converter must be used between the computer and the control elements to provide the required analog signal.

A block diagram for a popular type of D/A converter is shown in Fig. 4.36. The digital word, in this case consisting of 16 bits, is parallel bussed into an input register. The digital word in the register activates a series of switches that connect a precisely controlled voltage $E_r$ to a precision resistor network. The

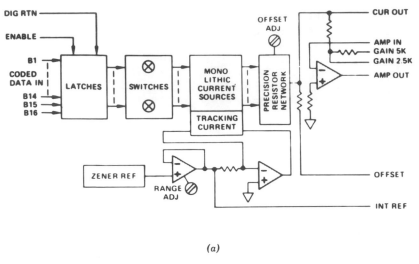

(a)

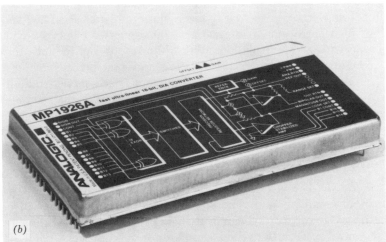

(b)

**Figure 4.36** A 16-bit D/A converter. (Courtesy of Analogic Corp.) (a) Block diagram. (b) Photograph of a D/A converter.

resistor network is connected to a summing amplifier (see Section 4.8) that sums the analog voltage contribution from each of the bits.

A simple schematic for the process for a 16-bit D/A converter is shown in Fig. 4.37. In this case, the analog output voltage $E_o$ is given by Eq. (4.54) as

$$E_o = -E_r R_f \left( \frac{1}{R_1} * \frac{1}{R_2} * \frac{1}{R_3} * \ldots * \frac{1}{R_{16}} \right) \tag{4.66}$$

where the symbol * indicates that the following term is to be added if the switch is closed and deleted if the switch is open. The resistors $R_1$ to $R_{16}$ are used to proportion the contribution of a particular bit to the output voltage according

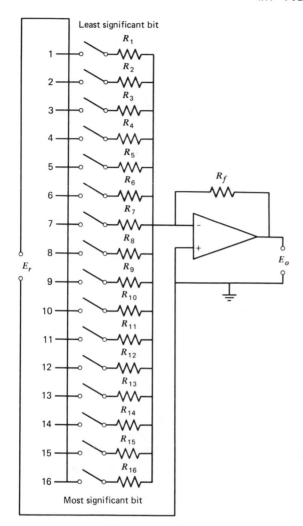

**Figure 4.37**  Schematic diagram for a summing amplifier in a D/A converter.

to the weight of the bit. For instance, the least significant bit in a 16-bit digital word has a weight of $1/2^{16} = 1/65{,}536$ and would require a resistance ratio

$$\frac{R_f}{R_1} = \frac{k}{65{,}536}$$

where $k$ is a proportionality constant. For the most significant bit in the digital word

$$\frac{R_f}{R_{16}} = k$$

The performance of high-quality D/A converters is outstanding. The devices are linear to within 0.001 percent of full scale, are virtually noise-free, and have conversion times of a few microseconds.

## 4.12   TIME-, FREQUENCY-, COUNT-, AND PHASE-MEASUREMENT CIRCUITS

Time, frequency, count, and phase are important engineering measurements. The digital electronic counter, which is used to make several of these measurements, has become a common instrument in all laboratories and on many production lines. These counters are versatile, accurate, and relatively inexpensive; thus, their use is being extended to many new applications as more and more digital instrumentation is incorporated into new and existing systems. The different kinds of time-related quantities that can be measured with a digital electronic counter and their symbolic names are:

1.  EPUT (events per unit time) is the number of events that occur in a precisely determined interval of time. FREQUENCY is a special case of EPUT where the events are equally spaced over the time interval.
2.  TIM (time-interval measurement) is the time between two events. PERIOD is a special case of TIM where the time interval is between two identical points on a periodic signal, such as the time required for a sinusoidal signal to complete one cycle.
3.  GATE (totalize and count) is a basic measurement technique where events are counted during a time interval that is started by one event and stopped by a second event. RATIO is a special case of GATE where the ratio of two input frequencies is measured directly (the number of A events per B event).

The digital electronic counter consists of a number of fundamental components that are arranged in different ways to perform the different measurements. The major components are:

1.  A digital counting unit (DCU) that counts the events and displays the results.
2.  A quartz controlled oscillator (CLOCK) that provides a precise time base.
3.  An electronic device (GATE) that controls when counting is to take place.
4.  Electronic units (AMPLIFIERS AND SCHMITT TRIGGERS) that condition the signal and generate pulses to either control the gate unit or register that an event has occurred.

In the subsections that follow, the operation of each of these major components and arrangements for performing the different measurements are described.

## Digital Counting Unit

A typical digital counting unit is shown schematically in Fig. 4.38. The event pulses pass through the gate that is connected to the clock (CLK) input of the counter unit. The counter unit shown is a binary ÷ 16 type that can be reset to zero whenever the reset line is toggled. Each counter has four outputs that correspond to dividing the input pulses by 2, 4, 8, and 16, as shown. The divide-by-16 line of the first counter is connected to the clock input of the second counter and so forth down the line. Thus, a binary sequence of 1, 2, 4, 8, 16, 32, . . ., 2048 is formed by using this cascaded arrangement of counters. When the gate stops passing the event pulses, the number of events is present on the counter output lines in binary form. This number, ranging from the least significant bit (LSB) to the most significant bit (MSB), can be held by the latch unit when the latch switch is toggled. The data, in 12-bit binary form, are then available for use by a computer or to drive a printer or visual display.

The major difference among digital counting units is in the type of counter used to process the input data. Currently, binary counters, binary-coded-decimal (BCD) counters, and decade (divide-by-10) counters are widely used. For ex-

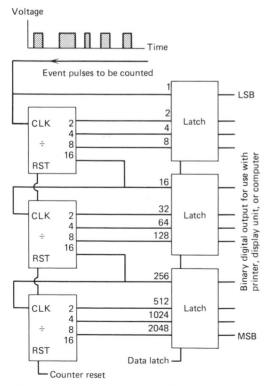

*Figure 4.38*   Typical binary digital counting unit (DCU).

ample, decade counters are often cascaded and used to control the gate function for EPUT-type measurements. In this application, the time base clock pulses (either 1 MHz or 10 MHz are common) are counted to give 0.001-, 0.01-, 0.1-, 1-, and 10-s gate times.

## The Gate Process

The *gate process* controls the time during which pulses can be counted by the digital counting unit (DCU). First, the DCU is cleared by toggling the counter reset line. The gate (an AND gate) can then be activated by switching the gate signal to the high state (Pin A at 1) so that the event pulses are passed to the DCU in accordance with the truth table and timing diagram illustrated in Fig. 4.39. When the measuring time elapses, the gate signal returns to the low state (Pin A at 0) and the event pulses are blocked from the DCU. Toggling the latch line holds the data in the latch unit where they are available for display. A display time control normally holds the display for a controlled amount of time or, in some instances, until the process repeats itself.

As illustrated in Fig. 4.39, the event duration does not have to be constant. Also, the DCU can be set to count ether the positive edge of a pulse (the O's) or the negative edge of a pulse (the X's of Fig. 4.39). In either case, the unit would count six events during the gate period shown in the figure. In all counting measurements, the count uncertainty is ±1 count.

## Input Amplifiers and Schmitt Triggers

Analog-type input signals must be converted to pulses before they can be processed with a digital electronic counter. A circuit that is often used to make this conversion consists of a Schmitt trigger (inverting type) and several operational amplifiers (op-amps) as shown in Fig. 4.40.

The Schmitt trigger is an electronic device that snaps from a high state to a low state when the input voltage exceeds an upper trip level and snaps from a low state to a high state when the input level falls below a lower trip level.

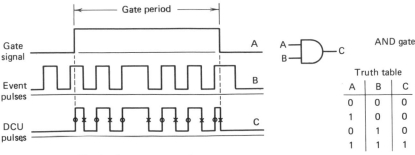

*Figure 4.39*   The basic gate operation.

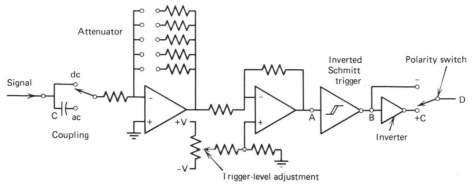

**Figure 4.40** Input amplifier and Schmitt trigger circuit.

The characteristics of a Schmitt trigger are illustrated in Fig. 4.41. The difference in trip voltages, known as hysteresis, prevents signal noise from generating pulses that should not exist.

The first op-amp shown in Fig. 4.40 is used to attenuate the input signal to a range suitable for use with the Schmitt trigger. This attenuation is accomplished by providing a wide range of feedback resistors for use with the op-amp, as shown in the figure. The second op-amp inverts the signal and provides a means (through the trigger-level adjustment) for offsetting (positioning) the input signal voltage levels with respect to the trigger levels of the Schmitt trigger.

A typical conversion of an analog input signal to pulses is illustrated in Fig. 4.42. The trigger trips from the high to the low state at points $A_1$ and $A_2$, and from the low to the high state at points $B_1$ and $B_2$. This snap action of the trigger produces the voltage pulses $V_B$ shown in the figure. The voltage pulses $V_C$ are obtained by passing the signals from the Schmitt trigger through the second inverting amplifier. In some applications, the polarity of the output pulses is important. In these instances, the polarity switch D of Fig. 4.40 provides the option for selecting a signal with the proper polarity.

The effects of trigger-level adjustment and hysteresis on pulse formation is illustrated in Fig. 4.43. With the trigger levels set for a low level of hysteresis, the trigger will trip at points $A_1$, $B_1$, $A_2$, $B_2$, $A_3$, $B_3$, $A_4$, and $B_4$; therefore, four pulses will form for the digital counting unit to count. If the trigger levels are

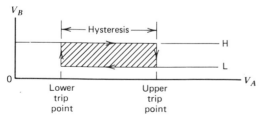

**Figure 4.41** Schmitt trigger characteristics.

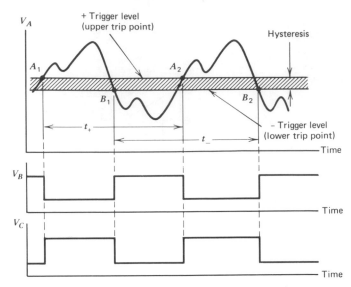

**Figure 4.42**   Conversion of an analog signal to pulses.

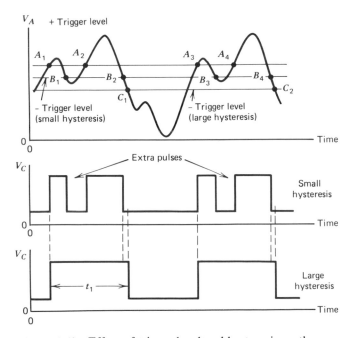

**Figure 4.43**   Effect of trigger level and hysteresis on the conversion of analog signals to digital pulses.

set for a higher level of hysteresis, the trigger will trip only at points $A_1$, $C_1$, $A_3$, and $C_2$; therefore, only two pulses will be formed for the DCU to count. Thus, it is obvious that increasing hysteresis reduces the effects of noise in the signal, but it also increases the uncertainty of pulse formation due to small but significant voltage variations in the input signal.

## EPUT, TIM, and GATE Measurements

The components required for EPUT, TIM, and GATE measurements include input amplifiers, a gate, a time base, and a digital counting unit. The input amplifers will be referred to in the different circuit arrangements that follow as the A-Amplifier, the B-Amplifier, and the STOP-Amplifier. The B-Amplifier and the STOP-Amplifier are identical and have all of the features described in the previous subsection ("Input Amplifiers and Schmitt Triggers") that are needed to convert analog input signals to pulses. Usually, the A-Amplifier does not have the polarity switch, since it is used only to generate start and stop pulses for the gate.

### EPUT Measurements

The arrangement for EPUT (events per unit time or frequency) measurements is shown in Fig. 4.44. In this case, the time base controls the gate duration. Pulses from the time base clock drive a decade counter that holds the gate open for a predetermined number of counts corresponding to times of 0.001, 0.01, 0.1, 1, or 10 s. The A-Amplifier converts the analog input signal into a pulse train that is counted by the DCU while the gate is open. The count is accurate to ±1 count.

### TIM Measurements

Component arrangement for TIM (time-interval or period) measurements is shown in Fig. 4.45. In this case, signal 1, through the B-Amplifier, generates the start pulse. The time base provides precision pulses that are counted by the digital counting unit (DCU). Signal 2, through the STOP-Amplifier, generates the stop pulse. The attenuators and trigger levels of the B- and STOP-Amplifiers must be set the same in order to obtain accurate TIM measurements. The DCU reads directly in microseconds when the clock rate is 1 MHz.

The period measurement arrangement, shown in Fig. 4.46, is a special case

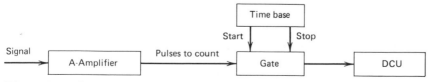

*Figure 4.44* Component arrangement for EPUT measurements.

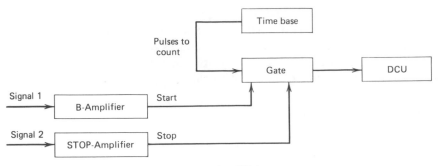

*Figure 4.45*   Component arrangement for TIM measurements.

of TIM measurement where both the start and the stop pulses come from the same input signal through the B-Amplifier. In this case, signal noise can cause significant period measurement error as illustrated in Fig. 4.47. In Fig. 4.47, the envelopes for input signal ± peak noise are shown for a typical period of the signal. With the trigger level set near the midrange of the signal, the start and stop pulses may each be formed in such a way (start early and stop late) that the maximum error shown in the figure is produced. The exact amount of error depends on the precise character of the signal and its noise.

### GATE Measurements

Component arrangements for GATE measurements and the two simpler cases of RATIO and COUNT are shown in Fig. 4.48. In the GATE arrangement, signal 2, through the B-Amplifier, generates the start pulse while signal 3, through the STOP-Amplifier, generates the stop pulse. Signal 1, through the A-Amplifier, provides the pulses to be counted. A precision time base is not used in any of the GATE measurements. The start and stop attenuation and trigger levels must be carefully adjusted in order to obtain an accurate GATE time interval.

Component arrangement for the simpler case of RATIO measurement is shown in Fig. 4.48b. For this measurement, signal 2, through the B-Amplifier, provides both the start pulse and the stop pulse. Signal 1, through the A-Am-

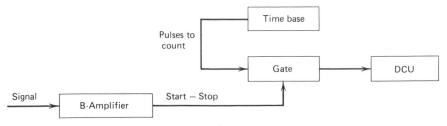

*Figure 4.46*   Component arrangement for period measurements.

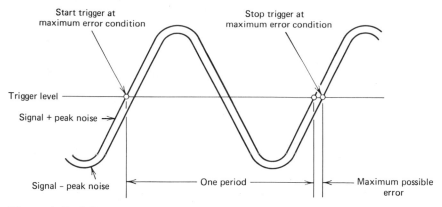

**Figure 4.47** Effect of noise on period measurement when the trigger level is near mid-point and hysteresis is small.

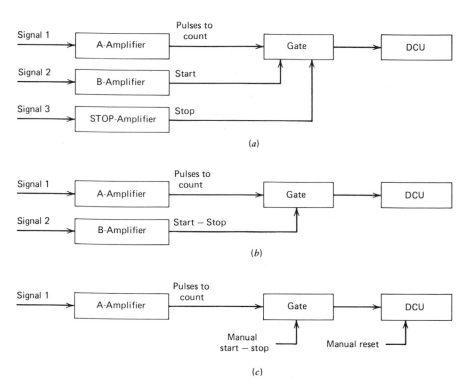

**Figure 4.48** Component arrangement for GATE, RATIO, and COUNT measurement. (*a*) General GATE arrangement; (*b*) RATIO arrangement; (*c*) COUNT arrangement.

plifier, provides the pulses to be counted. Thus, the digital counting unit measures the number of A (signal 1) events per B (signal 2) event (i.e., the ratio A/B). This measurement can be used for digital-data-transmission monitoring and for reliability assurance studies. For example, during digital data transmission, sending and receiving stations can each count the number of bits sent and received. Agreement indicates a high probability that the data were properly received.

The COUNT mode of operation, shown in Fig. 4.48c, is a completely manual mode of operation. A manual start–stop switch both initiates and terminates the counting process. The digital counting unit is also manually reset to zero.

## Period and Frequency Measurement by Comparison Methods

When a universal electronic counter is not available, the engineer often makes period and frequency measurements by comparing his signal with the output from a calibrated oscillator or clock. The clock rate of any oscillator can be subdivided by using decade counters to generate a lower clock rate (as low as $10^{-6}$ times the clock rate of the oscillator).

Typical examples of the use of timing signals for period and frequency measurements are illustrated in Fig. 4.49. The timing method often used with strip-chart recorders is shown in Fig. 4.49a. Here, the timing signal consists of a series of rectangular or triangular pulses having a known period. The timing signal provides a time base for the data and a means for checking the paper speed of the recorder.

A second method, illustrated in Fig. 4.49b, utilizes an electronic switch to periodically switch the signal being recorded to ground. Thus, the segmented signal contains both the information being recorded and the time base.

A third method, also illustrated in Fig. 4.49b, applies the square-pulse output signal from a calibrated oscillator or clock to the Z-modulated input terminal of the oscilloscope. Periodic trace intensity variations induced by the clock signal cause the oscilloscope trace to appear dashed, as indicated on the lower trace of Fig. 4.49b. Since the time from dash to dash is equal to the period of the clock signal, the oscilloscope trace contains both the information being recorded and the time base.

A major disadvantage of the second two methods is loss of signal. In order to adequately define a sinusoidal input, the clock rate should be at least 10 times the frequency of the input signal. With transient signals, the possibility of losing a signal peak also exists. Fortunately, improved accuracy and reliability of oscilloscope time bases has virtually eliminated the need for this type of timing signal. The methods currently find some use for trace identification in complex multitrace applications and in the periodic calibration of the time base of the oscilloscope.

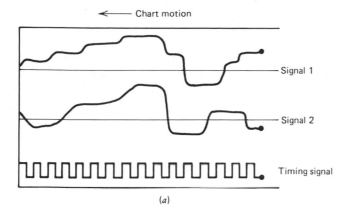

(a)

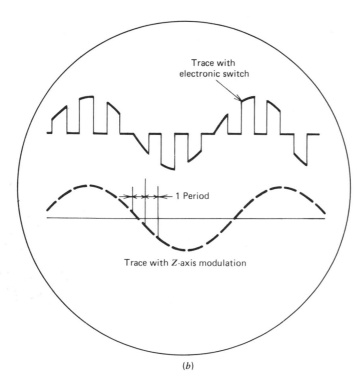

(b)

***Figure 4.49*** Use of timing signals for period and frequency measurement. (*a*) Strip chart record with timing signal. (*b*) Oscilloscope trace with electronic switch and *Z*-axis modulation.

## Phase Measurement with Lissajous Figures

The phase angle between two sinusoidal signals of the same frequency can be obtained from an oscilloscope trace known as a *Lissajous figure* or diagram. Such figures are obtained by connecting the input (reference) signal to the horizontal amplifier input of the oscilloscope and the phase-shifted output signal to the vertical amplifier input, as shown schematically in Fig. 4.50. The resulting figure can be analyzed as follows.

For the reference signal, let

$$x = A \cos \omega t \tag{4.67}$$

For the phase-shifted output signal

$$y = B \cos(\omega t + \theta) \tag{4.68}$$

$$= B \cos \theta \cos \omega t - B \sin \theta \sin \omega t$$

When $\theta = 0$, Eqs. (4.67) and (4.68) become

$$x = A \cos \omega t$$

$$y = B \cos \omega t$$

or

$$y = \frac{B}{A} x \tag{4.69}$$

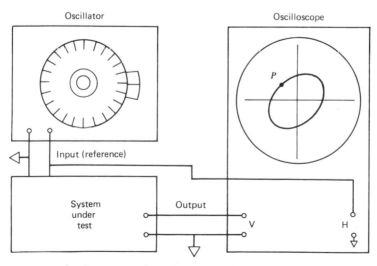

**Figure 4.50**   System used to obtain Lissajous figures.

which is the equation of the straight line shown in Fig. 4.51a. When $\theta = \pi/2$, Eqs. (4.67) and (4.68) become

$$x = A \cos \omega t$$

$$y = -B \sin \omega t$$

or

$$\frac{x^2}{A^2} + \frac{y^2}{B^2} = 1 \tag{4.70}$$

which is the equation of the ellipse shown in Fig. 4.51b. For this case, point $P$ travels around the elliptical path in a clockwise direction. Similarly, when $\theta = -(\pi/2)$, point $P$ travels around the elliptical path in a counterclockwise direction. Finally, when $\theta = \pm \pi$, Eqs. (4.67) and (4.68) reduce to

$$x = A \cos \omega t$$

$$y = -B \cos \omega t$$

or

$$y = -\frac{B}{A} x \tag{4.71}$$

which is the equation of the straight line shown in Fig. 4.51c. Equations (4.69), (4.70), and (4.71), together with the oscilloscope traces of Fig. 4.51, suggest that the phase angle $\theta$ can be measured with Lissajous figures provided $-\pi \leqslant \theta \leqslant \pi$.

For the general case of an arbitrary phase angle $\theta$, Eqs. (4.67) and (4.68) can be combined to yield

$$\frac{x^2}{A^2} + \frac{y^2}{B^2} - \frac{2 \cos \theta}{AB} xy = \sin^2 \theta \tag{4.72}$$

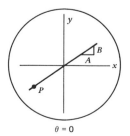
$\theta = 0$

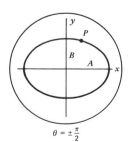
$\theta = \pm \dfrac{\pi}{2}$

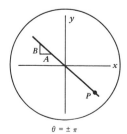
$\theta = \pm \pi$

**Figure 4.51** Three basis Lissajous figures.

which is the equation of the ellipse shown in Fig. 4.52. When $x = 0$, Eq. (4.72) indicates that

$$\sin \theta = \frac{y_0}{B} \tag{4.73}$$

The construction shown in Fig. 4.52 illustrates the method used to obtain $y_0$ and $B$, which are required for the determination of $\theta$. The sign of the phase angle is determined by observing the motion of $P$ ($+$ clockwise and $-$ counterclockwse) as it generates the elliptical trace of Fig. 4.52.

### Direct Phase-Angle Measurement

The phase angle between two sinusoidal signals of the same frequency can most easily be measured by observing the two signals on a common time base, as illustrated in Fig. 4.53. The phase angle $\theta$ in terms of the measurements indicated on Fig. 4.53 is

$$\theta = \frac{2\pi}{T_P} T_S \tag{4.74}$$

**where**    $T_P$ is the period of the reference signal.
$\phantom{where}$    $T_S$ is the time shift associated with the phase-shifted signal.

When a time shift $T_S$ to the left is considered positive (a shift to the right is then considered negative), Eq. (4.74) provides both the magnitude and direction (leading $+$ or lagging $-$) of the phase angle.

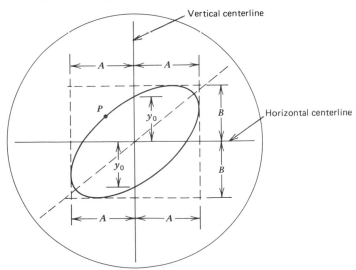

*Figure 4.52*    Construction used for the determination of an arbitrary phase angle.

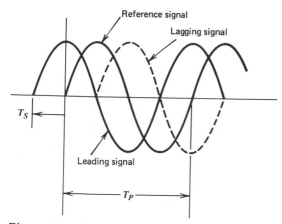

**Figure 4.53** Phase-angle determination between two sinusoidal signals.

## 4.13 SUMMARY

Many different signal conditioning circuits are employed in instrumentation systems and the performance of a system can be markedly affected by the behavior of any or all of these circuits. Power supplies, both constant voltage and constant current, are commonly used in several different elements of an instrumentation system. It is imperative that the power supplies be stable over long periods of time and that noise and/or ripple be suppressed.

Both the potentiometer circuit and the Wheatstone bridge circuit convert resistance change to a voltage variation. The potentimeter circuit can utilize either a constant-voltage or a constant-current power supply. The significant characteristics of each type of circuit are described by the following equations:

**Constant-Voltage Potentiometer Circuit**

Output voltage:
$$\Delta E_o = \frac{r}{(1 + r)^2} \left( \frac{\Delta R_1}{R_1} - \frac{\Delta R_2}{R_2} \right) (1 - \eta) E_i \qquad (4.4)$$

Nonlinear term:
$$\eta = 1 - \cfrac{1}{1 + \cfrac{1}{1 + r} \left( \cfrac{\Delta R_1}{R_1} + r \cfrac{\Delta R_2}{R_2} \right)} \qquad (4.3)$$

Circuit sensitivity:
$$S_{cv} = \frac{r}{1 + r} \sqrt{P_T R_T} \qquad (4.10)$$

Loss factor:
$$\mathscr{L} = \cfrac{2r \dfrac{R_T}{R_M} \left[ 1 + r \left( 1 + \dfrac{1}{2} \dfrac{R_T}{R_M} \right) \right]}{\left[ 1 + r \left( 1 + \dfrac{R_T}{R_M} \right) \right]^2} \qquad (4.14)$$

**Constant-Current Potentiometer Circuit**

Output voltage: $\quad E_o = IR_1 \dfrac{\Delta R_1}{R_1}$ $\hspace{3cm}$ (4.16)

Circuit sensitivity: $S_{cc} = \sqrt{P_T R_T}$ $\hspace{3cm}$ (4.18)

The Wheatstone bridge is widely used for converting resistance change to voltage, since it can be employed for both static and dynamic measurements. The Wheatstone bridge can also be driven with either a constant-voltage or a constant-current power supply. The equations describing the behavior of the Wheatstone bridge are:

**Constant-Voltage Wheatstone Bridge**

Output voltage: $\quad \Delta E_0 = \dfrac{r}{(1+r)^2} \left( \dfrac{\Delta R_1}{R_1} - \dfrac{\Delta R_2}{R_2} + \dfrac{\Delta R_3}{R_3} - \dfrac{\Delta R_4}{R_4} \right)(1 - \eta)E_i$ $\hspace{0.5cm}$ (4.23)

Nonlinear term: $\quad \eta = \dfrac{1}{1 + \dfrac{r+1}{\dfrac{\Delta R_1}{R_1} + \dfrac{\Delta R_4}{R_4} + r\left( \dfrac{\Delta R_2}{R_2} + \dfrac{\Delta R_3}{R_3} \right)}}$ $\hspace{1cm}$ (4.24)

Circuit sensitivity: $\quad S_{cv} = \dfrac{r}{1+r} \sqrt{P_T R_T}$ $\hspace{2cm}$ (4.28)

**Constant-Current Wheatstone Bridge**

Output voltage: $\quad \Delta E_o = \dfrac{IR_T r}{2(1+r)} \dfrac{\Delta R_T}{R_T} (1 - \eta)$ $\hspace{2cm}$ (4.35)

Nonlinear term: $\quad \eta = \dfrac{\dfrac{\Delta R_T}{R_T}}{2(1+r) + \dfrac{\Delta R_T}{R_T}}$ $\hspace{2.5cm}$ (4.36)

Circuit sensitivity: $\quad S_{cc} = \dfrac{r}{1+r} \sqrt{P_T R_T}$ $\hspace{2cm}$ (4.38)

The constant-current Wheatstone bridge is superior because of its extended range; however, the availability of high-performance constant-current power supplies often limits its application.

Amplifiers are used in most instrumentation systems to increase the low-level output signal from a transducer to a level sufficient for recording with a voltage measuring instrument. Ideally, the output and input signals for the amplifier are related by the expression

$$E_o = GE_i \hspace{3cm} (4.39)$$

Frequency response and linearity are two important characteristics of instrument amplifiers that must be adequate if signal distortion is to be avoided. A popular amplifier in instrumentation systems is the differential amplifier because it rejects common-mode signals. Instrument amplifiers that employ op-amps with resistor feedback are commonly employed because of their stability, low cost, and favorable operating characteristics.

Operational amplifiers (op-amps) are the basic circuit element in signal conditioning circuits, such as voltage followers, summing amplifiers, integrating amplifiers, and differentiating amplifiers. The voltage follower exhibits a gain of unity and is used because of its high input impedance. Summing, integrating, and differentiating amplifiers, as the names imply, are used to add (or subtract) two or more input signals, integrate an input signal with respect to time, or differentiate an input signal with respect to time. Equations describing important characteristics of each are as follows:

**Voltage Follower**

Input resistance: $R_{ci} = (1 + G)R_i$      (4.52)

Output resistance: $R_{co} = \dfrac{R_o}{1 + G}$      (4.53)

**Summing Amplifier**

Output voltage: $E_o = -R_f\left(\dfrac{E_{i1}}{R_1} + \dfrac{E_{i2}}{R_2} + \dfrac{E_{i3}}{R_3}\right)$    (4.54)

**Integrating Amplifier**

Output voltage: $E_o = -\dfrac{1}{R_1 C_f}\displaystyle\int_0^t E_i \, dt$      (4.57)

**Differentiating Amplifier**

Output voltage: $E_o = -R_f C_1\dfrac{dE_i}{dt}$      (4.58)

Filters are used to eliminate undesirable signals such as noise from transducer signals. Voltage ratios $E_o/E_i$ for four commonly used filters are given by the following expressions:

*RL* **Filter—Low Pass**

Voltage ratio: $\dfrac{E_o}{E_i} = \dfrac{1}{\sqrt{1 + \left(\dfrac{\omega L}{R}\right)^2}}$      (4.59)

*RC* **Filter—High Pass**

Voltage ratio: $\dfrac{E_o}{E_i} = \dfrac{1}{\sqrt{1 + \left(\dfrac{1}{\omega RC}\right)^2}} = \dfrac{\omega RC}{\sqrt{1 + (\omega RC)^2}}$   (4.60)

### *RC* Filter—Low Pass

Voltage ratio:  $\dfrac{E_o}{E_i} = \dfrac{1}{\sqrt{1 + (\omega RC)^2}}$   (4.61)

### Wein-Bridge Filter—Notched

Voltage ratio:  $\dfrac{E_o}{E_i} = \dfrac{1}{3 + i\left(\dfrac{\omega}{\omega_f} - \dfrac{\omega_f}{\omega}\right)} - \dfrac{1}{1 + \dfrac{R_2}{R_1}}$   (4.62)

Filters must be selected very carefully; otherwise, the filter may attenuate both the noise signal and the transducer signal (if the frequencies are similar) and produce serious error.

Amplitude modulation is a signal conditioning process in which the signal from a transducer is multiplied by a carrier signal of much higher frequency and constant amplitude. The resulting output voltage is given by the expression

$$E_o = \frac{E_{it}E_{ic}}{2}\left[\cos(\omega_c - \omega_t)t - \cos(\omega_c + \omega_t)t\right]$$   (4.64)

The higher frequencies associated with Eq. (4.64) require less power for long-distance transmissions and permit use of high-pass filters to eliminate low-frequency noise.

Analog-to-digital (A/D) converters are electronic devices that digitize an analog transducer signal at a very high conversion rate so that it can be processed with a computer or microprocessor. Conversion rates of 2.5 μs are possible with 10-bit resolution and an accuracy of 0.05 percent. Two types of A/D converters are widely used. The successive-approximation type is used when a high conversion speed is essential. The dual-slope integrating converter is used for most other applications because of its lower cost, better accuracy, and linear stability.

Digital-to-analog (D/A) converters are electronic devices that accept digital signals at their input terminals and then generate an analog output voltage corresponding to the digital input word. Conversion rates of a few microseconds are possible and accuracies of 0.001 percent of full scale can be achieved.

Time, frequency, count, and phase are important engineering measurements. The digital electronic counter, which is used to make several of these measurements, has become a common instrument in all laboratories and on many production lines. These counters are versatile, accurate, and relatively inexpensive; therefore, their use is being extended to many new applications as more and more digital instrumentation is being incorporated into new and existing systems.

# REFERENCES

**1.** Lenk, J. D.: *Manual for Operational Amplifier Users*, Reston Publishing Company, Reston, Virginia, 1976.

**2.** Ahmed, H., and P. J. Spreadbury: *Electronics for Engineers*, Cambridge University Press, London, 1973.

**3.** Brophy, J. J.: *Basic Electronics for Scientists*, 3rd ed., McGraw-Hill, New York, 1977.

**4.** Malmstadt, H. V., and C. G. Enke: *Digital Electronics for Scientists*, Benjamin, Menlo Park, California, 1969.

**5.** Doebelin, E. O.: *Measurement Systems: Application and Design*, 3rd ed., McGraw-Hill, New York, 1983.

**6.** *1980 Design Engineers Handbook & Selection Guide for A/D and D/A Converter Modules, Signal Conditioners, and Control Modules,* Analogic Corp., Wakefield, Massachusetts.

**7.** Teledyne Philbrik Product Guide, Dedham, Massachusetts, 1978.

---

## EXERCISES

**4.1** A strain gage with $R_g = 350$ $\Omega$ and $S_g = 2.00$ is used to monitor a sinusoidal signal with an amplitude of 1500 $\mu$in./in. and a frequency of 200 Hz. Determine the output voltage $E_o$ if a constant-voltage potentiometer circuit is used to convert the resistance change to voltage. Assume $E_i = 22$ V and $r = 5$.

**4.2** Determine the magnitude of the nonlinear term $\eta$ for the data of Exercise 4.1.

**4.3** If the strain gage described in Exercise 4.1 can dissipate 0.5 W, determine the input voltage $E_i$ required to maximize the output voltage $E_o$.

**4.4** Determine the circuit sensitivity $S_{cv}$ for the constant-voltage potentiometer circuit described in Exercise 4.1.

**4.5** Determine the load error $\mathscr{L}$ if the output voltage $E_o$ of Exercise 4.1 is monitored with:

(a) An oscilloscope having an input impedance of $10^6$ $\Omega$
(b) An oscillograph having an input impedance of 350 $\Omega$

**4.6** If a constant-current potentiometer circuit was used in Exercise 4.1 in place of the constant-voltage potentiometer circuit, determine the output voltage $E_o$ if $I = 5$ mA.

**4.7**   Determine the magnitude of the nonlinear term $\eta$ for the data of Exercise 4.6.

**4.8**   If the strain gage described in Exercise 4.1 can dissipate 0.5 W, determine the current $I$ that should be used with a constant-current potentiometer circuit to maximize the output voltage $E_o$.

**4.9**   Determine the circuit sensitivity $S_{cc}$ for the constant-current potentiometer circuit of Exercise 4.6.

**4.10**   Determine the circuit sensitivity $S_{cc}$ for the constant-current potentiometer circuit of Exercise 4.8.

**4.11**   Determine the load error $\mathcal{L}$ if the output voltage $E_o$ of Exercise 4.6 is monitored with:

(a)   An oscilloscope having an input impedance of $10^6$ $\Omega$
(b)   An oscillograph having an input impedance of 350 $\Omega$

**4.12**   A constant-voltage Wheatstone-bridge circuit is employed with a displacement transducer (potentiometer type) to convert resistance change to output voltage. If the displacement transducer has a total resistance of 1000 $\Omega$, then $\Delta R = \pm 500$ $\Omega$ if the wiper is moved from the center position to either end. If the transducer is placed in arm $R_1$ of the bridge and, if $R_1 = R_2 = R_3 = R_4 = 500$ $\Omega$:

(a)   Determine the magnitude of the nonlinear term $\eta$ as a function of $\Delta R$.
(b)   Prepare a graph of $\eta$ versus $\Delta R$ as $\Delta R$ varies from $-500$ $\Omega$ to $+500$ $\Omega$.

**4.13**   Determine the output voltage $E_o$ as a function of $\Delta R$ for the displacement transducer and Wheatstone bridge described in Exercise 4.12 if $E_i = 12$ V.

**4.14**   The nonlinear output voltage of Exercise 4.13 makes data interpretation difficult. How can the Wheatstone-bridge circuit be modified to improve the linearity of the output voltage $E_o$?

**4.15**   A strain gage with $R_g = 120$ $\Omega$, $P_T = 0.2$ W, and $S_g = 2.00$ is used in arm $R_1$ of a constant-voltage Wheatstone bridge. Determine:

(a)   Values of $R_2$, $R_3$, and $R_4$ needed to maximize $E_o$ if the available power supply is limited to 36 V
(b)   The circuit sensitivity of the bridge of Part (a)

**4.16**   If the strain gage of Exercise 4.15 is subjected to a strain of 900 $\mu$in./in., determine the output voltage $E_o$.

**4.17**   Four strain gages are installed on a cantilever beam as shown in Fig. E4.17 to produce a displacement transducer.

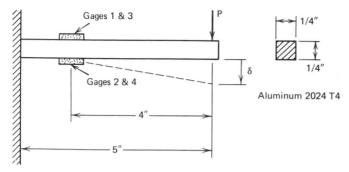

*Figure E4.17*

(a) Indicate how the gages should be wired into a Wheatstone bridge to produce maximum signal output.

(b) Determine the circuit sensitivity if $R_g = 350\ \Omega$, $P_T = 0.10$ W, and $S_g = 2.00$.

(c) Determine the calibration constant $c = \delta/E_o$ for the transducer.

**4.18** If the cantilever beam of Exercise 4.17 is used as a load transducer, determine the calibration constant $C = P/E_o$.

**4.19** A strain gage with $R_g = 120\ \Omega$, $P_T = 0.2$ W, and $S_g = 2.00$ is used in arm $R_1$ of a constant-current Wheatstone bridge. Determine:

(a) Values of $R_2$, $R_3$, and $R_4$ needed to maximize $E_o$ if the available power supply can deliver a maximum of 20 mA.

(b) The circuit sensitivity of the bridge of Part (a)

(c) The output voltage $E_o$ if the gage is subjected to a strain of 900 $\mu$in./in.

**4.20** If the displacement transducer of Exercise 4.12 is used with a constant-current Wheatstone bridge:

(a) Determine the magnitude of the nonlinear term $\eta$ as a function of $\Delta R$.

(b) Prepare a graph of $\eta$ versus $\Delta R$ as $\Delta R$ varies from $-500\ \Omega$ to $+500\ \Omega$.

**4.21** Determine the output voltage $E_o$ as a function of $\Delta R$ for the displacement transducer and Wheatstone bridge described in Exercise 4.20 if $I = 20$ mA.

**4.22** Use an op-amp with a gain of 100 dB and $R_a = 7\ \text{M}\Omega$ to design an inverting amplifier with a gain of 20.

**4.23** Use an op-amp with a gain of 100 dB and $R_a = 7\ \text{M}\Omega$ to design a noninverting amplifier with a gain of 20.

**4.24**  Use an op-amp with a gain of 100 dB and $R_a = 7\ \text{M}\Omega$ to design a differential amplifier with a gain of 20.

**4.25**  Determine the input and output impedances for a voltage follower that incorporates an op-amp having a gain of 120 dB and $R_i = 7\ \text{M}\Omega$.

**4.26**  Verify Eq. (4.56).

**4.27**  Three signals $E_{i1}$, $E_{i2}$, and $E_{i3}$ are to be summed so that the output voltage $E_o$ is proportional to $E_{i1} + 2E_{i2} + \frac{1}{2}E_{i3}$. Select resistances $R_1$, $R_2$, $R_3$, and $R_f$ to accomplish this operation.

**4.28**  Show that the op-amp circuit shown in Fig. E4.28 is a combined adding/scaling and subtracting/scaling amplifier by deriving the following equation for the output voltage $E_o$:

$$E_o = \frac{R_f^*}{R_4}E_{i4} + \frac{R_f^*}{R_5}E_{i5} - \frac{R_f}{R_1}E_{i1} - \frac{R_f}{R_2}E_{i2} - \frac{R_f}{R_3}E_{i3}$$

$$\text{where } R_f^* = R_f \left[ \frac{\dfrac{1}{R_1} + \dfrac{1}{R_2} + \dfrac{1}{R_3} + \dfrac{1}{R_f}}{\dfrac{1}{R_4} + \dfrac{1}{R_5} + \dfrac{1}{R}} \right]$$

**4.29**  The signals shown in Fig. E4.29 are to be used as input to an integrating amplifier having $R_1 = 1\ \text{M}\Omega$ and $C_1 = 0.5\ \mu\text{F}$. Sketch the output signal corresponding to each of the input signals.

**4.30**  Discuss potential problem areas associated with the output voltages from signals (*a*) and (*c*) of Exercise 4.29.

**4.31**  Repeat Exercise 4.29 with a differentiating amplifier in place of the integrating amplifier.

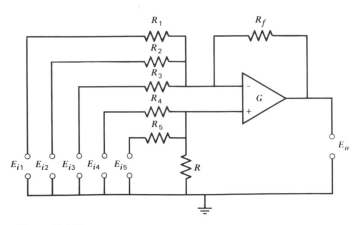

*Figure E4.28*

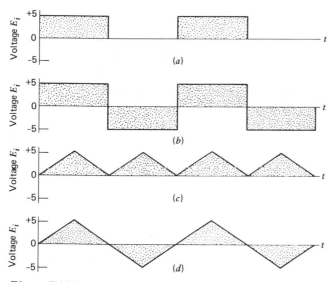

**Figure E4.29**

**4.32** Select $L$ and $R$ for a low-pass filter, so that a 10-Hz transducer signal will be transmitted with less than 2 percent attenuation, while a 60-Hz noise signal will undergo at least a 90 percent attenuation.

**4.33** Select $R$ and $C$ for a high-pass filter so that the dc component of the output from a potentiometer circuit will be blocked, while the 20-Hz ac signal from a transducer will be transmitted with less than 2 percent attenuation.

**4.34** Determine the voltage ratio $E_o/E_i$ as a function of the frequency ratio $\omega/\omega_f$ for a Wein-bridge filter having $R_2/R_1 = 2$.

**4.35** If the data signal is a triangular wave and the carrier signal is a square wave, sketch the amplitude modulated signal.

**4.36** Prepare a table showing resolution as a function of the number of bits in a digital word.

**4.37** Select values for the resistances shown in Fig. 4.37 if $R_f = 1 \, M\Omega$, $E_r = 10$ V, and the maximum output voltage for the 16-bit digital word 1111111111111111 is 20 V.

**4.38** Review binary code and word representation in an appropriate text-book. Determine the binary representation for the following numbers:

(a)  4                         (d)  137
(b)  26                       (e)  2752
(c)  43

**4.39** Show that Eqs. (4.67) and (4.68) can be combined to yield Eq. (4.72).

**4.40** Show that Eq. (4.72) reduces to Eq. (4.70) when $\theta = \pi/2$ and to Eq. (4.69) when $\theta = 0$.

# FIVE

# RESISTANCE-TYPE STRAIN GAGES

## 5.1 INTRODUCTION

Historically, the development of strain gages has followed many different approaches, and gages have been based on mechanical, optical, electrical, acoustical, and pneumatic principles. A strain gage has several characteristics that should be considered in judging its adequacy for a particular application. These characteristics are:

1.  The calibration constant for the gage should be stable with respect to both temperature and time.
2.  The gage should be capable of measuring strains with an accuracy of $\pm 1$ $\mu$in./in. ($\mu$m/m) over a range of $\pm 5$ percent strain ($\pm 50,000$ $\mu$in./in.).
3.  The gage length and width should be small so that the measurement approximates strain at a point.
4.  The inertia of the gage should be minimal to permit the recording of high-frequency dynamic strains.

5. The response or output of the gage should be linear over the entire strain range of the gage.
6. The gage and associated electronics should be economical.
7. Installation and readout of the gage should require minimal skills and understanding.

While no single gage system can be considered optimum, the electrical resistance strain gage very nearly meets all of the required characteristics listed above.

## 5.2  ETCHED-FOIL STRAIN GAGES

The sensitivity $S_A$ of a metallic conductor to strain was developed in Section 3.4 and it is evident from Eq. (3.4)

$$ S_A = \frac{dR/R}{\varepsilon} = \frac{d\rho/\rho}{\varepsilon} + (1 + 2v) \qquad \text{(3.4 bis)} $$

that it is possible to measure strain with a straight length of wire if the change in resistance is monitored as the wire is subjected to a strain. However, the circuits required to measure $dR$ (in practice $\Delta R$) have power supplies with limited current capabilities and the power dissipated by the gage itself must be limited. As a result, strain gages are usually manufactured with a resistance of 120 $\Omega$ or more. These high values of gage resistance, in most cases, preclude fabrication from a straight length of wire, since the gage becomes too long.

When electrical resistance strain gages were first introduced (1936–1956), the gage element was produced by winding a grid with very-fine-diameter wire. Since the late 1950s, most gages have been fabricated from ultra-thin metal foil by using an advanced photoetching process. Since this process is quite versatile, a wide variety of gage sizes and grid shapes are produced (see Fig. 3.12). Gages as small as 0.20 mm in length are commercially available. Standard gage resistances are 120 $\Omega$ and 350 $\Omega$; but in some configurations, resistances of 500 $\Omega$ and 1000 $\Omega$ are available. The foil gages are normally fabricated from Advance, Karma, or Isoelastic alloys (see Table 3.1). In addition, high-temperature gages are available in several of the heat-resistant alloys.

The etched metal-film grids are very fragile and easy to distort or tear. To avoid these difficulties, the metal film is bonded to a thin sheet of plastic (see Fig. 3.10), which serves as a backing material and carrier before the photoetching process is performed. The carrier contains markings for the centerlines of the gage length and width to facilitate installation and serves to electrically insulate the metal grid from the specimen once it is installed.

For general-purpose strain-gage applications, a polyimide plastic that is tough and flexible is used for the carrier. For transducer applications, where precision and linearity are extremely important, a very thin, brittle, high-modulus epoxy is used for the carrier. Glass-reinforced epoxy is used when the gage will be exposed to high-level cyclic strains or when the gage will be employed at temperatures as high as 750°F (400°C). For very-high-temperature applications, a gage with a strippable carrier is available. The carrier is removed during installation of the gage. A ceramic adhesive is used to maintain the grid config- uration and to electrically insulate the grid from the specimen.

## 5.3  STRAIN-GAGE INSTALLATION

The bonded type of electrical resistance strain gage is a high-quality pre- cision resistor that must be attached to a specimen by utilizing the correct ad- hesive and by employing proper mounting procedures. The adhesive serves a vital function in the strain-measuring system, since it must transmit the surface displacement from the specimen to the gage grid without distortion. At first it may appear that this function can be accomplished with almost any strong ad- hesive; however, experience has shown that improperly selected and cured ad- hesives can seriously degrade a gage installation by changing the gage factor and/or the initial resistance of the gage. Improperly cured or viscoelastic ad- hesives also produce gage hysteresis and signal loss due to stress relaxation. Best results are obtained with a strong, viscous-free, well-cured adhesive that forms a very thin bond line.

The surface of the component in the area where gages are to be positioned must be carefully prepared before the gages are installed. This preparation consists of paint and/or rust removal followed by sanding to obtain a smooth but not highly polished surface. Solvents are then used to eliminate all traces of grease and oil. Finally, the surface should be treated with a basic solution to give it the proper chemical affinity for the adhesive.

Next, the gage location is marked on the specimen with a very light scribe line and the gage, without adhesive, is positioned by using a rigid transparent tape in the manner illustrated in Fig. 5.1. The position and orientation of the gage are maintained by the tape as the adhesive is applied and as the gage is pressed into place by squeezing out the excess adhesive.

Once the gage is positioned, the adhesive must be subjected to a proper combination of pressure and temperature for the period of time needed to ensure a complete cure. The curing process is quite critical since the adhesive will expand during heating, experience a volume change during polymerization, exhibit a contraction while cooling, and sometimes exhibit a postcure shrinkage. Since the adhesive is strong enough to control deformation of the gage, changes in the volume of the adhesive influence the resistance of the gage. Of particular importance is postcure shrinkage, which can influence gage resistance long after the adhesive is supposed to be completely cured. If a long-term measurement of strain is made with a gage having an adhesive that has not completely poly- merized, the signal from the gage will drift with time and accuracy of the data will be seriously impaired.

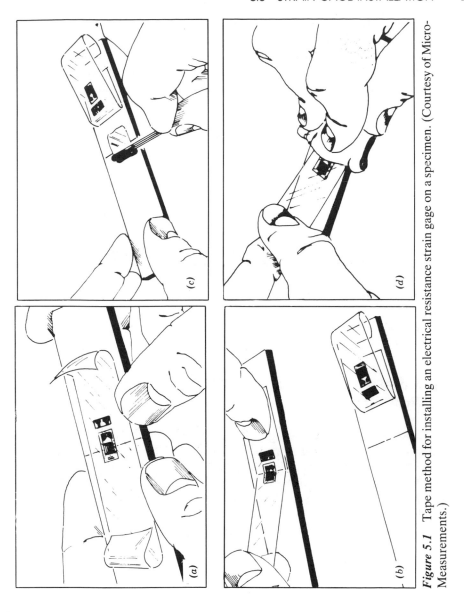

*Figure 5.1* Tape method for installing an electrical resistance strain gage on a specimen. (Courtesy of Micro-Measurements.)

For most strain-gage applications, either cyanoacrylate or epoxy adhesives are used. The cyanoacrylate adhesive (Permabond 910) has the advantage of being easier to apply, since it requires no heat, requires only a gentle pressure that can be applied with one's thumb, and requires only about 10 min for complete polymerization. Its disadvantages include a deterioration with time, water absorption, and elevated temperatures. The epoxy adhesives are superior to cyanoacrylates; however, they are more difficult to apply since they require a pressure of 5 to 20 psi (35 to 140 kPa) and often require application of heat for an hour or more while the pressure is applied. After the adhesive is completely cured, the gage should be waterproofed with a light overcoating of crystalline wax or a polyurethane.

Lead wires are attached to the terminals of the gage so that the change in resistance can be monitored with a suitable instrumentation system. Since the foil strain gages are fragile even when bonded to a structure, care must be exercised as the lead wires are attached to the soldering tabs. Intermediate anchor terminals, which are much more rugged than the strain gage tabs, are used to protect the gage from damage, as shown in Fig. 5.2. A small-diameter wire (32 to 36 gage) is used to connect the gage terminal to the anchor terminal. Three lead wires are soldered to the anchor terminal, as shown in Fig. 5.2, to provide for temperature compensation of the lead wires in the Wheatstone bridge (see Section 5.4).

## 5.4 THE WHEATSTONE BRIDGE FOR STRAIN-GAGE SIGNAL CONDITIONING

The basic equations governing balance condition, output voltage, nonlinearity, and sensitivity of Wheatstone bridges with constant-voltage and constant-current power supplies were developed in Sections 4.4 and 4.5. Since the Wheatstone bridge is the most commonly employed circuit to convert the resistance change $\Delta R/R$ from a strain gage to an output voltage $E_o$, its application for this

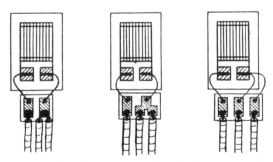

***Figure 5.2*** A strain-gage installation with anchor terminals. (Courtesy of Micro-Measurements.)

purpose is considered in detail in this section. One of the first questions that arises pertaining to use of the Wheatstone bridge for strain measurements concerns location of the gage or gages within the bridge. An answer to this question can be provided by considering the four common bridge arrangements shown in Fig. 5.3.

*Case 1:*  This bridge arrangement utilizes a single active gage in position $R_1$ and is often employed for both static and dynamic strain-gage measurements where temperature compensation is not required. The resistance $R_1$ equals $R_g$ and the other three resistances are selected to maximize the circuit sensitivity while maintaining the balance condition $R_1 R_3 = R_2 R_4$.

The sensitivity $S_s$ of the strain-gage Wheatstone-bridge system is defined as the product of the sensitivity of the gage $S_g$ and the sensitivity of the bridge circuit $S_c$. Thus,

$$S_s = S_g S_c = \frac{\Delta R_g / R_g}{\varepsilon} \frac{\Delta E_o}{\Delta R_g / R_g} \tag{5.1}$$

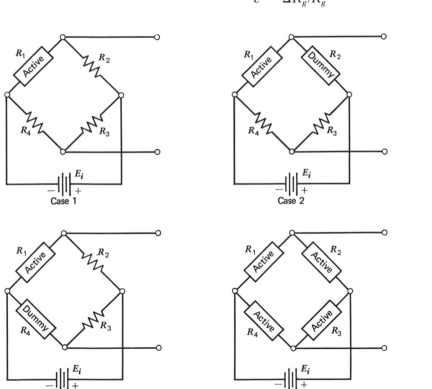

*Figure 5.3*    Four common strain-gage arrangements in a Wheatstone bridge.

From Eqs. (3.5) and (4.28),

$$S_s = \frac{r}{1 + r}S_g\sqrt{P_gR_g} \tag{5.2}$$

Equation (5.2) indicates that the sensitivity of the system is controlled by the circuit efficiency $r/(1 + r)$ and the characteristics of the strain gage $S_g$, $P_g$, and $R_g$. The most important of the two factors is the characteristics of the strain gage that vary widely with gage selection. The gage factor $S_g$ is about 2 for gages fabricated from Advance or Karma alloys and about 3.6 for Isoelastic alloys. Resistances of 120 and 350 $\Omega$ are available for most grid configurations; resistances of 500 and 1000 $\Omega$ can be obtained for a few configurations. Power dissipation $P_g$ is more difficult to specify since it depends upon the conductivity and heat-sink capacity of the specimen to which the gage is bonded. Power density $P_D$ is defined as

$$P_D = \frac{P_g}{A} \tag{5.3}$$

**where**   $P_g$ is the power that can be dissipated by the gage.
$A$ is the area of the grid of the gage.

Recommended power densities for different materials and different test conditions are given in Table 5.1.

A graph showing bridge supply or input voltage $E_i$ as a function of grid area for a large number of different gage configurations is shown in Fig. 5.4. The bridge voltage $E_i$ specified in Fig. 5.4 is for a four-equal-arm bridge with $r = 1$. In this case, the bridge voltage is given by

$$E_i = 2\sqrt{AP_DR_g} \tag{5.4}$$

When $r \neq 1$, the bridge voltage is given by

$$E_i = (1 + r)\sqrt{AP_DR_g} \tag{5.5}$$

**TABLE 5.1**   Recommended Power Densities

| Power Density $P_D$ | | |
|---|---|---|
| W/in.² | W/mm² | Specimen Conditions |
| 5–10 | 0.008–0.016 | Heavy aluminum or copper sections |
| 2–5 | 0.003–0.008 | Heavy steel sections |
| 1–2 | 0.0015–0.003 | Thin steel sections |
| 0.2–0.5 | 0.0003–0.0008 | Fiberglass, glass, ceramics |
| 0.02–0.05 | 0.00003–0.00008 | Unfilled plastics |

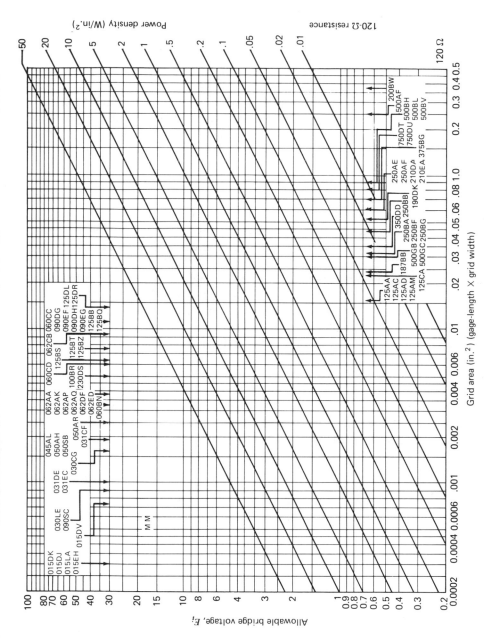

*Figure 5.4*   Allowable bridge voltage as a function of grid area for different power densities. (Courtesy of Micro-Measurements.)

The power that can be dissipated by a gage will vary over very wide limits. A small gage with a grid area of 0.001 in.$^2$ bonded to an insulating material such as a ceramic ($P_D$ = 0.2 W/in.$^2$) can dissipate 0.2 mW. On the other hand, a large strain gage with $A$ = 0.2 in.$^2$ mounted on a heavy aluminum section ($P_D$ = 10 W/in.$^2$) can dissipate 2 W.

System sensitivity can be maximized by selecting high-resistance gages with the largest grid area consistent with allowable errors due to gage-length and gage-width effects. Specification of Isoelastic alloys to obtain $S_g$ = 3.6 should be limited to dynamic strain measurements where temperature stability of the gage is not a consideration.

The second factor controlling system sensitivity is circuit efficiency $r/(1 + r)$. The value of $r$ should be selected to increase circuit efficiency, but not so high that the bridge voltage given by Eq. (5.5) increases beyond reasonable limits. Values of $r$ between 4 and 9 give circuit efficiencies between 80 and 90 percent; therefore, most bridges should be designed to fall within this range.

*Case 2:*   This bridge arrangement contains a single active gage in arm $R_1$, a dummy gage in arm $R_2$, and fixed-value resistors in arms $R_3$ and $R_4$. The active gage and the dummy gage must be identical (preferably two gages from the same package), must be applied with the same adhesive, and must be subjected to the same curing cycle. The dummy gage can be mounted in a stress-free region of the specimen or on a small block of specimen material that is placed in the same thermal environment as the specimen. In the Wheatstone bridge, the dummy gage output serves to cancel any active gage output due to temperature fluctuations during the test interval. The manner in which this bridge arrangement compensates for temperature changes can be illustrated by considering the resistance changes experienced by the active and dummy gages during a test. Thus

$$\left(\frac{\Delta R_g}{R_g}\right)_a = \left(\frac{\Delta R_g}{R_g}\right)_\varepsilon + \left(\frac{\Delta R_g}{R_g}\right)_{\Delta T} \tag{a}$$

$$\left(\frac{\Delta R_g}{R_g}\right)_d = \left(\frac{\Delta R_g}{R_g}\right)_{\Delta T} \tag{b}$$

In Eqs. (a) and (b) the subscripts $a$ ands $d$ refer to the active and dummy gages, respectively, while the subscripts $\varepsilon$ and $\Delta T$ refer to the effects of strain and temperature. Substituting Eqs. (a) and (b) into Eq. (4.22) and noting that $\Delta R_3$ = $\Delta R_4$ = 0 (fixed-value resistors) gives

$$\Delta E_o = E_i \frac{r}{(1 + r)^2} \left[\left(\frac{\Delta R_g}{R_g}\right)_\varepsilon + \left(\frac{\Delta R_g}{R_g}\right)_{\Delta T} - \left(\frac{\Delta R_g}{R_g}\right)_{\Delta T}\right] \tag{5.6}$$

Since the last two terms in the bracketed quantity cancel, the output $\Delta E_o$ is due only to the strain applied to the active gage, and temperature compensation is achieved.

With this bridge arrangement, $r$ must equal 1 to satisfy the bridge balance requirement; therefore, the system sensitivity obtained from Eq. (5.2) is

$$S_s = \frac{1}{2} S_g \sqrt{P_g R_g} \tag{5.7}$$

Equation (5.7) indicates that placement of a dummy gage in arm $R_2$ of the Wheatstone bridge to effect temperature compensation reduces the circuit efficiency to 50 percent. This undesirable feature can be avoided by use of the bridge arrangement described under Case 3.

*Case 3:*  In this bridge arrangement, the dummy gage is inserted in arm $R_4$ of the bridge instead of in arm $R_2$. The active gage remains in arm $R_1$ and fixed-value resistors are used in arms $R_2$ and $R_3$. With this positioning of the dummy gage $r$ is not restricted by the balance condition and the system sensitivity is the same as that given by Eq. (5.2). Temperature compensation is achieved in the same manner that was illustrated in Case 2, but without loss of circuit efficiency. Thus, if a dummy gage is to be used to effect temperature compensation, arm $R_4$ of the bridge is the perferred location for the gage.

*Case 4:*  Four active gages are used in this Wheatstone-bridge arrangement: one active gage in each arm of the bridge (thus, $r = 1$). When the gages are placed on a specimen such as a cantilever beam in bending with tensile strains on gages 1 and 3 (top surface of the beam) and compressive strains on gages 2 and 4 (bottom surface of the beam), then

$$\frac{\Delta R_1}{R_1} = \frac{\Delta R_3}{R_3} = -\frac{\Delta R_2}{R_2} = -\frac{\Delta R_4}{R_4} \tag{c}$$

Substituting Eqs. (c) into Eq. (4.22) gives

$$\Delta E_o = \frac{1}{4} E_i \left(4 \frac{\Delta R_g}{R_g}\right) = E_i \frac{\Delta R_g}{R_g} \tag{5.8}$$

The Wheatstone bridge has added the four resistance changes to increase the output voltage; therefore, the system sensitivity is

$$S_c = \frac{1}{2} (4 S_g \sqrt{P_g R_g}) = 2 S_g \sqrt{P_g R_g} \tag{5.9}$$

This arrangement (with four active gages) has doubled the system sensitivity of Cases 1 and 3 and has quadrupled the sensitivity of Case 2. Also, this bridge is temperature compensated. The use of multiple gages to gain sensitivity is not usually recommended because of the costs involved in the installation of the extra gages. High-quality, high-gain differential amplifiers can be used more economically to increase the output signal.

Examination of the four bridge arrangements shows that the system sensitivity can be varied from 1/2 to 2 times $S_g\sqrt{P_g R_g}$. Temperature compensation is best achieved by placing the dummy gage in position $R_4$ to avoid loss of system sensitivity. System sensitivity can be improved by using multiple gages; however, the costs involved for the added gaging is usually not warranted except for transducer applications, where the additional gages serve other purposes (see Chapter Six).

## 5.5 RECORDING INSTRUMENTS FOR STRAIN-GAGE APPLICATIONS

The selection of a recording system for strain-gage applications depends primarily upon the nature of the strain to be measured (static or dynamic) and upon the number of strain gages to be monitored. Static recording of short-term strain data is generally the easiest and least expensive. Static recording of long-term strain data can be very difficult when it is impossible to return to a bench-mark zero strain condition and long-term stability is required. Dynamic recording is difficult and much more expensive because of noise problems that arise as a result of the higher levels of signal amplification needed for dynamic recording devices and because of the increased complexity of multichannel dynamic recorders.

Many different recording instruments can be used to monitor the output of the Wheatstone bridge. The reader is referred to Chapter Two where these recorders are described in considerable detail. In this section, four different instrumentation systems that have been adapted for strain-gage applications and are frequently used in industry will be described.

### Direct-Reading Strain Indicator

A strain indicator that employs an integrating digital voltmeter to record the system output is shown in Fig. 5.5. This system contains a Wheatstone bridge that is initially balanced by a potentiometer that serves as a parallel-balance resistor. The voltage output from the bridge is amplified and then displayed on a digital voltmeter. A constant-current power supply that can be adjusted to match the power limit $P_g$ of the gage is used to drive the bridge. System calibration is accomplished with shunt resistors. The output of the amplifier is attenuated so that the digital voltmeter directly displays the strain.

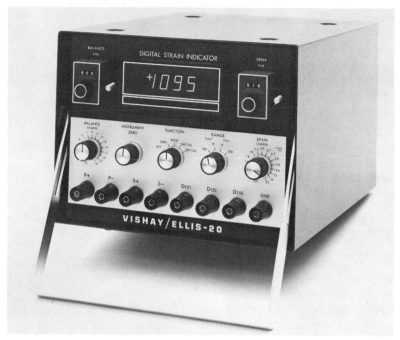

**Figure 5.5** A direct-reading digital strain indicator. (Courtesy of Vishay Instruments.)

Several strain gages can be monitored with this indicator if a separate parallel-balance resistor is provided for each gage and if a switch is provided so that each gage and balance resistor can be switched, in sequence, into the bridge. The switching operation can be performed either manually or automatically with a multiplexer. A digital voltmeter with a multiplexer is capable of providing 10 to 20 readings per second.

The direct-reading indicators can also be used to log data automatically, since the output from the digital voltmeter can be either printed or recorded in digital form on punched tape or magnetic tape. A typical data logger for recording the output of 25 gages automatically is shown in Fig. 5.6. The main advantage of recording in a digital format is the capability provided for computer processing of the data that results in a considerable savings in both costs and time when a large number of gages must be monitored.

## Null-Balance Bridges

The Wheatstone bridges described previously have all been direct-reading types where a recording device is used to measure the output voltage $\Delta E_o$ of the bridge. For static measurements of strain, it is possible to employ a null-balance bridge where the resistance in a nonactive arm is changed to match the resistance change $\Delta R/R$ of the active gage. The null-balance approach is much

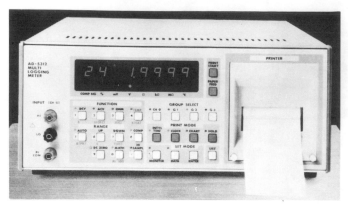

**Figure 5.6** Data logger for automatic recording of static strain-gage signals and the outputs from other sensors. (Courtesy of Soltec Corp.)

slower than the direct-reading methods because of the time required to balance the bridge for each reading; however, this approach provides accurate strain data with low-cost instruments.

The reference bridge, illustrated in Fig. 5.7, is a measuring device that utilizes the null-balance principle. Here, two bridges are used in combination to achieve the null-balance condition. The strain-gage bridge on the left is used for the active gages (one to four), while the reference bridge on the right is used for the variable resistors needed to effect the balance condition. When gages and fixed-value bridge-completion resistors are inserted in the strain-gage bridge, an initial unbalance occurs between the two bridges. This unbalance produces a voltage difference that, after amplification, causes the galvanometer to deflect. The variable resistor in the reference bridge is then adjusted to effect a balance between the bridges (indicated by a zero or null reading on the galvanometer). A reading on the scale associated with the variable resistor (after initial balance is achieved) provides a datum or base reading for subsequent strain measurements. If strains are applied to the gages, an unbalance again occurs and is eliminated by changing the setting of the variable resistor. The difference between the two readings from the scale of the variable resistor provides a measure of the resistance change $\Delta R/R$ experienced by the active gage as a result of the applied strain. The scale on the variable resistor can be constructed to read strain directly.

In actual instruments, the circuits are more complex than the schematic arrangement shown in Fig. 5.7. The bridges are powered by a common oscillator with a 1000-Hz square-wave output of 1.5 V (rms). The voltage to the reference bridge is adjusted with a potentiometer (calibrated by introducing a gage factor setting) so that readout from the scale of the variable resistor is direct in terms of strain.

The null-balance type of strain indicator will function over a range of gage resistances from 50 $\Omega$ to 2000 $\Omega$. With gage resistances less than 50 $\Omega$, the

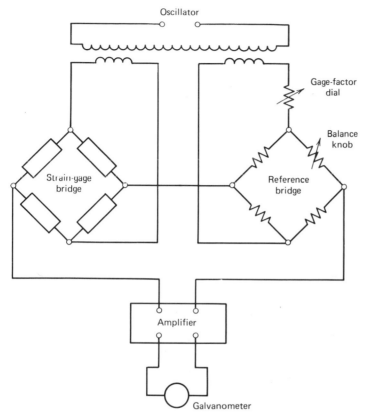

**Figure 5.7**  Schematic diagram of a reference bridge that is commonly used for the measurement of static strains.

oscillator is overloaded; with gage resistances greater than 2000 $\Omega$, the load on the amplifier becomes excessive. The gage factor adjustment will accommodate gages with the range $1.5 \leqslant S_g \leqslant 4.5$. The scale on the variable resistor can be read to $\pm 2$ $\mu$m/m, and is accurate to $\pm 0.1$ percent of the reading or 5 $\mu$m/m whichever is greater. The range of strain that can be measured is $\pm 50,000$ $\mu$m/m. The null-balance instrument shown in Fig. 5.8 is small, light weight, and portable. It is easy to operate and is adequate for all static measurements of strain except those requiring a very large number of gages with extensive data analysis.

## The Wheatstone Bridge and the Oscilloscope

When strain gages are used to measure dynamic strains at only a very few locations, the oscilloscope is probably the best recording instrument. A typical Wheatstone bridge–oscilloscope arrangement is shown schematically in Fig. 5.9.

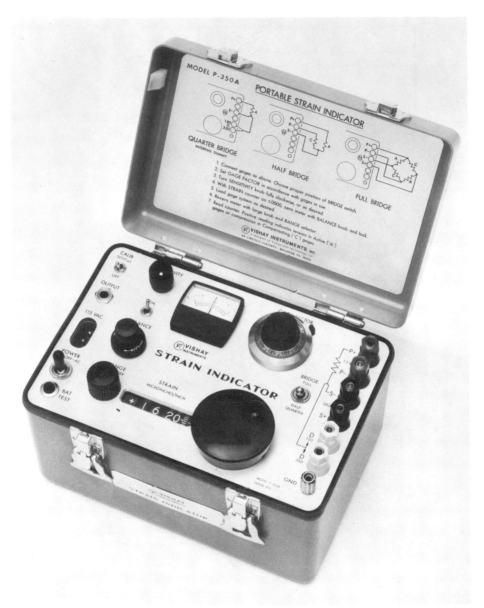

*Figure 5.8* A null-balance type of static strain indicator. (Courtesy of Vishay Instruments.)

The connection from the bridge to the oscilloscope can be direct if a differential amplifier with sufficient gain is available as a plug-in unit for the oscilloscope. Some single-ended amplifiers and power supplies cannot be used, since they ground point *D* of the bridge. This grounding seriously affects the output voltage of the bridge and thus introduces errors in the strain measurements.

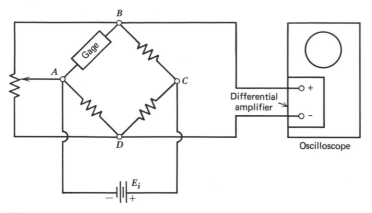

**Figure 5.9** A Wheatstone bridge–oscilloscope strain-measurement system.

The input impedance of an oscilloscope is quite high (about 1 MΩ) and as a consequence, $R_s/R_m < 0.001$ for the Wheatstone bridge–oscilloscope combination and loading errors (see Fig. 2.2) are negligible. The frequency response of an oscilloscope is extremely high and even low-frequency models (1-MHz bandwidth) greatly exceed the requirements for mechanical strain measurements, which rarely exceed 50 kHz. The observation interval depends upon the sweep rate and can range from about 1 μs to 50 s.

Strain as a function of time is displayed as a trace on the face of the cathode-ray tube (CRT). The trace can be photographed, or if a storage or memory oscilloscope is used, readings of voltage and/or strain can be taken directly from the CRT. With a conventional oscilloscope, strain ε is computed from the height $d_s$ of the strain-time pulse, as illustrated in Fig. 5.10, and the distance $d_c$ between two calibration lines produced by shunt calibration (see Section 5.6) of the bridge. The strain experienced by the gage is

$$\varepsilon = \frac{d_s}{d_c}\varepsilon_c \qquad (5.10)$$

where $\varepsilon_c$ is the equivalent strain produced by shunt calibration.

If the shunt calibration technique is not used, the strain must be computed by using the output voltage from the bridge. For example, consider a single

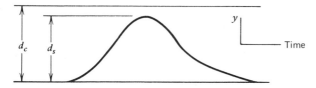

**Figure 5.10** Determining strain from an oscilloscope trace.

gage. From Eqs. (4.22) and (3.5), the output voltage from the bridge will be

$$\Delta E_o = \frac{r}{(1 + r)^2} E_i S_g \varepsilon \qquad (a)$$

This output voltage can be expressed in terms of oscilloscope parameters as

$$\Delta E_o = S_R d_s \qquad (b)$$

**where**    $d_s$ is the height of the strain-time pulse (in CRT divisions, see Fig. 5.10).

$S_R$ is the sensitivity of the oscilloscope in volts per division.

Substituting Eq. (b) into Eq. (a) and solving for the strain gives

$$\varepsilon = \frac{(1 + r)^2}{r} \frac{S_R d_s}{E_i S_g} \qquad (5.11)$$

## The Wheatstone Bridge and the Oscillograph

The oscillograph (see Section 2.5) is also used with the Wheatstone bridge for dynamic strain measurements. The oscillograph is preferred over the oscilloscope[1] when large numbers of strain gages must be monitored and when the observation period is relatively long. When an oscillograph is used for dynamic strain measurements, care must be exercised in selection of the galvanometers, since many galvanometers are satisfactory only for low-frequency signals.

When a galvanometer of the type used in an oscillograph is connected directly to a Wheatstone bridge, the output voltage of the bridge is seriously affected. These galvanometers have a very low input impedance (between 30 and 300 Ω); therefore, the equations developed previously, which were based upon use of a high-impedance recorder, must be modified to account for the characteristics of the galvanometer. A circuit diagram for an oscillograph connected directly to a Wheatstone bridge is shown in Fig. 5.11a. An equivalent circuit (obtained by using Thevenin's theorem), consisting of an equivalent resistance $R_B$ and a voltage source $\Delta E_o$ replacing the bridge, is shown in Fig. 5.11b. The Wheatstone bridge–oscillograph circuit must be designed such that the equivalent resistance $R_B$ provides the external damping resistance $R_x$ required by the galvanometer to maintain frequency response. Also, the bridge should be designed to produce the largest possible output current $I_G$ per unit strain in order to maximize the deflection θ of the galvanometer.

---

[1] Memory oscilloscopes with magnetic storage also provide very long observation periods together with a capability for expanding the time scale to improve resolution; however, these oscilloscopes are relatively new and quite expensive. Until costs are reduced, oscillographs will continue to be widely used for dynamic recording of strain.

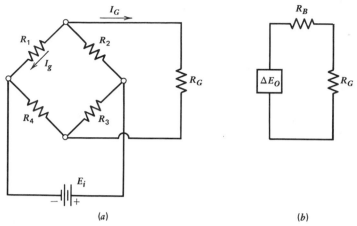

**Figure 5.11**  A Wheatstone bridge–oscillograph strain measurement system. (*a*) Schematic diagram with $R_G$ representing the input impedance of the galvanometer. (*b*) Equivalent circuit for determining $I_G$.

The equivalent circuit shown in Fig. 5.11*b* can be used to determine the current $I_G$ passing through the galvanometer in terms of strain gage and circuit parameters. Consider an initially balanced bridge where $R_1 R_3 = R_2 R_4$, and define resistance ratios $r$ and $q$ as

$$r = \frac{R_2}{R_1} \quad \text{and} \quad q = \frac{R_2}{R_3} \tag{a}$$

The value of the equivalent resistance $R_B$ can be obtained from Eq. (4.29) if the resistance of the voltage supply is small ($R_V \approx 0$). Substituting Eq. (a) into Eq. (4.29) gives

$$R_B = \left(\frac{r}{1 + r}\right)\left(\frac{1 + q}{q}\right) R_1 \tag{5.12}$$

If the bridge contains one active gage in arm $R_1$, Eqs. (4.22) and (3.5) can be combined to give

$$\Delta E_o = \frac{r}{(1 + r)^2} E_i S_g \varepsilon \tag{5.13}$$

Since the voltage $E_i$ is limited by the power $P_g$, that can be dissipated by the gage, $E_i$ is given by Eqs. (5.3) and (5.5) as

$$E_i = (1 + r) \sqrt{P_g R_g} \tag{b}$$

Substituting Eq. (b) into Eq. (5.13) gives

$$\Delta E_o = \frac{r}{1 + r} \sqrt{P_g R_g} S_g \varepsilon \tag{5.14}$$

For the equivalent circuit of Fig. 5.11$b$, the current $I_G$ is given by

$$I_G = \frac{\Delta E_o}{R_B + R_G} = \frac{\Delta E_o}{R_B} \frac{R_B}{R_B + R_G} \tag{c}$$

If Eqs. (5.12) and (5.14) are now substituted into Eq. (c), the current $I_G$ is given by the expression

$$I_G = \frac{q}{1 + q} \sqrt{\frac{P_g}{R_g}} S_g \varepsilon \frac{R_B}{R_B + R_G} \tag{5.15}$$

The deflection $\theta$ of the galvanometer is given by Eq. (2.14) as

$$\theta = S_G I_G \tag{2.14}$$

where $S_G$ is the galvanometer sensitivity. System sensitivity is defined as

$$S_s = \frac{\theta}{\varepsilon} \tag{5.16}$$

Therefore,

$$S_s = \frac{q}{1 + q} \sqrt{\frac{P_g}{R_g}} S_g \frac{R_B}{R_B + R_G} S_G$$

or

$$S_s = \frac{q}{1 + q} \sqrt{\frac{P_g}{R_g}} S_g \frac{1}{1 + \dfrac{R_G}{R_B}} S_G \tag{5.17}$$

Equation (5.17) shows that the system sensitivity is controlled by

1. Circuit efficiency [$q/(1 + q)$].
2. Gage selection ($\sqrt{P_g/R_g} S_g$).

3. The ratio of galvanometer resistance to equivalent bridge resistance ($R_G/R_B$).
4. Galvanometer sensitivity ($S_G$).

The constraint that $R_B$ must provide the required external resistance for proper damping of the galvanometer limits the options available for maximizing system sensitivity. Most galvanometers used for strain measurement are designed (see Table 2.2) so that the required external resistance $R_x$ equals 120 or 350 $\Omega$. The equivalent resistance of a four-equal-arm bridge (i.e., $R_1 = R_2 = R_3 = R_4$) with 120- or 350-$\Omega$ gages is 120 and 350 $\Omega$, respectively. Thus, $R_B = R_g$ and $q$ is limited to a value of 1. Under these conditions, Eq. (5.17) reduces to

$$S_s = \frac{1}{2}\sqrt{\frac{P_g}{R_g}}S_g \frac{1}{1 + \frac{R_G}{R_g}}S_G \tag{5.18}$$

If the equivalent resistance of the bridge is less than the required external resistance for the galvanometer, which is often the case, a series resistor $R_s$ must be added to the circuit between the bridge and the galvanometer. The value of the series resistor is given by

$$R_s = R_x - R_B$$

When the series resistor is added, the circuit sensitivity given by Eq. (5.17) becomes

$$S_s = \frac{q}{1+q}\sqrt{\frac{P_g}{R_g}}S_g \frac{R_B}{R_B + R_G + R_s}S_g \tag{5.19}$$

If the external resistance is maintained at the value specified in Table 2.2, the frequency response of the system will be flat within a $\pm 5$ percent accuracy band over the frequency range $0 \leqslant \omega \leqslant 0.87\, \omega_n$.

In many cases, the system sensitivity achieved by connecting a galvanometer with adequate frequency response to the Wheatstone bridge is too low to provide sufficient galvanometer deflection. In these instances, a current amplifier must be inserted between the bridge and the galvanometer to provide the current required for adequate response. Such amplifiers are designed to provide an output impedance equal to or less than the required external resistance for most galvanometers. When the output impedance is less than $R_x$, a series resistor is used on the output of the amplifier for proper matching.

Strain is determined from the oscillograph record by using Eq. (5.10). In all instances, the system is calibrated either by applying a known strain to a gage in arm $R_1$ of the Wheatstone bridge or by applying an equivalent strain with a shunt resistor across arm $R_2$ of the bridge. Calibration methods are described in Section 5.6.

## 5.6  CALIBRATION METHODS

A strain-measurement system (see Fig. 5.12) usually includes one or more strain gages, a power supply, circuit-completion resistors, an amplifier, and a voltage or current measuring instrument. It is possible to calibrate such a system by precisely measuring $R_1$, $R_2$, $R_3$, $R_4$, $E_i$, sensitivity or gain of the amplifier, and sensitivity of the recorder. The calibration constant $C$ for the system can then be computed by using the expression

$$C = \frac{(1 + r)^2 S_R}{r E_i S_g G} \tag{5.20}$$

**where**   $G$  is the gain of the amplifier.
         $S_R$  is the recorder sensitivity (volts per division).

The strain recorded with the system is given in terms of the calibration constant as

$$\varepsilon = C d_s \tag{5.21}$$

where $d_s$ is the deflection of the recorder in divisions. This procedure is time consuming and is subject to measurement errors in each of the quantities appearing in Eq. (5.20). A more direct, less time-consuming, and more accurate

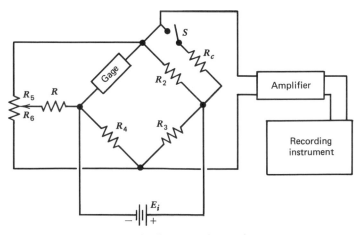

***Figure 5.12***   A schematic diagram of a strain-measurement system.

may be accomplished by introducing a known strain in the bridge (either mechanically or electrically), measuring $d_s$ resulting from this strain, and computing $C$ by using Eq. (5.21).

Mechanical calibration is performed by mounting a strain gage (which must have the same gage factor $S_g$ as the gages being employed for the measurements) on a calibration specimen (usually a cantilever beam), connecting this calibration gage into arm $R_1$ of the Wheatstone bridge, and observing the deflection of the trace on the recorder as a known strain is applied to the gage. If the free end of a cantilever beam is deflected a distance $\delta$, the calibration strain $\varepsilon_c$ induced in the calibration gage is

$$\varepsilon_c = \frac{3\delta hx}{2l^3} \tag{5.22}$$

**where**   $h$ is the depth of the cantilever beam.
  $l$ is the length of the cantilever beam.
  $x$ is the distance from the load point to the center of the gage.

The voltage output from an initially balanced bridge is recorded before and after the beam is deflected (for example, note the two horizontal traces shown in Fig. 5.10). The distance between these two lines $d_c$ is used with the calibration strain $\varepsilon_c$ to determine the calibration constant $C$. Thus,

$$C = \frac{\varepsilon_c}{d_c} \tag{5.23}$$

Electrical calibration is performed in a similar manner, except that the calibration strain is induced by shunting a calibration resistor $R_c$ across arm $R_2$ of the Wheatstone bridge, as shown in Fig. 5.12. The effective resistance of arm $R_2$ with $R_c$ in place is

$$R_{2e} = \frac{R_2 R_c}{R_2 + R_c} \tag{5.24}$$

The change of resistance $\Delta R_2 / R_2$ is then given by

$$\frac{\Delta R_2}{R_2} = \frac{R_{2e} - R_2}{R_2} = -\frac{R_2}{R_2 + R_c} \tag{5.25}$$

The output voltage produced by shunting $R_c$ across $R_2$ of the bridge is obtained by substituting Eq. (5.25) into Eq. (4.21). Thus

$$\Delta E_o = E_i \frac{R_1 R_2}{(R_1 + R_2)^2} \frac{R_2}{R_2 + R_c} \tag{a}$$

The output from a single active gage in arm $R_1$ of a bridge due to a strain equal to the calibration strain is given by Eq. (4.21) as

$$\Delta E_o = E_i \frac{R_1 R_2}{(R_1 + R_2)^2} S_g \varepsilon_c \tag{b}$$

Equating Eqs. (a) and (b) and solving for $\varepsilon_c$ gives

$$\varepsilon_c = \frac{R_2}{S_g(R_2 + R_c)} \tag{5.26}$$

Once $\varepsilon_c$ is known (calculated), the calibration constant $C$ can be determined by following the procedure outlined previously for mechanical calibration.

This technique of shunt calibration is accurate and simple to use. It provides a single calibration constant for the complete system, which incorporates the sensitivities of all components. The calibration strain produces a deflection $d_c$ on the recording instrument that is used to determine the calibration constant $C$. All other deflections on the recording instrument are linearly related to the unknown strains that produced them by this calibration constant as indicated in Eq. (5.21).

## 5.7  EFFECTS OF LEAD WIRES, SWITCHES, AND SLIP RINGS

The resistance change from a strain gage is small; therefore, any disturbance that produces a resistance change within the bridge circuit is extremely important, since it also affects the output voltage. Components within the bridge typically include gages, soldered joints, terminals, lead wires, and binding posts. Frequently, switches and slip rings are also included. The effects of lead wires, switches, and slip rings are the most important; therefore, they will be covered individually in this section. The effects of soldered joints, terminals, and binding posts must not be neglected since they can also produce significant errors; however, if cold-soldered connections are avoided and if binding posts are tight, joint resistance will be constant and negligibly small.

### Lead Wires

Frequently, a strain gage is mounted on a component that is located a considerable distance from the bridge and recording system. The gage must be connected to the bridge with two long lead wires, as shown in Fig. 5.13. With this arrangement, two detrimental effects occur: signal attenuation and loss of temperature compensation. Both can seriously compromise the accuracy of the measurements.

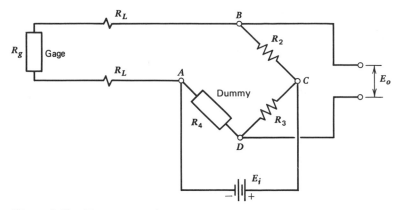

**Figure 5.13** Gage connection to the bridge with two long lead wires.

Signal attenuation or loss due to the resistance of the two lead wires can be determined by noting in Fig. 5.13 that

$$R_1 = R_g + 2R_L \qquad (a)$$

where $R_L$ is the resistance of a single lead wire. The added resistance in arm $R_1$ of the bridge (due to the lead wires) leads to the expression

$$\frac{\Delta R_1}{R_1} = \frac{\Delta R_g}{R_g + 2R_L} = \frac{\Delta R_g}{R_g} \frac{R_g}{R_g + 2R_L}$$

$$= \frac{\Delta R_g}{R_g} \frac{1}{1 + (2R_L/R_g)} \qquad (b)$$

Equation (b) can be rewritten in terms of a signal loss factor $\mathcal{L}$ as

$$\frac{\Delta R_1}{R_1} = \frac{\Delta R_g}{R_g}(1 - \mathcal{L}) \qquad (c)$$

where

$$\mathcal{L} = \frac{2R_L}{R_g + 2R_L} = \frac{2R_L/R_g}{1 + (2R_L/R_g)} \qquad (5.27)$$

Signal loss factor $\mathcal{L}$ is shown as a function of resistance ratio $R_L/R_g$ in Fig. 5.14. Error due to lead wires can be reduced to less than 1 percent if $R_L/R_g \leq 0.005$. The resistance of a 100-ft (30.5-m) length of solid copper lead wire as a function of gage size is listed in Table 5.2. It is obvious from the data in Table 5.2 that long lengths of small-diameter wire (large gage numbers) must be avoided in strain-gage measurements.

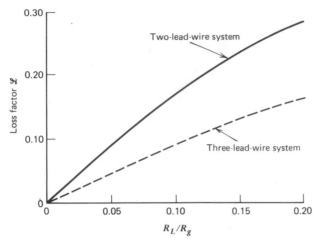

**Figure 5.14**  Loss factor $\mathscr{L}$ as a function of resistance ratio $R_L/R_g$ for two- and three-lead-wire systems.

The second detrimental effect resulting from long lead wires is loss of temperature compensation. As an example, consider a Wheatstone bridge with an active gage and two long lead wires in arm $R_1$ and a dummy gage with two short lead wires in arm $R_4$. If both gages and all lead wires are subjected to the same temperature change $\Delta T$ during the time interval when strain is being monitored, the output of the bridge is given by Eq. (4.22) as

$$\Delta E_o = E_i \frac{r}{(1 + r)^2} \left[ \left( \frac{\Delta R_g}{R_g + 2R_L} \right)_\varepsilon + \left( \frac{\Delta R_g}{R_g + 2R_L} \right)_{\Delta T} \right.$$

$$\left. + \left( \frac{2\Delta R_L}{R_g + 2R_L} \right)_{\Delta T} - \left( \frac{\Delta R_g}{R_g} \right)_{\Delta T} \right] \tag{5.28}$$

**TABLE 5.2**  Resistance of Solid-Conductor Copper Wire Ohms per 100 ft (30.5 m)

| Gage Size | Resistance | Gage Size | Resistance |
|-----------|-----------|-----------|-----------|
| 12 | 0.159 | 28 | 6.490 |
| 14 | 0.253 | 30 | 10.310 |
| 16 | 0.402 | 32 | 16.41 |
| 18 | 0.639 | 34 | 26.09 |
| 20 | 1.015 | 36 | 41.48 |
| 22 | 1.614 | 38 | 65.96 |
| 24 | 2.567 | 40 | 104.90 |
| 26 | 4.081 | | |

The first term in the brackets is the resistance change in the active gage due to the strain.

The second term is the resistance change in the active gage resulting from the temperature change.

The third term is the resistance change in the lead wires of arm $R_1$ resulting from the temperature change.

The fourth term is the resistance change in the dummy gage resulting from the temperature change.

The resistance change in the lead wires of arm $R_4$ is negligible.

In this example, temperature compensation is not achieved, since the second and fourth terms do not cancel. Significant additional error due to resistance changes in the lead wires is represented by the third term in Eq. (5.28).

The detrimental effects of long lead wires can be reduced by employing the simple three-wire system illustrated in Fig. 5.15. With this three-wire arrangement, both the active gage and the dummy gage are located at the remote site. One of the three wires is not considered a lead wire, since it is not within the bridge (not in arm $R_1$ or $R_4$) and serves only to transfer point $A$ of the bridge to the remote location. The active and dummy gages each have one long lead wire with resistance $R_L$ and one very short lead wire with negligible resistance. The signal loss factor $\mathcal{L}$ for the three-wire system is

$$\mathcal{L} = \frac{R_L/R_g}{1 + (R_L/R_g)} \tag{5.29}$$

A comparison of Eqs. (5.27) and (5.29) indicates that signal attenuation due to lead-wire resistance is reduced by a factor of approximately 2 by using the three-wire system (also see Fig. 5.14).

The temperature-compensating feature of the Wheatstone bridge is retained

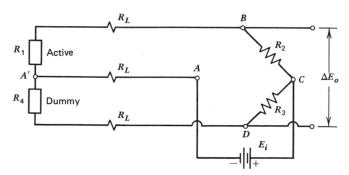

*Figure 5.15*   Gage connections to the bridge with the three-lead-wire system.

when the three-wire system is used. In this case, Eq. (5.28) becomes

$$
\begin{aligned}
\Delta E_o = E_i \frac{r}{(1 + r)^2} &\left[ \left( \frac{\Delta R_g}{R_g + R_L} \right)_\varepsilon + \left( \frac{\Delta R_g}{R_g + R_L} \right)_{\Delta T} \right. \\
&\left. + \left( \frac{\Delta R_L}{R_g + R_L} \right)_{\Delta T} - \left( \frac{\Delta R_g}{R_g + R_L} \right)_{\Delta T} - \left( \frac{\Delta R_L}{R_g + R_L} \right)_{\Delta T} \right]
\end{aligned} \quad (5.30)
$$

It is clear from Eq. (5.30) that temperature compensation is achieved since all of the temperature-dependent terms in the bracketed quantity cancel.

In all cases where lead-wire resistance causes measurable signal attenuation, the calibration resistor should be applied at the remote active gage in order to establish experimentally the effects of the lead wires. Once this effect is established, the calibration resistor can be inserted as shown in Fig. 5.12 to check overall system performance.

## Switches

Frequently, a large number of gages are used to evaluate a component or structure and the output of each gage is read several times during a typical test. In this type of application, the number of gages is too large to employ a separate recording system for each gage. Instead, a single recording system is used and the gages are switched in and out of the system according to some schedule. Two different switching arrangements are commonly used today for multiple-gage installations.

The most common and least expensive arrangement is illustrated in Fig. 5.16. Here, one side of each active gage is switched, in turn, into arm $R_1$ of the bridge, while the other side of each active gage is connected to terminal $A$ of the bridge with a common lead wire. This arrangement places the switch in arm $R_1$ of the bridge; therefore, a high-quality switch with a small reproducible resistance (less than 500 $\mu\Omega$) must be employed. Low resistance is achieved by using silver-tipped contacts and two or more parallel contacts per switch. If the switch resistance is not reproducible, the change in switch resistance $\Delta R_s$ adds to the strain-induced change in gage resistance $\Delta R_g$ to produce an apparent strain $\varepsilon'$, which can be expressed as

$$
\varepsilon' = \frac{\Delta R_s / R_g}{S_g} \quad (5.31)
$$

The quality of a switch can be easily checked, since a nonreproducible switch resistance results in a shifting of the zero reading. Switches must be cleaned regularly, since even high-quality switches will begin to perform erratically when the contacts become dirty or oxidized.

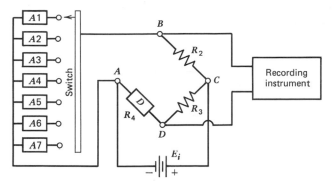

**Figure 5.16**  Switching a large number of individual gages into arm $R_1$ of the Wheatstone bridge with a single-pole switch.

A second switching arrangement is shown in Fig. 5.17. Here, a three-pole switch is used to transfer terminals $A$, $B$, and $D$ to the power supply and the recording instrument. Terminal $C$ of each bridge is grounded in common with the power supply with a single common lead wire. Since none of the switches are located within the bridge, switch resistance is not important; however, switching the complete bridge is more expensive, since separate dummy gages and two bridge completion resistors are required for each bridge.

A major disadvantage of all switching schemes is the thermal drift induced by heating of the gages and resistors when power is suddenly applied to the system. Depending upon the application, this drift may continue for a minute or more after the switch is closed.

## Slip Rings

When strain gages are used on rotating members, slip rings are often used to complete the lead-wire connections, as shown in Fig. 5.18. The slip rings are usually mounted on a shaft that can be attached to the rotating member so that the axes of rotation of the shaft and member coincide. The outer shell of the slip ring assembly is stationary and carries several brushes per ring to transfer the signal from the rotating rings to terminals on the stationary shell. Satisfactory operation up to speeds of 24,000 rpm is possible with a properly designed slip ring assembly.

Brush movement and dirt collecting on the slip rings due to brush wear tend to produce a change in resistance that can be reduced by using multiple brushes in parallel. Even with multiple brushes, however, changes in resistance between rings and brushes tend to be large; therefore, slip rings should not be placed within the arms of the bridge. Instead, a complete bridge should be assembled for each active gage on the rotating member, as shown in Fig. 5.18. The slip rings should be used only to connect the bridge to the power supply and the recording instrument. This arrangement minimizes the effect of resist-

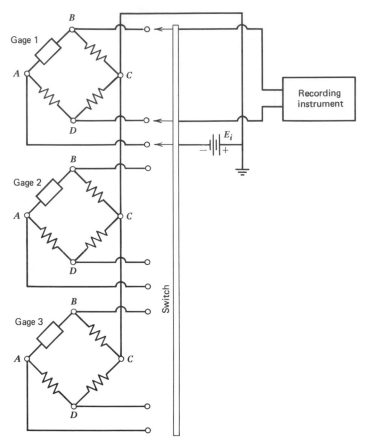

***Figure 5.17*** Switching several complete bridges into the power supply and recording instrument with a three-pole switch.

ance change due to the slip rings $\Delta R_{sr}$ and provides a means for accurately recording strain-gage signals from rotating members.

## 5.8  ELECTRICAL NOISE

The output voltage from a Wheatstone bridge due to the resistance change $\Delta R/R$ of a strain gage (or other transducer) is usually quite small (a few millivolts). Because of this very small output voltage $\Delta E_o$, electrical noise is frequently a problem. Electrical noise occurs as a result of magnetic fields generated by current flow in wires in close proximity to the lead wires or bridge, as shown in Fig. 5.19. When an alternating current flows in an adjacent wire, a cyclic magnetic field (frequently 60 Hz) is produced, which cuts both wires of the signal circuit and induces a voltage (noise) in the signal loop. The magnitude of this induced voltage (noise) is proportional to the current $I$ flowing in the disturbing

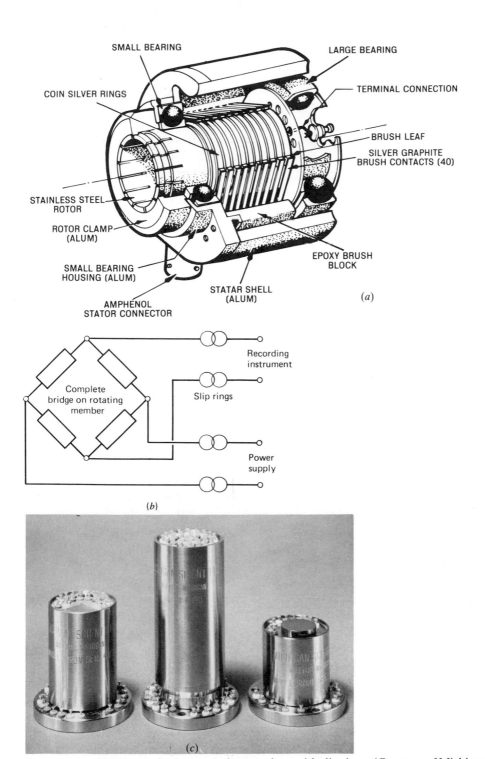

**Figure 5.18** Signal transfer from rotating members with slip rings. (Courtesy of Michigan Scientific Corp.) (*a*) Construction details of a slip ring unit. (*b*) Schematic diagram of a strain measurement system with slip rings. (*c*) Several slip-ring units.

**221**

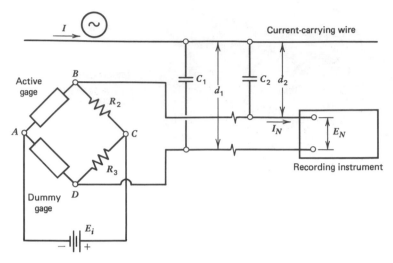

*Figure 5.19* Schematic diagram showing generation of electrical noise.

wire and the area enclosed by the signal loop, and inversely proportional to the distance between the disturbing wire and the strain-gage lead wires (see Fig. 5.19). In some cases, the voltage induced by the magnetic field is so large that it is difficult to separate the noise from the strain-gage signal.

Three precautions can be taken to minimize noise. First, all lead wires should be twisted or placed adjacent to each other to minimize the area of the signal loop. Second, only shielded cables should be used, and the shields should be grounded only at the negative terminal of the power supply to the bridge, as shown in Fig. 5.20. With this arrangement, the shield is grounded without forming a ground loop and any noise voltage generated in the shield is maintained at nearly zero potential. The power supply itself should be floated relative to

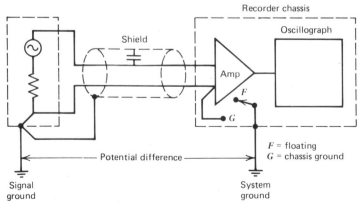

*Figure 5.20* Shielding and single-point grounding to eliminate noise.

the system ground (the third conductor in the power cord) to avoid a ground loop at the supply. The third method for reducing noise is by common-mode rejection. If the lead wires are twisted, any noise developed will be equal and will occur simultaneously in both lead wires. If a differential amplifier is used with the Wheatstone bridge, the noise signals are rejected (canceled by the amplifier) and only the strain signal is amplified. Unfortunately, common-mode rejection even for very-high-quality differential amplifiers is not perfect, and a small portion of the noise voltage is transmitted by the amplifier. Common-mode rejection for good-quality, low-level data amplifiers is about $10^6$ to 1 at 60 Hz; therefore, most of the noise is suppressed.

If the three previously listed precautions are followed, the signal-to-noise ratio can be maximized and clean data can be recorded even under adverse electrical conditions.

## 5.9 TEMPERATURE-COMPENSATED GAGES

Temperature compensation in the Wheatstone bridge was discussed in Section 5.4; however, temperature compensation of the gage itself is possible and in practice both the bridge and the gage should be compensated to nullify any signal due to temperature variations during the readout interval. When the ambient temperature changes, four effects occur that influence the signal $\Delta R/R$ from the gage:

1. The gage factor $S_g$ changes with temperature.
2. The grid undergoes an elongation or contraction ($\Delta l/l = \alpha \Delta T$).
3. The specimen elongates or contracts ($\Delta l/l = \beta \Delta T$).
4. The resistance of the gage changes ($\Delta R/R = \gamma \Delta T$).

The strain sensitivities $S_A$ of the two most commonly used alloys (Advance and Karma) are linear functions of temperature as shown in Fig. 5.21. These plots indicate that $\Delta S_A/\Delta T$ equals 0.00735 and $-0.00975$ percent per Celsius degree for Advance and Karma alloys, respectively. Since these changes are small (less than 1 percent for $\Delta T = 100°C$), variations in $S_A$ with temperature are usually neglected in routine stress analysis work; however, in thermal stress studies where temperature variations of several hundred degrees are common, changes in $S_A$ become significant and must be considered.

Effects 2, 3, and 4 are much more significant and combine to produce a change in resistance of the gage with temperature change $(\Delta R/R)_{\Delta T}$, which can be expressed as

$$\left(\frac{\Delta R}{R}\right)_{\Delta T} = (\beta - \alpha)S_g\Delta T + \gamma\Delta T \tag{5.32}$$

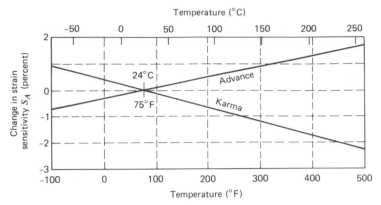

**Figure 5.21** Alloy sensitivity as a function of temperature for Advance and Karma alloys.

**where**   $\alpha$  is the thermal coefficient of expansion of the gage alloy.
          $\beta$  is the thermal coefficient of expansion of the specimen material.
          $S_g$ is the gage factor.
          $\gamma$  is the temperature coefficient of resistivity of the gage alloy.

A differential expansion between the gage grid and the specimen due to a temperature change ($\alpha \neq \beta$) subjects the gage to a thermally induced mechanical strain $\varepsilon_T = (\beta - \alpha)\Delta T$, which does not occur in the specimen. The gage responds to the strain $\varepsilon_T$ in the same way that it responds to a load-induced strain $\varepsilon$ in the specimen. Unfortunately, it is impossible to separate the component of the response due to temperature change from the response due to the load.

If the gage alloy is matched to the specimen ($\alpha = \beta$), the first term in Eq. (5.32) will not produce a response; however, the second term will produce a response that indicates an apparent strain that does not exist in the specimen. A temperature-compensated gage is obtained only if both terms in Eq. (5.32) are zero or if they cancel.

The values of $\alpha$ and $\gamma$ are quite sensitive to the composition of the alloy and to the degree of cold working imparted during the rolling of the foil. It is common practice for the strain-gage manufacturers to measure the thermal-response characteristics of a few gages from each roll of foil that they use in manufacturing the gages. Because of variations in $\alpha$ and $\gamma$ between melts and rolls of foil, it is possible to select gages that are temperature compensated for almost any specimen material. These gages are known as selected-melt or temperature-compensated gages.

Unfortunately, selected-melt gages are not perfectly compensated over a wide range of temperature because of nonlinear terms that were omitted in Eq. (5.32). A typical selected-melt strain gage exhibits an apparent strain with temperature as shown in Fig. 5.22. The apparent strain produced by a temperature

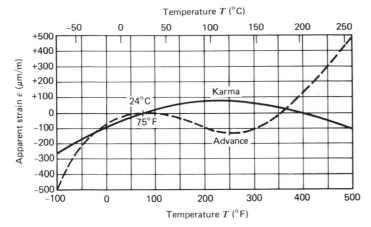

**Figure 5.22** Apparent strain as a function of temperature for Karma and Advance alloys.

change of a few degrees in the neighborhood of 75°F (24°C) is quite small (less than 0.5 μm/m/°C); however, when the temperature change is large, the apparent strain generated by the gage becomes large and corrections to account for apparent strain must be made.

## 5.10 ALLOY SENSITIVITY, GAGE FACTOR, AND CROSS-SENSITIVITY FACTOR

The sensitivity of a single, uniform length of conductor to strain was defined (see Section 3.4) as

$$S_A = \frac{dR/R}{\varepsilon} = \frac{\Delta R/R}{\varepsilon} \tag{3.4 bis}$$

where $S_A$ is the alloy sensitivity. In a typical strain gage, the conductor is formed into a pattern (commonly referred to as the grid) to keep the gage length short. Also, the conductor is usually not uniform over its entire length. As a result, the alloy sensitivity $S_A$ is not a true calibration constant for a strain gage.

A better understanding of the response of a grid-type strain gage can be obtained by considering a gage mounted on a specimen that is subjected to a biaxial strain field. For this situation,

$$\frac{\Delta R}{R} = S_a \varepsilon_a + S_t \varepsilon_t + S_s \gamma_{at} \tag{5.33}$$

**where**    $\varepsilon_a$  is the normal strain along the axial direction of the gage.
     $\varepsilon_t$  is the normal strain along the transverse direction of the gage.

$\gamma_{at}$ is the shearing strain associated with the $a$ and $t$ directions.
$S_a$ is the sensitivity of the gage to axial strain.
$S_t$ is the sensitivity of the gage to transverse strain.
$S_s$ is the sensitivity of the gage to shearing strain.

In general, the gage sensitivity to shearing strain is small and can be neglected. The gage sensitivity to transverse strain may or may not be small. A transverse sensitivity factor $K_t$ is provided by the manufacturer for each gage and is defined as

$$K_t = \frac{S_t}{S_a} \tag{5.34}$$

If Eq. (5.34) is substituted into Eq. (5.33) and if $S_s = 0$,

$$\frac{\Delta R}{R} = S_a(\varepsilon_a + K_t\varepsilon_t) \tag{5.35}$$

The sensitivity of strain gages is usually expressed in terms of a gage factor $S_g$ as previously indicated in Eq. (3.5).

$$\frac{\Delta R}{R} = S_g\varepsilon_a \tag{3.5 bis}$$

This gage factor $S_g$ is determined by the manufacturer for each lot of gages by mounting sample gages drawn from the lot on a calibration beam having a Poisson's ratio $\nu_o = 0.285$. With this method of calibration, the transverse strain present during the determination of $S_g$ is

$$\varepsilon_t = -\nu_o\varepsilon_a \tag{5.36}$$

The response of the gage in calibration is obtained by substituting Eq. (5.36) into Eq. (5.35). Thus

$$\frac{\Delta R}{R} = S_a\varepsilon_a(1 - \nu_o K_t) \tag{5.37}$$

A comparison of Eqs. (5.37) and (3.5) indicates that the gage factor $S_g$ can be expressed in terms of $S_a$ and $K_t$ as

$$S_g = S_a(1 - \nu_o K_t) \tag{5.38}$$

The simplified form of the $\Delta R/R$ versus $\varepsilon_a$ relationship given by Eq. (3.5) is usually used to interpret strain-gage response. It is very important to recognize that this equation is approximate unless either $K_t = 0$ or $\varepsilon_t = 0$. The magnitude of the error incurred by using Eq. (3.5) (the approximate relationship) can be determined by considering the response of a gage in a general biaxial strain field. If Eq. (5.38) is substituted into Eq. (5.35), the gage response is given as

$$\frac{\Delta R}{R} = \frac{S_g \varepsilon_a}{1 - v_o K_t} \left(1 + K_t \frac{\varepsilon_t}{\varepsilon_a}\right) \tag{5.39}$$

The true value of strain $\varepsilon_a$ can then be written as

$$\varepsilon_a = \frac{\Delta R/R}{S_g} \left[\frac{1 - v_o K_t}{1 + K_t(\varepsilon_t/\varepsilon_a)}\right] \tag{5.40}$$

The apparent strain $\varepsilon_a'$ obtained by using Eq. (3.5) (approximate relationship) is

$$\varepsilon_a' = \frac{\Delta R/R}{S_g} \tag{a}$$

Substituting Eq. (a) into Eq. (5.40) gives

$$\varepsilon_a = \varepsilon_a' \left[\frac{1 - v_o K_t}{1 + K_t(\varepsilon_t/\varepsilon_a)}\right] \tag{5.41}$$

The percent error $\mathscr{E}$ incurred by neglecting the transverse sensitivity of a strain gage in a general biaxial strain field is obtained from Eqs. (5.40) and (a) as

$$\mathscr{E} = \frac{\varepsilon_a' - \varepsilon_a}{\varepsilon_a}(100) = \frac{K_t(\varepsilon_t/\varepsilon_a + v_o)}{1 - v_o K_t}(100) \tag{5.42}$$

Some representative values of $\mathscr{E}$ as a function of $K_t$ for different biaxiality ratios $\varepsilon_t/\varepsilon_a$ are illustrated in Fig. 5.23.

Two different procedures are used to correct for the error involved with the use of Eq. (3.5). First, if the biaxiality ratio $\varepsilon_t/\varepsilon_a$ is known (in a thin-walled cylinder or sphere under internal pressure, for example), the bracketed term in Eq. (5.41) can be viewed as a correction factor $C_f$ that modifies the apparent strain $\varepsilon_a'$ to give the true strain $\varepsilon_a$. The factor by which all apparent strain values must be multiplied to give true strains is

$$C_f = \frac{1 - v_o K_t}{1 + K_t(\varepsilon_t/\varepsilon_a)} \tag{5.43}$$

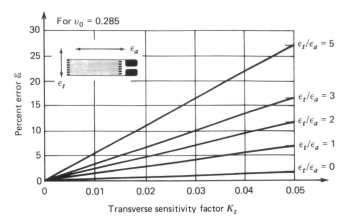

**Figure 5.23**   Percent error as a function of transverse sensitivity factor for different ratios of transverse to axial strain.

Alternatively, a corrected gage factor $S_g^*$ can be used in place of $S_g$ to adjust the bridge before readings are taken. The corrected gage factor as determined from Eqs. (3.5) and (5.41) is

$$S_g^* = S_g \left[ \frac{1 + K_t(\varepsilon_t/\varepsilon_a)}{1 - v_o K_t} \right] \tag{5.44}$$

The second correction procedure is used when the biaxiality ratio $\varepsilon_t/\varepsilon_a$ is not known. If apparent strains $\varepsilon'_{xx}$ and $\varepsilon'_{yy}$ are recorded in orthogonal directions, then from Eqs. (5.39) and (3.5)

$$\varepsilon'_{xx} = \frac{1}{1 - v_o K_t} (\varepsilon_{xx} + K_t \varepsilon_{yy})$$

$$\varepsilon'_{yy} = \frac{1}{1 - v_o K_t} (\varepsilon_{yy} + K_t \varepsilon_{xx}) \tag{5.45}$$

Solving Eqs. (5.45) for the true strains $\varepsilon_{xx}$ and $\varepsilon_{yy}$ yields

$$\varepsilon_{xx} = \frac{1 - v_o K_t}{1 - K_t^2} (\varepsilon'_{xx} - K_t \varepsilon'_{yy})$$

$$\varepsilon_{yy} = \frac{1 - v_o K_t}{1 - K_t^2} (\varepsilon'_{yy} - K_t \varepsilon'_{xx}) \tag{5.46}$$

## 5.11   THE STRESS GAGE

While the transverse sensitivity of a strain gage is detrimental in that it complicates the analysis of strain-gage response, it can be used to advantage in the design of another related transducer known as the stress gage. The stress gage, shown in Fig. 5.24, looks like a strain gage; however, its grid is designed to give a predetermined value of transverse sensitivity $K_t$ so that the gage output $\Delta R/R$ is proportional to the stress along the axis of the gage rather than the strain.

The principle of operation of the stress gage can be illustrated by considering the output of a grid-type strain gage bonded to the free surface of a specimen that is being subjected to a biaxial state of stress. The relationship between stress and strain at the free surface of a specimen is given by generalized Hooke's law as

$$\varepsilon_a = \frac{1}{E}(\sigma_a - \nu\sigma_t)$$

$$\varepsilon_t = \frac{1}{E}(\sigma_t - \nu\sigma_a) \tag{5.47}$$

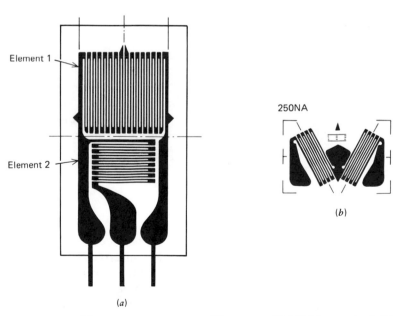

*Figure 5.24*   (a) A stress–strain gage. (Courtesy of BLH Electronics.) (b) A stress gage. (Courtesy of Micro-Measurements.)

The output of the gage in terms of the stresses $\sigma_a$ and $\sigma_t$ is obtained by substituting Eqs. (5.47) into Eq. (5.35). Thus,

$$\frac{\Delta R}{R} = \frac{\sigma_a S_a}{E} (1 - vK_t) + \frac{\sigma_t S_a}{E} (K_t - v) \tag{5.48}$$

The second term of Eq. (5.48) will vanish if $K_t = v$; therefore, the output of the gage $\Delta R/R$ will be proportional only to $\sigma_a$. For this special type of gage (stress gage with $K_t = v$), Eq. (5.48) can be written as

$$\frac{\Delta R}{R} = S_{sg} \sigma_a \tag{5.49}$$

where $S_{sg}$ is the sensitivity or gage factor for the stress gage. Equation (5.48) indicates that the stress-gage sensitivity (gage factor) depends upon the gage alloy, the gage configuration, and the elastic constants $E$ and $v$ of the specimen material; and since $K_t = v$, can be expressed as

$$S_{sg} = \frac{S_a}{E} (1 - v^2) \tag{5.50}$$

## 5.12  DATA REDUCTION METHODS

Strain gages are normally used on the free surface ($\sigma_{zz} = \tau_{zx} = \tau_{zy} = 0$) of a specimen to determine the stresses at a particular point (or points) when the specimen is subjected to a specified system of loads. The conversion from strain data to stresses requires knowledge of the elastic constants $E$ and $v$ of the material and, depending upon the state of stress at the point, from one to three normal strains. Three different stress states are considered in the following subsections. Data analysis methods and special-purpose strain gages are described for each stress state.

### The Uniaxial State of Stress ($\sigma_{xx} \neq 0$)

In a uniaxial state of stress (encountered in tension members and in beams in pure bending, for example), the stress $\sigma_{xx}$ is the only nonzero component and the direction of $\sigma_{xx}$ is known. In this case, a single-element strain gage (see Fig. 3.12a, b, or c) mounted with its axis in the x direction can be used to determine the strain $\varepsilon_{xx}$. The stress is then given by the uniaxial form of generalized Hooke's law as

$$\sigma_{xx} = E\varepsilon_{xx} \tag{5.51}$$

One principal direction is coincident with the $x$ axis for this special case of uniaxial stress. Any direction perpendicular to the $x$ axis is also a principal direction for the uniaxial state of stress.

## The Biaxial State of Stress ($\sigma_{xx}$ and $\sigma_{yy} \neq 0$)

If the directions of the principal stresses are known at the point of interest (on an axis of load and geometric symmetry, for example), two strain measurements in perpendicular directions provide sufficient data to determine the stresses at the point. Special strain gages known as two-element rectangular rosettes (see Fig. 5.25) are available for this purpose. The two-element rosette should be mounted on the specimen with its axes coincident with the principal stress directions in order to determine the two principal strains $\varepsilon_1$ and $\varepsilon_2$. The stresses are then given by the biaxial form of generalized Hooke's law as

$$\sigma_1 = \frac{E}{1 - v^2} (\varepsilon_1 + v\varepsilon_2)$$

$$\sigma_2 = \frac{E}{1 - v^2} (\varepsilon_2 + v\varepsilon_1) \qquad (5.52)$$

Recall that principal planes are free of shear stress.

## The General State of Stress ($\sigma_{xx}, \sigma_{yy}, \tau_{xy} \neq 0$)

In the most general case, the principal stress directions are not known; therefore, three unknowns $\sigma_1$, $\sigma_2$, and the principal angle $\phi_1$ must be determined in order to specify the state of stress at the point. Three-element rosettes (see Fig. 5.26) are used in these cases to obtain the required strain data. The fact that three strain measurements are sufficient to determine the state of strain at

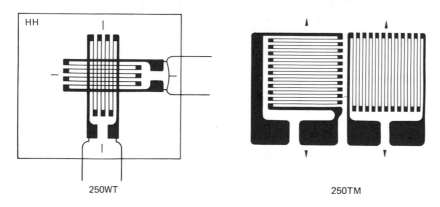

**Figure 5.25**   Two-element rectangular rosettes. (Courtesy of Micro-Measurements.)

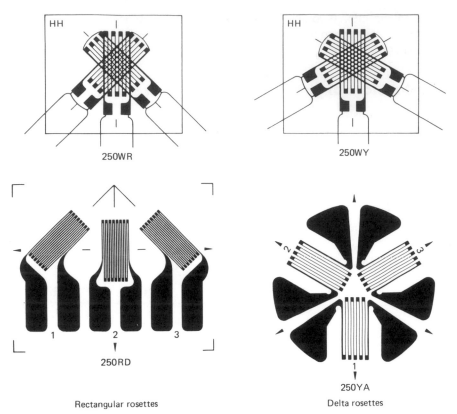

Rectangular rosettes                                    Delta rosettes

*Figure 5.26*   Three-element rectangular and delta rosettes. (Courtesy of Micro-Measurements.)

a point on the free surface of a specimen can be demonstrated by considering three gages aligned along axes $A$, $B$, and $C$, as shown in Fig. 5.27. From the equations of strain transformation

$$\varepsilon_A = \varepsilon_{xx} \cos^2 \theta_A + \varepsilon_{yy} \sin^2 \theta_A + \gamma_{xy} \sin \theta_A \cos \theta_A$$

$$\varepsilon_B = \varepsilon_{xx} \cos^2 \theta_B + \varepsilon_{yy} \sin^2 \theta_B + \gamma_{xy} \sin \theta_B \cos \theta_B$$

$$\varepsilon_C = \varepsilon_{xx} \cos^2 \theta_C + \varepsilon_{yy} \sin^2 \theta_C + \gamma_{xy} \sin \theta_C \cos \theta_C \tag{5.53}$$

The cartesian components of strain $\varepsilon_{xx}$, $\varepsilon_{yy}$, and $\gamma_{xy}$ can be determined by solving Eqs. (5.53) if $\varepsilon_A$, $\varepsilon_B$, and $\varepsilon_C$ are known. The principal strains $\varepsilon_1$ and $\varepsilon_2$ and the principal direction $\phi_1$ can then be determined from

$$\varepsilon_1 = \frac{1}{2} (\varepsilon_{xx} + \varepsilon_{yy}) + \frac{1}{2} \sqrt{(\varepsilon_{xx} - \varepsilon_{yy})^2 + \gamma_{xy}^2}$$

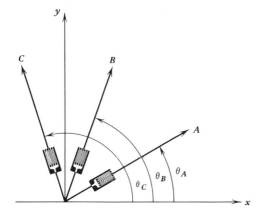

**Figure 5.27** Three gages oriented at angles $\theta_A$, $\theta_B$, and $\theta_C$ with respect to the x axis.

$$\varepsilon_2 = \frac{1}{2}(\varepsilon_{xx} + \varepsilon_{yy}) - \frac{1}{2}\sqrt{(\varepsilon_{xx} - \varepsilon_{yy})^2 + \gamma_{xy}^2}$$

$$2\phi = \tan^{-1}\frac{\gamma_{xy}}{\varepsilon_{xx} - \varepsilon_{yy}} \tag{5.54}$$

where $\phi$ is the angle between the principal direction for $\varepsilon_1$ and the x axis.

Two of the most commonly employed rosettes are the delta rosette and the three-element rectangular rosette. The three-element rectangular rosette will be discussed here; the analysis of the delta rosette is left as an exercise for the student (see Exercise 5.29). The three-element rectangular rosette is constructed with

$$\theta_A = 0°, \qquad \theta_B = 45°, \qquad \theta_C = 90° \tag{5.55}$$

With these fixed angles, Eqs. (5.53) reduce to

$$\varepsilon_A = \varepsilon_{xx}$$

$$\varepsilon_B = \frac{1}{2}(\varepsilon_{xx} + \varepsilon_{yy} + \gamma_{xy})$$

$$\varepsilon_C = \varepsilon_{yy} \tag{5.56}$$

From Eqs. (5.56)

$$\gamma_{xy} = 2\varepsilon_B - \varepsilon_A - \varepsilon_C \tag{5.57}$$

The principal strains $\varepsilon_1$ and $\varepsilon_2$ and the principal angle $\phi$ are obtained in terms of $\varepsilon_A$, $\varepsilon_B$, and $\varepsilon_C$ by substituting Eqs. (5.56) into Eqs. (5.54). Thus

$$\varepsilon_1 = \frac{1}{2}(\varepsilon_A + \varepsilon_C) + \frac{1}{2}\sqrt{(\varepsilon_A - \varepsilon_C)^2 + (2\varepsilon_B - \varepsilon_A - \varepsilon_C)^2}$$

$$\varepsilon_2 = \frac{1}{2}(\varepsilon_A + \varepsilon_C) - \frac{1}{2}\sqrt{(\varepsilon_A - \varepsilon_C)^2 + (2\varepsilon_B - \varepsilon_A - \varepsilon_C)^2}$$

$$\phi = \frac{1}{2}\tan^{-1}\frac{2\varepsilon_B - \varepsilon_A - \varepsilon_C}{\varepsilon_A - \varepsilon_C} \tag{5.58}$$

Equation (5.58) yields two values for the angle $\phi$. One value $\phi_1$ refers to the angle between the $x$ axis and the axis of $\varepsilon_1$, while the second value $\phi_2$ refers to the angle between the $x$ axis and the axis of $\varepsilon_2$. It can be shown (see Exercise 5.38) that the following classification procedure will define the angle $\phi_1$:

$$0° < \phi_1 < 90° \qquad \text{when } \varepsilon_B > \frac{1}{2}(\varepsilon_A + \varepsilon_C)$$

$$-90° < \phi_1 < 0° \qquad \text{when } \varepsilon_B < \frac{1}{2}(\varepsilon_A + \varepsilon_C)$$

$$\phi_1 = 0 \qquad \text{when } \varepsilon_A > \varepsilon_C \text{ and } \varepsilon_A = \varepsilon_1$$

$$\phi_1 = \pm 90 \qquad \text{when } \varepsilon_A < \varepsilon_C \text{ and } \varepsilon_A = \varepsilon_2 \tag{5.59}$$

Finally, the principal stresses can be expressed in terms of $\varepsilon_A$, $\varepsilon_B$, and $\varepsilon_C$ by substituting Eqs. (5.58) into Eqs. (5.52). The results are

$$\sigma_1 = E\left[\frac{\varepsilon_A + \varepsilon_C}{2(1 - v)} + \frac{1}{2(1 + v)}\sqrt{(\varepsilon_A - \varepsilon_C)^2 + (2\varepsilon_B - \varepsilon_A - \varepsilon_C)^2}\right]$$

$$\sigma_2 = E\left[\frac{\varepsilon_A + \varepsilon_C}{2(1 - v)} - \frac{1}{2(1 + v)}\sqrt{(\varepsilon_A - \varepsilon_C)^2 + (2\varepsilon_B - \varepsilon_A - \varepsilon_C)^2}\right] \tag{5.60}$$

While derivation of these equations may appear tedious, their application is simple and rapid. As a result, rosettes are widely used to establish the complete state of stress at a point on the free surface of a general three-dimensional body.

## 5.13  SUMMARY

The electrical resistance strain gage nearly meets the optimum requirements for a strain gage; therefore, it is widely employed in stress analysis and as the sensing element in other transducers, such as load cells and pressure gages.

While the gage is inexpensive and relatively easy to use, care must be exercised in its installation to ensure that it is properly bonded to the specimen, water-proofed, and wired correctly into the Wheatstone bridge.

The voltage that can be applied to a Wheatstone bridge having a single active gage is limited by the power the gage can dissipate. Proper input voltage is obtained from Eq. (5.5). Thus,

$$E_i = (1 + r) \sqrt{AP_D R_g} \qquad (5.5)$$

With this voltage applied, the system sensitivity is given by Eq. (5.2) as

$$S_s = \frac{r}{1 + r} S_g \sqrt{P_g R_g} \qquad (5.2)$$

The bridge can provide temperature compensation if a temperature-compensating gage (dummy gage) is used in arm $R_4$ of the bridge.

Digital voltmeters and oscilloscopes are high-impedance recording instruments that can be used with the Wheatstone bridge to measure the output voltage $\Delta E_o$ without introducing significant loading errors. When the Wheatstone bridge is used with a low-impedance oscillographic recording system, however, loading effects are significant and system sensitivity is reduced to

$$S_s = \frac{1}{2} \sqrt{\frac{P_g}{R_g}} \frac{S_g S_G}{1 + (R_G/R_g)} \qquad (5.18)$$

Also, care must be exercised to ensure that the specified external resistance is provided in the circuit in order for the galvanometer to maintain its range of frequency response.

Both electrical and mechanical procedures are used to calibrate a strain-measuring system. For electrical calibration, the calibration strain $\varepsilon_c$ is simulated by shunting a calibration resistor $R_c$ across arm $R_2$ of the bridge. The magnitude of the calibration strain is given by Eq. (5.26) as

$$\varepsilon_c = \frac{R_2}{S_g(R_2 + R_c)} \qquad (5.26)$$

The strain being measured is then obtained by comparing deflections of the recorder trace induced by the calibration strain and the load-induced strain. From Eq. (5.21),

$$\varepsilon = \frac{d_s}{d_c} \varepsilon_c = C d_s \qquad (5.21)$$

Lead wires, slip rings, and switches, which are commonly employed with strain gages, can in some cases seriously degrade the instrumentation system. The detrimental effects of long lead wires can be significantly reduced by using a three-wire system. Signal loss $\mathscr{L}$, due to long lead wires, is given by the expression

$$\mathscr{L} = \frac{R_L/R_g}{1 + (R_L/R_g)} \tag{5.29}$$

The signal loss $\mathscr{L}$ can be accounted for during calibration if the calibration resistor is introduced into the circuit at the remote strain gage.

Only high-quality switches should be used in the arms of a Wheatstone bridge; otherwise, errors due to changes in switch resistance will occur. Slip rings should not be used within the arms of the bridge to transmit signals from rotating members. Instead, a complete bridge should be assembled on the rotating member and the supply voltage and output voltage should be transmitted with the slip rings.

Noise in strain-gage circuits is common and can be minimized by employing twisted leads with a properly grounded shield. Also, common-mode rejection by well-designed amplifiers further reduces the noise-to-signal ratio.

Temperature-compensating gages are available for a wide range of specimen materials and should be used for all tests where large temperature changes are expected to occur.

Strain gages exhibit a sensitivity to both axial and transverse strains, which can be expressed as

$$\frac{\Delta R}{R} = \frac{S_g \varepsilon_a}{1 - v_o K_t} \left( 1 + K_t \frac{\varepsilon_t}{\varepsilon_a} \right) \tag{5.39}$$

If the transverse sensitivity of the gage is neglected, the response from the gage is related to the axial strain by the simple (approximate) expression

$$\frac{\Delta R}{R} = S_g \varepsilon_a \tag{3.5}$$

The percent error resulting from use of the approximate equation is

$$\mathscr{E} = \frac{K_t(\varepsilon_t/\varepsilon_a + v_o)}{1 - v_o K_t}(100) \tag{5.42}$$

If the ratio $\varepsilon_t/\varepsilon_a$ is known, a corrected gage factor can be used to eliminate error due to the transverse sensitivity of the gage. The corrected gage factor

$S_g^*$ is given by the expression

$$S_g^* = \left[ \frac{1 + K_t(\varepsilon_t/\varepsilon_a)}{1 - v_o K_t} \right] S_g \tag{5.44}$$

If the ratio $\varepsilon_t/\varepsilon_a$ is not known, transverse sensitivity errors are eliminated by measuring two orthogonal apparent strains $\varepsilon_{xx}'$ and $\varepsilon_{yy}'$ with a two-element rectangular rosette and computing true strains by using the equations

$$\varepsilon_{xx} = \frac{1 - v_o K_t}{1 - K_t^2} (\varepsilon_{xx}' - K_t \varepsilon_{yy}')$$

$$\varepsilon_{yy} = \frac{1 - v_o K_t}{1 - K_t^2} (\varepsilon_{yy}' - K_t \varepsilon_{xx}') \tag{5.46}$$

A stress gage (a strain gage designed with a transverse sensitivity such that the output is proportional to the stress along the axis of the gage) is available for special situations where a single stress component in a specific direction is required.

Strain measurements can be converted to stresses for the uniaxial state of stress by using the simple expression

$$\sigma_{xx} = E\varepsilon_{xx} \tag{5.51}$$

In more complex situations, where a three-element rectangular rosette is used to record $\varepsilon_A$, $\varepsilon_B$, and $\varepsilon_C$, the principal stresses $\sigma_1$ and $\sigma_2$ and their directions $\phi$ are obtained from these three strains by using the equations

$$\sigma_1 = E \left[ \frac{\varepsilon_A + \varepsilon_C}{2(1 - v)} \right.$$

$$\left. + \frac{1}{2(1 + v)} \sqrt{(\varepsilon_A - \varepsilon_C)^2 + (2\varepsilon_B - \varepsilon_A - \varepsilon_C)^2} \right]$$

$$\sigma_2 = E \left[ \frac{\varepsilon_A + \varepsilon_C}{2(1 - v)} \right. \tag{5.60}$$

$$\left. - \frac{1}{2(1 + v)} \sqrt{(\varepsilon_A - \varepsilon_C)^2 + (2\varepsilon_B - \varepsilon_A - \varepsilon_C)^2} \right]$$

$$\phi = \frac{1}{2} \tan^{-1} \frac{2\varepsilon_B - \varepsilon_A - \varepsilon_C}{\varepsilon_A - \varepsilon_C} \tag{5.58}$$

## REFERENCES

1. Thomson, W. (Lord Kelvin): On the Electrodynamic Qualities of Metals, *Proceedings of the Royal Society,* 1856.

2. Tomlinson, H.: The Influence of Stress and Strain on the Action of Physical Forces, *Philosophical Transactions of the Royal Society (London)*, vol. 174, 1883, pp. 1–172.

3. Simmons, E. E., Jr.: Material Testing Apparatus, U.S. Patent 2,292,549, February 23, 1940.

4. Maslen, K. R., and I. G. Scott: Some Characteristics of Foil Strain Gauges, *Royal Aircraft Establishment Technical Note Instruction 134,* 1953.

5. Micro-measurements, Strain Gage Installations, *Instruction Bulletin B-130-2,* July 1972.

6. Stein, P. K.: *Advanced Strain Gage Techniques,* Chapter 2, Stein Engineering Services, Phoenix, Arizona, 1962.

7. Perry, C. C., and H. R. Lissner: *The Strain Gage Primer,* 2nd ed., McGraw-Hill, New York, 1962, pp. 200–217.

8. Measurements Group, Inc., Noise Control in Strain Gage Measurements, *Technical Note* 501, 1980, pp. 1–5.

9. Measurements Group, Inc., Optimizing Strain Gage Excitation Levels, *Technical Note* 502, 1979, pp. 1–5.

10. Measurements Group, Inc., Temperature-Induced Apparent Strain and Gage Factor Variation in Strain Gages, *Technical Note* 504, 1976, pp. 1–9.

11. Measurements Group, Inc., Strain Gage Selection Criteria, Procedures, & Recommendations, *Technical Note* 505, 1976, pp. 1–12.

12. Measurements Group, Inc., Fatigue Characteristics of Micro-Measurements Strain Gages, *Technical Note* 508, 1982, pp. 1–4.

13. Measurements Group, Inc., Errors Due to Transverse Sensitivity in Strain Gages, *Technical Note* 509, 1982, 1–8.

14. Measurements Group, Inc., Errors Due to Misalignment of Strain Gages, *Technical Note* 138, 1979, pp. 1–7.

15. Measurements Group, Inc., Errors Due to Wheatstone Bridge Nonlinearity, *Technical Note* 139, 1974, pp. 1–4.

16. Dally, J. W., and W. F. Riley: *Experimental Stress Analysis,* 2nd ed., McGraw-Hill, New York, 1978, pp. 153–336.

## EXERCISES

**5.1** A strain gage is to be fabricated from Advance wire having a diameter of 0.001 in. and a resistance of 25 $\Omega$/in. The gage is to have a gage length of $\frac{1}{4}$ in. and a resistance of 200 $\Omega$. Design a grid configuration.

**5.2** Plot a family of curves showing bridge voltage $E_i$ as a function of $r$ for a Wheatstone bridge with a single active 350-$\Omega$ gage on a material with $P_D = 0.002$ W/mm$^2$. Use the area of the gage as a parameter and vary $A$ in increments between 0.0005 mm$^2$ to 0.5 mm$^2$. From these results, indicate the value of $r$ that should be used in the design of the bridge.

**5.3** Repeat Exercise 5.2 for a 120-$\Omega$ gage.

**5.4** Determine the system sensitivity for a bridge with a single active gage having $R_g = 350$ $\Omega$ and $S_g = 2.05$, if $r = 4$ and if the bridge voltage is 6 V.

**5.5** If the gage in Exercise 5.4 can dissipate 0.05 W, is the bridge voltage correct? If not, what is the correct voltage?

**5.6** Determine the voltage output from a Wheatstone bridge if a single active gage is used in an initially balanced bridge to measure a strain of 600 $\mu$m/m. Assume that a digital voltmeter will be used to measure the voltage and that $S_g = 2.06$, $r = 1$, and $E_i = 9$ V.

**5.7** Determine the loading error produced by connecting an oscilloscope with an input impedance of 10$^6$ $\Omega$ to a Wheatstone bridge with one active gage. Use $R_g = 350$ $\Omega$, $r = 5$, and $q = 1$.

**5.8** The bridge in Exercise 5.7 is powered with a 9-V constant-voltage supply and the strain gage has a gage factor $S_g = 3.35$. If the gage responds to a dynamic strain pulse having a magnitude of 1200 $\mu$m/m, determine the sensitivity setting on the oscilloscope that will give a trace deflection of four divisions.

**5.9** If the bridge and gage of Exercise 5.8 respond to a strain of 1000 $\mu$m/m, determine the trace deflection if an oscilloscope having a sensitivity of 1 mV/div is used for the measurement.

**5.10** If the bridge, gage, and oscilloscope of Exercise 5.9 record a trace deflection of 3.7 divisions, determine the strain at the gage location.

**5.11** Determine the sensitivity of a gage–bridge–galvanometer system if a four-equal-arm bridge ($R_1 = R_2 = R_3 = R_4 = R_g$) is used and if $R_g = 350$ $\Omega$, $S_g = 2.07$, $P_g = 0.1$ W, $R_G = 100$ $\Omega$, and $S_G = 0.003$ mm/$\mu$A.

**5.12** Would the sensitivity of the system described in Exercise 5.11 be improved by replacing the 350-$\Omega$ gage with a 120-$\Omega$ gage? Explain.

**5.13** An oscillograph chart shows a calibration displacement $d_c = 30$ mm.

Determine the strain corresponding to a trace displacement $d_s = 26$ mm if the calibration constant $C = 50$ μm/m per millimeter of displacement.

**5.14**  Determine the value of the calibration constant $C$ for a gage–bridge–amplifier–recorder system if $S_g = 2.04$, $r = 4$, $E_i = 12$ V, $G = 50$, and $S_R = 10$ mV/div.

**5.15**  Determine the resistance $R_c$ that must be shunted across arm $R_2$ of a Wheatstone bridge to produce a calibration strain $\varepsilon_c = 600$ μm/m if $r = 4$, $R_g$ (in arm $R_1$) $= 350$ Ω, and $S_g = 2.05$.

**5.16**  Design a displacement fixture to be used with a strain gage mounted on a cantilever beam to mechanically produce a calibration strain $\varepsilon_c$ ranging from 0 to 1500 μm/m in 500 μm/m increments.

**5.17**  Determine the error that results if a strain gage compensated for aluminum ($\alpha = 13 \times 10^{-6}/°F$) is used on steel ($\alpha = 6 \times 10^{-6}/°F$). The total response of the gage was 200 μm/m and the temperature change between the zero and final reading was 40°F. Assume that a dummy gage was not used in the bridge.

**5.18**  Determine the axial sensitivity $S_a$ of a strain gage if $S_g = 2.04$ and $K_t = 0.03$.

**5.19**  Determine the error involved if transverse sensitivity is neglected in a measurement of hoop strain in a thin-walled steel cylindrical pressure vessel when $K_t$ for the gage is 0.02.

**5.20**  If strain gages with $S_g = 2.03$ and $K_t = 0.04$ are used to determine the apparent strains $\varepsilon'_{xx}$ and $\varepsilon'_{yy}$, determine the true strains $\varepsilon_{xx}$ and $\varepsilon_{yy}$ if:

| Case | $\varepsilon'_{xx}$ | $\varepsilon'_{yy}$ |
|---|---|---|
| 1 | 1000 μin./in. | 1000 μin./in. |
| 2 | 800 | −600 |
| 3 | 1400 | −200 |
| 4 | −700 | 2000 |
| 5 | 300 | 1200 |

**5.21**  Determine the error produced by ignoring transverse sensitivity effects if strain gages with $K_t = 0.03$ are employed in a simple tension test to measure Poisson's ratio of a material.

**5.22**  Assume that $K_t$ for an ordinary strain gage is zero. Show how this gage could be used to measure stress in a specified direction. Clearly indicate all assumptions made in your derivation.

**5.23**  A stress gage is fabricated with two grids each of which exhibit an axial sensitivity $S_a = 2.00$. Determine the output from the gage in terms of $\Delta R/R$ if it is mounted on a steel specimen and subjected to a stress $\sigma_a = 60,000$ psi.

**5.24** Determine the uniaxial state of stress associated with the following strain measurements:

| $\varepsilon_1$ | Material | |
|---|---|---|
| 650 μin./in. | Steel | $(E = 29{,}000{,}000 \text{ psi}, \nu = 0.29)$ |
| 900 | Aluminum | $(E = 10{,}000{,}000 \text{ psi}, \nu = 0.33)$ |
| 1410 | Titanium | $(E = 14{,}000{,}000 \text{ psi}, \nu = 0.25)$ |

**5.25** Determine the biaxial state of stress associated with the following strain measurements:

| $\varepsilon_1$ | $\varepsilon_2$ | Material |
|---|---|---|
| − 100 μin./in. | − 400 μin./in. | Aluminum |
| 900 | − 300 | Titanium |
| 700 | 600 | Steel |

**5.26** Determine the general state of stress associated with the following strain measurements made with a 0°, 45°, 90° rosette.

| $\varepsilon_A$ | $\varepsilon_B$ | $\varepsilon_C$ | Material |
|---|---|---|---|
| 400 μin./in. | 800 μin./in. | − 200 μin./in. | Aluminum |
| 700 | 700 | 700 | Steel |
| − 300 | − 600 | 900 | Titanium |

**5.27** Verify Eqs. (5.56), (5.57), and (5.58) for a three-element rectangular rosette with $\theta_A = 0°$, $\theta_B = 45°$, and $\theta_C = 90°$.

**5.28** Use Mohr's strain circle to verify the classification procedure for a three-element rectangular rosette described in Eqs. (5.59).

**5.29** Consider a delta rosette with $\theta_A = 0°$, $\theta_B = 120°$, and $\theta_C = 240°$. Derive:

(a) The equations for the cartesian components of strain—similar to Eqs. (5.56) and (5.57)
(b) The equations for the principal strains—similar to Eqs. (5.58)
(c) The classification procedure for the principal angles—similar to Eqs. (5.59)
(d) The equations for the principal stresses—similar to Eqs. (5.60)

# SIX

# FORCE, TORQUE, AND PRESSURE MEASUREMENTS

## 6.1 INTRODUCTION

Transducers that measure force, torque, or pressure usually contain an elastic member that converts the quantity to be measured to a deflection or strain. A deflection sensor or, alternatively, a set of strain gages can then be used to measure the quantity of interest (force, torque, or pressure) indirectly. Characteristics of the transducer, such as range, linearity, and sensitivity, are determined by the size and shape of the elastic member, the material used in its fabrication, and the sensor.

A wide variety of transducers are commercially available for measuring force (load cells), torque (torque cells), and pressure. The different elastic members employed in the design of these transducers include links, columns, rings, beams, cylinders, tubes, washers, diaphragms, shear webs, and numerous other shapes for special-purpose applications. Strain gages are usually used as the sensor; however, linear variable-differential transformers (LVDTs) and linear potentiometers are sometimes used for static or quasi-static measurements. A selection of force transducers (load cells) and a torque transducer are shown in Fig. 6.1.

(a)

(b)

*Figure 6.1*  Transducers for measuring force and torque.

(a) A selection of force transducers. (Courtesy of Hottinger-Baldwin Measurements, Inc.)

(b) A shaft torque sensor for general applications. (Courtesy of Lebow Products, Eaton Corp.)

## 6.2  FORCE MEASUREMENTS (Load Cells)

The elastic members commonly used in load cells are links, beams, rings, and shear webs. The operating characteristics for each of these transducer types are developed in the following subsections.

## Link-Type Load Cell

A simple uniaxial link-type load cell with strain gages as the sensor is shown in Fig. 6.2*a*. The load $P$ can be either a tensile load or a compressive load. The four strain gages are bonded to the link such that two are in the axial direction and two are in the transverse direction. The four gages are wired into a Wheatstone bridge with the axial gages in arms 1 and 3 and the transverse gages in arms 2 and 4, as shown in Fig. 6.2*b*.

When the load $P$ is applied to the link, axial and transverse strains $\varepsilon_a$ and $\varepsilon_t$ develop in the link and are related to the load by the expressions

$$\varepsilon_a = \frac{P}{AE} \qquad \varepsilon_t = -\frac{vP}{AE} \tag{a}$$

**where**   $A$ is the cross-sectional area of the link.
   $E$ is the modulus of elasticity of the link material.
   $v$ is Poisson's ratio of the link material.

The response of the gages to the applied load $P$ is given by Eqs. (3.5) and (a) as

$$\frac{\Delta R_1}{R_1} = \frac{\Delta R_3}{R_3} = S_g \varepsilon_a = \frac{S_g P}{AE}$$

$$\frac{\Delta R_2}{R_2} = \frac{\Delta R_4}{R_4} = S_g \varepsilon_t = -\frac{v S_g P}{AE} \tag{b}$$

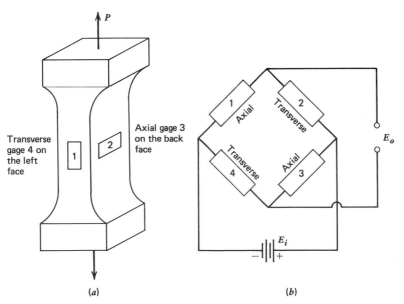

*(a)*  *(b)*

***Figure 6.2***   Link-type load cell. (*a*) Elastic element with strain gages. (*b*) Gage positions in the Wheatstone bridge.

The output voltage $E_o$ from the Wheatstone bridge can be expressed in terms of the load $P$ by substituting Eqs. (b) into Eq. (4.22). If it is assumed that the four strain gages on the link are identical, then $R_1 = R_2$ and Eq. (4.22) yields

$$E_o = \frac{S_g P(1 + v)E_i}{2AE} \tag{6.1}$$

or

$$P - \frac{2AE}{S_g(1 + v)E_i} E_o = CE_o \tag{6.2}$$

Equation (6.2) indicates that the load $P$ is linearly proportional to the output voltage $E_o$ and that the constant of proportionality or calibration constant $C$ is

$$C = \frac{2AE}{S_g(1 + v)E_i} \tag{6.3}$$

The sensitivity of the load cell–Wheatstone bridge combination is given by Eq. (1.2) as $S = E_o/P$; therefore, from Eq. (6.3)

$$S = \frac{E_o}{P} = \frac{1}{C} = \frac{S_g(1 + v)E_i}{2AE} \tag{6.4}$$

Equation (6.4) indicates that the sensitivity of the link-type load cell depends upon the cross-sectional area of the link ($A$), the elastic constants of the material used in fabricating the link ($E$ and $v$), the strain gages used as sensors ($S_g$), and the input voltage applied to the Wheatstone bridge ($E_i$).

The range of a link-type load cell is determined by the cross-sectional area of the link and by the fatigue strength $S_f$ of the material used in its fabrication. Thus

$$P_{max} = S_f A \tag{6.5}$$

Since both sensitivity and range depend upon the cross-sectional area $A$ of the link, high sensitivities are associated with low-capacity load cells, while low sensitivities are associated with high-capacity load cells.

The voltage ratio at maximum load $(E_o/E_i)_{max}$ for the link-type load cell is obtained by substituting Eq. (6.5) into Eq. (6.1). Thus

$$\left(\frac{E_o}{E_i}\right)_{max} = \frac{S_g S_f(1 + v)}{2E} \tag{6.6}$$

Most load-cell links are fabricated from AISI 4340 steel ($E = 30,000,000$ psi and $v = 0.30$), which is heat treated to give a fatigue strength $S_f \approx 80,000$ psi.

Since $S_g \approx 2$ for the strain gages used in load cells, Eq. (6.6) indicates that

$$\left(\frac{E_o}{E_i}\right)_{max} = \frac{2(80{,}000)(1 + 0.30)}{2(30{,}000{,}000)} = 3.47 \text{ mV/V}$$

Most link-type load cells are rated at $(E_o/E_i)^* = 3$ mV/V at the full-scale value of the load $(P = P_{max})$. With this full-scale specification of voltage ratio $(E_o/E_i)^*$, the load $P$ on the load cell is given by

$$P = \frac{E_o/E_i}{(E_o/E_i)^*} P_{max} \tag{6.7}$$

The voltage $E_i$ is typically about 10 V; therefore, the output voltage of a link-type load cell at the maximum rated load is approximately 30 mV. This output can be monitored with a digital voltmeter or if the signal is dynamic, it can be displayed on an oscillographic recorder or an oscilloscope.

## Beam-Type Load Cell

Beam-type load cells (see Fig. 6.3a) are commonly employed for measuring low-level loads where the link-type load cell is not effective. A simple cantilever beam (see Fig. 6.3b) with two strain gages on the top surface and two strain gages on the bottom surface (all oriented along the axis of the beam) serves as the elastic member and sensor for this type of load cell. The gages are wired into a Wheatstone bridge as shown in Fig. 6.3c.

The load $P$ produces a moment $M = Px$ at the gage location $x$ that results in the following strains:

$$\varepsilon_1 = -\varepsilon_2 = \varepsilon_3 = -\varepsilon_4 = \frac{6M}{Ebh^2} = \frac{6Px}{Ebh^2} \tag{a}$$

**where**    $b$ is the width of the cross section of the beam.
$h$ is the height of the cross section of the beam.

The response of the strain gages is obtained from Eqs. (3.5) and (a). Thus

$$\frac{\Delta R_1}{R_1} = -\frac{\Delta R_2}{R_2} = \frac{\Delta R_3}{R_3} = -\frac{\Delta R_4}{R_4} = \frac{6S_g Px}{Ebh^2} \tag{b}$$

The output voltage $E_o$ from the Wheatstone bridge, resulting from application of the load $P$, is obtained by substituting Eq. (b) into Eq. (4.22). If the four

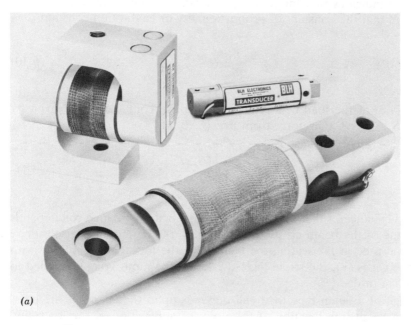

(a)

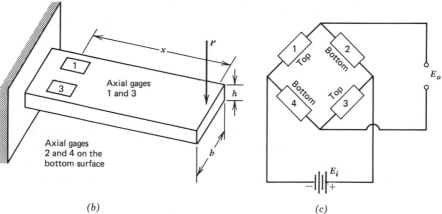

(b)                                          (c)

*Figure 6.3*   Beam-type load cells. (*a*) A selection of beam-type load cells. (Courtesy of BLH Electronics.) (*b*) Elastic element with strain gages. (*c*) Gage positions in the Wheatstone bridge.

strain gages on the beam are assumed to be identical

$$E_o = \frac{6S_g PxE_i}{Ebh^2}$$   (6.8)

or

$$P = \frac{Ebh^2}{6S_g xE_i} E_o = CE_o$$   (6.9)

Equation (6.9) indicates that the load $P$ is linearly proportional to the output voltage $E_o$ and that the constant of proportionality or calibration constant $C$ is

$$C = \frac{Ebh^2}{6S_gxE_i} \tag{6.10}$$

The sensitivity of the load cell–Wheatstone bridge combination is given by Eq. (1.2) as $S = E_o/P$; therefore, from Eq. (6.10)

$$S = \frac{E_o}{P} = \frac{1}{C} = \frac{6S_gxE_i}{Ebh^2} \tag{6.11}$$

Equation (6.11) indicates that the sensitivity of the beam-type load cell depends upon the shape of the beam cross section ($b$ and $h$), the modulus of elasticity of the material used in fabricating the beam ($E$), the location of the load with respect to the gages ($x$), the strain gages ($S_g$), and the input voltage applied to the Wheatstone bridge ($E_i$).

The range of a beam-type load cell depends upon the shape of the cross section of the beam, the location of the point of application of the load, and the fatigue strength of the material from which the beam is fabricated. If it is assumed that the gages are located at or near the beam support, then $M_{\text{gage}} \approx M_{\text{max}}$ and

$$P_{\text{max}} = \frac{S_f bh^2}{6x} \tag{6.12}$$

Equations (6.11) and (6.12) indicate that both the range and the sensitivity of a beam-type load cell can be changed by varying the point of load application. Maximum sensitivity and minimum range occurs as $x$ approaches the length of the beam. The sensitivity decreases and the range increases as the point of load application moves nearer the gages.

The voltage ratio at maximum load $(E_o/E_i)_{\text{max}}$ is obtained by substituting Eq. (6.12) into Eq. (6.8). Thus,

$$\left(\frac{E_o}{E_i}\right)_{\text{max}} = \frac{S_g S_f}{E} \tag{6.13}$$

A comparison of Eq. (6.13) with Eq. (6.6) indicates that the beam-type load cell is approximately 50 percent more sensitive than the link-type load cell. Beam-type load cells are commercially available with ratings of $(E_o/E_i)^*$ between 4 and 5 mV/V at full-scale load.

## Ring-Type Load Cell

Ring-type load cells incorporate a proving ring (see Fig. 6.4) as the elastic element. The ring element can be designed to cover a very wide range of loads by varying the diameter $D$ (or radius $R$), the thickness $t$, or the depth $w$ of the

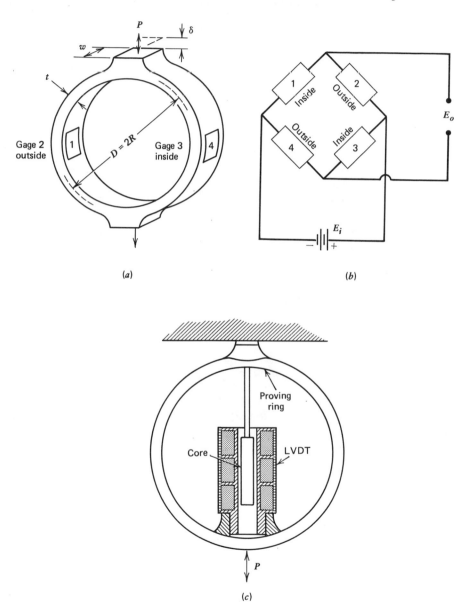

*Figure 6.4*   Ring-type load cell. (*a*) Elastic element with strain-gage sensors. (*b*) Gage positions in the Wheatstone bridge. (*c*) Elastic element with an LVDT sensor.

ring. Either strain gages or a linear variable-differential transformer (LVDT) can be used as the sensor.

If an LVDT is used to measure the diametric compression or extension $\delta$ of the ring, the relationship between displacement $\delta$ and load $P$ is given by the following approximate expression:

$$\delta = 1.79 \frac{PR^3}{Ewt^3} \tag{6.14}$$

Equation (6.14) is approximate since the reinforced areas at the top and bottom of the ring that accommodate the loading attachments have not been considered in its development. The output voltage $E_o$ of an LVDT can be expressed as

$$E_o = S\delta E_i \tag{6.15}$$

**where**   $S$  is the sensitivity of the LVDT.
       $E_i$ is the voltage applied to the primary winding of the LVDT.

Expressions relating output voltage $E_o$ and load $P$ are obtained by substituting Eq. (6.14) into Eq. (6.15). Thus

$$E_o = 1.79 \frac{SPR^3 E_i}{Ewt^3} \tag{6.16}$$

or

$$P = 0.56 \frac{Ewt^3}{SR^3 E_i} E_o = CE_o \tag{6.17}$$

Equation (6.17) shows that the load $P$ is linearly proportional to the output voltage $E_o$ and that the constant of proportionality or calibration constant $C$ will be approximately equal to

$$C = 0.56 \frac{Ewt^3}{SR^3 E_i} \tag{6.18}$$

The sensitivity of the ring–LVDT combination $S_t$ is given by Eqs. (1.2) and (6.18) as

$$S_t = \frac{E_o}{P} = \frac{1}{C} = 1.79 \frac{SR^3 E_i}{Ewt^3} \tag{6.19}$$

Thus, it is seen that the sensitivity of the ring-type load cell with an LVDT sensor depends upon the geometry of the ring ($R$, $t$, and $w$), the material from which the ring is fabricated ($E$), and the characteristics of the LVDT ($S$ and $E_i$).

The range of a ring-type load cell is controlled by the strength of the material used in fabricating the ring. If the load cell is to be used to measure cyclic loads, the fatigue strength $S_f$ is important. If the load cell will be used only to measure static loads, the proportional limit of the material can be used to establish the range of the load cell. The maximum stress in a ring element, reinforced at the top and the bottom, is highest on the inside surface of the ring on a diameter perpendicular to the line of the loads. The approximate value for the stress at this location is

$$\sigma_\theta = 1.09 \frac{PR}{wt^2} \tag{6.20}$$

Thus, from Eq. (6.20), for cyclic load measurements

$$P_{max} = 0.92 \frac{wt^2 S_f}{R} \tag{6.21}$$

The voltage ratio at maximum load $(E_o/E_i)_{max}$ is obtained by substituting Eq. (6.21) into Eq. (6.16). Thus

$$\left(\frac{E_o}{E_i}\right)_{max} = 1.64 \frac{SR^2 S_f}{Et} \tag{6.22}$$

The rated voltage ratio $(E_o/E_i)^*$ for most ring-type load cells will be slightly less than the value established by Eq. (6.22), since the ring will not be operated at a stress level equal to the fatigue strength of the material. Once the rated voltage ratio $(E_o/E_i)^*$ and maximum load $P_{max}$ for a particular load cell are known, Eq. (6.7) can be used to establish the load corresponding to a measured output voltage.

$$P = \frac{E_o/E_i}{(E_o/E_i)^*} P_{max} \tag{6.7}$$

A typical short-range LVDT ($\pm 1.25$ mm), used for the load-cell sensor, will exhibit a sensitivity of 250 mV/V $\cdot$ mm. If the ring element of the load cell is designed to have a maximum deflection $\delta_{max} = 1.25$ mm at $P_{max}$, then Eq. (6.15) indicates that

$$\left(\frac{E_o}{E_i}\right)_{max} = S\delta_{max} = 250(1.25) = 313 \text{ mV/V}$$

Ring-type load cells rated at $(E_o/E_i)^* \approx 300$ mV/V are available and have the capability of measuring both tensile and compressive loads (universal load cells). The rated output of a ring-type load cell with an LVDT sensor is significantly higher than the output achieved with load cells with strain-gage sensors.

## Shear-Web-Type Load Cell

The shear-web-type load cell (also known as a low-profile or a flat load cell) is useful for applications where space is limited along the line of application of the load. The flat load cell consists of an inner loading hub and an outer supporting flange connected by a continuous shear web (see Fig. 6.5). Shear-type strain gages installed in small holes drilled into the neutral surface of the web (see Fig. 6.5b) are used as the sensor. Some characteristics of a line of flat load cells are shown in Table 6.1.

The flat load cell is compact and stiff; therefore, it can be used for dynamic applications where measurement of load at a high frequency is required. The effective weight $w_e$ of a flat load cell is relatively small (see Table 6.1), since only the inner hub and the inboard parts of the shear web vibrate at or near full amplitude when the transducer is dynamically excited. The outer rim and the outer region of the shear web do not contribute to the effective weight since they remain essentially motionless. The frequency limit of a load cell in a dynamic application is determined by the effective weight $w_e$, the weight of any moving attachments $w_x$, and the stiffness of the load cell. The natural frequencies of the flat load cells listed in Table 6.1 have been determined by the manufacturer to

**TABLE 6.1** Mechanical Properties of Flat Load Cells[a]

| Force Capacity $P$ (lb) | Spring Rate $k$ (lb/in.) | Effective Weight $w_e$ (lb) | Natural Frequency $f_n$ (kHz) |
|---|---|---|---|
| 250 | 920,000 | 0.028 | 18 |
| 500 | 920,000 | 0.028 | 18 |
| 1,000 | 1,220,000 | 0.023 | 22.8 |
| 2,500 | 2,200,000 | 0.040 | 23.3 |
| 5,000 | 6,600,000 | 0.135 | 22.0 |
| 10,000 | 8,500,000 | 0.34 | 15.7 |
| 25,000 | 11,700,000 | 0.69 | 13.0 |
| 50,000 | 20,400,000 | 1.58 | 11.3 |
| 100,000 | 28,600,000 | 4.50 | 7.9 |
| 200,000 | 35,800,000 | 10.0 | 5.9 |
| 500,000 | 65,000,000 | 33.0 | 4.4 |

[a] Courtesy of Strainsert Company (from technical bulletin No. 365-4MP).

be given by the following expression:

$$f_n = 3.13 \sqrt{\frac{k}{w_e + w_x}} \qquad (6.23)$$

Supporting structure

(*a*)

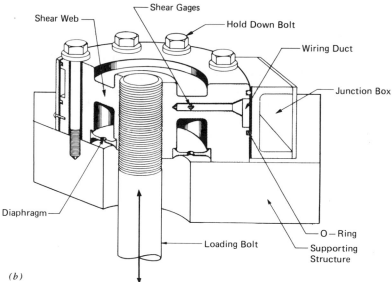

Shear Gages

Shear Web

Hold Down Bolt

Wiring Duct

Junction Box

Diaphragm

Loading Bolt

O – Ring

Supporting Structure

(*b*)

***Figure 6.5*** Universal flat load cell. (*a*) Universal flat load cell. (*b*) Construction details. (Courtesy of Strainsert Company.)

**where**    $k$   is the spring rate or stiffness of the load cell.

$w_e$ is the effective weight of the active portion of the load cell.

$w_x$ is the external weight attached to the moving portion of the load cell.

The natural frequencies listed in Table 6.1 were calculated under the assumption that no external weight ($w_x = 0$) was attached to the load cell. With external weight attached to the load cell, Eq. (6.23) indicates that the frequency limit will be lower than the values listed in Table 6.1. A complete discussion of the performance characteristics of dynamic force transducers and their application to dynamic loading situations is presented in Section 7.6.

## 6.3   TORQUE MEASUREMENT (Torque Cells)

*Torque cells* are transducers that convert torque to an electrical signal. The two types of torque cells in common usage include those installed on fixed shafts and those installed on rotating shafts. The latter type is more difficult to utilize, since the electrical signal must be transmitted from the rotating shaft to a stationary assembly of recording instruments. The problem of signal transmission will be considered after design concepts associated with torque cells are discussed.

### Torque Cells—Design Concepts

Torque cells are very similar to load cells; they consist of a mechanical element (usually a shaft with a circular cross section) and a sensor (usually electrical resistance strain gages). A circular shaft with four strain gages mounted on two perpendicular 45-degree helixes is shown in Fig. 6.6a. Gages 1 and 3, mounted on the right-hand helix, sense a positive strain, while gages 2 and 4, mounted on the left-hand helix, sense a negative strain. The two 45-degree helixes define the principal stress and strain directions for a circular shaft subjected to pure torsion.

The shearing stress $\tau$ in the circular shaft is related to the applied torque $T$ by the equation

$$\tau_{xz} = \frac{TD}{2J} = \frac{16T}{\pi D^3} \tag{6.24}$$

**where**    $D$ is the diameter of the shaft.

$J$ is the polar moment of inertia of the circular cross section.

Since the normal stresses $\sigma_x = \sigma_y = \sigma_z = 0$ for a circular shaft subjected to pure torsion, Mohr's circle, shown in Fig. 6.7, indicates that

$$\sigma_1 = -\sigma_2 = \tau_{xz} = \frac{16T}{\pi D^3} \tag{6.25}$$

Principal strains $\varepsilon_1$ and $\varepsilon_2$ are obtained by using Eqs. (6.25) and Hooke's law for the plane state of stress. Thus,

$$\varepsilon_1 = \frac{1}{E}(\sigma_1 - \nu\sigma_2) = \frac{16T}{\pi D^3}\left(\frac{1 + \nu}{E}\right)$$

$$\varepsilon_2 = \frac{1}{E}(\sigma_2 - \nu\sigma_1) = -\frac{16T}{\pi D^3}\left(\frac{1 + \nu}{E}\right) \tag{6.26}$$

The response of the strain gages is obtained by substituting Eqs. (6.26) into Eq. (3.5).

$$\frac{\Delta R_1}{R_1} = -\frac{\Delta R_2}{R_2} = \frac{\Delta R_3}{R_3} = -\frac{\Delta R_4}{R_4} = \frac{16T}{\pi D^3}\left(\frac{1 + \nu}{E}\right)S_g \tag{a}$$

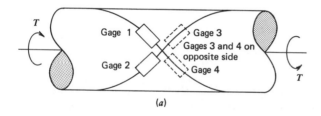

(a)

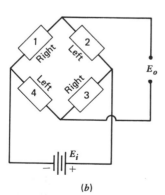

(b)

**Figure 6.6**   Torque cell. (*a*) Elastic element with strain-gage sensors. (*b*) Gage positions in the Wheatstone bridge.

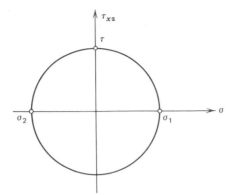

**Figure 6.7** Mohr's circle for the state of stress in a circular shaft subjected to a pure torque.

If the gages are connected into a Wheatstone bridge, as illustrated in Fig. 6.6b, the relationship between output voltage $E_o$ and torque $T$ is obtained by substituting Eq. (a) into Eq. (4.22). The results are

$$E_o = \frac{16T}{\pi D^3} \left( \frac{1 + \nu}{E} \right) S_g E_i \tag{6.27}$$

or

$$T = \frac{\pi D^3 E}{16(1 + \nu)S_g E_i} E_o = CE_o \tag{6.28}$$

Equation (6.28) indicates that the torque $T$ is linearly proportional to the output voltage $E_o$ and that the constant of proportionality or calibration constant $C$ is given by

$$C = \frac{\pi D^3 E}{16(1 + \nu)S_g E_i} \tag{6.29}$$

The sensitivity of the torque cell–Wheatstone bridge combination is given by Eq. (1.2) as

$$S = \frac{E_o}{T} = \frac{1}{C} = \frac{16(1 + \nu)S_g E_i}{\pi D^3 E} \tag{6.30}$$

Equation (6.30) shows that the sensitivity of a torque cell depends upon the diameter of the shaft ($D$), the material used in fabricating the shaft ($E$ and $\nu$), the strain gages ($S_g$), and the voltage applied to the Wheatstone bridge ($E_i$).

The range of the torque cell depends upon the diameter $D$ of the shaft and the yield strength of the material in torsion $S_\tau$. For static applications, the range is given by Eq. (6.24) as

$$T_{max} = \frac{\pi D^3 S_\tau}{16} \tag{6.31}$$

The voltage ratio at maximum torque $(E_o/E_i)_{max}$ is obtained by substituting Eq. (6.31) into Eq. (6.27). Thus,

$$\left(\frac{E_o}{E_i}\right)_{max} = \frac{S_\tau S_g(1 + v)}{E} \tag{6.32}$$

If the torque cell is fabricated from heat-treated steel ($S_\tau \approx 60{,}000$ psi),

$$\left(\frac{E_o}{E_i}\right)_{max} = \frac{60{,}000(2)(1 + 0.30)}{30{,}000{,}000} = 5.2 \text{ mV/V}$$

Typically, torque cells are rated at values of $(E_o/E_i)^*$ between 4 and 5 mV/V. The torque $T$ corresponding to an output voltage $E_o$ is then given by Eq. (6.7) as

$$T = \frac{E_o/E_i}{(E_o/E_i)^*} T_{max} \tag{6.7}$$

## Torque Cells—Data Transmission

Frequently, torque must be measured on a rotating shaft, which necessitates signal transmission between a Wheatstone bridge on the rotating shaft and a stationary instrumentation center. Signal transmission under these circumstances is usually accomplished with either slip rings or telemetry.

### Signal Transmission with Slip Rings

A schematic illustration of a slip-ring connection between a Wheatstone bridge on a rotating shaft and a recording instrument at a stationary location is shown in Fig. 6.8. The slip-ring assembly contains a series of insulated rings mounted on a shaft and a companion series of insulated brushes mounted in a case. High-speed bearings between the shaft and the case permit the case to remain stationary while the shaft rotates with the torque cell. A commercial slip-ring assembly is shown in Fig. 6.9.

The major problem associated with slip-ring usage is noise (generated by contact resistance variations between the rings and brushes). These contact

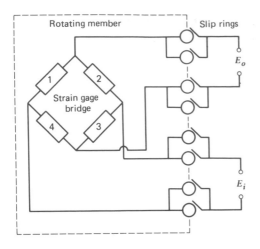

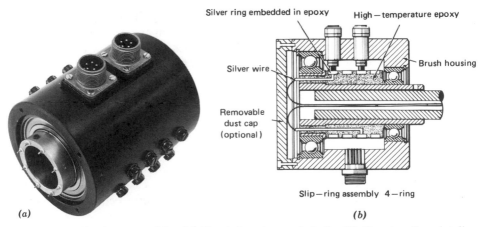

*Figure 6.8* Schematic illustration of a slip-ring connection between a rotating member and a fixed instrumentation station.

resistance variations can be kept within acceptable limits if the rings are fabricated from monel metal (a copper–nickel alloy), if the brushes are fabricated from a silver–graphite mixture, and if the ring-brush contact pressure is maintained between 50 and 100 psi. Rotational speed limits of slip-ring assemblies are determined by the concentricity that can be maintained between the shaft and the case and by the quality of the bearings. Slip-ring units with speed ratings of 6000 rpm are available.

### Signal Transmission with Telemetry

In many applications, the end of the shaft is not accessible for mounting of the slip-ring unit; therefore, direct connection through slip rings is not possible and telemetry must be used to transmit the signal from the rotating shaft to the

*Figure 6.9* Slip-ring assembly. (*a*) Brush housing and shaft. (*b*) Construction details. (Courtesy of Lebow Products, Eaton Corp.)

recording instrument. In a relatively simple telemetry system, the output from the Wheatstone bridge is used to modulate a radio signal. The strain gages, bridge, power supply, and radio transmitter are mounted on the rotating shaft, while the receiver and recorder are located at a stationary instrumentation center. In most applications, the distance over which the signal must be transmitted is only a few feet; therefore, low-power transmitters, which do not need to be licensed, can be used.

A commercially available short-range telemetry system, designed to measure torque on a rotating shaft, is shown in Fig. 6.10. The system utilizes a split collar that fits over the shaft. The collar contains a power supply for the bridge, a modulator, a voltage-controlled oscillator (VCO), and an antenna. The signal from the bridge is used to pulse-width modulate a constant-amplitude 5-kHz square wave (i.e., the width of the square wave is proportional to the voltage output from the bridge). The square wave is used to vary the frequency of the VCO, which is centered at 10.7 MHz. The VCO signal is transmitted at a very low power level by the rotating antenna in the split collar. The signal is received by a stationary loop antenna that encircles the split collar, as shown in Fig. 6.10. The transmitting unit is completely self-contained, since the power to drive the bridge, modulator, and voltage-controlled oscillator is obtained by inductively coupling a 160-kHz signal to the power supply through the stationary loop antenna.

For longer-range, multiple-transducer applications, a much more complex telemetry system must be used. Also, licensing is required, since the frequencies available for transmissions are limited and crowded. In the United States, telemetry transmissions are restricted to two frequency bands: 1435 to 1535 MHz and 2200 to 2300 MHz. As the number of transducers is increased in an application where telemetry must be employed for signal transmission, it becomes less and less practical to have a separate transmitter and receiver for each signal. Instead, the long-range, multiple-transducer systems combine signals from a number of transducers into a single signal for transmission. The process for combining individual signals from several transducers into a composite signal is called *multiplexing;* the two types of multiplexing commonly used are *frequency-division multiplexing* and *time-division multiplexing.* At the radio reciever, the individual signals are separated from the composite signal and recorded on individual recording instruments.

As an example of frequency-division multiplexing, consider the system shown schematically in Fig. 6.11 where three transducer outputs are used to frequency modulate three subcarrier frequencies that are then combined and transmitted as a single signal. As shown in Fig. 6.11, the oscillator for one transducer is centered at 400 Hz, and the output signal from the transducer produces a maximum deviation from this frequency of $\pm 30$ Hz. Similarly, the oscillator for transducer 2 is centered at 560 Hz with a maximum deviation of $\pm 42$ Hz, and the oscillator for transducer 3 is centered at 730 Hz with a maximum deviation of $\pm 55$ Hz. Note that there is no overlapping between channels and that guard

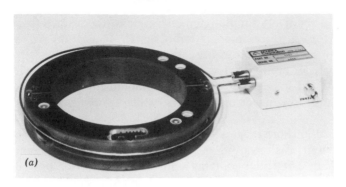

(a)

(b)

(c)

*Figure 6.10* Telemetry system for data transmission. (*a*) Rotating collar and stationary loop antenna. (*b*) Read-out and display unit. (*c*) Torsion-measuring system on a shaft. (Courtesy of Acurex Corporation.)

bands lie between them to ensure separation. When the signals from the three channels are mixed, they form a composite signal having a frequency range from 370 Hz to 785 Hz. The composite signal is then transmitted over a radio link at a transmitting frequency of 2200 Mhz. At the receiving station, bandpass filters are used to separate the channels. The recovered frequency bands are then sent to individual discriminators for demodulation and recovery of the individual transducer signals for display on recording instruments.

With time-division multiplexing, all channels use the same portion of the frequency spectrum, but not at the same time. Each channel is sampled in a repeated sequence to give a composite signal consisting of time-spaced segments of the signals from each transducer. Since the individual channels are not monitored continuously, the sampling rate must be sufficient to ensure that the individual signal amplitudes do not change significantly during the time between samples. Sampling rates about five times greater than the highest frequency component in any transducer signal are used in most telemetry systems that utilize time-division multiplexing.

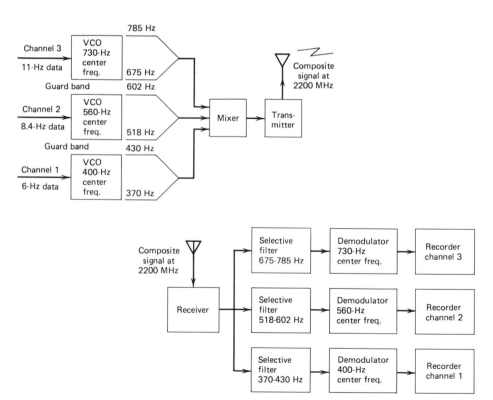

*Figure 6.11*   Schematic diagram of a data transmission system that utilizes frequency-division multiplexing.

## 6.4   PRESSURE MEASUREMENTS (Pressure Transducers)

*Pressure transducers* are devices that convert pressure into an electrical signal through a measurement of either displacement, strain, or piezoelectric response. The operating characteristics for each of these transducer types are covered in the following subsections.

### Displacement-Type Pressure Transducer

A common type of pressure transducer in which a measurement of displacement is used to convert pressure to an electrical output voltage is illustrated in Fig. 6.12. This transducer utilizes a bourdon tube as the elastic element and a linear variable-differential transformer (LVDT) as the sensor. The bourdon tube is a C-shaped pressure vessel with a flat-oval cross section that tends to straighten as internal pressure is applied. In the displacement-type pressure transducer, one end of the bourdon tube is fixed, while the other end is free to displace. The core of the LVDT is attached to the free end of the bourdon tube and to a small cantilever spring that maintains tension on the core assembly. The coil of the LVDT is attached to the housing that anchors the fixed end of the tube (see Fig. 6.12).

As pressure is applied to the bourdon tube, the core of the LVDT is pulled through the coil and an output voltage develops. The output voltage is a linear function of the pressure provided the displacement of the bourdon tube is kept

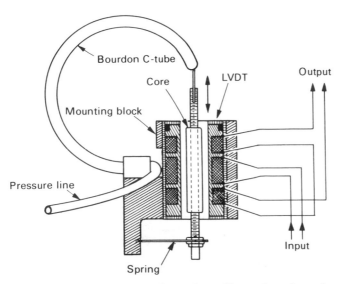

*Figure 6.12*   Pressure transducer that utilizes a bourdon tube as the elastic element and a linear variable-differential transformer as the sensor.

small. Displacement-type transducers of this type provide stable and reliable measurements of pressure over extended periods of time. Such transducers are excellent for static or quasi-static applications; however, they are not suitable for dynamic measurements of pressure, since the mass of the tube and core limits frequency response to approximately 10 Hz.

## Diaphragm-Type Pressure Transducers

A second general type of pressure transducer utilizes either a clamped circular plate (diaphragm) or a hollow cylinder as the elastic element and electrical resistance strain gages as the sensor. Diaphragms are used for the low- and middle-pressure ranges (0 to 30,000 psi), while cylinders are mostly used for the high- and very-high-pressure ranges (30,000 to 100,000 psi). The strain distribution resulting from a uniform pressure on the face of a clamped circular plate of constant thickness is given by the following expressions:

$$\varepsilon_{rr} = \frac{3p(1 - v^2)}{8Et^2} (R_o^2 - 3r^2)$$

$$\varepsilon_{\theta\theta} = \frac{3p(1 - v^2)}{8Et^2} (R_o^2 - r^2) \tag{6.33}$$

**where**  $p$  is the pressure.
  $t$  is the thickness of the diaphragm.
  $R_o$ is the outside radius of the diaphragm.
  $r$  is a position parameter.

Examination of Eqs. (6.33) indicates that the circumferential strain $\varepsilon_{\theta\theta}$ is always positive and assumes its maximum value at $r = 0$. The radial strain $\varepsilon_{rr}$ is positive in some regions but negative in others and assumes its maximum negative value at $r = R_o$. Both distributions are shown in Fig. 6.13.

A special-purpose diaphragm strain gage, which has been designed to take advantage of this strain distribution, is widely used in diaphragm-type pressure transducers. Circumferential elements are employed in the central region of the diaphragm where $\varepsilon_{\theta\theta}$ is a maximum. Similarly, radial elements are employed near the edge of the diaphragm where $\varepsilon_{rr}$ is a maximum. Also, the circumferential and radial elements are each divided into two parts, as shown in Fig. 6.14, so that the special-purpose gage actually consists of four separate gages. Terminals are provided that permit the individual gages to be connected into a Wheatstone bridge with the circumferential elements in arms $R_1$ and $R_3$ and the radial elements in arms $R_2$ and $R_4$. If the strains are averaged over the areas of the circumferential and radial elements, and if the average values of $\Delta R/R$ (with a gage factor $S_g = 2$) obtained from Eq. (3.5) are substituted into Eq. (4.22), the output voltage $E_o$ is given by

$$E_o = 0.82 \frac{pR_o^2(1 - v^2)}{Et^2} E_i \qquad (6.34)$$

Similarly, the pressure $p$ is given in terms of the output voltage $E_o$ as

$$p = 1.22 \frac{Et^2}{R_o^2(1 - v^2)E_i} E_o = CE_o \qquad (6.35)$$

Equation (6.35) indicates that the pressure $p$ is linearly proportional to the output voltage $E_o$ and that the constant of proportionality or calibration constant $C$ is given by

$$C = 1.22 \frac{Et^2}{R_o^2(1 - v^2)E_i} \qquad (6.36)$$

Special-purpose diaphragm strain gages are commercially available in seven sizes, ranging from 0.187 to 1.25 in. (4.75 to 31.8 mm). The gage factors for these gages is approximately 2.

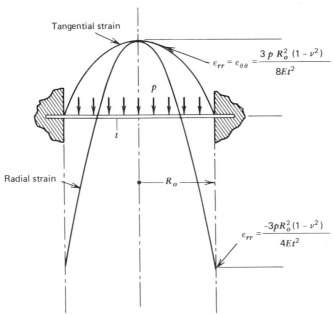

**Figure 6.13** Strain distribution in a thin clamped circular plate (diaphragm) due to a uniform lateral pressure.

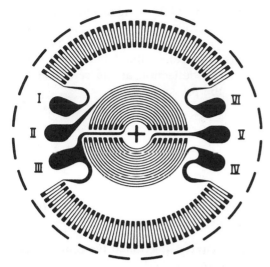

**Figure 6.14** Special-purpose four-element strain gage for diaphragm-type pressure transducers. (Courtesy of Micro-Measurements.)

The sensitivity of the diaphragm pressure transducer–Wheatstone bridge combination is given by Eq. (1.2) as

$$S = \frac{E_o}{P} = \frac{1}{C} = 0.82 \frac{R_o^2(1 - v^2)E_i}{Et^2} \tag{6.37}$$

The voltage $E_i$ that can be applied to the Wheatstone bridge is controlled by the power $P_T$, which can be dissipated by the gage elements. The voltage–power relationship for a four-equal-arm Wheatstone bridge is given by Eq. (4.27) as

$$E_i = 2\sqrt{P_T R_T} \tag{a}$$

Substituting Eq. (a) into Eq. (6.37) gives

$$S = 1.64 \frac{R_o^2(1 - v^2)\sqrt{P_T R_T}}{Et^2} \tag{6.38}$$

It is clear from Eq. (6.38) that the sensitivity of a diaphragm-type pressure gage can be varied over a very wide range by adjusting the geometry $(R_o/t)$ of the diaphragm. Maximum sensitivity will occur at $(R_o/t)_{max}$. Unlike most other transducers, diaphragm deflection rather than yield strength determines the limit of $(R_o/t)_{max}$.

The relationship between pressure $p$ and output voltage $E_o$, given by Eq. (6.35) is linear; however, this relationship is valid only if the deflection of the diaphragm is small. As the deflection of the diaphragm become large, the diaphragm begins to act like a pressurized shell rather than a laterally loaded plate, and strain is no longer a linear function of the pressure. The relationship between pressure and output voltage will be linear to within 0.3 percent if the deflection $\delta_c$ at the center of the diaphragm is less than $t/4$. The deflection at the center of the diaphragm $\delta_c$ can be expressed in terms of the pressure as

$$\delta_c = \frac{3pR_o^4(1 - \nu^2)}{16t^3E} \tag{6.39}$$

With the restriction that $\delta_c \leq t/4$,

$$P_{max} \leq \frac{4}{3}\left(\frac{t}{R_o}\right)^4 \frac{E}{1 - \nu^2} \tag{6.40}$$

The maximum sensitivity is also controlled by the deflection at the center of the diaphragm. Substituting Eq. (6.40) into Eq. (6.38) yields

$$S_{max} = 2.19 \left(\frac{t}{R_o}\right)^2 \frac{\sqrt{P_T R_T}}{P_{max}} \tag{6.41}$$

Several diaphragm-type pressure transducers are shown in Fig. 6.15. The units are all small and compact and exhibit a relatively high frequency response.

(a) 0 – 5000 psi          (b) 0 – 10,000 psi

*Figure 6.15*  Diaphragm-type pressure transducers. (Courtesy of BLH Electronics.)

The frequency limits of the diaphragm-type pressure transducer depends primarily upon the degree of damping provided by the fluid in contact with the diaphragm. A general rule often followed in determining the frequency limit is that the resonant frequency of the diaphragm should be three to five times higher than the highest frequency associated with the applied dynamic pressure. The resonant frequency $f_r$ for a diaphragm is expressed as

$$f_r = 0.471 \frac{t}{R_o^2} \sqrt{\frac{gE}{w(1 - v^2)}}$$ (6.42)

**where** $f_r$ is the resonant frequency in hertz.

$g$ is the gravitational constant (386.4 in./s² or 9815 mm/s²).

$w$ is the specific weight of the diaphragm material (lb/in.³ or N/mm³).

Typical values of $f_r$ range from 10 to 50 kHz depending on the material and the ratio $t/R_o$. Thus, the diaphragm-type pressure transducer can be utilized over a wide range of frequencies from static measurements to dynamic measurements involving frequencies as high as 10 kHz.

## Piezoelectric-Type Pressure Transducers

The piezoelectric-type of pressure transducer uses a piezoelectric crystal (see Section 3.7) as both the elastic element and the sensor. Quartz is the most widely used piezoelectric material because of its high modulus of elasticity, high resonant frequency, linearity over several decades, and very low hysteresis. Resonant frequencies of 0.25 to 0.50 MHz can be achieved while maintaining relatively high sensitivity.

Construction details of a piezoelectric pressure transducer are shown in Fig. 6.16. The quartz crystal is enclosed in a cylindrical shell that has a thin pressure-transmitting diaphragm on one end and a rigid support base for the crystal on the other end. As pressure is applied to the face of the crystal in contact with the diaphragm, an electrostatic charge is generated. The magnitude of the charge depends upon the pressure, the size of the crystal, and the orientation of the crystal axes as indicated by Eq. (3.16). Miniature pressure transducers that utilize a quartz crystal having a diameter of 6 mm and a length of 6 mm exhibit a sensitivity of approximately 1 pC/psi. Pressure transducers with larger crystals (11-mm diameter and 12-mm length) often exhibit sensitivities of 5 pC/psi.

Piezoelectric transducers all exhibit an extremely high output impedance that depends upon the frequency of the applied pressure as indicated by Eq. (3.20). Because of this high output impedance, a charge amplifier must be inserted between the transducer and any conventional voltage measuring instrument. The charge amplifier converts the charge to a voltage, amplifies the voltage,

*(a)*

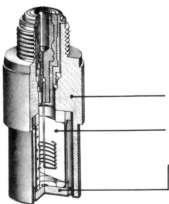

Quartz pressure transducers consist basically of three parts (see sectional drawing):

1. Transducer housing, which serves for
   mounting and encloses the quartz elements hermetically at the same time.
2. Quartz elements, yielding an electrical
   charge proportional to the pressure.
3. Diaphragm, welded tightly with the transducer housing and transmitting to the
   quartz elements the pressure exerted by
   the medium.

*(b)*

*Figure 6.16* A piezoelectric pressure transducer. (Courtesy of Kistler Instrument Corp.)
(*a*) Pressure transducer with its connector adaptor. (*b*) Construction details.

and provides an output impedance of approximately 100 $\Omega$, which is satisfactory for most voltage measuring instruments.

A schematic diagram that illustrates the operating principle of a charge amplifier is shown in Fig. 6.17. The first component in the circuit is a high-impedance operational amplifier (op-amp) with capacitive feedback. This component serves as an integrator and converts the charge to a voltage. The second op-amp, with an adjustable feedback resistor, is used to amplify the voltage by specified amounts. The input voltage $E_i$ is very small because of the high open-loop gain of the first op-amp and the capacitive feedback; therefore, the charge from the crystal is simply transferred to the capacitor $C_1$. The feedback resistor $R_1$ is placed in parallel wth capacitor $C_1$ to control the time constant and eliminate drift. The charge amplifier is zeroed by shorting (closing the switch) across the integrating capacitor $C_1$.

Charge amplifiers are commercially available with input impedances of $10^{14}$ $\Omega$ and maximum output voltages of 10 V. Charge ranges can vary from 10 to 500,000 pC. Time constants depend upon the resistance $R_1$ and for high values of $R_1$, time constants of 100,000 s are possible. Thus, piezoelectric transducers can be used for both dynamic and quasi-static pressure measurements. Since the voltage $E_i$ is small, the capacitance of the cable (up to 1000 pF) does not affect the measurement of the charge $q$.

A complete discussion of the performance characteristics of dynamic pressure transducers and the circuits used in their application is presented in Sections 7.6 and 7.7.

Piezoelectric transducers can also be used for very-high-pressure (up to 100,000 psi) measurements and for pressure measurements at temperatures as high as 350°C. If water cooling is used to protect the crystal and its insulation, the temperature range can be extended.

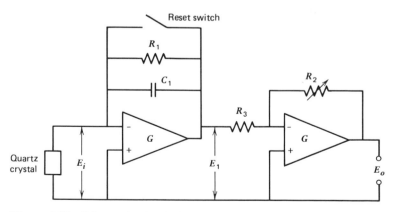

*Figure 6.17*  Schematic circuit for a charge amplifier.

## 6.5  COMBINED MEASUREMENTS

In some applications, two or more quantities must be measured simultaneously. This can be accomplished by using two or more separate strain-gage bridges on a single elastic element or by using selected combinations of gages from a single bridge on the elastic element. The combinations that can be designed are unlimited (for example, six component systems are commercially available for wind tunnel measurements of the three force components $P_x$, $P_y$, and $P_z$ simultaneous with the three moment components $M_x$, $M_y$, and $M_z$). In order to illustrate the concepts while restricting the treatment to a reasonable length, only two different combined-measurement transducers will be discussed; namely, the force–moment transducer and the force–torque transducer.

### Force–Moment Measurements

A transducer for measuring both force and moment can be designed by using a simple link as the elastic member as shown in Fig. 6.18. For simplicity, consider the link to have a square cross section ($A = h^2$) and assume that the strain gages are mounted on the centerline of each side in the longitudinal or load $P_z$ direction. The force $P_z$ is measured by wiring gages A and C into positions 1 and 3 of a Wheatstone bridge as shown in Fig. 6.19a. Resistances $R_2$ and $R_4$ are fixed-value resistors, with $R_2 = R_4 = R_g$. Under these conditions, Eq. (4.22) reduces to

$$E_o = \frac{1}{4} \left[ \frac{\Delta R_1}{R_1} + \frac{\Delta R_3}{R_3} \right] E_i \tag{a}$$

The response of the strain gages is given by Eq. (3.5) as

$$\frac{\Delta R_1}{R_1} = \frac{\Delta R_3}{R_3} = S_g \varepsilon = \frac{S_g P_z}{AE} \tag{b}$$

Substituting Eq. (b) into Eq. (a) and solving for $P_z$ yields

$$P_z = \frac{2AE}{S_g E_i} E_o = C E_o \tag{6.43}$$

The sensitivity of this combined force–moment transducer to the force $P_z$ is given by Eq. (1.2) as

$$S = \frac{E_o}{P_z} = \frac{1}{C} = \frac{S_g E_i}{2AE} \tag{6.44}$$

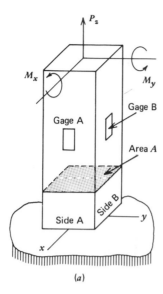

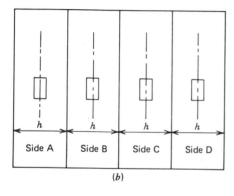

**Figure 6.18** Combined-measurement transducer used to measure axial load $P_z$ and moments $M_x$ and $M_y$. (a) Elastic element with strain gages. (b) Developed surface showing strain-gage orientations.

A comparison of this sensitivity with that of the four-gage force transducer given by Eq. (6.4) shows that there has been a loss in sensitivity of $v/(1 + v)$ or approximately 25 percent. This loss of sensitivity is the price paid for reserving gages B and D for moment $M_x$ measurements.

The moment $M_x$ is measured by wiring gages B and D into the Wheatstone bridge as shown in Fig. 6.19b. With gages B and D in arms $R_1$ and $R_4$ of the bridge and with fixed resistors of equal value in arms $R_2$ and $R_3$, the output of the bridge is given by Eq. (4.22) as

$$E_o = \frac{1}{4}\left[\frac{\Delta R_1}{R_1} - \frac{\Delta R_4}{R_4}\right]E_i \qquad (c)$$

The response of the strain gages is given by Eq. (3.5) as

$$\frac{\Delta R_1}{R_1} = -\frac{\Delta R_4}{R_4} = S_g \varepsilon = \frac{6 S_g M_x}{E h^3} \tag{d}$$

Substituting Eq. (d) into Eq. (c) and solving for $M_x$ gives

$$M_x = \frac{E h^3}{3 S_g E_i} E_o = C E_o \tag{6.45}$$

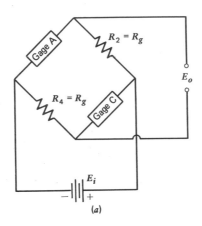

(a)

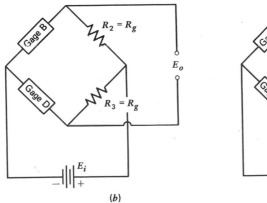

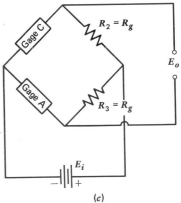

(b)     (c)

**Figure 6.19** Wheatstone-bridge arrangements used with a combined-measurement transducer to measure an axial force $P_z$ and moments $M_x$ and $M_y$. (a) Bridge arrangement for measuring force $P_z$. (b) Bridge arrangement for measuring moment $M_x$. (c) Bridge arrangement for measuring moment $M_y$.

Similarly, the moment $M_y$ can be measured by wiring gages C and A into arms $R_1$ and $R_4$, respectively, and using fixed-value resistors of equal value in arms $R_2$ and $R_3$, as indicated in Fig. 6.19c. Thus

$$M_y = \frac{Eh^3}{3S_g E_i} E_o = CE_o \tag{6.46}$$

Transducers designed for combined measurements are usually equipped with switch boxes that contain the bridge completion resistors and the wiring needed to position the gages in the proper bridge positions. Care must be exercised during certain measurements, since temperature compensation is not maintained within the Wheatstone bridge.

## Force–Torque Measurements

A transducer designed to measure both force $P_z$ and torque $M_z$ is shown in Fig. 6.20. For the force measurment, gages A and C are wired into arms $R_1$ and $R_3$ of the Wheatstone bridge, and fixed resistors of equal value are used in the other two positions. This arrangement is identical to that shown in Fig. 6.19a; therefore, Eq. (6.43) applies.

A torque measurement is made by connecting gages B and D into arms $R_1$ and $R_4$ of the Wheatstone bridge as shown in Fig. 6.21. With fixed resistors of equal value in the other two arms of the bridge, Eq. (4.22) reduces to

$$E_o = \frac{1}{4} \left[ \frac{\Delta R_1}{R_1} - \frac{\Delta R_4}{R_4} \right] E_i \tag{e}$$

The gage response for this transducer can be expressed as follows:

$$\frac{\Delta R_1}{R_1} = -\frac{\Delta R_4}{R_4} = S_g \varepsilon = \frac{16(1 + v)S_g M_z}{\pi D^3 E} \tag{f}$$

Substituting Eq. (f) into Eq. (e) and solving for $M_z$ yields

$$M_z = \frac{\pi D^3 E}{8 S_g (1 + v) E_i} E_o = CE_o \tag{6.47}$$

A comparison of Eq. (6.47) with Eq. (6.28) for the standard torque cell indicates that the sensitivity of the combination transducer is lower by a factor of 2. The loss of sensitivity is due to the fact that gages A and C have been reserved for measurement of the axial load $P_z$.

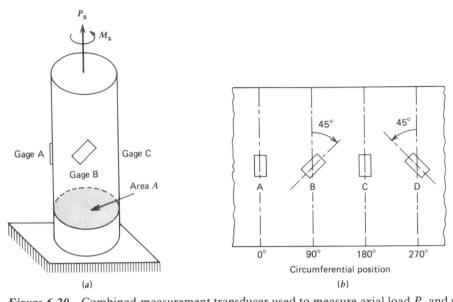

*(a)*

*(b)*

***Figure 6.20*** Combined-measurement transducer used to measure axial load $P_z$ and moment $M_z$. (*a*) Elastic element with strain gages. (*b*) Developed surface showing strain-gage orientations.

## 6.6 MINIMIZING ERRORS IN TRANSDUCERS

Most transducers, designed to measure force, torque, or pressure, utilize electrical resistance strain gages as sensors because they are inexpensive, easy to install, and provide an output voltage $E_o$ (when used as elements of a Wheatstone bridge) that can be related easily to the load, torque, or pressure. In applications of strain gages to stress analysis, accuracies of $\pm 2$ percent are acceptable. When strain gages are used as sensors in transducers, however,

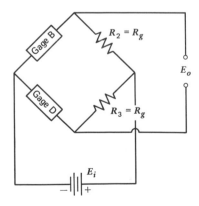

***Figure 6.21*** Wheatstone-bridge arrangement for torque measurements with a force–torque transducer.

accuracy requirements are an order of magnitude more stringent; therefore, more care must be exercised in the selection and installation of the gages and in the design of the Wheatstone bridge. Typical performance specifications for general-purpose, improved-accuracy, and high-accuracy load cells are listed in Table 6.2.

Errors that degrade the accuracy of a transducer include dual sensitivity, zero shift with temperature change, bridge balance, span adjust, and span change with temperature change. Each of these sources of error is discussed in the following subsections together with procedures for minimizing error.

## Dual Sensitivity

All transducers exhibit a *dual sensitivity*, to some small degree, which means that the output voltage is due to both a primary quantity, such as load, torque, or pressure, and a secondary quantity, such as temperature or a secondary load. Provision must be made during design of the transducer to minimize the secondary sensitivity.

## Dual Sensitivity—Temperature

As an example of dual sensitivity due to temperature, consider a link-type load cell subjected to both a load $P_z$ and a temperature change $\Delta T$ during the readout period. The strain gages on the link will respond to both the strain $\varepsilon$

**TABLE 6.2** Specifications for Load-Cell Accuracies

| Characteristic | General-Purpose Load Cell | Improved-Accuracy Load Cell | High-Accuracy Load Cell |
|---|---|---|---|
| Calibration inaccuracy | 0.5% FS[a] | 0.25% FS | 0.1% FS |
| Temperature effect on zero | ±0.005%/°F FS | ±0.0025%/°F FS | ±0.0015%/°F FS |
| Zero-balance error | ±5% FS | ±2½% FS | ±1% FS |
| Temperature effect on span | ±0.01%/°F OL[b] | ±0.005%/°F OL | ±0.008%/°F OL |
| Nonlinearity | 0.25% FS | 0.1% FS | 0.05% FS |
| Hysteresis | 0.1% FS | 0.05% FS | 0.02% FS |
| Nonrepeatability | 0.1% FS | 0.05% FS | 0.02% FS |
| System inaccuracy[c] | 1% FS | ½% FS | 0.15% FS |

[a] Full scale.
[b] Of load.
[c] Combined effects, but not including temperature.

produced by the load and the apparent strain $\varepsilon'$ produced by the temperature change. The total response of each gage will appear as

$$\frac{\Delta R_1}{R_1} = \left.\frac{\Delta R_g}{R_g}\right)_{P_z} + \left.\frac{\Delta R_g}{R_g}\right)_{\Delta T} = S_g \varepsilon_1 + S_g \varepsilon_1' \tag{a}$$

Similar expressions will apply for the other three gages. If the strain gages are identical and if the temperature change for each gage is the same, Eq. (4.22) indicates that the response of the gages due to temperature will cancel and the output of the Wheatstone bridge will be a function only of the load-induced strains in the elastic element (link) of the transducer. In this example, the signal summing property of the Wheatstone bridge provides temperature compensation of the load cell.

### Dual Sensitivity—Secondary Load

When link-type load cells are used for force measurements, it is usually difficult to apply the load $P_z$ exactly on and parallel to the centroidal axis of the link. As a consequence, both the load $P_z$ and a bending moment $M$ are imposed on the link and it is necessary to design the transducer with a very low sensitivity to $M$ while maintaining a high sensitivity to $P_z$. This objective is accomplished in the link-type load cell by proper placement of the strain gages.

As an example, consider that an arbitrary moment $M$ is being applied to the cross section of the elastic element of the load cell as shown in Fig. 6.22. If the moment $M$ is resolved into cartesian components $M_x$ and $M_y$, the effect of $M_x$ will be to bend the element about the $x$ axis such that

$$\varepsilon_{a1} = -\varepsilon_{a3} \tag{a}$$

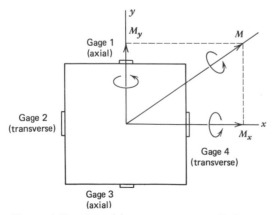

**Figure 6.22**   An arbitrary moment applied to a cross section of the elastic element in a link-type load cell.

Since the transverse gages are on the neutral axis for bending about the $x$ axis,

$$\varepsilon_{t2} = \varepsilon_{t4} = 0 \tag{b}$$

Equations (a) and (b) indicate that the response of the gages to a moment $M_x$ will be

$$\frac{\Delta R_1}{R_1} = -\frac{\Delta R_3}{R_3} \tag{c}$$

$$\frac{\Delta R_2}{R_2} = \frac{\Delta R_4}{R_4} = 0 \tag{d}$$

When Eqs. (c) and (d) are substituted into Eq. (4.22), the output resulting from strain-gage response to the moment $M_x$ vanishes. It is easy to show in a similar way that output resulting from the moment $M_y$ also vanishes. Since neither $M_x$ nor $M_y$ produce an output, any arbitrary moment $M$ can be applied to the load cell without influencing the measurement of the load $P_z$. In this example, proper placement of the strain gages eliminates any sensitivity to the secondary load.

## Zero Shift with Temperature Change

It was shown in Section 5.9 that some electrical resistance strain gages are temperature compensated (resistance changes due to temperature change are minimized through proper selection of the gage alloy) over a limited range of temperature. When temperature variations are large, small changes in resistance occur and zero output under zero load is not maintained.

Zero shift with temperature change is reduced by using either half or full Wheatstone bridges as discussed previously. Here, temperature-induced resistance changes are partially canceled by the summing properties of the Wheatstone bridge. However, since the strain gages are never identical, some zero shift persists.

A third compensation procedure for reducing zero shift in transducers is illustrated in Fig. 6.23. Here, a low-resistance copper ladder gage is inserted between arms 3 and 4 of the Wheatstone bridge. Since the ladder gage is a part of both arms, it makes a positive contribution to both $\Delta R_3$ and $\Delta R_4$ when the temperature is increased. During calibration (which involves temperature cycling of the transducer over its specified range of operation), the ladder gages are trimmed ($\Delta R_3$ and $\Delta R_4$ due to temperature change are adjusted) until the zero shift for this particular transducer is within acceptable limits.

## Bridge Balance

In general, transducers should exhibit zero output under no-load conditions. Unfortunately, the strain gages employed as sensors are not perfectly matched (do not have exactly the same resistance); therefore, the Wheatstone bridge is

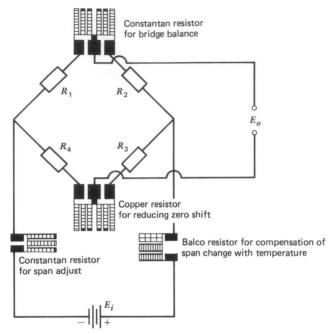

*Figure 6.23*   Compensation resistors that can be introduced into the Wheatstone bridge of a transducer to minimize the effects of temperature change. (After Dorsey.)

usually out-of-balance under the no-load condition. Balance can be achieved by inserting a compensation resistor between arms 1 and 2 of the bridge, as shown in Fig. 6.23. The compensation resistor is a double-ladder gage that can be trimmed to add either $\Delta R_1$ or $\Delta R_2$ until nearly perfect balance is achieved.

## Span Adjust

*Span* refers to the sensitivity of the transducer. In large instrumentation systems, where transducers are often interchanged or replaced, it is important to have transducers whose span can be adjusted to a preselected level. The span or sensitivity is usually adjusted by using a temperature-insensitive resistor (ladder gage) in series with the voltage supply, as indicated in Fig. 6.23. As the ladder gage reduces the supply voltage applied to the bridge, the span of the transducer is reduced to the specified value (usually 3 mV/V full scale).

## Span Change with Temperature

Compensation of span change with temperature is difficult because the procedure involves simultaneous application of load and cycling of temperature. Usually, temperature compensation of span is achieved by inserting a resistor that changes with temperature (a nickel–iron alloy known as Balco) in the second lead from the voltage supply, as shown in Fig. 6.23. This resistor (also of the ladder type) can be trimmed to give a resistance change with temperature that compensates for changes in sensitivity with temperature.

Considerable fine tuning of a transducer is required to achieve the accuracies specified in Table 6.2. Also, periodic recalibration is needed to ensure that the transducer is operating within specified limits of accuracy. If these high accuracies are not required, then some of the compensation procedures will not be necessary and lower cost transducers can be specified.

## 6.7 FREQUENCY RESPONSE OF TRANSDUCERS

Static or quasi-static (vary slowly with time) measurements are relatively simple to make when compared to dynamic measurements. Dynamic measurements are difficult to make because of the effects of frequency response of each element in the instrumentation system on the amplitude and phase of the quantity being measured. Previously, the dynamic response of a galvanometer was considered (see Section 2.5) to illustrate the importance of response characteristics to the recording process. The response characteristics of a transducer are equally important since serious errors can be introduced in dynamic measurements if the frequency response of the transducer is not adequate.

Transducers for measuring load, torque, and pressure are all second-order systems; therefore, their dynamic behavior can be described by a second-order differential equation similar to Eq. (2.33). The application of second-order theory to transducers will be illustrated by considering a link-type load cell with a uniaxial elastic member such as the one illustrated in Fig. 6.2. The dynamic response of this load cell can be described by the differential equation of motion for the mass–spring–dashpot combination shown in Fig. 6.24. The elastic member of the load cell is represented by the spring. The spring modulus or spring constant $k$ is given by the expression

$$k = \frac{P}{\delta} = \frac{AE}{L} \tag{6.48}$$

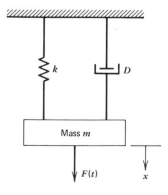

*Figure 6.24* A spring–mass–dashpot representation of a link-type load cell.

**where** $P$ is the load.

$\delta$ is the extension or contraction of the elastic member.

$A$ is the cross-sectional area of the elastic member.

$E$ is the modulus of elasticity of the material from which the elastic member is fabricated.

$L$ is the length of the elastic member.

The dashpot represents the parameters in the transducer system (such as internal friction) that produce damping. In load and torque cells, damping is usually a very small quantity; however, in pressure transducers, damping is larger since the fluid interacts with the diaphragm. The mass is a lumped mass consisting of the mass of the object to which the transducer is fastened plus the effective mass of the elastic element of the transducer. The force $F(t)$ acts on the moving object and is the quantity being measured. The position parameter $x$ describes the motion (displacement) of the mass $m$ as a function of time. The dynamic response of this second-order system is described by the following differential equation:

$$\frac{1}{\omega_n^2}\frac{d^2x}{dt^2} + \frac{2d}{\omega_n}\frac{dx}{dt} + x = \frac{F(t)}{k} \tag{6.49}$$

where $\omega_n$ is the natural frequency of the system ($\omega_n = \sqrt{k/m}$)

The natural frequency $\omega_n$ of the system depends upon the mass of the object and the effective mass of the transducer that for the link-type load cell can be approximated as

$$m = m_o + (m_t/3)$$

**where** $m_o$ is the mass of the object.

$m_t$ is the mass of the elastic element in the transducer.

Thus, both the mass of the transducer and the mass of the object affect the fidelity of the measurement of $F(t)$.

The fidelity of the measurement of $F(t)$ depends primarily on the rise time associated with $F(t)$. In the treatment of galvanometers and other electronic recording instruments, $F(t)$ is usually considered to be a step function, since electrical signals can be applied almost instantaneously. In mechanical systems, however, it is not realistic to consider $F(t)$ as a step function since application of $F(t)$ requires some finite time even in the most severe dynamic application. For this reason, a more realistic forcing function for mechanical systems is the terminated ramp function shown in Fig. 6.25. This terminated ramp function

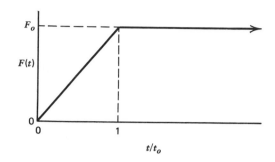

**Figure 6.25** A terminated ramp function type of input to a load cell.

can be expressed in equation form as

$$F(t) = \frac{F_o t}{t_o} \qquad \text{for } 0 \le t \le t_o$$

$$F(t) = F_o \qquad \text{for } t > t_o \qquad (6.50)$$

Since the degree of damping in the link-type load cell is very small, only the underdamped solution to Eq. (6.49) needs to be considered. After Eq. (6.50) is substituted into Eq. (6.49), the differential equation can be solved to yield

**Homogeneous solution:**

$$x_h = e^{-d\omega_n t}[C_1 \sin\sqrt{1 - d^2}\,\omega_n t + C_2 \cos\sqrt{1 - d^2}\,\omega_n t] \qquad (a)$$

**Particular solution:**

$$x_p = \frac{F_o}{kt_o}\left(t - \frac{2d}{\omega_n}\right) \qquad \text{for } 0 \le t \le t_o$$

$$x_p = \frac{F_o}{k} \qquad \text{for } t > t_o \qquad (b)$$

The coefficients $C_1$ and $C_2$ in Eq. (a) are obtained by using the initial conditions for the system. These conditions are

$$x = 0 \qquad \text{and} \qquad \frac{dx}{dt} = 0 \qquad \text{at } t = 0$$

$$x(t_o^+) = x(t_o^-) \qquad \text{and} \qquad \frac{dx}{dt}(t_o^+) = \frac{dx}{dt}(t_o^-) \qquad \text{at } t = t_o \qquad (c)$$

The general solution $(x = x_h + x_p)$ obtained from Eqs. (a), (b), and (c) is

$$\frac{x}{x_o} = \frac{-e^{-d\omega_n t}}{\omega_n t_o \sqrt{1 - d^2}}\left[\sin(\sqrt{1 - d^2}\,\omega_n t \oplus \phi)\right]$$

$$+ \frac{1}{t_o}\left[t - \frac{2d}{\omega_n}\right] \qquad \text{for } 0 \le t \le t_o \qquad (6.51)$$

$$\frac{x}{x_o} = 1 - \frac{e^{-d\omega_n t}}{\omega_n t_o \sqrt{1 - d^2}} \Big[ \sin(\sqrt{1 - d^2}\,\omega_n t - \phi)$$

$$- e^{-d\omega_n t_o} \sin\left(\sqrt{1 - d^2}\,\omega_n (t - t_o) - \phi\right) \Big] \text{ for } t > t_o \tag{6.52}$$

where

$$x_o = \frac{F_o}{k}, \quad \tan\phi = \frac{2d\sqrt{1 - d^2}}{1 - 2d^2} \tag{6.53}$$

Equations (6.51), (6.52), and (6.53) provide the information needed to determine the error introduced by a load cell while tracking a terminated ramp function if the degree of damping $d$ is known.

In many transducers, the degree of damping $d$ is very low (less than 0.02) and Eqs. (6.51), (6.52), and (6.53) reduce to

$$\frac{x}{x_o} = \frac{t}{t_o} - \frac{\sin(\omega_n t)}{\omega_n t_o} \qquad \text{for } 0 \le t \le t_o \tag{6.54}$$

$$\frac{x}{x_o} = 1 - \left[ \frac{\sin(\omega_n t) - \sin(\omega_n (t - t_o))}{\omega_n t_o} \right] = 1 - \frac{D \sin(\omega_n t - \beta)}{\omega_n t_o} \tag{6.55}$$

where

$$D = \sqrt{2(1 - \cos(\omega_n t_o))}, \quad \tan\beta = \frac{\sin(\omega_n t_o)}{(1 - \cos(\omega_n t_o))} \tag{6.56}$$

The first term in Eq. (6.54) represents the desired ramp response. The second term represents an oscillation about the desired ramp with an amplitude given by

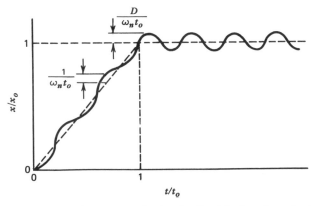

**Figure 6.26**  Response of a load cell with $d = 0$ to the terminated ramp function type of input.

$1/\omega_n t_o$ as shown in Fig. 6.26. The deviation from the ramp is minimized by ensuring that $1/\omega_n t_o$ is small compared to unity. The maximum error during the ramp portion of the curve occurs at the smallest loads when $t$ is small compared to $t_o$. The maximum error that can occur at the peak, $t = t_o$, is given by

$$\mathcal{E} = \frac{1}{\omega_n t_o} \sin \omega_n t_o \tag{6.57}$$

It is evident from Eq. (6.57) that the error is zero when $\omega_n t_o$ is a multiple of $\pi$. The error is a maximum when $\omega_n t_o$ is an odd multiple of $\pi/2$.

The transducer response after $t > t_o$ shows an oscillation at its natural frequency with an amplitude of D as given in Eq. (6.56) and shown in Fig. 6.56. This oscillation amplitude varies from zero to $2/\omega_n t_o$ depending on the value of $\omega_n t_o$. When $\omega_n t_o$ is an odd multiple of $\pi$, the maximum error of

$$\mathcal{E} = 2/\omega_n t_o \tag{6.58}$$

occurs. This error is twice as large as that which can occur when $t = t_o$. To limit the error to a specified amount, the natural frequency of the transducer and connecting mass must be selected such that

$$\omega_n = 2/\mathcal{E} t_o \tag{6.59}$$

A practical consequence of these equations is that five transducer oscillations must occur during the ramp period if the deviation is to be limited to 6.4 per cent or less near the peak. It is seen that ten transducer oscillations are required if the deviation is to be less than 3.2 per cent. Thus, the ramp rise time $t_o$ that can be measured by a given transducer is limited by the natural frequency of the transducer. The general rule of thumb is that $t_o >$ than five to ten natural periods $(2\pi/\omega_n)$. However, if the error is too high, a transducer with a higher natural frequency must be utilized in the instrumentation system.

The dynamic response of transducers to a periodic forcing function can be studied by letting $F(t) = F_o \sin \omega t$ in Eq. (6.49). Thus

$$\frac{1}{\omega_n^2} \frac{d^2 x}{dt^2} + \frac{2d}{\omega_n} \frac{dx}{dt} + x = \frac{F_o}{k} \sin \omega t \tag{6.60}$$

where $\omega$ is the circular frequency of the applied force. If the damping coefficient $d$ is small (as it is in most transducers), Eq. (6.60) can be reduced to

$$\frac{1}{\omega_n^2} \frac{d^2 x}{dt^2} + x = \frac{F_o}{k} \sin \omega t \tag{6.61}$$

Since the forcing function is periodic, the complementary solution to Eq. (6.61) has no significance and the particular solution alone gives the steady-state response of the transducer to the sinusoidal forcing function. The particular solution is

$$x_p = \frac{1}{1 - (\omega/\omega_n)^2} \frac{F_o}{k} \sin \omega t \qquad (6.62)$$

Equation (6.62) can be expressed in terms of the periodic forcing function $F(t)$ as

$$kx_p = \frac{1}{1 - (\omega/\omega_n)^2} F_o \sin \omega t = AF(t) \qquad (6.63)$$

where $A$ is an amplification factor that relates the steady-state response to the periodic forcing function.

The error $\mathscr{E}$ associated with a measurement of $F(t)$ can be expressed as

$$\mathscr{E} = \frac{kx_p - F(t)}{F(t)} \qquad (a)$$

Substituting Eq. (6.63) into Eq. (a) gives an expression for the error in terms of frequency ratio as

$$\mathscr{E} = A - 1 = \frac{(\omega/\omega_n)^2}{1 - (\omega/\omega_n)^2} \qquad (6.64)$$

Equation (6.64) indicates that substantial error can occur in a measurement of force, torque, or pressure unless the frequency ratio $\omega/\omega_n$ is very small. For example, if $\omega/\omega_n = 0.142$, an error of 2 percent occurs. Similarly, if $\omega/\omega_n = 0.229$, the error is 5 percent. Thus, if the error is to be kept within reasonable limits, the natural frequency of the transducer system must be 5 to 10 times higher than the frequency of the forcing function.

The dynamic response of transducers to other common inputs, such as the impulse function and the ramp function, should also be studied. These cases are covered in the exercises at the end of the chapter. Additional information on the dynamic response characteristics of transducers is presented in Sections 7.6, 7.7, and 7.8.

## 6.8 CALIBRATION OF TRANSDUCERS

All transducers must be calibrated periodically to ensure that the calibration constant has not changed. Calibration of load cells and torque cells is usually accomplished with a testing machine whose accuracy has been certified to be within specified limits. After the transducer is mounted in the testing machine, load is applied in increments that cover the complete range of the transducer. The output from the transducer is compared to the load indicated by the testing machine at each level of load, and differences (errors) are recorded. If the error is small, the calibration constant for the transducer is verified and the transducer can be used with confidence. If the error is excessive but consistent (i.e., the response is linear, but the slope is not correct), the calibration constant can be adjusted to correct the error. In cases where the calibration constant requires correction, the calibration test should be repeated to ensure that the new calibration constant is reproducible and correct. In some cases it will be observed that the error is not consistent and the output from the transducer is erratic. If the instrumentation system is checked and found to be operating properly, then the transducer is malfunctioning and cannot be calibrated. Such transducers should be removed from service immediately and returned to the manufacturer for repair.

A second method of calibrating load cells and torque cells is commonly employed in instances where a testing machine, certified to the required limits of accuracy, is not available. This method utilizes two transducers connected in series. One transducer is a standard transducer that is used only for calibration purposes, the other is the working transducer. With this method, the calibration loads are given by the standard transducer instead of the testing machine.

For low-capacity load cells and torque cells, deadweight loads are frequently used in the calibration process. The standard weights (traceable to the Bureau of Standards or certified by weighing with a calibrated scale) are applied directly to the transducer to provide the known input. The transducer output is compared to the known input as discussed previously. While deadweight loading in calibration has many advantages and is usually preferred, it is not practical when the range or capacity of the load cell exceeds a few hundred pounds (about 1 kN).

Pressure transducers are usually calibrated with a deadweight pressure source, such as the one illustrated schematically in Fig. 6.27. The calibration pressures are generated in the deadweight tester by adding standard weights to the piston tray. The calibration pressure is related to the weight by the expression

$$p = \frac{W}{A} \tag{6.65}$$

**where** $W$ is the total weight of the piston, tray, and standard weights.
$A$ is the cross-sectional area of the piston.

After the weights are placed on the piston tray, a screw-driven plunger is forced into the hydraulic oil chamber to reduce its volume and thus lift the piston-weight assembly. The piston-weight assembly is then rotated to eliminate frictional forces between the piston and the cylinder. By adding weights incrementally to the piston tray, it is possible to generate 8 or 10 calibration pressures

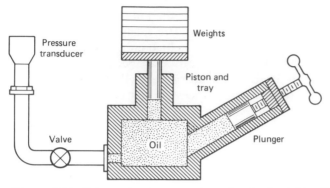

*Figure 6.27*  Schematic representation of a deadweight calibration system for pressure transducers.

that cover the operating range of the transducer. Comparisons are made between the calculated calibration pressures and the pressures indicated by the transducer in order to certify the calibration constant. Since deadweight testers are relatively inexpensive and operate over a very wide range of pressures, they are preferred over methods that utilize a standard transducer. The cost associated with purchase of a number of standard transducers to cover a wide range of pressures exceeds the cost of a deadweight tester to cover the same range.

Dynamic calibration of pressure transducers is usually accomplished with a shock tube. A shock tube is simply a closed section of smooth-walled tubing that is divided by a diaphragm into a short high-pressure chamber and a long low-pressure chamber. When the diaphragm is ruptured, a shock wave propagates into the low-pressure chamber as illustrated schematically in Fig. 6.28. The pressure associated with the shock wave (the dynamic calibration pressure $p_c$), with air as the gas in the shock tube, is given by the expression

$$p_c = p_h \left( \frac{5}{6} - \frac{v}{c} \right) \tag{6.66}$$

**where**   $p_h$ is the static pressure in the high-pressure chamber.
$v$   is the velocity of the shock wave in the low-pressure chamber.
$c$   is the velocity of sound at the static pressure $p_l$ in the low-pressure chamber.

The velocity $v$ of the shock wave is determined by placing a number of pressure transducers along the length of the low-pressure chamber in the shock tube and measuring the time of arrival of the shock front at the various locations. With a shock tube it is possible to apply a sharp-fronted pressure pulse to a transducer so that its dynamic response can be characterized.

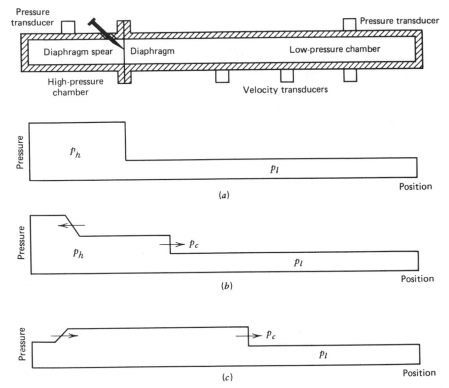

*Figure 6.28* Schematic illustration of the use of a shock tube to generate dynamic pressure pulses for transducer calibration. (*a*) Pressure distribution with the diaphragm intact. (*b*) Pressure distribution before reflection of the rarefraction wave. (*c*) Pressure distribution after reflection of the rarefraction wave.

## 6.9 SUMMARY

A wide variety of transducers are commercially available for measuring load, torque, or pressure. All of these transducers incorporate an elastic member and a sensor to convert the deformation of the elastic member into an electrical signal. Accuracies of 0.1 to 0.2 percent are often specified.

Most load cells use strain gages as the sensor; however, for static measurements, where long-term stability is important, the linear variable differential transformer sensors are more suitable. For dynamic measurements, where a very high natural frequency is required, piezoelectric sensors are recommended. Load cells are covered in sufficient detail in Section 6.2 to give the reader adequate background to design special-purpose transducers and to thoroughly understand the sensitivities that are possible with many of the commonly employed elastic elements. It should be noted, however, that it is usually less expensive to buy a transducer than to build one, and the reader is encouraged to design and fabricate transducers only in those instances when it is not possible to purchase the required transducer.

Torque cells are very similar to load cells. Strain gages are the most common sensors and a simple circular shaft is the most common elastic member. The most significant difference between load and torque measurements arises when torque measurements must be made on a rotating shaft. In these measurements, either slip rings or telemetry must be used to transmit the signal from the rotating member to the stationary instrumentation station. Slip rings are usually preferred if the end of the shaft is accessible for mounting of the slip-ring assembly. If the shaft ends are not accessible, telemetry must be used for signal transmission.

Pressure transducers are available in a wide variety of designs and capacities. The diaphragm-type pressure transducer, with electrical resistance strain-gage sensors, is probably the most common type because of ease of manufacturing. Selection of a pressure transducer for a given application is usually made on the basis of stability and frequency response. For long-term stability, transducers with linear variable-differential transformer (LVDT) sensors are usually preferred. For quasi-static and medium-frequency measurements, the diaphragm-type pressure transducer with strain-gage sensors has some advantages. For extremely high-frequency measurements, transducers with piezoelectric sensors should be used.

Considerable care must be exercised in all dynamic measurements to ensure that the desired quantities are recorded with the required accuracy. The capability of a transducer to record a dynamic signal depends primarily upon the natural frequency of the transducer since damping is very small. Since error increases in proportion to the frequency ratio $\omega/\omega_n$, piezoelectric sensors with natural frequencies of 100 kHz have significant advantages. Also, the form of the dynamic signal is important. It is shown in Section 6.7 that the terminated ramp function can be recorded with greater fidelity than the periodic input function.

---

### EXERCISES

**6.1**    Determine the sensitivity of a load cell–Wheatstone bridge combination if $S_g = 2$, $v = 0.30$, $E_i = 6$ V, $A = 0.5$ in.$^2$, and $E = 30,000,000$ psi.

**6.2**    The sensitivity of the transducer of Exercise 6.1 can be increased if the input voltage $E_i$ is increased. If each gage in the bridge can dissipate 0.5 W of power, determine the maximum sensitivity that can be achieved without endangering the strain-gage ($R_g = 350\ \Omega$) sensors.

**6.3**    Determine the voltage ratio $E_o/E_i$ for the load cell of Exercise 6.1 if the fatigue strength of the elastic member is 75,000 psi.

**6.4**    If the load cell of Exercise 6.3 is used in a static load application, what maximum load could be placed on the transducer? What voltage ratio $E_o/E_i$ would result?

**6.5**   The calibration constant of a transducer procured from a commercial supplier is listed as 2 mV/V. Determine the sensitivity $S$ of the transducer if $P_{max} = 40,000$ lb and $E_i = 10$ V.

**6.6**   Design a beam-type load cell with variable range and sensitivity. Use aluminum ($E = 10,000,000$ psi, $v = 0.33$, and $S_f = 20,000$ psi) as the beam material and four electrical resistance strain gages ($S_g = 2$ and $R_g = 120 \, \Omega$) as the sensors. Design the load cell to give the following sensitivities and corresponding range:

| $(E_o/E_i)^*$ (mV/V) | Range (lb) |
|:---:|:---:|
| 1 | 1000 |
| 2 | 500 |
| 5 | 200 |

**6.7**   Design a ring-type load cell with a linear-variable-differential-transformer (LVDT) sensor. The load cell should have a capacity of 2000 lb. The radius to thickness ratio of the ring $R/t$ should be 10. Select an LVDT for this application from Table 3.1. Use steel ($E = 30,000,000$ psi and $v = 0.30$) for the ring. Determine the sensitivity $S_t$ for your transducer.

**6.8**   For the transducer designed in Exercise 6.7, determine $(E_o/E_i)^*$ if the fatigue strength of the steel $S_f = 60,000$ psi.

**6.9**   Show that the torque cell shown in Fig. 6.6 is insensitive to both axial load $P$ and moments $M_x$ and $M_y$.

**6.10**   Determine the sensitivity of a torque cell if $E = 30,000,000$ psi, $v = 0.30$, $E_i = 8$ V, $D = 1$ in., $S_g = 2$, and $R_g = 120 \, \Omega$.

**6.11**   The sensitivity of the torque cell described in Exercise 6.10 can be increased if the input voltage $E_i$ is increased. If each gage in the bridge can dissipate 0.8 W of power, determine the maximum sensitivity that can be achieved without endangering the strain-gage sensors.

**6.12**   Determine the sensitivity of the torque cell of Exercise 6.11 if strain gages having $R_g = 500 \, \Omega$ are used in place of the 120-$\Omega$ gages.

**6.13**   A torque cell with a capacity of 500 ft · lbs is supplied with a calibration constant of $(E_o/E_i)^* = 4$ mV/V and a recommendation that the input voltage $E_i = 10$ V. If the cell is used with $E_i = 8$ V and a measurement of $E_o$ yields 24 mV, determine the torque $T$.

**6.14**   Determine the sensitivity of the torque cell described in Exercise 6.13.

**6.15**   Why are at least four slip rings used to transmit the voltages associated with a torque cell on a rotating shaft?

**6.16**   Outline the advantages associated with the use of telemetry for data transmission from a rotating shaft.

**6.17**   A solid circular shaft having a diameter of 2 in. is rotating at 800 rpm and is transmitting 100 hp. Show how four strain gages can be used to convert the shaft itself into a torque cell. Determine the sensitivity of this shaft–torque transducer if the shaft is made of steel having $E = 30,000,000$ psi and $v = 0.30$.

**6.18**    Design a static pressure transducer having a 2-in.-diameter diaphragm fabricated using steel with a fatigue strength of 60,000 psi. Select a linear variable-differential transformer (LVDT) from Table 3.1 to use as a sensor to convert the center point deflection of the diaphragm to an output voltage $E_o$. The capacity of the transducer is to be 1000 psi and linearity must be maintained within 0.3 percent. Determine the sensitivity of your transducer.

**6.19**    Repeat Exercise 6.18 by using a special-purpose four-element diaphragm strain gage as the sensor in place of the LVDT. Assume that for each element of the strain gage $P_g = 2$ W, $R_g = 350$ Ω, and $S_g = 2$.

**6.20**    Determine the natural frequency of the pressure transducer of Exercise 6.18.

**6.21**    Determine the natural frequency of the pressure transducer of Exercise 6.19.

**6.22**    A cylindrical elastic element for a very-high-pressure transducer is shown in Fig. E6.22. If the capacity of the pressure transducer is to be 50,000 psi and if the fatigue strength of the steel used in fabricating the cylindrical elastic element is 80,000 psi, determine the diameters $D_i$ and $D_o$.

**6.23**    Determine the sensitivity of the pressure transducer described in Exercise 6.22 if the strain gages used as sensors have $P_g = 0.5$ W, $S_g = 2$, and $R_g = 350$ Ω.

**6.24**    Explain why dummy gages are mounted at the positions shown in Fig. E6.22.

**6.25**    The load cell shown in Fig. 6.18 was shown to be insensitive to moments $M_x$ and $M_y$. Show, in addition, that it is not sensitive to the torque $M_z$.

**6.26**    A load cell with a natural frequency $f_n = 10$ kHz is to be used to measure a terminated ramp function that exhibits a rise time of 1 ms. Determine the maximum error due to the response characteristics of the transducer.

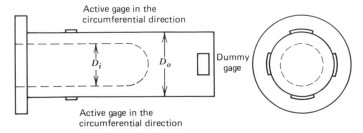

*Figure E6.22*

**6.27**  Repeat Exercise 6.26 by assuming that the rise time of the terminated ramp is reduced to:

(a)  500 μs
(b)  200 μs
(c)  100 μs

Plot a curve showing the maximum possible error as a function of rise time for the transducer.

**6.28**  A transducer having a natural frequency $\omega_n$ will be used to monitor a sinusoidal forcing function having a frequency $\omega$. Determine the error if $\omega/\omega_n$ equals:

(a)  0.05
(b)  0.10
(c)  0.20
(d)  0.50

Plot a curve showing error as a function of frequency ratio $\omega/\omega_n$.

**6.29**  Derive the response equation for a transducer with $d = 0$ if the input function is a ramp that can be expressed as $F(t) = \dot{F}_o t$.

**6.30**  Interpret the results of Exercise 6.29 and determine the magnitude of any error that may result.

**6.31**  Derive the response equation for a transducer with damping $(d \neq 0)$ if the input function is an impulse $I_o$ as shown in Fig. E6.31.

**6.32**  Use the results of Exercise 6.31 to show the response of the transducer if

(a)  $d = 0.01$
(b)  $d = 0.10$
(c)  $d = 0.20$
(d)  $d = 0.50$
(e)  $d = 0.70$

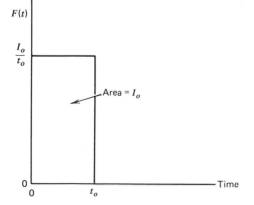

*Figure E6.31*

# SEVEN
# DISPLACEMENT, VELOCITY, AND ACCELERATION MEASUREMENTS

## 7.1 INTRODUCTION

Many methods have been developed to measure linear and angular displacements ($s$ and $\theta$), velocities ($v$ and $\omega$), and accelerations ($a$ and $\alpha$). Displacements and accelerations are usually measured directly, while velocities are often obtained by integrating acceleration signals. The definitions of velocity ($v = ds/dt$ or $\omega = d\theta/dt$) and acceleration ($a = dv/dt = d^2s/dt^2$ or $\alpha = d\omega/dt = d^2\theta/dt^2$) suggest that any convenient quantity can be measured and the others can be obtained by integrating or differentiating the recorded signal. Since the integration process is an error-smoothing process, while the differentiation process is an error-amplifying process, only the integration process is widely used for practical applications. Inexpensive and accurate systems are available for such applications. Displacement measurements are most frequently made in manufacturing and process-control applications, while acceleration measurements are made in vibration, shock, or motion-measurement situations.

Measurements of kinematic quantities, such as displacement, velocity, and acceleration, must be made with respect to a system of reference axes. The basic frame of reference used in mechanics is known as the *primary inertial system* (or astronomical frame of reference) and consists of an imaginary set of rectangular axes that neither translate nor rotate in space. Measurements made with respect to this primary inertial system are said to be absolute. Measurements show that the laws of Newtonian mechanics are valid for this reference system as long as velocities are negligible with respect to the speed of light (300,000 km/s or 186,000 mi/s). For velocities of the same order as the speed of light, the theory of relativity must be applied.

A reference frame attached to the surface of the earth exhibits motion in the inertial reference system; therefore, corrections to the basic equations of mechanics may be required when measurements are made relative to an earth-based reference frame. For example, the absolute motion of the earth must be considered in calculations related to rocket-flight trajectories. For engineering calculations involving machines and structures that remain on the surface of the earth, corrections are extremely small and can usually be neglected. Thus, measurements made relative to the earth in most earth-bound engineering applications can be considered absolute.

There are many engineering problems for which the analysis of motion is simplified by making measurements with respect to a moving coordinate system. These measurements, when combined with the observed motion of the moving coordinate system with respect to some fixed system, permits the determination of the absolute motion. This approach is known as a *relative motion analysis*. For most earth-bound engineering applications, a set of fixed axes attached to the surface of the earth gives satisfactory results, since the effects of the motion of the earth in space for these applications are insignificant and can be neglected. For problems involving earth satellites, a nonrotating coordinate system with its origin on the earth's axis of rotation is a convenient fixed reference system. For problems involving interplanetary travel, a nonrotating coordinate system fixed on the axis of rotation of the sun would be appropriate. From the above it can be concluded that the choice of a fixed reference system depends on the type of problem being considered in the relative motion analysis.

## 7.2 DISPLACEMENT, VELOCITY, AND ACCELERATION RELATIONSHIPS

The position of a particle $P$ moving along a straight line, as shown in Fig. 7.1, can be specified at any instant of time $t$ by its displacement $s$ from some convenient fixed reference point 0 on the line. The average velocity $v_{avg}$ of the

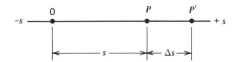

*Figure 7.1* Rectilinear motion of a particle.

particle during any time interval $\Delta t$ is simply the distance moved $\Delta s$ divided by the time interval $\Delta t$. The instantaneous velocity $v$ of the particle at any time $t$ is given by the expression

$$v = \lim_{\Delta t \to 0} \frac{\Delta s}{\Delta t} = \frac{ds}{dt} = \dot{s} \tag{7.1}$$

where a dot over a quantity indicates differentiation with respect to time. Similarly, the instantaneous acceleration $a$ of the particle is the instantaneous time rate of change of the velocity. Thus,

$$a = \lim_{\Delta t \to 0} \frac{\Delta v}{\Delta t} = \frac{dv}{dt} = \dot{v} \tag{7.2}$$

An expression for acceleration $a$ in terms of displacement $s$ is obtained by substituting Eq. (7.1) into Eq. (7.2). Thus,

$$a = \frac{dv}{dt} = \frac{d^2 s}{dt^2} = \ddot{s} \tag{7.3}$$

Equations involving only displacement $s$, velocity $v$, and acceleration $a$ are obtained by eliminating time $dt$ from Eqs. (7.1) and (7.2). Thus,

$$v \, dv = a \, ds \qquad \text{or} \qquad \dot{s} \, d\dot{s} = \ddot{s} \, ds \tag{7.4}$$

Since displacement $s$, velocity $v$, and acceleration $a$ are algebraic quantities, their signs may be either positive or negative. In Eqs. (7.1)–(7.4) the positive direction for velocity $v$ or accelertion $a$ is the same as that chosen for displacement $s$.

Angular motion of a line in a plane can be described by using an angular displacement $\theta$ with respect to a convenient fixed reference direction, as shown in Fig. 7.2. For the case illustrated in Fig. 7.2, angular displacement in a counterclockwise direction has been defined as positive. The choice of reference axis

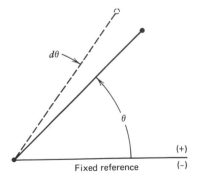

Fixed reference    (+)
(−)    *Figure 7.2*    Angular motion of a line.

and sense for positive measurements is arbitrary. Angular velocity $\omega$ and angular acceleration $\alpha$ of a line are, respectively, the first and second time derivatives of the angular displacement $\theta$. Thus,

$$\omega = \frac{d\theta}{dt} = \dot{\theta} \qquad (7.5)$$

and

$$\alpha = \frac{d\omega}{dt} = \dot{\omega} = \frac{d^2\theta}{dt^2} = \ddot{\theta} \qquad (7.6)$$

By eliminating time $dt$ from Eqs. (7.5) and (7.6), it is possible to obtain equations for angular motion of a line that are similar to Eqs. (7.4) for rectilinear motion of a particle. Thus,

$$\omega \, d\omega = \alpha \, d\theta \qquad \text{and} \qquad \dot{\theta} \, d\dot{\theta} = \ddot{\theta} \, d\theta \qquad (7.7)$$

In the previous equations for angular motion of a line, the positive direction for angular velocity $\omega$ and angular acceleration $\alpha$, clockwise or counterclockwise, is the same as that chosen for angular displacement $\theta$.

Motion of a particle along a curved path is called *curvilinear motion.* The vast majority of problems encountered in engineering practice involve plane curvilinear motion. A particle moving along a plane curved path $s$ is shown in Fig. 7.3. At position $A$, the particle can be located by a position vector $\mathbf{r}$ measured from a convenient fixed origin 0. Similarly, at position $B$, the particle can be located by a vector $\mathbf{r} + \Delta\mathbf{r}$. The vector change of position $\Delta\mathbf{r}$ is known as a *linear displacement,* while the scalar length $\Delta s$ measured along the path $s$ is the distance actually traveled. The velocity $\mathbf{v}$ of the particle at position $A$ is defined as

$$\mathbf{v} = \lim_{\Delta t \to 0} \frac{\Delta\mathbf{r}}{\Delta t} = \frac{d\mathbf{r}}{dt} = \dot{\mathbf{r}} \qquad (7.8)$$

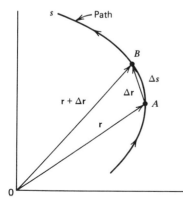

**Figure 7.3** Plane curvilinear motion.

The magnitude of the velocity is usually referred to as the *speed* of the particle. The instantaneous acceleration **a** of the particle at position $A$ is defined as

$$\mathbf{a} = \lim_{\Delta t \to 0} \frac{\Delta \mathbf{v}}{\Delta t} = \frac{d\mathbf{v}}{dt} = \dot{\mathbf{v}} = \ddot{\mathbf{r}} \tag{7.9}$$

Three different coordinate systems (rectangular, normal and tangential, and polar) are commonly used to describe plane curvilinear motion. The choice of system is determined by the problem being studied. Useful forms of Eqs. (7.8) and (7.9) for each of these coordinate systems are as follows.

**Rectangular Coordinates** (see Fig. 7.4):

$$\mathbf{r} = x\mathbf{i} + y\mathbf{j}$$
$$\mathbf{v} = \dot{\mathbf{r}} = \dot{x}\mathbf{i} + \dot{y}\mathbf{j} \tag{7.10}$$
$$\mathbf{a} = \dot{\mathbf{v}} = \ddot{\mathbf{r}} = \ddot{x}\mathbf{i} + \ddot{y}\mathbf{j}$$

where

$$v_x = \dot{x}$$
$$v_y = \dot{y}$$
$$a_x = \dot{v}_x = \ddot{x}$$
$$a_y = \dot{v}_y = \ddot{y}$$
$$v^2 = v_x^2 + v_y^2$$

$$\tan \theta = \frac{v_y}{v_x}$$

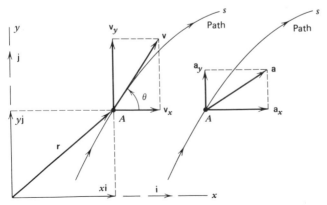

*Figure 7.4*   Plane curvilinear motion in rectangular coordinates.

**Normal and Tangential Coordinates** (see Fig. 7.5):

$$\mathbf{v} = v\mathbf{t}_1 \tag{7.11}$$

$$\mathbf{a} = \dot{\mathbf{v}} = v\dot{\mathbf{t}}_1 + \dot{v}\mathbf{t}_1$$

however,

$$\dot{\mathbf{t}}_1 = \frac{d\mathbf{t}_1}{dt} = \frac{d\theta}{dt}\,\mathbf{n}_1 = \dot{\theta}\mathbf{n}_1$$

therefore,

$$\mathbf{a} = \dot{\mathbf{v}} = \dot{\theta}v\mathbf{n}_1 + \dot{v}\mathbf{t}_1$$

$$= \mathbf{a}_n + \mathbf{a}_t$$

where

$$v = |\mathbf{v}| = \dot{s} = \frac{ds}{dt} = \frac{\rho\,d\theta}{dt} = \rho\dot{\theta}$$

$$a_n = v\dot{\theta} = \rho\dot{\theta}^2 = \frac{v^2}{\rho}$$

$$a_t = \dot{v} = \ddot{s}$$

**Polar Coordinates** (see Fig. 7.6):

$$\mathbf{r} = r\mathbf{r}_1$$

$$\mathbf{v} = \dot{\mathbf{r}} = \dot{r}\mathbf{r}_1 + r\dot{\mathbf{r}}_1 \tag{7.12}$$

$$\mathbf{a} = \dot{\mathbf{v}} = \ddot{\mathbf{r}} = \ddot{r}\mathbf{r}_1 + 2\dot{r}\dot{\mathbf{r}}_1 + r\ddot{\mathbf{r}}_1$$

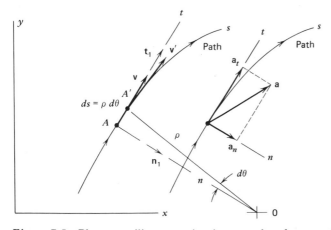

***Figure 7.5*** Plane curvilinear motion in normal and tangential coordinates.

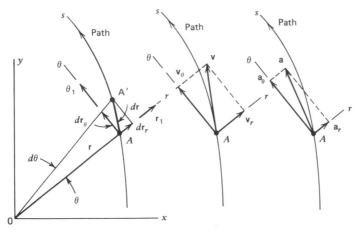

**Figure 7.6**   Plane curvilinear motion in polar coordinates.

however,

$$\dot{\mathbf{r}}_1 = \dot{\theta}\boldsymbol{\theta}_1$$

$$\dot{\boldsymbol{\theta}}_1 = -\dot{\theta}\mathbf{r}_1$$

therefore,

$$\mathbf{v} = \dot{\mathbf{r}} = \dot{r}\mathbf{r}_1 + r\dot{\theta}\boldsymbol{\theta}_1$$

$$= \mathbf{v}_r + \mathbf{v}_\theta$$

$$\mathbf{a} = \dot{\mathbf{v}} = \ddot{\mathbf{r}} = (\ddot{r} - r\dot{\theta}^2)\mathbf{r}_1 + (r\ddot{\theta} + 2\dot{r}\dot{\theta})\boldsymbol{\theta}_1$$

$$= \mathbf{a}_r + \mathbf{a}_\theta$$

where

$$v_r = \dot{r}$$

$$v_\theta = r\dot{\theta}$$

$$a_r = \ddot{r} - r\dot{\theta}^2$$

$$a_\theta = r\ddot{\theta} + 2\dot{r}\dot{\theta}$$

Motion of a particle along a curved path in space, as shown in Fig. 7.7, is known as *space curvilinear motion*. Extension of Eqs. (7.10) for plane curvilinear motion in rectangular coordinates and Eqs. (7.12) for plane curvilinear motion in polar coordinates to problems involving particle motion along space rather than plane curves requires addition of a $z$-coordinate to the system. Results for rectangular and cylindrical coordinate systems are as follows.

**Rectangular Coordinates** $(x$–$y$–$z)$ (see Fig. 7.7):

$$\mathbf{R} = x\mathbf{i} + y\mathbf{j} + z\mathbf{k}$$

$$\mathbf{v} = \dot{\mathbf{R}} = \dot{x}\mathbf{i} + \dot{y}\mathbf{j} + \dot{z}\mathbf{k} \tag{7.13}$$

$$\mathbf{a} = \dot{\mathbf{v}} = \ddot{\mathbf{R}} = \ddot{x}\mathbf{i} + \ddot{y}\mathbf{j} + \ddot{z}\mathbf{k}$$

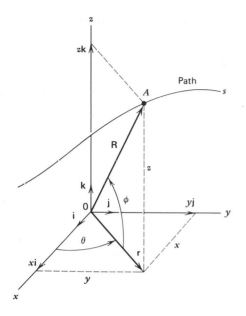

***Figure 7.7***  Space curvilinear motion.

where

$v_x = \dot{x}$

$v_y = \dot{y}$

$v_z = \dot{z}$

$a_x = \dot{v}_x = \ddot{x}$

$a_y = \dot{v}_y = \ddot{y}$

$a_z = \dot{v}_z = \ddot{z}$

**Cylindrical Coordinates** $(r$–$\theta$–$z)$ (see Fig. 7.7):

$$\mathbf{v} = \mathbf{v}_r + \mathbf{v}_\theta + \mathbf{v}_z \qquad\qquad (7.14)$$

$$\mathbf{a} = \mathbf{a}_r + \mathbf{a}_\theta + \mathbf{a}_z$$

where

$v_r = \dot{r}$

$v_\theta = r\dot{\theta}$

$v_z = \dot{z}$

$a_r = \ddot{r} - r\dot{\theta}^2$

$a_\theta = r\ddot{\theta} + 2\dot{r}\dot{\theta}$

$a_z = \ddot{z}$

Spherical coordinates can also be used to describe space curvilinear motion. Expressions for velocity and acceleration in spherical coordinates are

**Spherical Coordinates** $(R–\theta–\phi)$ (see Fig. 7.7):

$$\mathbf{v} = \mathbf{v}_R + \mathbf{v}_\theta + \mathbf{v}_\phi \tag{7.15}$$

$$\mathbf{a} = \mathbf{a}_R + \mathbf{a}_\theta + \mathbf{a}_\phi$$

where

$$v_R = \dot{R}$$

$$v_\theta = R\dot{\theta} \cos \phi$$

$$v_\phi = R\dot{\phi}$$

$$a_R = \ddot{R} - R\dot{\phi}^2 - R\dot{\theta}^2 \cos^2 \phi$$

$$a_\theta = \frac{1}{R} \cos \phi \frac{d}{dt}(R^2\dot{\theta}) - 2R\dot{\theta}\dot{\phi} \sin \phi$$

$$a_\phi = \frac{1}{R}\frac{d}{dt}(R^2\dot{\phi}) + R\dot{\theta}^2 \sin \phi \cos \phi$$

The choice of rectangular, cylindrical, or spherical coordinates for a particular problem, involving space curvilinear motion, will depend upon the manner in which the measurements are made or upon the manner in which the motion is generated. The most convenient system is usually fairly obvious.

## 7.3  VIBRATORY MOTION AND ITS REPRESENTATION

Vibratory motion can occur in all types of machines and structures. Vibrations may result from a slight unbalance of forces in rotating machine components or from the action of wind loadings in transmission lines and suspension bridges. Loss of efficiency, objectional noise or motion, increased bearing loads, and failure are some of the effects of unwanted vibrations.

A *vibration* can be described as a cyclic or periodically repeated motion about a position of equilibrium. The *amplitude A* of a vibration is defined as the distance from the equilibrium position to the point of maximum displacement. The *period T* of the vibration is the minimum amount of elapsed time before the motion starts to repeat itself. The motion completed in one period is a *cycle*. The number of times the motion repeats itself in a unit of time is the *frequency f of the motion*.

The periodic motion that occurs when an elastic system is displaced from its equilibrium position and released is known as a *free vibration*. The frequency of a free vibration is called the *natural frequency $f_n$ of the system*. When the vibration results from application of an external periodic force, it is called a *forced vibration*. If the frequency of the external periodic force is the same as the natural frequency of the system, the amplitude of vibration becomes very large and the system is said to be in a *state of resonance*. When motion of a body is constrained so that its position can be completely specified by one coordinate, it is said to have a *single degree of freedom*. If the system can vibrate in two directions or if the system is composed of two bodies that can vibrate independently in one direction, it is said to have *two degrees of freedom* since two coordinates are required to specify the position of the system at any instant of time. A single rigid body has, in general, six degrees of freedom, since it can translate in three coordinate directions and can rotate about three coordinate axes.

A common type of vibratory motion, which applies to many physical systems, is shown in Fig. 7.8. The line $0R$ of magnitude $A_o$ rotates in a counterclockwise direction about point $0$ with a constant angular velocity $\omega$. The projection of line $0R$ onto the horizontal $x$ axis can be used to represent a special type of motion for a particle $P$ having a single degree of freedom in the $x$ direction. The position $x$ of particle $P$ at any time $t$ is given by the expression

$$x = A_o \cos \omega t \qquad (7.16)$$

Similarly, the projection of line $0R$ onto the vertical $y$ axis can be used to represent motion of a particle $Q$ having a single degree of freedom for motion in the $y$ direction. The position $y$ for particle $Q$ as a function of time is

$$y = A_o \sin \omega t \qquad (7.17)$$

where $A_o$ is the magnitude of $0R$ and the amplitude of each of its sinusoidal components. The velocities of particles $P$ and $Q$ , obtained by taking derivatives

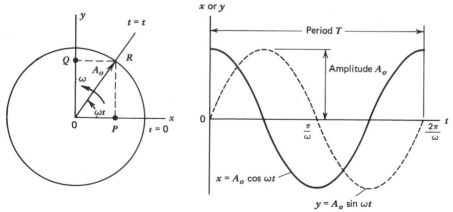

**Figure 7.8**   A common rotating line representation of a simple type of vibratory motion.

with respect to time of the expressions for displacement, are

$$v_P = \dot{x} = -A_o\omega \sin \omega t = A_o\omega \cos\left(\omega t + \frac{\pi}{2}\right)$$

$$v_Q = \dot{y} = A_o\omega \cos \omega t \quad = A_o\omega \sin\left(\omega t + \frac{\pi}{2}\right) \tag{7.18}$$

Likewise, the accelerations are

$$a_P = \dot{v}_P = \ddot{x} = -A_o\omega^2 \cos \omega t = A_o\omega^2 \cos(\omega t + \pi)$$

$$a_Q = \dot{v}_Q = \ddot{y} = -A_o\omega^2 \sin \omega t = A_o\omega^2 \sin(\omega t + \pi) \tag{7.19}$$

Equations (7.18) and (7.19) show that velocities and accelerations for this special case of motion can be obtained from the expressions for displacement by multiplying by $\omega$ and $\omega^2$ and increasing the angles by $\pi/2$ and $\pi$ radians, respectively. The angles $\pi/2$ and $\pi$ are known as phase angles and indicate that velocities and accelerations are out of phase with the displacements by 90 and 180 degrees, respectively, as shown in Fig. 7.9.
    If Eqs. (7.16) and (7.17) are substituted into Eqs. (7.19), the accelerations $a_P$ and $a_Q$ can be expressed as

$$a_P = -\omega^2 x \tag{7.20}$$

$$a_Q = -\omega^2 y$$

Both accelerations are seen to be proportional to distance from the equilibrium position. Any motion for which the acceleration is proportional to the displacement from a fixed point on the path of motion and always directed toward that

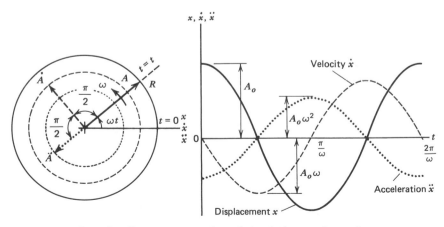

**Figure 7.9**   Rotating line representation of simple harmonic motion.

fixed point is defined as simple harmonic motion. Any periodic motion which is not *simple harmonic motion* can be considered as being a sum of simple harmonic motions of frequencies that are multiples of the frequency of the fundamental motion. In general, any periodic motion can be expressed by using the Fourier form as follows:

$$x = A_0 + A_1 \sin \omega t + A_2 \sin 2\omega t + \cdots$$

$$+ B_1 \cos \omega t + B_2 \cos 2\omega t + \cdots \tag{7.21}$$

The motion described by the rotating line $0R$ of Fig. 7.8 can also be represented by using phasors in the complex plane. The phasor is written using an exponential expression of the form

$$A = A_o e^{i\omega t} = \underbrace{A_o \cos \omega t}_{\text{Real}} + \underbrace{iA_o \sin \omega t}_{\text{Imaginary}} \tag{7.22}$$

where $A$ is the phasor and $i = \sqrt{-1}$. Each phasor has a magnitude $A_o$ and a phase angle measured relative to a reference phasor which also rotates with the same angular frequency $\omega$. In Eq. (7.22), $A$ is the reference phasor; therefore, its phase angle is zero. The first and second derivatives of $A$ with respect to time are

$$\dot{A} = \frac{dA}{dt} = iA_o\omega e^{i\omega t} = A_o\omega e^{i(\omega t + \pi/2)}$$

$$= A_o\omega \cos\left(\omega t + \frac{\pi}{2}\right) + iA_o\omega \sin\left(\omega t + \frac{\pi}{2}\right) \tag{7.23}$$

$$\ddot{A} = \frac{d^2A}{dt^2} = -A_o\omega^2 e^{i\omega t} = A_o\omega^2 e^{i(\omega t + \pi)}$$

$$= A_o\omega^2 \cos(\omega t + \pi) + iA_o\omega^2 \sin(\omega t + \pi) \tag{7.24}$$

since $i = e^{i\pi/2}$ and $-1 = e^{i\pi}$. It is clear from Eqs. (7.22), (7.23), and (7.24) that $A$ is the reference phasor from which the phase angles for $\dot{A}$ and $\ddot{A}$ are measured as shown in Fig. 7.9.

With this exponential (phasor) representation of the rotating line $0R$, the position, velocity, and acceleration of point $P$ can be interpreted to be the real parts of the complex expressions for phasor $A$ and its first and second derivatives with respect to time. The advantages of using the exponential representation of motion include the ease associated with differentiation and integration, the elimination of extensive trigonometric identity manipulation to reduce expressions to a useful form, and the presence of both magnitude and phase information in a simple form. Equations (7.22), (7.23), and (7.24) and Fig. 7.9 indicate that taking a derivative simply involves multiplying by $\omega$ and adding $\pi/2$ radians to the phase angle. Similarly, performing an integration requires dividing by $\omega$ and subtracting $\pi/2$ radians from the phase angle. The phase angles associated with velocity and acceleration (with displacement as the reference vector) are positive (counterclockwise) and are referred to as leading phase angles. Similarly, a lagging phase angle would be negative (clockwise). The exponential or complex representation of sinusoidal time histories will be used for all further developments in this chapter.

## 7.4   DIMENSIONAL MEASUREMENTS

One of the most fundamental of all measurements is the determination of linear displacement. Measurements related to the size of an object are usually referred to as *dimensional measurements*; whereas, measurements related to the extent of movement of an object are usually referred to as *displacement measurements*. In either case, the measurement process generally requires only the simple process of comparing the size or movement of the object with some standard (basis for length measurements).

The primary standard for length measurements is a platinum–iridium bar, maintained under very accurate environmental conditions, at the International Bureau of Weights and Measures in Sèvres, France. In 1960, The General Conference on Weights and Measures defined the standard meter in terms of the wavelength of the orange-red light of a krypton-86 lamp. The standard meter was defined as

$$1 \text{ meter} = 1,650,763.73 \text{ wavelengths}$$

The National Bureau of Standards of the United States has adopted this standard and the inch is now defined in terms of the krypton light as

$$1 \text{ in.} = 41,929.39854 \text{ wavelengths}$$

The relationship between English and International (SI) systems of units for length measurement, which has been used in industry and engineering since the SI system was adopted, is

$$1 \text{ in.} = 25.4 \text{ mm (exactly)}$$

Determinations of the size of an object are frequently made with graduated metal or wood scales (secondary standards). For large dimensional measurements, metal tapes are used. Such devices can provide accuracies of approximately $\pm 0.02$ in. A modification of the metal scale that can provide dimensional information with an accuracy of approximately $\pm 0.002$ in. is a vernier caliper of the type shown schematically in Fig. 7.10a. A more precise measurement device is the micrometer of Fig. 7.10b, which can provide dimensional information with an accuracy of the order of $\pm 0.0002$ in. Dial indicators represent a similar class of measurement devices that have an accuracy of the order of $\pm 0.001$ in.

## Optical Measurement Methods

Optical instruments in wide use for dimensional measurements include fixed-scale and filar microscopes for small dimensional magnitudes (0.05 to 0.200 in.), traveling and traveling-stage microscopes for larger dimensional magnitudes (4 to 6 in.), and drawtube microscopes for small displacement measurements (0 to $1\frac{1}{2}$ in.) in a directional parallel to the optical axis of the instrument. Accuracies for the various microscopes range from $\pm 0.001$ to $\pm 0.00005$ in. Optical instruments for very large dimensional measurements (50 to 1000 ft) are represented by the conventional surveyor's transit. Accuracies for this type of instrument are approximately $\pm 1/2$ percent.

The modification of intensity of light by superposition of light waves is discussed in any undergraduate physics text and is defined as an *interference effect*. An optical device that can be used to measure lengths or changes in length with great accuracy by means of interference effects is known as an *interferometer*. A fundamental requirement for the existence of well-defined fringes (lines of zero intensity of light) is that the light waves producing the fringes have a sharply defined phase difference that remains constant with time. When light beams from two independent sources are superimposed, interference fringes are not observed, since the phase difference between the beams varies in a random way (the beams are incoherent). Two beams from the same source, on the other hand, interfere, since the individual wave trains in the two beams have the same phase initially (the beams are coherent) and any difference in phase at the point of superposition results solely from differences in optical paths. The concept of optical-path difference and its effect on the production of interference fringes

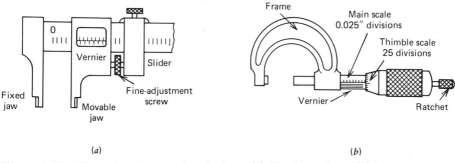

(a)                                        (b)

*Figure 7.10*   Dimensional measuring devices. (a) Vernier caliper. (b) Micrometer.

can be illustrated by considering the reflection and refraction of light rays from a transparent plate and a reflecting surface, as shown in Fig. 7.11. Consider a plane wavefront associated with light ray A that strikes the plate at an angle of incidence $\alpha$. Ray B results from reflection at the front surface of the plate. A second ray is refracted at the front surface, reflected at the back surface, and refracted from the front surface before emerging from the front surface of the plate as ray C. The optical-path difference between rays B and C can be expressed as

$$\delta_1 = \frac{2h}{\cos \gamma} n_{21} - \frac{2h}{\cos \gamma} n_{21} \sin^2 \gamma = 2hn_{21} \cos \gamma \tag{7.25}$$

where $n_{21}$ is the index of refraction of material 2 with respect to material 1. Since ray B suffers a phase change of $\lambda/2$ on reflection, rays B and C will interfere destructively (produce minimum intensity or extinction when brought together) whenever

$$\delta_1 = 2hn_{21} \cos \gamma = m\lambda \qquad m = 0, 1, 2, 3, \ldots$$

If the light beam illuminates an extended area of the plate, and if the thickness of the plate varies slightly with position, the locus of points experiencing the same order of extinction will combine to form an interference fringe. The fringe spacings will represent thickness variations of approximately 7 $\mu$in., or 180 nm (in glass with mercury light and a small angle $\gamma$).

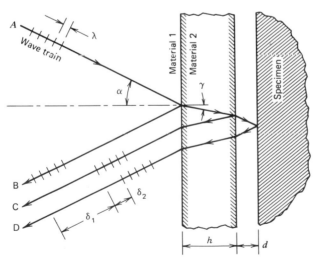

**Figure 7.11**  Schematic illustration of the operating principles associated with measurements employing interference effects.

Rays emerging from the back surface of the plate and reflecting from a working surface will also produce interference effects as indicated by ray D. If the glass plate is a strain-free optical flat (a plate that has been polished flat within a few microinches), beam D will have traveled farther than beam C by a distance $2d/\cos\alpha$. Complete extinction will occur (fringe will form) when

$$\delta_2 = \frac{2d}{\cos\alpha} = \frac{(2m+1)\lambda}{2} \qquad m = 0, 1, 2, 3, \ldots$$

The previous discussion serves to illustrate the principles associated with measurements employing interference effects. For the system illustrated schematically in Fig. 7.11, the optical-path difference $\delta_1$ associated with thickness measurements would be a large number of wavelengths ($m \to \infty$). Such a system would involve high-order interference; therefore, it would require extreme coherence and long wave trains, such as those provided by a laser for successful operation. Other systems, such as the one for the determination of the spacing $d$, utilize low-order interference ($m = 0, 1, 2$, etc.); therefore, less stringent requirements are imposed on the light source. Interference methods are frequently used to calibrate dimensional standards.

## Pneumatic Measurement Methods

Air or pneumatic gaging has been developed in the metal working industry and is widely employed for determining the dimensions of machined parts. The method is relatively simple and can be applied to a wide variety of gaging problems. The pneumatic gage, shown schematically in Fig. 7.12, is used in the discussion that follows to illustrate the principles involved in pneumatic gaging.

As shown in Fig. 7.12, air at a constant pressure head $H$ passes from a reservoir to a plenum through an orifice $R$ (which may be only a short length of pipe). The pressure drop across this orifice is proportional to the pressure head $H - h$. The flow rate through the orifice is proportional to $A_R\sqrt{2g(H-h)}$ for air and a given manometer fluid. The air then flows through a pipe from the plenum to an opening $G$ in the pneumatic device, where it strikes the surface opposite the opening, is deflected, and is forced to flow radially outward until vented to the atmosphere. The cylindrical area at $S$ forms a second orifice $S$ in series with the orifice at $R$. The pressure drop across this second orifice is proportional to the pressure head $h$. Since the orifices are in series, the flow can be expressed in terms of the pressure heads as

$$A_R\sqrt{2g(H-h)} = A_S\sqrt{2gh} \qquad (7.26)$$

**where**  $A_R$ and $A_S$ are the cross-sectional areas of orifices $R$ and $S$.
  $H$  is the constant head maintained by the reservoir.
  $h$  is the head maintained by the manometer.

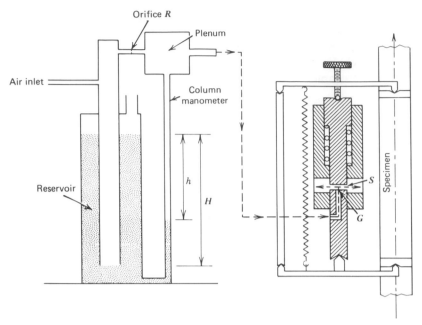

***Figure 7.12*** Schematic illustration of a pneumatic measurement device.

In Eq. (7.26) it has been assumed that the flow is incompressible. This assumption is valid, and density changes can be neglected so long as the pressures used with the system are small (less than 10 to 15 in. of water).

When the specimen shown in Fig. 7.12 is loaded, the distance between the two gage points changes. This elongation is transmitted through the lever system to the pneumatic gage where it changes the gap between opening $G$ and the top surface. Thus, the area $A_S$ changes in direct proportion to the change in length of the specimen. The manometer reading $h$ varies as a quadratic function of this change in length. With proper design, this nonlinear characteristic of the gage can be minimized and a nearly linear response can be obtained over short ranges. Multiplication factors of 100,000 are possible with this type of pneumatic amplification.

In many types of industrial applications, the system is simplified considerably by utilizing the work piece as the upper surface (at $S$) of the gage shown in Fig. 7.12. The dimensional changes in the part being gaged controls the area $A_S$ of orifice $S$. The principle of operation (two orifices in series) is identical for this simplified system.

## Calibration Methods

Calibration of dimensional measuring devices by industry is accomplished by using gage-block sets similar to the one shown in Fig. 7.13. Gage blocks, as the name implies, are small blocks of heat-treated steel having parallel faces and a thickness that is accurate to within a given tolerance. Gage blocks are available in the following classes:

### Gage Blocks for the English System of Units

| Class | Type | Tolerance |
|-------|------|-----------|
| B | "Working" blocks | $\pm 8$ $\mu$in. |
| A | "Reference" blocks | $\pm 4$ $\mu$in. |
| AA | "Master" blocks | $\pm 2$ $\mu$in. for all blocks up to 1 in. and $\pm 2$ $\mu$in./in. for larger blocks |

A commercially available set of 83 blocks consists of the following:

    9 blocks with 0.0001-in. increments from 0.1001 to 0.1009 in. inclusive.

    49 blocks with 0.001-in increments from 0.101 to 0.149 in. inclusive.

    19 blocks with 0.050-in. increments from 0.050 to 0.950 in. inclusive.

    4 blocks with 1-in. increments from 1 to 4 in. inclusive.

    2 tungsten–steel wear blocks having a thickness of 0.050 in.

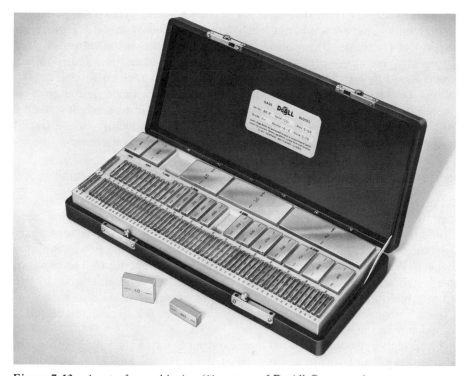

*Figure 7.13*    A set of gage blocks. (Courtesy of DoAll Company.)

The previously listed set of 83 blocks can provide approximately 120,000 dimensional values in steps of 0.0001 in. The blocks must be assembled by using a process known as wringing to obtain a specific dimension. This assembly process involves eliminating all but the thinnest oil film (approximately 0.2 μin.) between the individual blocks. Properly wrung blocks will exhibit a significant resistance to separation. Extreme care must be exercised in maintaining the temperature of both the blocks and the machine component during any gaging operation. The reader should refer to an *Industrial Metrology Handbook* or text for additional information regarding calibration of industrial measurement standards.

## 7.5   DISPLACEMENT AND VELOCITY MEASUREMENTS IN A FIXED REFERENCE FRAME

### Displacement Measurements

Measurements related to the extent of movement of an object are usually referred to as displacement measurements. Most of the sensor devices described in Chapter Three can be used for displacement determinations when a fixed reference frame is available. Variable-resistance (Section 3.4) and capacitance (Section 3.5) sensors are widely used for small static and dynamic displacement measurements (from a few microinches to small fractions of an inch). Differential transformers (Section 3.3) are used for larger displacement magnitudes (from fractions of an inch to several inches). Resistance potentiometers are used where less accuracy but greater range (small fractions of an inch to several feet) is required. In the paragraphs that follow, the use of resistance potentiometers, photosensing transducers, and microswitch position indicators are described in detail, since they provide a convenient means for introducing several circuits that have been designed and perfected for displacement measurement.

### Displacement Measurements with Resistance Potentiometers

Potentiometer devices, such as those shown in Figs. 3.1 and 3.2, are commonly used to measure linear and angular motions. A typical potentiometer circuit is shown in Fig. 7.14. The circuit consists of a potentiometer sensor with a resistance $R_P$, a recording instrument with a resistance $R_M$, a power supply to provide an input voltage $E_i$, and in certain instances a capacitor $C$. The capacitor is used with wire-wound potentiometers to smooth the output signal by momentarily maintaining the voltage as the wiper moves from wire to wire along the spiral resistance coil. In the following analysis of the circuit, the effects of the capacitor are neglected.

The effects of the load imposed on the output signal by the voltage measuring instrument is obtained from an analysis of the circuit of Fig. 7.14. The output

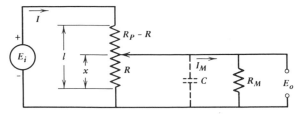

**Figure 7.14** Displacement measuring circuit with a potentiometer as a sensor.

voltage $E_o$ can be expressed in terms of the various resistances as

$$E_o = I_M R_M = (I - I_M)R = E_i - I(R_P - R) \qquad (7.27)$$

**where** $R = (x/l)R_P$ is a resistance that is proportional to position $x$ of the wiper on the potentiometer coil.

$R_M$ is the load imposed on the output signal $E_o$ by the measuring instrument.

Solving Eq. (7.27) for the output voltage $E_o$ yields

$$E_o = \left[ \frac{R_M/R_P}{(R_M/R_P) + (R/R_P) - (R/R_P)^2} \right] (R/R_P) E_i \qquad (7.28)$$

Equation (7.28) clearly indicates that the output voltage $E_o$ of this circuit is a nonlinear function of resistance $R$ (and thus position $x$) unless the resistance $R_M$ of the measuring instrument is large with respect to the potentiometer resistance $R_P$. This nonlinear behavior can be expressed in terms of a nonlinear factor $\eta$ such that Eq. (7.28) becomes

$$E_o = (1 - \eta)(R/R_P) E_i \qquad (7.29)$$

Values of the nonlinear term $\eta$ as a function of resistance ratio $R/R_P$ (wiper position) for different values of $R_M/R_P$ are shown in Fig. 7.15a. The significance of the nonlinear effect is further illustrated in Fig. 7.15b, where normalized output voltage $E_o/E_i$ for the circuit is plotted as a function of resistance ratio $R/R_P$ (wiper position) for different values of $R_M/R_P$. These results clearly indicate that the resistance ratio $R_M/R_P$ must be at least 10 if significant deviations from linearity are to be avoided.

Figure 7.15a also indicates that the maximum deviation from linearity will occur at a position $x$ where $R/R_P = 0.5$ for all measuring instruments. At such a position, the error $\mathscr{E}$ due to nonlinear effects can be obtained from Eq. (7.28) as

$$\mathscr{E} = \frac{E_o - E_o'}{E_o} = \frac{1}{1 + 4(R_M/R_P)} \qquad (7.30)$$

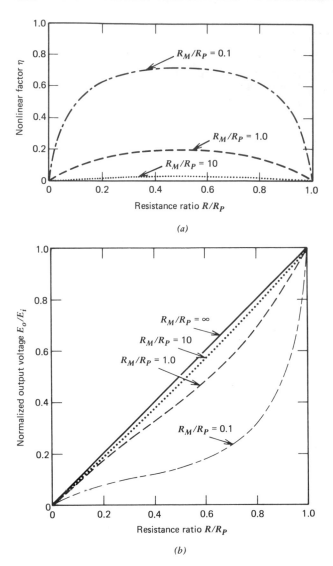

**Figure 7.15**   (a) Nonlinear factor $\eta$ and (b) Normalized output voltage $E_o/E_i$ as a function of resistance ratio $R/R_P$ (position) for different values of $R_M/R_P$.

where $E_o$ and $E_o'$ are the open circuit and indicated output voltages from a measuring instrument when $R/R_P = 0.5$. Thus, for a measuring instrument with $R_M/R_P$ equal to 10, the maximum error due to nonlinear effects $\mathscr{E}_{max} = 0.0244 = 2.44$ percent.

A second circuit that utilizes a resistance potentiometer as a displacement sensor is shown in Fig. 7.16. A special feature of this circuit is incorporation of an operational amplifier (op-amp) as a voltage follower interface between the

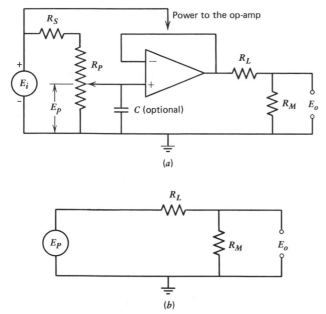

**Figure 7.16** (*a*) Displacement measuring circuit with a potentiometer as a sensor and a voltage follower interface between the sensor and indicator. (*b*) Equivalent circuit.

sensor and the indicating instrument. Use of the voltage follower in the circuit converts the high-resistance voltage source $E_P$ of the potentiometer into a low-resistance (about 50 $\Omega$) source device with sufficient power to drive a measuring instrument with a low input resistance $R_M$. This type of circuit is useful in situations where long lead wires must be used between the sensor and the voltage indicating or recording instrument. In such cases, the circuit is very susceptible to 60-Hz electrical noise or other parasitic signals. Such noise can be significantly reduced by using an indicating or recording instrument with a lower input resistance $R_M$.[1]

As shown in Fig. 7.16, the supply voltage $E_i$ can be used to power both the potentiometer and the op-amp by inserting a resistor $R_S$ in series with the potentiometer. The resistor $R_S$ should be sized to yield a voltage drop of approximately 2 V. Several op-amps are available (such as the LM 324 series) that

---

[1] The student should perform the following experiment to verify the effect of lowering the load resistance on the attenuation of 60-Hz electrical noise. Connect a 3- to 5-ft length of unshielded cable to the input terminals of an oscilloscope (input resistance > 1 M$\Omega$); grasp the cable and observe the magnitude of the 60-Hz signal. Now shunt a 100- to 500-$\Omega$ resistor across the input terminals of the oscilloscope and observe the reduction in signal. This simple experiment shows that most magnetically induced noise signals contain very little power; hence, a lowering of the terminating resistance will produce a dramatic increase in the desired signal-to-noise ratio.

exhibit linear output with single-sided inputs from zero to supply voltage less 1.5 V.

Use of the voltage follower interface between the potentiometer sensor and the recording instrument permits the use of long lead wires (with significant resistance $R_L$), and provides reduced susceptibility to electrical noise provided the op-amp is located near the potentiometer. This circuit also eliminates all of the undesirable nonlinear potentiometer characteristics, since the potentiometer has been converted to a low-resistance source device that has the effect of increasing the $R_M/R_P$ ratio. Drift and thermal instability of the op-amp, on the other hand, frequently create other types of problems.

Calibration of a potentiometer-type, displacement-measuring device can be accomplished by using a micrometer as the source of accurate displacements. The process of calibration can also be automated by comparing the output of the instrument to be calibrated with that from a previously calibrated instrument. Resolutions of the order of $\pm 0.001$ in. are possible. Sensitivities of 30 V/in. can be achieved by using potentiometers with resistances of 2000 $\Omega$/in. and the ability to dissipate a power of 0.5 W/in. of coil.

### Displacement Measurement with Multiple-Resistor Devices

Another type of variable-resistance, displacement-measuring device consists of a sequence of resistors in parallel, as shown in the circuit of Fig. 7.17a. With a fixed input voltage $E_i$, the initial output voltage $E_0$ of the circuit is given by the expression

$$E_0 = \frac{R_0}{R_0 + R_e} E_i \qquad (7.31)$$

where

$$\frac{1}{R_e} = \frac{1}{R_1} + \frac{1}{R_2} + \frac{1}{R_3} + \cdots + \frac{1}{R_n} \qquad (7.32)$$

As the resistors are successively removed from the circuit (the moving object either breaks the series of wires or opens the series of switches), the output voltage $E_o$ varies in the descending step fashion illustrated in Fig. 7.17b. Resolution of this simple system depends upon the spacing of the wires or switches.

### Unbonded-Wire, Variable-Resistance Displacement Gages

A high-resolution, variable resistance, displacement gage can be fabricated by using strain-sensitive loops of wire as the sensing elements. The operating principle of this type of displacement gage is illustrated schematically in Fig. 7.18. Components A and B are rigid frames connected by pretensioned loops of strain-sensitive wire. Any relative motion between the fixed frame (reference) and the movable frame (displacement measuring) increases the tension in two

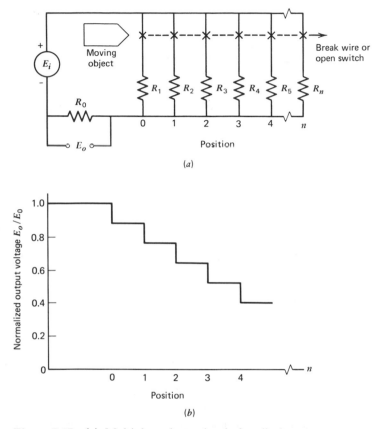

**Figure 7.17** (a) Multiple-resistor circuit for displacement measurements. (b) Output voltage $E_o$ from the multiple-resistor circuit.

of the wire loops and decreases the tension in the other two, thus producing two positive and two negative changes in resistance. When the four individual elements are connected in an appropriate fashion to form a four-equal-arm Wheatstone-bridge circuit, the output voltage $E_o$ of the bridge is proportional to $4 \, \Delta R/R$. Because of the stepless character of the output, the ultimate resolution of this type of displacement transducer is limited primarily by the characteristics of the voltage measuring instrument used to indicate the output voltage.

## Variable-Inductance Displacement Gages

Linear and angular displacement gages based on the variable-inductance principle are known as linear variable-differential transformers (LVDTs) and rotary variable-differential transformers (RVDTs). Application of these transducers to displacement measurements was discussed in Section 3.3 when the variable-inductance sensor was introduced.

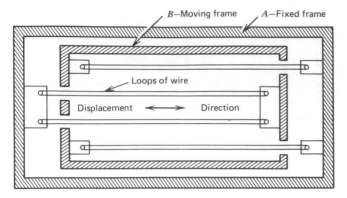

*Figure 7.18*   Unbonded wire, variable-resistance displacement gage.

## Variable-Capacitance Displacement Gages

Displacement gages based on the variable-capacitance principle are discussed in Section 3.5.

## Photoelectric Displacement Transducers

A light source, an opaque object, and a photoelectric sensor (see Section 3.9) can be combined to produce a very useful displacement measuring system that requires no contact with the object in motion. Exact implementation depends on the type of photoelectric sensor to be used. For photoemissive and photoconductive sensors, which generate a current $I$ that is proportional to the illumination $\psi$ imposed on the sensitive area of the sensor, a system similar to the one illustrated schematically in Fig. 7.19 is used. In this system, a parallel beam of light is generated by using a point-light source and a collimating lens. The size and shape of the parallel beam is established with a reticle opening. An opaque flag, whose position in the parallel beam is related to the displacement being measured, interrupts the light beam and controls the illumination $\psi$ falling on the focusing lens and ultimately on the sensitive face of the sensor.

The focusing lens and photoelectric sensor of Fig. 7.19 are often replaced by a flat photovoltaic cell that exhibits the behavior of a current generator in parallel with a capacitor as shown in Fig. 7.20. The voltage output $E_o$ from the circuit of Fig. 7.20 depends on the load resistance $R_M$ of the instrument used to

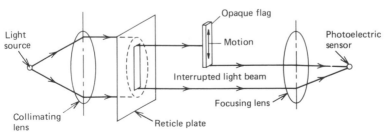

*Figure 7.19*   Schematic diagram of a photoelectric displacement transducer.

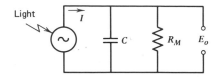

**Figure 7.20**  Displacement measuring circuit with a photovoltaic sensor.

make the measurement. For an open-circuit condition ($R_M \rightarrow \infty$) the output $E_o$ varies logarithmically with illumination $\psi$; however, nearly linear output can be obtained with the appropriate load resistance $R_M$. Dynamic response and linearity can be improved by introducing an op-amp into the circuit as a *current amplifier*, as shown in Fig. 7.21. In this circuit, the feedback resistor $R$ is adjusted to give the best linear output with the desired signal range. The inherent capacitance of the sensor is effectively removed with this circuit, since the voltage across the capacitor is very small ($E = E_o/A$) due to the large open-loop gain A of the op-amp.

Errors with this system result from light source and lens imperfections that tend to produce a nonparallel beam. Light source power variations and external light variations and/or reflections can also introduce additional errors.

## Microswitch Position Sensors

A microswitch is often used in production lines to indicate passage of an object or to record the number of items produced. Features of microswitches that have contributed to their widespread use include low cost, rugged construction, and low activating force. Microswitches normally have three terminals labeled COM (common), NO (normally open), and NC (normally closed) for use in different electrical circuits. The simple circuit of Fig. 7.22, which contains a power supply, a microswitch, and an electronic counter, is used for many counting applications. The series resistor $R_s$ is inserted into the circuit to protect the power supply in case of a short in the circuit.

The major problem encountered with mechanical switches in counting operations is contact bounce. When an electrical switch with mechanical contacts is closed, numerous contacts are made and broken before solid contact is finally achieved. A typcial voltage output $E_o$ from a circuit containing a switch with

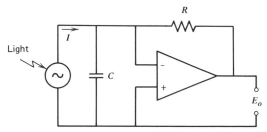

**Figure 7.21**  Displacement measuring circuit with a photovoltaic sensor and an op-amp as a current amplifier.

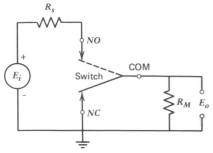

**Figure 7.22** Simple microswitch circuit for counting applications.

mechanical contacts is shown in Fig. 7.23. An electronic counter responding to positive voltage slopes would count each of the spikes as a legitimate event. Such errors can be eliminated by inserting two digital-logic elements known as NAND gates into the switching circuit to form what is known as a latch. The latch will respond to the first momentary contact but will ignore any other contacts until the circuit is reset.

A dual NAND-gate, switch-debouncer circuit together with a NAND-gate logic or truth table is shown in Fig. 7.24. As shown in the figure, the COM terminal of the switch and the negative terminal of the power supply ($E_i = 5$ VDC for the TTL chip referred to in the figure) are connected to ground, while the NO and NC terminals of the switch and the A1 and B2 inputs of the NAND gates are connected to the positive terminal of the power supply through series resistors $R_s$ (20 k$\Omega$ to 100 k$\Omega$ for the 5-V TTL chips). With the switch initially in the NC position, as shown in Fig. 7.24, NAND input A1 is held low (L) by the closed switch. The logic (truth) table for the NAND gate indicates that output Y1 will be high (H) regardless of the value of the input B1. Since input A2 is connected to Y1 and since input B2 is connected to the NO terminal of the switch, both A2 and B2 will be high (H); therefore, according to the NAND-gate truth table, output Y2 will be low (L). With B1 forced low (L) by the Y2 condition, the NC contact can be opened and closed repeatedly without altering the logic state of output Y1 high (H) and output Y2 low (L). In other words,

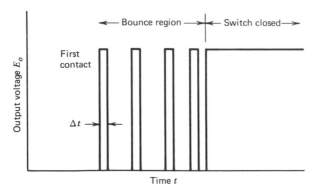

**Figure 7.23** Typical voltage output from a switch with mechanical contacts.

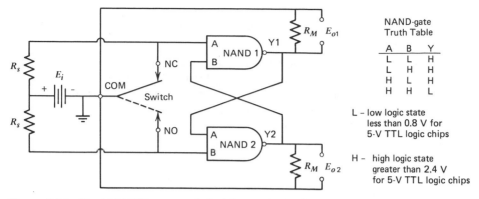

*Figure 7.24* Dual NAND-gate, switch-debouncer circuit.

the logic state would not be altered by any NC contact bounce as long as the NO contact does not make a momentary closure.

When a NO contact closure occurs, input B2 is forced low (L) by the new switch position. When this occurs, output Y2 goes high (H) as indicated in the truth table, and forces Y1 to go low (L) with the accompanying result that output Y2 remains high (H) regardless of NO contact bounce. The time required to effect this logic shift is approximately 20 ns. The duration $\Delta t$ of a first contact closure (see the first voltage spike of Fig. 7.23) for a mechanical switch is several orders of magnitude longer (millisecond duration) than this activation time; therefore, the response of the NAND circuit is more than adequate for this switch–contact–bounce application. Either output voltage can be used to record the switch activation.

## Full-Field Displacement Measuring Methods

All of the displacement-measuring methods presented in previous sections of this chapter (except for interference methods) were concerned with relative movements between discrete points. On occasion, for the purpose of performing an experimental stress or strain analysis,[2] relative movements associated with an array of points or complete maps of displacement fields are required. Such information can be provided by moire and grid methods.

The word *moire* is the French name for a fabric known as watered silk, which exhibits patterns of light and dark bands. This moire effect occurs whenever two similar but not quite identical arrays of equally spaced lines or dots are positioned so that one array can be viewed through the other. Almost everyone has seen the effect in two parallel snow fences or when two layers of window screen are placed in contact.

The first practical application of the moire effect may have been its use in judging the quality of line rulings used for diffraction gratings or halftone screens.

---

[2] For more information on the subject of experimental stress analysis, see the text *Experimental Stress Analysis,* 2nd ed., by J. W. Dally and W. F. Riley, McGraw-Hill, New York, 1978.

In this application, the moire fringes provide information on errors in spacing, parallelism, and straightness of the lines in the ruling. All of these factors contribute to the quality of the ruling.

Elimination of the moire effect has always been a major problem associated with screen photography in the printing industry. In multicolor printing, for example, where several screened images must be superimposed, the direction of screening must be carefully controlled to minimize moire effects.

Two line arrays must be overlaid to produce moire fringes. The overlaying can be accomplished by mechanical or optical means. In the following discussion, the two line arrays (gratings) will be referred to as the model (or specimen) grating and the master (or reference) grating. Quite frequently, the model grating is applied by coating the specimen with a photographic emulsion and contact-printing through the master grating. In this way, the model and master gratings are essentially identical (matched) when the specimen is in the undeformed state. Model arrays can also be applied by bonding, etching, ruling, etc. In one method of analysis, the shadow of the master grating on the model serves as the model array.

A typical moire fringe pattern, obtained using transmitted light through the model and master gratings, is shown in Fig. 7.25. In this instance, both the master grating and the model grating before deformation had 1000 lines per inch. The number of lines per unit length is frequently referred to as the *density*

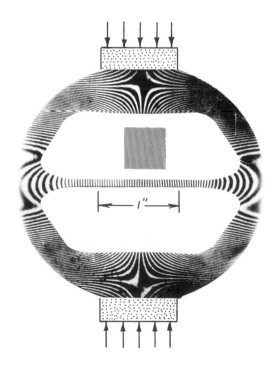

*Figure 7.25* Moire fringe pattern.

of the grating. The master grating was oriented with its lines running vertically, as shown in the inset.

The mechanism of formation of moire fringes can be illustrated by considering the transmission of a beam of light through model and reference arrays, as shown in Fig. 7.26. If model and master gratings are identical, and if they are aligned such that the opaque bars of one grating coincide exactly with the opaque bars of the other grating, the light will be transmitted as a series of bands having a width equal to one-half the pitch $p$ (reciprocal of the density) of the gratings. However, owing to diffraction and the resolution capabilities of the eye, this series of bands will appear as a uniform gray field with an intensity equal to approximately one-half the intensity of the incident beam when the pitch of the gratings is small.

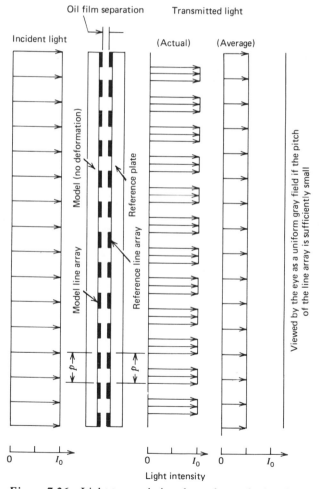

**Figure 7.26** Light transmission through matched and aligned model and master gratings.

If the model is then subjected to a uniform deformation (such as the one in the central tensile bar of Fig. 7.25), the model grating will exhibit a deformed pitch $p'$ as shown in Fig. 7.27. The transmission of light through the two gratings will now occur as a series of bands of different width, the width of the band depending on the overlap of an opaque bar with a transparent interspace. If the intensity of the emerging light is averaged over the pitch length of the master grating, to account for diffraction effects and the resolution capabilities of the eye, the intensity is observed to vary as a staircase function of position. The peaks of the function occur at positions where the transparent interspaces of the two gratings are aligned. A light band is perceived by the eye in these regions. When an opaque bar of one grating is aligned with the transparent interspace of the other grating, the light transmitted is minimum and a dark band known as a *moire fringe* is formed.

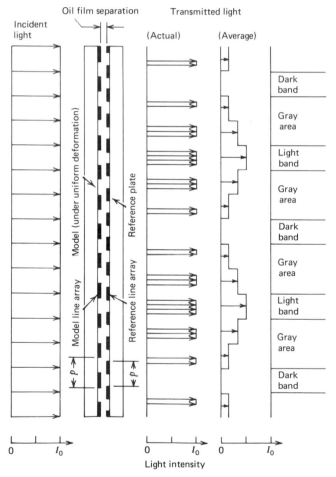

*Figure 7.27*  Formation of moire fringes in a uniformly deformed specimen.

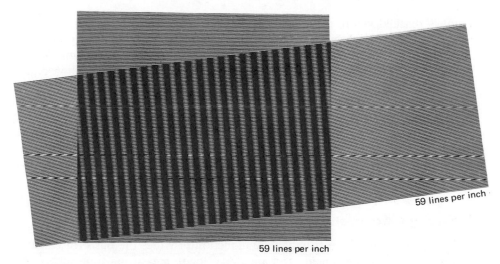

59 lines per inch

59 lines per inch

*Figure 7.28*   Moire fringes formed by rotation of one grating with respect to the other.

Inspection of the opaque bars in Fig. 7.27 indicates that a moire fringe is formed, within a given length, each time the model grating within the length undergoes a deformation equal to the pitch $p$ of the master grating. Deformations in a direction parallel to the lines do not produce fringes. In the case illustrated in Fig. 7.25, 32 fringes have formed in the 1-in. length indicated on the specimen. Thus, the change in length of the specimen in this 1-in. interval is

$$\Delta l = np = 32(0.001) = 0.032 \text{ in.} \tag{7.33}$$

In the previous discussions, the moire fringes were formed by elongations or contractions of the specimen in a direction perpendicular to the lines of the master grating. Simple experiments with a pair of identical gratings indicate that moire fringes can also be formed by pure rotations (no elongations or contractions), as illustrated in Fig. 7.28. In this illustration, two gratings having a line density of 59 lines per inch have been rotated through an angle $\theta$ with respect to one another. Note that the moire fringes have formed in a direction that bisects the obtuse angle between the lines of the two gratings. The relationship between angle of rotation $\theta$ and angle of inclination $\phi$ of the moire fringes, both measured in the same direction and with respect to the lines of the master grating, can be expressed as

$$\theta = 2\phi - \pi \tag{7.34}$$

The moire fringe patterns illustrated in Figs. 7.25 and 7.28 were produced by simple mechanical interference between two coarse line arrays. Moire fringe formation was easily explained by simple ray or geometric optics. As the line spacing of the arrays are reduced to improve the sensitivity of the method, a point is reached where diffraction effects become important and the complete wave character of light must be considered. Under such conditions, interference effects can no longer be explained by simple ray optics and the much more complicated methods of physical optics must be used. Ordinary photography can record only the intensity (related to the amplitude) of such waves. Holography, however, makes possible the recording of both amplitude and phase information. Extensive current research in both moire interferometry and holographic interferometry shows promise of providing totally new measurement capabilities for stress analysis, vibration analysis, and nondestructive evaluation. Such methods in combination with digital image analysis and interactive computer technology offer unlimited advancement possibilities for the future.[3]

## Velocity Measurements

The principle of electromagnetic induction provides the basis for construction of direct-reading, linear- and angular-velocity measuring transducers. Application of the principle for the two transducer types is shown in Fig. 7.29. For linear-velocity measurements, a magnetic field associated with the velocity to be measured moves with respect to a fixed conductor. For angular-velocity measurements, a moving conductor associated with the velocity to be measured moves with respect to a fixed magnetic field. In either case, a voltage is generated that can be related to the desired velocity. The basic equation relating voltage generated to velocity of a conductor in a magnetic field can be expressed as

$$E_T = Blv \tag{7.35}$$

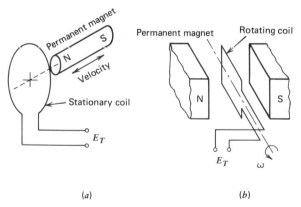

(a)                              (b)

*Figure 7.29*  Schematic representations of linear- and angular-velocity transducers. (*a*) Linear-velocity measurements. (*b*) Angular-velocity measurements.

---

[3] For more information on the subject of moire fringe analysis, see the text *Moire Analysis of Strain*, by A. J. Durelli and V. J. Parks, Prentice-Hall, Englewood Cliffs, NJ, 1970.

**where**  $E_T$ is the voltage generated by the transducer.
  $B$  is the component of the flux density (magnetic field strength) normal to the velocity.
  $l$  is the length of the conductor.
  $v$  is the velocity.

### Linear Velocity Measurements

A schematic representation of a self-generating linear-velocity transducer (LVT) is shown in Fig. 7.30. The windings are installed in series opposition so that the induced voltages add when the permanent magnet moves through the coil. Construction details of a commercially available linear-velocity transducer are shown in Fig. 7.31. As indicated, the core consists of an Alnico-V permanent magnet encased in a stainless steel sleeve. The coil form and windings may also be enclosed in a steel shell to prevent external magnetic materials from affecting the calibration.

The LVT is equivalent to a voltage generator connected in series with an inductance $L_T$ and a resistance $R_T$ as shown in Fig. 7.32. The governing differential equation for a velocity-measuring circuit with an LVT sensor and a recording instrument having an input resistance $R_M$ is

$$L_T \frac{dI}{dt} + (R_T + R_M)I = S_v v \tag{7.36}$$

**where**  $S_v$ is the voltage sensitivity of the transducer (mV/ips).
  $v$  is the time-dependent velocity being measured (ips).
  $I$  is the time-dependent current flowing in the circuit.

Response characteristics of an LVT can be obtained by considering the response of the transducer to a sinusoidal input velocity $v$ that can be expressed in phasor form as

$$v = v_o e^{i\omega t} = v_o(\cos \omega t + i \sin \omega t) \tag{a}$$

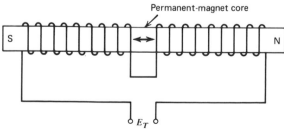

Permanent-magnet core

**Figure 7.30**  Schematic representation of a linear-velocity transducer.

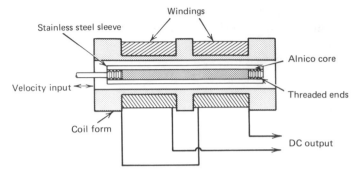

**Figure 7.31**   Cross section through a linear-velocity transducer.

**where**     $v_o$ is the magnitude of the velocity $v$.
$\omega$ is the angular frequency of the sinusoidal input.

Use of the phasor representation in the complex plane provides a convenient means for obtaining a solution to Eq. (7.36) that contains both magnitude and phase information. In a similar manner, the response $I$ of the LVT circuit to the sinusoidal input $v$ can be expressed in phasor form as

$$I = I_o e^{i\omega t} = I_o(\cos \omega t + i \sin \omega t) \tag{b}$$

where $I_o$ is the magnitude of the output $I$. Substituting Eqs. (a) and (b) into Eq. (7.36) yields

$$(R_M + R_T + iL_T\omega)I_o = S_v v_o \tag{c}$$

Solving Eq. (c) for $I_o$ yields

$$I_o = \frac{S_v v_o}{(R_M + R_T) + iL_T\omega} \tag{d}$$

The denominator of Eq. (d) is a complex number $(a + ib)$ with real $(a = R_M + R_T)$ and imaginary $(b = L_T\omega)$ parts that are perpendicular in the complex plane. The two parts can be combined by using the parallelogram law to yield

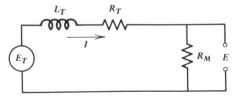

**Figure 7.32**   Linear-velocity transducer and circuit.

a resultant $r = \sqrt{a^2 + b^2}$. The real $a$ and imaginary $b$ components can be related to the resultant $r$ through a phase angle $\phi$ such that $a = r \cos \phi$ and $b = r \sin \phi$. Thus, the phase angle $\phi$ can be obtained from the expression $b/a = r \sin \phi / r \cos \phi = \tan \phi$. Equation (d) can be written in terms of $r$ and $\phi$ as

$$I_o = \left[ \frac{S_v v_o}{\sqrt{(R_M + R_T)^2 + (L_T \omega)^2}} \right] e^{-i\phi} \tag{e}$$

In Eq. (e), the quantity within the brackets (the real part of the expression) represents the magnitude of the response while the angle $\phi$ represents the phase of the response relative to the input. The negative sign in the exponential expression indicates that the current output lags the velocity input. The magnitude of the phase angle $\phi$ is given by the expression

$$\tan \phi = \frac{b}{a} = \frac{\text{Imag Part}}{\text{Real Part}} = \frac{L_T \omega}{R_M + R_T} \tag{f}$$

The output voltage $E$ from the circuit shown in Fig. 7.32 is given by the expression

$$E = IR_M = \frac{R_M S_v v_o}{(R_M + R_T) + iL_T \omega} e^{i\omega t} \tag{g}$$

The transfer function $H(\omega)$ for a circuit is defined as the ratio of the output to the input. Thus

$$H(\omega) = \frac{E}{v} = \frac{E}{v_o e^{i\omega t}} = \frac{R_M S_v}{(R_M + R_T) + iL_T \omega}$$

$$= \frac{R_M S_v e^{-i\phi}}{\sqrt{(R_T + R_M)^2 + (L_T \omega)^2}} \tag{7.37}$$

Equation (7.37) shows that the transfer function for an LVT circuit is a complex quantity and that both magnitude and phase of the output is frequency dependent. Equation (7.37) also indicates that the output will be attenuated at the higher frequencies. The break frequency $\omega_c$ for a circuit whose behavior is described by a first-order differential equation occurs when the real and imaginary parts of the transfer function are equal. Equation (7.37) indicates that this occurs when

$$\omega_c = \frac{R_M + R_T}{L_T} = \frac{R_M}{L_T}(1 + R_T/R_M) \tag{7.38}$$

Measurement error at the break frequency ($-3$ dB or 30 percent with a 45 degree phase shift) is too large for most practical applications. The more practical value of 2 percent error with a phase shift of 15.9 degrees occurs when $\omega = 0.28\omega_c$. Magnitude errors can be kept within 5 percent if instrument use is limited to frequencies below one-third of the break frequency.

Equations (7.37) and (7.38) clearly indicate that the sensitivity and frequency response of a given LVT is very dependent upon the input resistance $R_M$ of the recording instrument. The output voltage of an LVT sensor is relatively large (10 to 100 mV/in./s); therefore, signal amplification is not usually required. Linearity within $\pm 1$ percent can usually be achieved over the rated range of motion of the transducer. Common values for $R_T$ are less than 10 $\Omega$.

The projection of the rotating phasors $v$ and $I$ onto the "real" axis corresponds to the solution of Eq. (7.36) for the cosine part of the excitation of Eq. (a). Similarly, the projection of $v$ and $I$ onto the "imaginary" axis corresponds to the solution of Eq. (7.36) for the sine part of the excitation of Eq. (a). The principal advantage of using the phasor representation (complex notation) is that both sine and cosine solutions are obtained simultaneously and the phase angle is automatically obtained as a part of the complex algebra. Solutions of this type will be obtained for a number of seismic measuring instruments in later sections of this chapter.

## Angular Velocity Measurements

Angular velocities can be measured by using either dc or ac generators. With ac generators, which are used to measure average angular velocities, the number of cycles generated per revolution is an even integer; therefore, the readout device can be a simple frequency counter.

## Crack-Propagation Gages

A variable-resistance transducer, closely related to the multiple-resistor displacement-measuring system of Fig. 7.17, is the crack-propagation gage of Fig. 7.33. When a gage of this type is bonded to a structure, progression of a surface crack through the gage pattern causes successive open-circuiting of the strands, resulting in an increase in total resistance of the gage. Crack growth can be monitored by making a simple measurement of resistance with an ohmmeter, as shown schematically in Fig. 7.34, to establish the number of strands fractured and thus, the position of the crack tip. Typical step curves of resistance versus strands fractured for several gage types are shown in Fig. 7.35.

Crack-propagation-velocity determinations can also be made by using either a strip-chart recorder or an oscilloscope as the readout instrument in a circuit similar to the one shown in Fig. 7.36. The voltage-output record, which can be related to strands fractured as a function of time, provides the crack-tip position and time information needed for calculation of the average velocity of propagation between strands.

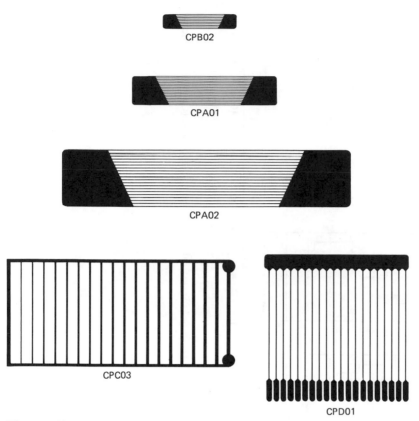

*Figure 7.33*   Alloy foil crack-propagation gages. (Courtesy of Micro-Measurements.)

## 7.6   **MOTION MEASUREMENT WITHOUT A FIXED REFERENCE**

Measurements of displacement, velocity, and acceleration must be made in many applications where it is not possible to use a fixed reference. Instances include earthquake-related phenomena and motion associated with aircraft or spacecraft in flight. Instruments that have been developed for such measurements are referred to as seismic instruments. Basically, such instruments detect relative motion between a base, which is attached to the structure of interest, and a seismic mass that, due to inertia, tends to resist any changes in velocity. Force

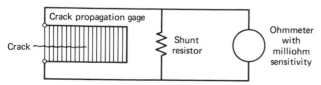

*Figure 7.34*   Circuit for monitoring crack growth.

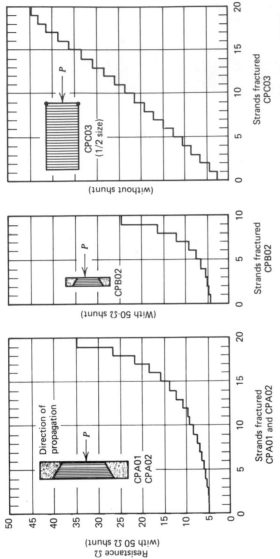

***Figure 7.35*** Resistance versus strands fractured for several crack-propagation gages. (Courtesy of Micro-Measurements.)

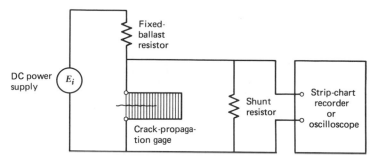

**Figure 7.36** Circuit for crack-propagation-velocity studies.

and pressure transducers also exhibit the seismic characteristics that are considered in this section.

## Seismic Transducer Theory

The mechanical behavior of motion-measuring instruments of the seismic type can be studied and their characteristics can be determined by considering a simple one-degree-of-freedom model with viscous damping as shown in Fig. 7.37a. A free-body diagram of the seismic mass is shown in Fig. 7.37b. The equation of motion of the seismic mass, obtained from an application of Newton's second law of motion, is

$$m\ddot{y} + C(\dot{y} - \dot{x}) + k(y - x) = F(t) \tag{7.39}$$

**where**  $m$  is the mass of the seismic body.
        $k$  is the deflection constant for the spring.
        $C$  is the viscous damping constant.
        $x$, $\dot{x}$, and $\ddot{x}$ are the displacement, velocity, and acceleration of the base.
        $y$, $\dot{y}$, and $\ddot{y}$ are the displacement, velocity, and acceleration of the seismic mass.
        $F(t)$ is the time-dependent, external, forcing function from either a force or pressure transducer.

The bolt force required to attach the transducer base to the structure can be expressed as

$$F_b = F(t) - m\ddot{y} - m_b\ddot{x}$$

The transducer does not measure this force directly.

Sensing elements in all types of seismic transducers respond to relative motion between the seismic mass and the base. As shown in Fig. 7.37a, this

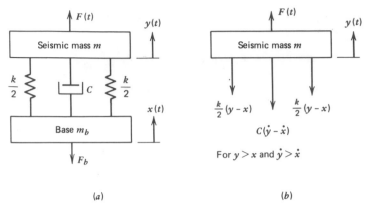

(a)                                                                    (b)

**Figure 7.37**   Single-degree-of-freedom model of a seismic instrument. (a) Schematic of a seismic transducer. (b) Free-body diagram of the seismic mass.

relative motion can be expressed as

$$z = y - x$$

$$\dot{z} = \dot{y} - \dot{x}$$

$$\ddot{z} = \ddot{y} - \ddot{x} \tag{7.40}$$

If Eqs. (7.40) are substituted into Eq. (7.39), the differential equation of motion becomes

$$m\ddot{z} + C\dot{z} + kz = F(t) - m\ddot{x} = R(t) \tag{7.41}$$

where $R(t) = F(t) - m\ddot{x}$ represents the external transducer excitation. Equation (7.41) indicates that the relative motion $z$ depends on both the external forcing function and base motion, in addition to instrument damping $(C)$ and the frequency of the input relative to the natural frequency $(\omega_n = \sqrt{k/m})$ of the transducer. Since Eq. (7.41) can be used to describe the response of displacement, velocity, and acceleration motion sensors as well as force and pressure transducers, a general solution of the equation for a sinusoidal type of excitation will be developed first and then applied to specific types of sensors. The solution for a sinusoidal excitation can be applied to a broad spectrum of inputs, and describes most of the mechanical characteristics of seismic sensors.

The sinusoidal excitation, expressed in very compact exponential (complex notation) form, is

$$R(t) = R_o e^{i\omega t} = R_o(\cos \omega t + i \sin \omega t) \tag{7.42}$$

where $R_o$ is the magnitude of the excitation. In a similar way, the relative motion response can be expressed in exponential form as

$$z = z_o e^{i\omega t}$$

from which

$$\dot{z} = i\omega z_o e^{i\omega t}$$

$$\ddot{z} = -\omega^2 z_o e^{i\omega t} \tag{7.43}$$

where $z_o$ is the complex amplitude (magnitude and phase) of the response phasor $z$ relative to the excitation phasor $R$. Substitution of Eqs. (7.42) and (7.43) into Eq. (7.41) yields

$$(k - m\omega^2 + iC\omega)z_o e^{i\omega t} = R_o e^{i\omega t} \tag{7.44}$$

From Eq. (7.44) it is obvious that the transfer function between the output $z$ and the input $R$ is

$$H(\omega) = \frac{1}{k - m\omega^2 + iC\omega} = \frac{1}{\sqrt{(k - m\omega^2)^2 + (C\omega)^2}} e^{-i\phi} \tag{7.45}$$

Thus, the magnitude of the response $z_o$ is given by the expression

$$z_o = \frac{R_o}{\sqrt{(k - m\omega^2)^2 + (C\omega)^2}} = \frac{R_o}{k\sqrt{(1 - r^2)^2 + (2rd)^2}} \tag{7.46}$$

and the phase angle $\phi$ is given by the expression

$$\tan \phi = \frac{\text{Imag Part}}{\text{Real Part}} = \frac{C\omega}{k - m\omega^2} = \frac{2rd}{1 - r^2} \tag{7.47}$$

**where**  $d = C/2\sqrt{km}$ is a dimensionless damping ratio.
   $r = \omega/\omega_n = f/f_n$ is a dimensionless frequency ratio.

In addition, the steady-state solution of Eq. (7.41) with the sinusoidal excitation of Eq. (7.42) is

$$z = \frac{R_o}{\sqrt{(k - m\omega^2)^2 + (C\omega)^2}} e^{i(\omega t - \phi)}$$

$$= \frac{R_o}{k\sqrt{(1 - r^2)^2 + (2rd)^2}} e^{i(\omega t - \phi)} \tag{7.48}$$

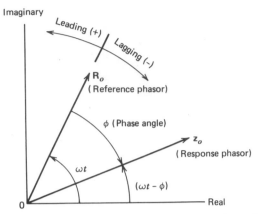

**Figure 7.38** Rotating phasor representation of steady-state instrument response.

The relationship between rotating phasors $R_o$ and $z_o$ is illustrated in Fig. 7.38. Note that phase angles associated with the response phasor $z_o$ are positive when counterclockwise and negative when clockwise with respect to the reference phasor $R_o$.

## Seismic Displacement Transducers

Seismic displacement transducers are instruments which have been designed to respond to base motion. If the base motion is assumed to be sinusoidal with amplitude $x_o$, then

$$x = x_o e^{i\omega t}$$

and

$$\dot{x} = i\omega x_o e^{i\omega t}$$

$$\ddot{x} = -\omega^2 x_o e^{i\omega t}$$

(7.49)

Since an external forcing function $F(t)$ is not present in these instruments, the excitation term $R(t)$ of Eq. (7.41) becomes

$$R(t) = R_o e^{i\omega t} = m\omega^2 x_o e^{i\omega t}$$

(7.50)

With this excitation, the steady-state solution of Eq. (7.48) becomes

$$z = \frac{m\omega^2 x_o}{\sqrt{(k - m\omega^2)^2 + (C\omega)^2}} e^{i(\omega t - \phi)}$$

$$= \frac{r^2 x_o}{\sqrt{(1 - r^2)^2 + (2dr)^2}} e^{i(\omega t - \phi)}$$

(7.51)

The transfer function $H(\omega)$ for the system represented by Eq. (7.51) is illustrated in Fig. 7.39 in the form of magnitude ratio $z/x_o$ and phase angle $\phi$ versus frequency ratio $\omega/\omega_n$ curves for different amounts of damping. It is evident from these curves that the ratio $z/x_o$ approaches value of unity and the phase angle $\phi$ approaches 180 degrees as the frequency ratio becomes large $(r > 4)$ irrespective of the amount of damping. The peak response decreases and occurs at a higher frequency ratio as the damping ratio is increased from 0 to 0.707. There are no peaks in the response curve for damping ratios greater than 0.707.

The results of Fig. 7.39 indicate that seismic displacement transducers should have a very low natural frequency so that the ratio $\omega/\omega_n$ can be large under most situations. Under this condition, $r \gg 1.0$, $\phi \to \pi$, and Eq. (7.51) reduces to

$$z = \frac{r^2 x_o}{\sqrt{(-r^2)^2}} e^{i(\omega t - \phi)} \approx -x_o e^{i\omega t} = -x \qquad (7.52)$$

Equation (7.52) indicates that the seismic mass remains essentially motionless and the response of the transducer results only from the motion of the base that is attached to the structure whose displacement is being measured. The amount of damping used in displacement sensing transducers is controlled by the peak amplitudes that can be tolerated during transients, the usable frequency range, and the phase shift that can be tolerated in the usable frequency range. These

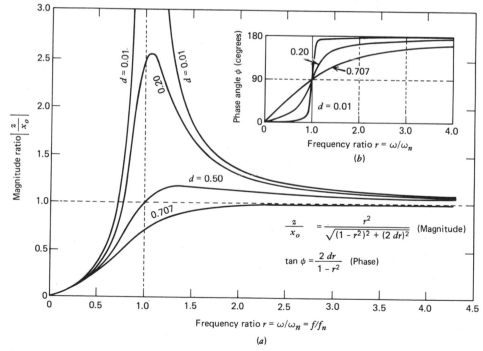

**Figure 7.39**  Transfer function (magnitude and phase) for a displacement transducer. (a) Magnitude. (b) Phase.

requirements cannot be satisfied simultaneously; therefore, some compromises will always be required. For example, a constant phase shift requires little or no damping, while the amplitude and frequency range requirements require some damping.

Seismic displacement transducers, because of the low natural frequency requirements, utilize soft springs with correspondingly large static deflections and instrument size. As a result, the common sensing elements used in these instruments are linear variable-differential transformers (LVDTs) and electrical resistance strain gages on flexible elastic members (beam-type elements).

## Seismic Velocity Transducers

The seismic displacement transducer can be converted to a seismic velocity transducer (velometer) by using a velocity-dependent magnetic sensing element to measure relative motion between the seismic mass and the base. The governing differential equation that describes the behavior of this type of instrument is obtained by differentiating Eq. (7.51) with respect to time. Thus,

$$\dot{z} = \frac{r^2 i\omega x_o}{\sqrt{(1 - r^2)^2 + (2dr)^2}} e^{i(\omega t - \phi)} \tag{7.53}$$

Under the conditions that $r \gg 1.0$ and $\phi \to \pi$, Eq. (7.53) reduces to

$$\dot{z} \approx -i\omega x_o e^{i\omega t} = -\dot{x} \tag{7.54}$$

Equation (7.54) indicates that the response of a seismic velocity transducer depends only on the velocity of the base if the instrument is designed with a very low natural frequency. Also, the transfer function is the same as that for the seismic displacement transducer.

## Seismic Acceleration Transducers

In the previous discussions involving seismic displacement and seismic velocity transducers, sensing elements, which respond to displacement and velocity, were available for use in the instruments. For the case of acceleration measurements, no such sensing element is available; therefore, other means must be used to accomplish the measurement. If Eq. (7.49) is substituted into Eq. (7.41), an expression is obtained that relates the relative displacement $z$ between the seismic mass and the base of the transducer to the acceleration $\ddot{x}$ of the base. Thus,

$$z = \frac{m\omega^2 x_o}{\sqrt{(k - m\omega^2)^2 + (C\omega)^2}} e^{i(\omega t - \phi)} \tag{7.51}$$

However from Eq. (7.49),

$$\ddot{x} = -\omega^2 x_o e^{i\omega t}$$

Therefore,

$$z = -\frac{e^{-i\phi}}{\omega_n^2 \sqrt{(1 - r^2)^2 + (2dr)^2}} \ddot{x} \tag{7.55}$$

The transfer function $H(\omega)$ for the system represented by Eq. (7.55) is illustrated in Fig. 7.40 in the form of magnitude ratios and phase angles versus frequency ratios for different amounts of damping. It is evident from these curves that the magnitude ratio approaches a value of unity and the phase angle approaches 0 degrees as the frequency ratio becomes very small ($r \to 0$), irrespective of the amount of damping. The peak response decreases and occurs at a lower frequency ratio as the damping ratio is increased from 0 to 0.707. There are no peaks in the magnitude response curve for damping ratios greater than 0.707.

Under the condition that $r \to 0$, Eq. (7.55) reduces to

$$z \approx -\frac{1}{\omega_n^2} \ddot{x} = -\frac{m}{k} \ddot{x} \tag{7.56}$$

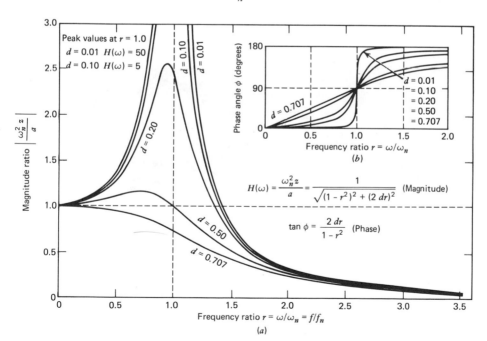

*Figure 7.40* Transfer function (magnitude and phase) for acceleration, force, and pressure transducers. (*a*) Magnitude. (*b*) Phase.

Equation (7.56) shows that the basic sensing mechanism of a seismic acceleration transducer consists of an inertia force $m\ddot{x}$ being resisted by a spring force $kz$. This type of instrument requires a very stiff spring and a very small seismic mass so that the instrument will have a very high natural frequency. Seismic acceleration transducers can be made very small; therefore, the presence of the transducer usually has little effect on the quantity being measured. An accelerometer with a piezoelectric sensing element that has been designed for shock measurements is shown in Fig. 7.41.

The condition that must exist if an accelerometer is to faithfully respond to a typical acceleration input can be determined by considering response of the instrument to a periodic input of the type shown in Fig. 7.42. Such an input signal can be represented by a Fourier series of the form

$$\ddot{x} = a_1 \sin \omega_1 t + a_2 \sin 2\omega_1 t + \cdots + a_q \sin q\omega_1 t \qquad (7.57)$$

**where** $\omega_1 = 2\pi/T_p$ is the fundamental frequency of the input signal.
$T_p$ is the period of the input signal.

The magnitude and phase of each component of the input signal will be modified as indicated by the transfer function for the accelerometer and as a result, the

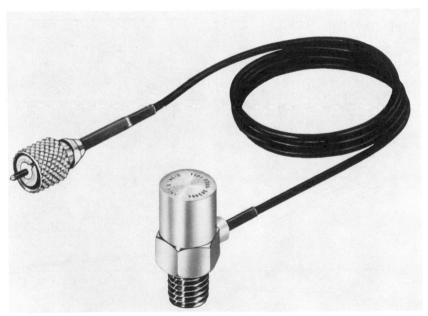

*Figure 7.41* An accelerometer with a piezoelectric sensing element. (Courtesy of Bruel and Kjaer.)

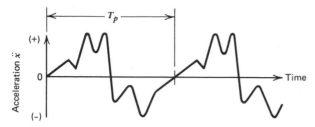

**Figure 7.42**  Periodic acceleration input.

output $z$ will take the form

$$z = b_1 \sin(\omega_1 t - \phi_1) + b_1 \sin(2\omega_1 t - \phi_2)$$
$$+ \cdots + b_q \sin(q\omega_1 t - \phi_q) \qquad (7.58)$$

A comparison of Eqs. (7.57) and (7.58) shows that they will have the same shape if:

1.  The $b_q$ coefficients are constant multiples of the $a_q$ coefficients.
2.  The phase angle is either zero or a linear function of frequency.

The plots shown in Fig. 7.40 indicate that the first requirement can be satisfied over a broad range of frequencies if either the damping ratio is near zero or if it is in the range between 0.59 to 0.707. With near-zero damping, the upper frequency limit is approximately 20 percent of the natural frequency of the instrument if errors are to be limited to $\pm 5$ percent. With a damping ratio of 0.59, errors can be limited to $\pm 5$ percent over a frequency range from 0 to 85 percent of the natural frequency of the instrument. Magnitude and phase errors as a function of frequency for different amounts of damping are shown in Fig. 7.43. The plots of Fig. 7.40 also indicate that the second requirement can be met by using either a damping ratio near zero to yield a very small phase angle over the useful range of frequencies of the instrument or a damping ratio between 0.59 to 0.707 to yield a phase shift that varies as a linear function of frequency. With a linear phase shift,

$$\phi_q = \alpha \omega_q = q\alpha\omega_1$$

The argument of the general term in Eq. (7.58) thus becomes

$$q\omega_1 t - \phi_q = q\omega_1(t - \alpha) \qquad (7.59)$$

Equation (7.59) indicates that a linear phase shift produces a simple time shift of the output signal with respect to the input signal, but does not change the shape of the signal.

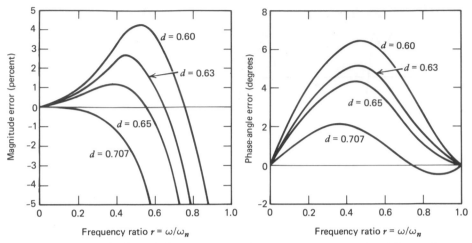

***Figure 7.43*** Magnitude and phase errors as a function of frequency for different degrees of damping.

At the present time, most seismic acceleration transducers are designed to have very high natural frequencies and very low damping ratios (of the order of 1 to 2 percent). The low damping ratios are used, since most damping schemes add significant mass to the transducer in the form of fluids or magnetic materials that would reduce their useful range of frequencies. Piezoelectric sensing elements are ideally suited for accelerometer applications, since they have a very small mass, exhibit a large spring constant (have a high modulus of elasticity), and exhibit little damping. The characteristics of several electrical circuits that are widely used with piezoelectric sensors are discussed in a later section of this chapter.

## Dynamic Force and Pressure Transducers

Force, torque, and pressure transducers were described in Chapter Six. In those discussions, the effects of acceleration on the output of the force or pressure transducer were neglected. Dynamic force and pressure transducers are constructed to have stiff structures with high natural frequencies; therefore, they respond to acceleration and can be considered a seismic type of instrument. Piezoelectric sensing elements are often used in order to obtain these high natural frequencies. The transfer function for these instruments is given by Eq. (7.45); therefore, the magnitude and phase curves of Fig. 7.40 are applicable. In the discussions that follow, problems unique to force and pressure transducers are discussed.

The manner in which a force or pressure transducer is used in a dynamic application can seriously affect its performance. For example, a connector such as a bolt with mass $m_c$ must usually be used between the transducer and the structure. This situation is illustrated in Fig. 7.44, where the force to be measured is $R(t)$ while the force sensed by the transducer is $F(t)$. Application of Newton's second law of motion to the connector mass indicates that

**For a force transducer**   $F(t) = R(t) - m_c \ddot{y}$

**For a pressure transducer** $F(t) = Ap(t) - m_c \ddot{y}$   (7.60)

**where**  $p(t)$ is the dynamic pressure.

   $A$  is the effective area of the face of the pressure transducer.

Equation (7.60) shows that the mounting hardware can significantly alter the force or pressure output of a transducer. The performance characteristics of a system consisting of a force transducer and mounting hardware can be obtained from the expression that results when Eq. (7.60) is substituted into Eq. (7.41). Thus

$$(m + m_c)\ddot{z} + C\dot{z} + kz = R(t) - (m + m_c)\ddot{x}$$   (7.61)

Equation (7.61) indicates that the natural frequency of the measurement system $\omega_{nm}$ is

$$\omega_{nm} = \sqrt{\frac{k}{m + m_c}} = \frac{1}{\sqrt{1 + (m_c/m)}} \sqrt{\frac{k}{m}}$$   (7.62)

Since the ratio $m_c/m$ can be quite large for instruments with piezoelectric sensors, the natural frequency of the measurement system (transducer and connector) is considerably less than the natural frequency of the transducer. Equation (7.61) also indicates that an increase in sensitivity to base acceleration will occur since the $m\ddot{x}$ term is multiplied by the factor $1 + m_c/m$. The response resulting from base acceleration can be eliminated electronically by subtracting a signal that is proportional to the base acceleration. Such a feature has been incorporated into several commercially marketed pressure transducers. The self-compensating feature is difficult to build into force transducers since the connector mass is determined by the user for his specific application. A discussion of acceleration

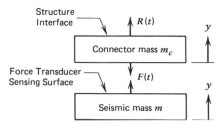

**Figure 7.44**  Schematic representation of a force transducer with a connector mass.

compensation circuits for force transducers can be found in a paper by McConnell and Park.[4]

The problem of force transducer response to base acceleration is especially important when the force transducer is used to measure the force being applied to a structure that is being excited at or near its resonant frequency. For such a situation, the motion of the structure is represented by the equation

$$m_s\ddot{x} + C_s\dot{x} + k_sx = R(t) = R_oe^{i\omega t} \tag{7.63}$$

**where**  $m_s$ is the mass of the structure.
$C_s$ is the structural damping coefficient.
$k_s$ is the structural spring constant.

The acceleration experienced by the force measuring system is

$$\ddot{x} = -\frac{R_o\omega^2}{k_s - m_s\omega^2 + iC_s\omega}e^{i\omega t}$$

As the excitation frequency $\omega$ approaches the structural resonance frequency $\omega_{ns} = \sqrt{k_s/m_s}$, the acceleration experienced by the force measuring system becomes

$$\ddot{x} = -\frac{R_o\omega_{ns}^2}{iC_s\omega_{ns}} = -\frac{R_o\omega_{ns}}{iC_s} \tag{7.64}$$

The force $F(t)$ sensed by the transducer is

$$F(t) = R(t) - (m + m_c)\ddot{x}$$

$$= R_o\left[1 + m\left(1 + \frac{m_c}{m}\right)\frac{\omega_{ns}}{iC_s}\right] \tag{7.65}$$

The term in the bracket of Eq. (7.65) can be considerably greater than unity when the connector mass is large with respect to the seismic mass of the transducer and the structural damping coefficient $C_s$ is small. Thus, the output of the transducer $F(t)$ can be much larger than the magnitude of the excitation force $R_o$. Equation (7.65) illustrates one of the many problems encountered in making dynamic force measurements.

An instrument known as an *impedance transducer* measures both the force transmitted through the instrument and the acceleration of the instrument. Since

---

[4] "Electronic Compensation of a Force Transducer for Measuring Fluid Forces Acting on an Accelerating Cylinder," by K. G. McConnell and Y. S. Park, *Experimental Mechanics,* vol. 21, no. 4, April 1981, pp. 169–172.

the two measurements are made simultaneously, compensation for the effects of connector mass can be made.

Typical characteristics for a selected list of seismic instruments (acceleration, force, and pressure transducers) are listed in Tables 7.1, 7.2, and 7.3, respectively.

## 7.7  PIEZOELECTRIC SENSOR CIRCUITS

Seismic acceleration transducers and dynamic force and pressure gages often use piezoelectric sensing elements in order to obtain the required frequency response. Since these piezoelectric sensing elements are charge generating devices, high-input-impedance signal conditioning instruments, such as cathode or voltage followers, charge amplifiers, or integrated-circuit amplifiers, built into the transducer package must be incorporated into the measurement circuit to convert the charge to an output voltage that can be measured and recorded. The operational and performance characteristics that are related to sensitivity and frequency response for each of these circuits are discussed in detail in the material that follows.

### Cathode-Follower Circuit

The general arrangement of a measurement circuit containing a piezoelectric sensor and a cathode follower (unity-gain buffer amplifier) is shown in Fig. 7.45a. The charge generated by the sensor can be expressed as

$$q = S_q a \tag{7.66}$$

**where**  $q$  is the charge generated by the piezoelectric sensor (pC).
  $S_q$  is the charge sensitivity of the transducer (pC/g, pC/lb, or pC/psi).
  $a$  is the quantity being measured (acceleration, force, or pressure).

An analysis of the circuit of Fig. 7.45a requires consideration of the capacitance $C_t$ of the transducer, the capacitance $C_c$ of the cable, the capacitance $C_s$ of the standardizing capacitor in the amplifier input portion of the circuit, the capacitance $C_b$ of the blocking capacitor used in some designs to protect the amplifier, and the resistance $R$ that represents the combined effect of the input impedance of the amplifier and any load resistance that may be placed in parallel with the amplifier. The circuit can be reduced for analysis to the form shown in Fig. 7.45b by combining the transducer, cable, and standardizing capacitances into a capacitance $C$ such that

$$C = C_t + C_c + C_s \tag{7.67}$$

**TABLE 7.1** Piezoelectric Accelerometer Characteristics

| | Manufacturer | | | |
|---|---|---|---|---|
| | **Endevco** | **Endevco** | **PCB** | **Kistler** |
| Model | 2215E | 2292 | 302A | 808A |
| Crystal material | Piezite P-8 | Piezite P-8 | Quartz | Quartz |
| Charge sensitivity (pC/g) | 170 | 0.14 | — | 1.0 |
| Voltage sensitivity (mV/g) | 16 | 0.4 | 10 | — |
| Capacitance (pF) | 10,000 | 80 | — | 90 |
| Internal time constant (s) | — | — | 0.5 | — |
| Frequency response (Hz) | 4 to 8000 | 50 to 20,000 | 1 to 5000 | 0 to 7000 |
| Mounted resonance frequency (Hz) | 32,000 | 125,000 | 45,000 | 40,000 |
| Transverse sensitivity | $\leqslant \pm 5\%$ | $\leqslant \pm 5\%$ | $\leqslant \pm 5\%$ | $\leqslant \pm 5\%$ |
| Range (g's) | 0 to 1000 | 0 to 20,000 | 0 to 500 | 0 to 10,000 |
| Size (in.) | $\frac{5}{8}$ Hex $\times$ 0.8 | 0.31 Hex $\times$ 0.3 | $\frac{1}{2}$ Hex $\times$ 1.25 | $\frac{1}{2}$ Hex $\times$ 0.90 |
| Weight (grams) | 32 | 1.3 | 23 | 20 |
| Vibration (maximum g's) | 1000 | 1000 | 2000 | — |
| Shock (maximum g's) | 2000 | 20,000 | 5000 | — |
| Temperature range (°F) | $-65$ to $+350$ | $-65$ to $+250$ | $-100$ to $+250$ | $-195$ to $+260$ |
| Application | General purpose | Shock | General purpose | General purpose |

**TABLE 7.2** Piezoelectric Dynamic Force Transducer Characteristics

| | Manufacturer | | | |
|---|---|---|---|---|
| | Endevco | PCB | PCB | Kistler |
| Model | 2103-500 | 208A03 | 200A05 | 912 |
| Crystal material | Piezite P-1 | Quartz | Quartz | Quartz |
| Charge sensitivity (pC/lb) | 45 | — | — | 50 |
| Voltage sensitivity (mV/lb) | 150 | 10 | 1.0 | — |
| Capacitance (pF) | 200 | — | — | 58 |
| Internal time constant (s) | — | 2000 | 2000 | — |
| Frequency response (Hz) | 2 to 4000 | 0.1 to 14,000 | 0.1 to 14,000 | To 14,000 |
| Mounted resonant frequency (Hz)[a] | 20,000 | 70,000 | 70,000 | 70,000 |
| Range (lb) | 0 to 500 | ±500 | 0 to −5000 | −5000 to +100 |
| Amplitude linearity | ±3% | ±1% | ±1% | ±1% |
| Size (in.) | $\frac{3}{4}$ Hex × 1.25 | $\frac{5}{8}$ Hex × 0.625 | 0.65 Dia × 0.36 | $\frac{5}{8}$ Hex × 0.500 |
| Weight (grams) | 57 | 25 | 14 | 17 |
| Stiffness (lb/in.) | $6 \times 10^6$ | $10 \times 10^6$ | $100 \times 10^6$ | $5 \times 10^6$ |
| Vibration (maximum g's) | — | 2000 | 2000 | — |
| Shock (maximum g's) | — | 10,000 | 10,000 | 10,000 |
| Temperature range (°F) | −30 to +200 | −100 to +250 | −100 to +250 | −240 to +150 |
| Application | Axial force link | General purpose | Impact compression | Impact |

[a] With no external mass attached.

**TABLE 7.3** Piezoelectric Dynamic Pressure Transducer Characteristics

| | Manufacturer | | | |
|---|---|---|---|---|
| | Endevco | PCB | Kistler | Kistler |
| Model | 2501-2000 | 109A | 601B1 | 606L |
| Crystal material | Piezite P-1 | Quartz | Quartz | Quartz |
| Charge sensitivity (pC/psi) | 14 | — | 1.0 | 5.5 |
| Voltage sensitivity (mV/psi) | 35 | 0.1 | — | — |
| Capacitance (pF) | 300 | — | 20 | 50 |
| Internal time constant (s) | — | 2000 | — | — |
| Frequency response (Hz) | 2 to 10,000 | 0 to 100,000 | 0 to 100,000 | 0 to 26,000 |
| Mounted resonant frequency (Hz) | 50,000 | 500,000 | 250,000 | 130,000 |
| Range (psi) | 0 to 2000 | 0 to 80,000 | 0 to 15,000 | 0 to 30 |
| Amplitude linearity | ±3% | ±2% | ±1% | ±1% |
| Size (in.) | $\frac{3}{4}$ Hex × 0.75 | $\frac{3}{8}$ Dia × 1.68 | $\frac{9}{32}$ Hex × 1.30 | $\frac{1}{2}$ Hex × 1.25 |
| Weight (grams) | 21 | — | 7 | 22 |
| Vibration (maximum g's) | — | 200 | 1000 | — |
| Shock (maximum g's) | — | 20,000 | 20,000 | 1000 |
| Acceleration sensitivity (psi/g) | 0.03 | 0.004 | 0.002 | 0.0005 |
| Temperature range (°F) | −22 to +230 | −100 to +275 | −450 to +500 | −350 to +450 |
| Applications | General purpose Flush mount | Shock wave Blast pressures | Transient pressure Rugged environment Acceleration compensated | Low pressure High sensitivity |

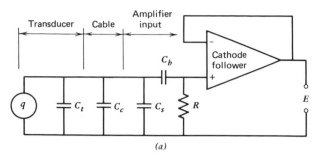

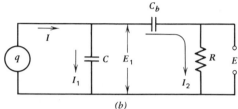

**Figure 7.45** (a) Measurement circuit with a piezoelectric sensor and a cathode follower. (b) Equivalent circuit for analysis.

The differential equation that describes the behavior of the circuit of Fig. 7.45b is obtained as follows:

$$I = I_1 + I_2 = \dot{q} = S_q\dot{a} \tag{7.68}$$

where

$$I_1 = C\dot{E}_1 \qquad \text{and} \qquad I_2 = C_b(\dot{E}_1 - \dot{E}) = \frac{E}{R} \tag{7.69}$$

Substituting Eqs. (7.69) into Eq. (7.68) and simplifying yields

$$\dot{E} + \frac{E}{RC_{eq}} = \frac{S_q}{C}\dot{a} \tag{7.70}$$

where

$$\frac{1}{C_{eq}} = \frac{1}{C} + \frac{1}{C_b} \tag{7.71}$$

Typical values of $C$ range from 300 to 10,000 pF while values of $C_b$ are usually of the order of 100,000 pF; therefore, the equivalent capacitance $C_{eq}$ of Eq. (7.71) is usually not significantly different from the combined capacitance $C$ of Eq. (7.67).

The transfer function associated with Eq. (7.70) can be obtained by assuming that the quantity being measured (acceleration, force, or pressure) experiences a sinusoidal variation that can be expressed as

$$a = a_o e^{i\omega t} = a_o(\cos \omega t + i \sin \omega t) \qquad (7.72)$$

The steady-state sinusoidal response can then be expressed as

$$E = E_o e^{i\omega t} = E_o(\cos \omega t + i \sin \omega t) \qquad (7.73)$$

Substitution of Eqs. (7.72) and (7.73) into Eq. (7.70) yields

$$E_o = \frac{iRC_{eq}\omega}{1 + iRC_{eq}\omega} \frac{S_q a_o}{C} \qquad (7.74)$$

Therefore, the voltage sensitivity of the measurement system is

$$S_v = \frac{E_o}{a_o} = \frac{S_q}{C} \qquad (7.75)$$

Equation (7.75) shows that the voltage sensitivity of the measurement system of Fig. 7.45a depends on all of the capacitances that contribute to Eq. (7.67). The specific cable used for the measurement, environmental factors such as temperature and humidity that can change the capacitance of the individual elements of the circuit, and dirt, grease, or sprays, which play no role in a clean laboratory but may be significant in an industrial setting, can produce a change in the voltage sensitivity of the system.

The electrical transfer function for the circuit of Fig. 7.45a is obtained from Eq. (7.74). Thus,

$$H(\omega) = \frac{CE_o}{S_q a_o} = \frac{iRC_{eq}\omega}{1 + iRC_{eq}\omega} = \frac{\omega T}{\sqrt{1 + (\omega T)^2}} e^{i\phi} \qquad (7.76)$$

$$T = RC_{eq} \qquad (7.77)$$

$$\phi = \frac{\pi}{2} - \tan^{-1} \omega T \qquad (7.78)$$

**where**    $T$ is the time constant of the circuit (seconds).

$\phi$ is the phase angle of the output relative to the input (radians).

The low-frequency characteristics of this circuit are shown in Fig. 7.46. The magnitude of the transfer function is plotted as a function of $\omega T$ in Fig. 7.46a. The curve shows that the response is very nonlinear in the range $0 \leqslant \omega T \leqslant 2$. The nonlinear characteristics of the magnitude response curve can be linearized

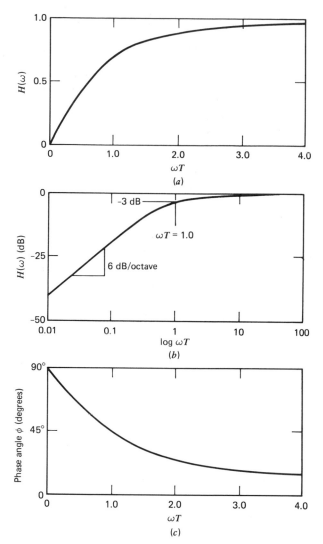

**Figure 7.46** Low-frequency transfer-function characteristics of piezoelectric sensors. (*a*) Magnitude of the transfer function $H(\omega)$ versus $\omega T$. (*b*) Magnitude of $H(\omega)$ in decibels versus log $\omega T$. (*c*) Phase angle $\phi$ versus $\omega T$.

for values of $\omega T$ less than 1.0 by plotting the magnitude in decibels as a function of log $\omega T$ (known as a *Bode plot*) as shown in Fig. 7.46*b*. This curve initially rises at a standard rate of 6 dB/octave (for $0 \leqslant \omega T \leqslant 0.7$), but then curves (the $-3$-dB point occurs at $\omega T = 1.0$) and approaches the 0-dB line asymptotically as the frequency is increased. Phase-angle values for $\omega T \leqslant 4$ are shown in Fig. 7.46*c*. Key values of $\omega T$ and $\phi$ for different levels of attenuation of the magnitude are listed in Table 7.4.

**TABLE 7.4**  Key Values of $\omega T$ and $\phi$ for Different Levels of Magnitude Attenuation

| Percent Attenuation | $\omega T$ | Phase Angle (Degrees) |
|:---:|:---:|:---:|
| 1 | 7.02 | 8.1 |
| 2 | 4.93 | 11.5 |
| 5 | 3.04 | 18.2 |
| 10 | 2.06 | 25.8 |
| 20 | 1.33 | 36.9 |
| 30 | 1.00 | 45.0 |

## Charge-Amplifier Circuit

The general arrangement of a measurement circuit containing a piezoelectric sensor and a charge amplifier is shown in Fig. 7.47. Two operational amplifiers in series are commonly used for describing the operational characteristics of such circuits. The first amplifier (the charge amplifier), which converts the charge $q$ into a voltage $E_2$, has both capacitive $C_f$ and resistive $R_f$ feedback. The second amplifier (an inverting amplifier), which is used to standardize the output sensitivity of the measurement system, has a variable resistive input $R_1 = bR$ ($0 \leqslant b \leqslant 1$) and a fixed resistive feedback $R$. The circuit can be reduced for analysis to the form shown in Fig. 7.48 by using Eq. (7.67) to combine the transducer capacitance $C_t$, the cable capacitance $C_c$, and the amplifier input capacitance $C_a$ into an effective amplifier input capacitance $C$. The differential equation that describes the behavior of this circuit is obtained as follows:

$$I = I_1 + I_2 + I_3 = \dot{q} = S_q \dot{a}$$

$$= C\dot{E}_1 + \frac{E_1 - E_2}{R_f} + C_f(\dot{E}_1 - \dot{E}_2) - S_q\dot{a} \qquad (7.79)$$

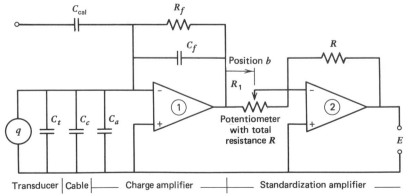

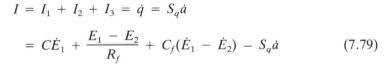

*Figure 7.47*  Measurement circuit with a piezoelectric sensor and a charge amplifier.

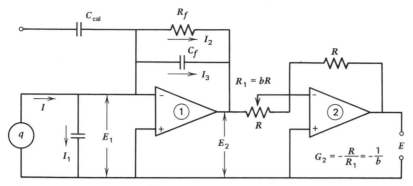

**Figure 7.48** Equivalent circuit for analysis.

Equation (7.79) can be expressed in terms of the output voltage $E$ of the circuit once it is observed that

$$E_2 = -G_1 E_1$$

$$E = -G_2 E_2 = -\frac{1}{b} E_2$$

**where**   $G_1$ is the open-loop gain of the first amplifier.
         $G_2$ is the circuit gain of the second (standardization) amplifier.
         $b$  is the fraction of the potentiometer that is active in the circuit.

Thus

$$\frac{1}{G_2}\left[\frac{C}{G_1} + C_f\left(1 + \frac{1}{G_1}\right)\right]\dot{E} + \frac{1}{G_2}\left(1 + \frac{1}{G_1}\right)\frac{E}{R_f} = S_q\dot{a} \qquad (7.80)$$

Since $G_1$ is very large (greater than $10^4$) for an op-amp in this configuration, Eq. (7.80) reduces to

$$\dot{E} + \frac{E}{R_f C_{eq}} = \frac{G_2 S_q}{C_{eq}}\dot{a} \qquad (7.81)$$

where

$$C_{eq} = \frac{C}{G_1} + C_f = C_f\left(1 + \frac{C}{C_f G_1}\right) \qquad (7.82)$$

Equation (7.82) shows that the total input capacitance $C$ has little effect on this measurement system, since the term $C/C_f G_1$ will usually be very small. The effect of this term on charge-amplifier performance is usually specified in terms of a maximum input capacitance $C$ permitted for each level of feedback capacitance $C_f$ available in the particular instrument. When the limitation on input capacitance is satisfied, the differential equation governing charge-amplifier performance becomes

$$\dot{E} + \frac{E}{R_f C_f} = \frac{G_2 S_q}{C_f} \dot{a} \tag{7.83}$$

Equation (7.83) is identical in form to Eq. (7.70) for the cathode-follower circuit; therefore, the performance characteristics (see Fig. 7.46) of the two circuits are identical.

The voltage sensitivity $S_v$ of the charge-amplifier circuit is given by the expression

$$S_v = \frac{G_2 S_q}{C_f} = \frac{S_q}{b C_f} = \frac{S_q^*}{C_f} \tag{7.84}$$

Equation (7.84) indicates that two parameters ($b$ and $C_f$) are available in the charge-amplifier circuit for controlling voltage sensitivity. The charge sensitivity $S_q$ of a particular transducer can be standardized to a value $S_q^*$ by adjusting the potentiometer to the required position $b$. In this way it is possible to have convenient charge sensitivities such as 1, 10, or 100 pC/unit of $a$. Once the charge sensitivity is standardized, instrument range is established by selecting the proper feedback capacitance. Typical instruments provide values from 10 to 50,000 pF in a 1–2–5–10 sequence.

Equation (7.83) indicates that the time constant $T$ of the charge-amplifier circuit is controlled by the feedback resistance $R_f$ and the feedback capacitance $C_f$ of the charge-amplifier, whereas the time constant of the cathode-follower circuit was controlled by the combined input capacitance $C$ of the transducer, cable, and amplifier and the input resistance $R$ of the amplifier. Charge amplifiers are available in two configurations. In the first, the time constant is fixed for each of the range positions. Typical values range from 0.1 to 2 s. In the second configuration, a short, medium, and long time constant is available for each range. Range, sensitivity, and time-constant information for a typical charge amplifier are listed in Table 7.5. The advantages of a charge-amplifier circuit over a cathode-follower circuit are

1. The time constant is controlled by charge-amplifier feedback resistance and capacitance instead of transducer and cable capacitance.
2. System performance is independent of transducer and cable capacitance so long as the total input capacitance is less than the maximum allowed for a given error limit.

**TABLE 7.5** Typical Sensitivities and Time Constants for a Charge Amplifier[a]

| Sensitivity (units of a/V) Range (pC/unit of $a$) | | | Feedback Capacitance $C_f$ (pF) | Time Constant (s) | | | Maximum Source Capacitance for 0.5% Error (μF) |
|---|---|---|---|---|---|---|---|
| 0.1 to 1.0 | 1.0 to 10 | 10 to 100 | | Short | Medium | Long | |
| 50,000 | 5,000 | 500 | 50,000 | 50 | 50,000 | 5,000,000 | 0.1 |
| 20,000 | 2,000 | 200 | 20,000 | 20 | 20,000 | 2,000,000 | 0.1 |
| 10,000 | 1,000 | 100 | 10,000 | 10 | 10,000 | 1,000,000 | 0.1 |
| 5,000 | 500 | 50 | 5,000 | 5.0 | 5,000 | 500,000 | 0.1 |
| 2,000 | 200 | 20 | 2,000 | 2.0 | 2,000 | 200,000 | 0.1 |
| 1,000 | 100 | 10 | 1,000 | 1.0 | 1,000 | 100,000 | 0.1 |
| 500 | 50 | 5.0 | 500 | 0.5 | 500 | 50,000 | 0.05 |
| 200 | 20 | 2.0 | 200 | 0.2 | 200 | 20,000 | 0.02 |
| 100 | 10 | 1.0 | 100 | 0.1 | 100 | 10,000 | 0.01 |
| 50 | 5 | 0.5 | 50 | 0.05 | 50 | 5,000 | 0.005 |
| 20 | 2 | 0.2 | 20 | 0.02 | 20 | 2,000 | 0.002 |
| 10 | 1 | 0.1 | 10 | 0.01 | 10 | 1,000 | 0.001 |

[a] The material in this table was adapted from the Kistler manual for the Model 504A charge amplifier. The sensitivity listed above is the reciprocal of the voltage sensitivity $S_v$ given by Eq. (7.84).

3. The charge sensitivity can be easily standardized by using the position parameter $b$ that controls the gain of the second (standardizing) amplifier.
4. A wide range of voltage sensitivities are available by simply changing the feedback capacitance.

## Transducers with Built-in Amplifiers

Recent developments in solid-state electronics have progressed to the point where it is possible to incorporate a miniature integrated-circuit amplifier into the transducer housing. A P-channel MOSFET (Metal–Oxide–Semiconductor Field-Effect Transistor) unity-gain amplifier (source follower) is used for this application. The additional components required for the measurement circuit are a power supply, a cable to connect the transducer to the power supply, and a recording instrument. A schematic diagram of a typical circuit is shown in Fig. 7.49.

In this type of circuit, the cable capacitance $C_c$ is removed from the charge-generating side of the circuit and placed on the low-impedance-output side of the circuit where it has negligible effect on the accuracy of the measurements. Also, the sensor capacitance $C$ and the amplifier input resistance $R$ are not influenced by changing environmental conditions, since these components are well protected inside the housing of the transducer. Resistance $R$ and capacitance $C$ are adjusted by the manufacturer during fabrication of the transducer to give a standard voltage sensitivity $S_v$ and a reasonable time constant $T$.

The power supply shown in Fig. 7.49, which consists of a dc supply voltage $E_i$ and a current regulating diode (CRD), provides a nominal $+11$ V at the transistor source ($S$) when there is no input signal. The blocking capacitor $C_1$ in the power supply is used to shield the recording instrument from this dc voltage. The meter $M$ in the power supply is used to adjust the voltage and monitor proper installation of the transducer and cable. If the meter reads zero, a short exists in the connecting cable; likewise, if the meter reads at the level of the supply voltage, the circuit is open.

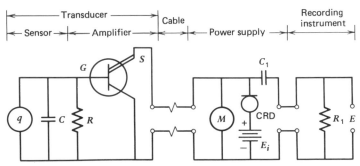

**Figure 7.49**  Schematic diagram of a measurement circuit that utilizes a piezoelectric transducer with a built-in amplifier.

Response of the circuit of Fig. 7.49 can be obtained by using the equivalent circuit shown in Fig. 7.50 for the analysis. Input to and output from the voltage follower amplifier yields two differential equations. From the input,

$$\dot{E}_1 + \frac{E_1}{RC} = \frac{S_q}{C}\dot{a} \qquad (7.85)$$

And since the MOSFET amplifier is a unity-gain ($G = 1$) amplifier, from the output,

$$\dot{E} + \frac{E}{R_1 C_1} = \dot{E}_2 = \dot{E}_1 \qquad (7.86)$$

The steady-state transfer function obtained from Eqs. (7.85) and (7.86) is

$$H(\omega) = \frac{CE_o}{S_q a_o} = \left(\frac{iR_1 C_1 \omega}{1 + iR_1 C_1 \omega}\right)\left(\frac{iRC\omega}{1 + iRC\omega}\right) \qquad (7.87)$$

Equation (7.87) indicates the presence of two time constants; namely,

$$T_1 = R_1 C_1 \qquad \text{and} \qquad T = RC$$

The internal time constant $T = RC$ is fixed by the manufacturer during fabrication of the transducer. Ideally, the value of $T$ should be very large; however, the value of $R$ is limited by MOSFET amplifier requirements and the value of $C$ is selected to yield the required voltage sensitivity. Typical values of $T$ range from 0.5 to 2000 s. The external time constant $T_1 = R_1 C_1$ is controlled by the blocking capacitor $C_1$ (fixed by the manufacturer of the power supply) and by the input resistance $R_1$ of the readout instrument (selected by the user). Since readout instruments have input resistances that range from 0.1 M$\Omega$ to 1 M$\Omega$, this external time constant usually becomes the controlling time constant. Magnitude and phase-angle response characteristics of the transducer with built-in amplifier are shown in Fig. 7.51.

The Bode plots of Fig. 7.51 clearly show the effects of the external time constant on the response characteristics of the piezoelectric transducer with a

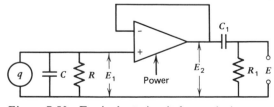

*Figure 7.50* Equivalent circuit for analysis.

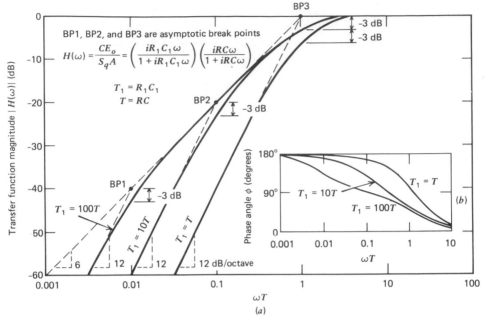

**Figure 7.51**   Transfer function (magnitude and phase) for a piezoelectric transducer with built-in amplifier. (*a*) Magnitude. (*b*) Phase angle ϕ versus ω*T*.

built-in amplifier. When $T_1 = T$, the low-frequency response drops off at a rate of 12 dB/octave and the phase angle quickly approaches 180 degrees. At the one break point (BP3) in the curve, the amplitude is attenuated by 6 dB (down 50 percent) and a 90-degree phase shift exists. When $T_1 = 100T$, the low-frequency response initially drops off at a rate of 6 dB/octave; but at the lower frequencies, the rate eventually increases to 12 dB/octave. At break point BP2 in this curve, the amplitude is attenuated by 3 dB and a 45-degree phase shift exists.

Practical error limits of 2 and 5 percent, with time constant ratios $T_1/T$ of 1, 10, and 100, are obtained when ω*T* is greater than the values listed in Table 7.6.

It is evident from Table 7.6 that errors can be limited to 2 percent when $T_1/T$ is greater than 10 if ω*T* is greater than 5. Similarly, errors can be limited to 5

**TABLE 7.6**   Minimum Values of ω*T* for Errors of 2 and 5 Percent

| $T_1/T$ | 2% | 5% |
|:---:|:---:|:---:|
| 1 | 7.00 | 4.96 |
| 10 | 4.95 | 3.06 |
| 100 | 4.93 | 3.04 |

percent when $T_1/T$ is greater than 10 if $\omega T$ is greater than 3. The values of $\omega T$ listed in Table 7.6 for $T_1/T$ equal to 100 are the same as those obtained for a single-time-constant system as described by Eq. (7.76).

The advantages of the system with the built-in amplifier are

1. The voltage sensitivity $S_v$ is fixed by the manufacturer.
2. Cable length and cable capacitance have little effect on the output.
3. A high-level voltage output signal with a low level of noise is obtained.
4. Low-cost power supplies are adequate.
5. Operates directly with most readout instruments.

## 7.8 MEASUREMENT OF TRANSIENT SIGNALS

In the previous two sections, mechanical and electrical steady-state (sinusoidal) transfer functions were developed for seismic transducers and their electrical circuits. In this section, the response of seismic transducers to transient inputs is investigated. For the purpose of these discussions, a *transient input* will be defined as one in which rapid changes in input occur over a short period of time that is both preceded and followed by a constant input for long periods of time. A typical transient input is illustrated in Fig. 7.52.

### Mechanical Response

Transient load-time histories that provide insight into the response limitations of mechanical systems include the terminated ramp function, the step function, the triangle function, and the half-sine function. The first of these inputs (the terminated ramp input) was considered in Chapter Six. The mechanical response of a load cell to a terminated ramp input (see Fig. 6.25) is given by Eqs. (6.51) through (6.55) and illustrated in Fig. 6.26. Errors are small when the quantity $1/\omega_n t_o$ is small with respect to unity ($\omega_n$ is the natural frequency of the transducer and $t_o$ is the rise time of the ramp input).

As the rise time $t_o$ of the ramp input becomes smaller and smaller, the input

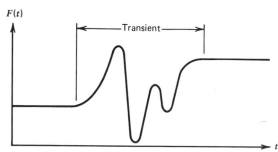

*Figure 7.52* Transient input with different constant values before and after the transient event.

approaches a step input, as shown in Fig. 7.53. No mechanical system can generate a step input; however, a pressure transducer subjected to shock or blast pressures can appraoch this ideal case. The response of a pressure transducer to a time-dependent loading is controlled by Eq. (7.61). For the case of a rigidly mounted (no connecting mass) pressure transducer responding to a step input, Eq. (7.61) becomes

$$m\ddot{z} + C\dot{z} + kz = Ap_o \tag{7.88}$$

**where**   $A$ is the effective area of the sensitive face of the transducer.
$p_o$ is the magnitude of the step pressure pulse.

Equation (7.88) has the solution

$$z = \frac{Ap_o}{k}\left[1 - \frac{1}{\sqrt{1 - d^2}}e^{-d\omega_n t}\cos(\omega_d t + \phi)\right] \tag{7.89}$$

$$\tan\phi = -d/\sqrt{1 - d^2} \tag{7.90}$$

$$\omega_d = \sqrt{1 - d^2}\,\omega_n \tag{7.91}$$

The first term of Eq. (7.89) is the particular solution (the step), while the second term is the homogeneous solution (transient), which decays exponentially with time. The terms inside the bracket of Eq. (7.89) are often referred to as the unit-step response function $h(t)$. Typical response of a lightly damped ($d = 0.05$ or 5 percent) transducer is shown in Fig. 7.54. This type of response is often referred to as *transducer ringing*, since the output oscillations occur at the damped frequency of oscillation $\omega_d$ of the transducer. Transducer damping can be estimated from the decay envelope of the response curve by using the log-decrement techniquc.

The *rise time* of the output signal from a transducer is frequently defined as the time required for the signal to rise from 10 to 90 percent of the final or static value as shown in Fig. 7.54. For lightly damped second-order mechanical

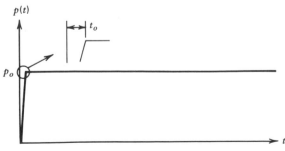

**Figure 7.53**   Dynamic pressure pulse from a shock tube.

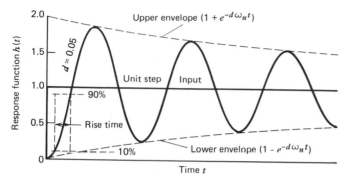

**Figure 7.54** Response of a lightly damped transducer to a unit-step input.

systems, which respond essentially as $(1 - \cos \omega_n t)$, the rise time is approximately 0.1623 times the natural period $T_n$ of the system. The rise time of any input can be used as a criterion for selecting a minimum natural frequency for the transducer to be used for the measurement. In order to prevent serious transducer ringing, at least five complete natural oscillations of the transducer should occur during the rise time of the input signal.

Transient half-sine and triangle inputs are characteristic of many impact problems. Mechanical response of a transducer to each of these inputs is shown in Figs. 7.55 and 7.56. In each of these figures, the response of the transducer for pulse durations of $T_n$ and $5T_n$ are shown where $T_n$ is the natural period of the transducer. It is evident from these plots that a reasonable output from the transducer requires a pulse duration that is at least five times the natural period of the transducer. For pulse durations less than $5T_n$, the output from the transducer is severely distorted as shown in Figs. 7.55 and 7.56 for the pulse of duration $T_n$.

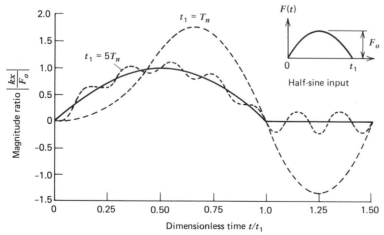

**Figure 7.55** Mechanical response of a transducer to a half-sine transient.

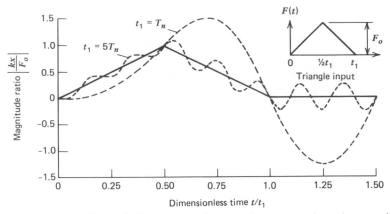

**Figure 7.56**  Mechanical response of a transducer to a triangular transient.

## Electrical Response

The electrical-response characteristics of the transducer circuit can also alter the output signal generated by transient inputs. This is particularly true for the ac-coupled circuits used with piezoelectric transducers. As an example, consider the electrical response of a cathode-follower circuit to the rectangular input pulse shown in Fig. 7.57a. While such a rectangular pulse is physically impossible to generate with a mechanical system, it provides a limiting case that can be used to judge the adequacy of the low-frequency response of a measuring system. The behavior of the cathode-follower circuit is governed by Eq. (7.70), which was developed in the previous section. Thus,

$$\dot{E} + \frac{E}{RC} = \frac{S_q}{C}\dot{a} \tag{7.70}$$

which has the solution

$$E = \frac{S_q}{C}e^{-t/RC}\int_0^t e^{t/RC}\dot{a}\,dt \tag{7.92}$$

The derivative with respect to time of the input $\dot{a}$ is shown in Fig. 7.57b and can be expressed as

$$\dot{a} = a_o\delta(t) - a_o\delta(t - t_1) \tag{7.93}$$

where $\delta(t)$ is the Dirac delta function. The properties of the Dirac delta function are such that the function is zero except when its argument is zero and the area under the function is unity. Substituting Eq. (7.93) into Eq. (7.92) yields

$$E = \frac{S_q a_o}{C}\left[e^{-t/RC} - u(t - t_1)e^{-(t-t_1)/RC}\right] \tag{7.94}$$

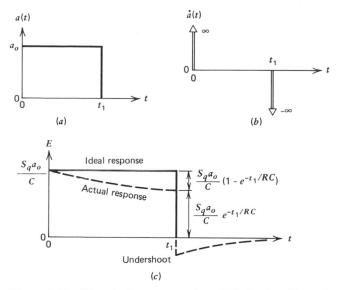

**Figure 7.57** Electrical response of an $RC$ circuit with a piezoelectric sensor to a rectangular pulse (transient) input. ($a$) Rectangular input pulse. ($b$) Time derivative of input pulse. ($c$) Output voltage.

where $u(t - t_1)$ is the unit-step function. The electrical response of the cathode-follower circuit, as represented by Eq. (7.94), is shown in Fig. 7.57c. Two characteristics of the low-frequency response of the circuit can be seen in this figure. First, the output signal decays exponentially during the pulse duration and generates an error of $(1 - e^{-t/RC})$ when compared to the input. Second, an undershoot occurs at the end of the pulse. This undershoot provides the maximum error associated with the rectangular pulse.

The maximum error $\mathscr{E}_{max}$ associated with this $RC$ circuit can be estimated from a series expansion of the function $(1 - e^{-t_1/RC})$. Thus

$$\mathscr{E}_{max} = \frac{t_1}{RC}\left[1 - \frac{1}{2}\frac{t_1}{RC} + \frac{1}{6}\left(\frac{t_1}{RC}\right)^2 - \cdots\right] \qquad (7.95)$$

When $t_1/RC$ is very small, Eq. (7.95) reduces to

$$\mathscr{E}_{max} = \frac{t_1}{RC} \qquad \text{or} \qquad T = RC = \frac{t_1}{\mathscr{E}_{max}} \qquad (7.96)$$

Equation (7.96) was used to determine the time constant requirements listed in Table 7.7 for the rectangular pulse.

The triangular and half-sine pulse shapes shown in Fig. 7.58 are also used to establish time constant requirements. Time constants for these pulse shapes, which will maintain measurements within specified error limits, are also listed

**TABLE 7.7**  Time Constant Requirements for Various Error Levels

| Pulse Shape | | Time Constant | | |
| --- | --- | --- | --- | --- |
| | | **2% Error** | **5% Error** | **10% Error** |
| Rectangular pulse | | $50t_1$ | $20t_1$ | $10t_1$ |
| Triangular pulse | | $25t_1$ | $10t_1$ | $5t_1$ |
| Half-sine pulse | $A$ | $16t_1$ | $6t_1$ | $3t_1$ |
| | $B$ | $31t_1$ | $12t_1$ | $6t_1$ |

in Table 7.7. It is evident from the data of Table 7.7 that the rectangular pulse places the greatest demand on the transducer time constant for a given level of error. The error present in a given measurement, which has been made with a particular transducer, can be estimated by comparing the shape of the output signal with one of the pulses listed in Table 7.7. Also, undershoot with its telltale exponential decay is a good indicator of time constant problems.

The transducer with built-in amplifier presents a special problem since it has two time constants ($T = RC$ and $T_1 = R_1C_1$) as indicated by Eq. (7.87).

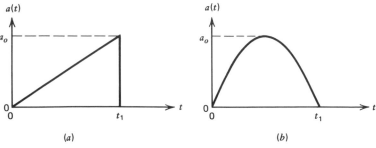

**Figure 7.58**  Common pulse forms used in studies of $RC$ time constants. (*a*) Triangular pulse. (*b*) Half-sine pulse.

The response of a transducer with built-in amplifier to a step input can be shown to be given by the following expression

$$E = \frac{S_v a_o}{T - T_1} (Te^{-t/T_1} - T_1 e^{-t/T}) \qquad (7.97)$$

**where** $a_o$ is the magnitude of the step input.
$S_v$ is the voltage sensitivity of the transducer.

The response of the circuit, as represented by Eq. (7.97), is plotted in Fig. 7.59 for several different ratios of the two time constants. When $T/T_1$ is large (of the order of 1000), the response is controlled entirely by the time constant $T_1$. When $T/T_1$ is small (near 1), both time constants play nearly equal roles, which results in a much faster decay and some undershoot. The effects of the two time constants can be combined into an equivalent time constant $T_e$, which is defined as

$$T_e = \frac{TT_1}{T + T_1} \qquad (7.98)$$

This effective time constant $T_e$ can be used with Table 7.7 to estimate maximum errors associated with any measurement by comparing the output pulse with the shapes listed in the table.

## Summary

A typical transfer function $H(\omega)$ for a measurement circuit with a piezoelectric transducer is shown in Fig. 7.60. At the low-frequency end of the spectrum (below $\omega_1$), the system exhibits a rapid drop in amplitude (6 dB/octave) and serious phase distortion of sinusoidal or periodic inputs as well as exponential

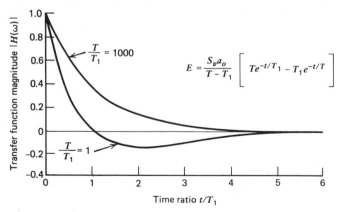

*Figure 7.59* Response of a dual time constant circuit to a step input.

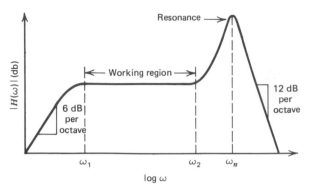

**Figure 7.60** Typical transfer function $H(\omega)$ for a piezoelectric transducer.

decay of transient inputs. Behavior at this end of the spectrum is controlled by the electrical $RC$ time constant of the system. Between $\omega_1$ and $\omega_2$, the system exhibits essentially zero phase shift and nearly constant output per unit input. Measurements should be limited to this range of the spectrum if possible. Above $\omega_2$, mechanical resonance becomes important and causes amplitude magnification and phase distortion of sinusoidal or periodic inputs and a ringing response to transient inputs if the rise time of the input is less than five times the natural period of the transducer. In every instrumentation application, it is the engineer's responsibility to select instruments that will provide the desired measurement with a minimum of error regardless of input type or shape; therefore, it is important that all of the effects illustrated in Fig. 7.60 be given careful consideration.

## 7.9 TRANSDUCER CALIBRATION

The charge and voltage sensitivities of a transducer are very important quantities. Transducer manufacturers provide calibration information, traceable to the National Bureau of Standards, when the instrument is purchased. Users must recalibrate the instrument occasionally to ensure that the original calibration has not changed with use. The National Bureau of Standards and most of the manufacturers use absolute calibration methods. Users generally use comparison methods to recalibrate the instruments. Calibration techniques for acceleration and force transducers, which utilize constant, sinusoidal, and transient inputs, will be examined in this section.

### Accelerometer Calibration with a Constant Input

The most widely used constant-acceleration calibration method requires only the simple act of rotating the sensitive axis of the accelerometer in the earth's gravity field, which produces a nominal change in acceleration of 2 g's ($-1\,\text{g}$ to $+1\,\text{g}$). The primary disadvantages of this simple method are the limited range and the absence of any check on the frequency response characteristics of the instrument due to the dc character of the test. The method provides a quick means to ensure that the transducer is at least nominally functional.

The range limitation problem can be solved by using a centrifuge. With this method of calibration, the sensitive axis of the accelerometer is aligned radially with respect to the axis of rotation of the centrifuge. Here it experiences an acceleration that is given by Eq. (7.12) as

$$a_r = -r\omega^2 \tag{7.12}$$

**where** $r$ is the radial position of the seismic mass of the accelerometer.

$\omega$ is the angular velocity (speed) of the centrifuge.

A range of 0 to 60,000 g's has been achieved by using this method of calibration. The primary problem encountered with this method of calibration is elimination of the electrical noise generated when the output signal is transmitted through slip rings from the transducer to the recording instrument.

## Accelerometer Calibration with a Sinusoidal Input

A sinusoidal type of input, such as that provided by a vibration generator, provides a means for calibrating an accelerometer at different frequencies and at different levels of acceleration. This form of loading is widely used for both comparison and absolute calibration.

When the comparison method is used, two accelerometers (test and standard) are mounted as close as possible on the exciter head and the output from the accelerometer being calibrated is compared with the output from the standard accelerometer, which is used only for calibration and whose calibration is traceable to the National Bureau of Standards. Standard accelerometers, together with charge amplifiers that have been adjusted to provide a constant sensitivity (say 10 mV/g) over a broad range of frequencies, are available for this use. A plot of test output versus standard output as the amplitude of vibration is varied at a given frequency establishes the linearity of the test instrument. The sensitivity as a function of frequency of the test instrument is obtained by varying the frequency while holding the standard output constant. Accuracies within ±2 percent are attainable with this comparison technique.

Several manufacturers have developed portable calibrators that can be used in the field to check an accelerometer. One such unit, shown in Fig. 7.61, uses a double magnet system with two mechanically coupled moving coils. One coil operates as the driver and the other serves as a velocity pickup. The coils can be connected to external instrumentation for performing reciprocity calibration.[5]

An accurate measure of the local acceleration of gravity can be obtained

---

[5] "A Portable Calibrator for Accelerometers," by Reinhard Kuhl, *Bruel and Kjaer Technical Review*, no. 1, 1971, pp. 26–32.

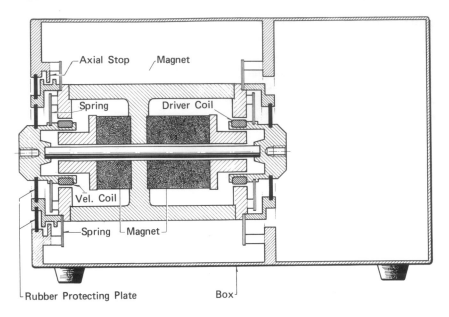

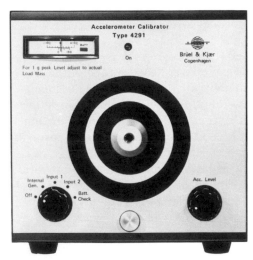

*Figure 7.61* Portable accelerometer calibrator. (Courtesy of Bruel and Kjaer.)

by mounting the calibrator with its axis vertical, mounting the test accelerometer on the calibrator, and placing a small (less than 1 gram) nonmagnetic object on the free face of the accelerometer. At a level of $\pm 1$ g of acceleration, the object will begin to "rattle." The onset of "rattle" can be detected in the output signal of the accelerometer within an accuracy of $\pm 1$ percent.

Absolute calibration of an accelerometer requires precise measurement of frequency $\omega$ and displacement $x$, since the magnitude of the peak acceleration

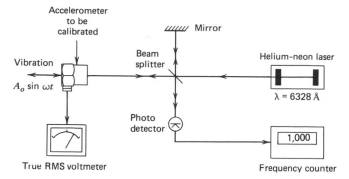

*Figure 7.62*  Absolute accelerometer calibration with laser interferometry.

$a_{max}$ for sinusoidal motion is given by Eq. (7.20) as

$$a_{max} = -\omega^2 x_{max} \tag{7.20}$$

As an example of the magnitudes of the quantities involved in such measurements, a sinusoidal displacement of 0.001 in. at a frequency of 100 Hz gives an acceleration of 1.02 g's. Thus, the peak-to-peak displacements that need to be measured are very small. For frequencies below 100 Hz, reasonable results can be obtained with a measuring microscope. For frequencies above 100 Hz, proximity gages and interferometry methods are commonly used.

A schematic diagram of a Michelson interferometer arranged for absolute calibration of an accelerometer is shown in Fig. 7.62. The helium–neon ($\lambda$ = 6328 Å) laser beam is split into two parts by the beamsplitter in such a way that one is reflected by the fixed mirror while the other is reflected by the accelerometer being calibrated. The two reflected beams, which are combined before arriving at the photodetector, exhibit a variation in intensity from minimum to maximum each time the accelerometer moves one-half of a wavelength. The number of changes in intensity per vibration cycle provides an accurate measure of the peak-to-peak displacement, since the wavelength of the helium–neon light is precisely known. Accuracies within $\pm 0.5$ percent have been obtained with this method over the frequency range from below 100 Hz to over 2000 Hz.

## Accelerometer Calibration with a Transient Input

A simple impulse method of accelerometer calibration known as gravimetric calibration[6] is based on Newton's Second Law of Motion, which can be written as

$$F = ma = mg\left(\frac{a}{g}\right) \tag{7.99}$$

---

[6] "Gravimetric Calibration of Accelerometers," by R. W. Lally, available from PCB Piezotronics, Inc., P.O. Box 33, Buffalo, NY.

A schematic diagram of the physical system used for this type of calibration is shown in Fig. 7.63. The system consists of a steel cylinder on which the accelerometer to be calibrated is mounted, a plastic tube to guide the motion of the cylinder, and a rigid base that supports a quartz force transducer. The calibration is performed in two steps and requires measurement of three voltages. First, the test mass (cylinder and accelerometer) is positioned on the force transducer and the output voltage $E_{mg}$ is measured as the mass is quickly removed. Next, the test mass is dropped on the force transducer and the output voltages $E_f$ and $E_a$ from the force transducer and accelerometer are simultaneously measured. By definition, the force and acceleration experienced by the force transducer and the accelerometer are their output voltages divided by their sensitivities. Thus

$$F_{mg} = mg = \frac{E_{mg}}{S_f}$$

$$F = \frac{E_f}{S_f}$$

$$\frac{a}{g} = \frac{E_a}{S_a} \tag{7.100}$$

**where**   $E_{mg}$   is the output voltage from the force transducer due to the weight ($mg$) of the cylinder and accelerometer.
   $E_f$   is the output voltage from the force transducer during impact.
   $E_a$   is the output voltage from the accelerometer during impact.
   $S_f$   is the voltage sensitivity of the force transducer.
   $S_a$   is the voltage sensitivity of the accelerometer (expressed in g's).

Substituting Eqs. (7.100) into Eq. (7.99) yields

$$\frac{E_f}{S_f} = \frac{E_{mg}}{S_f} \frac{E_a}{S_a} \quad \text{or} \quad S_a = \frac{E_{mg} E_a}{E_f} \tag{7.101}$$

Equation (7.101) shows that the unknown sensitivity $S_a$ of the accelerometer is simply the measured acceleration signal $E_a$ during impact divided by the ratio of the force signals $E_f/E_{mg}$. The sensitivity $S_f$ of the force transducer does not affect the calibration. Typical output signals from a gravimetric calibration of an accelerometer are shown in Fig. 7.64.

The cushion material shown in Fig. 7.63 controls the impact pulse duration. The drop height controls the pulse amplitude. A typical calibration is performed over a range of amplitudes and time durations by using different combinations of drop height and cushion material. The preamplifiers shown in Fig. 7.63 can be any one of the three standard types (high input impedance, charge, or built-

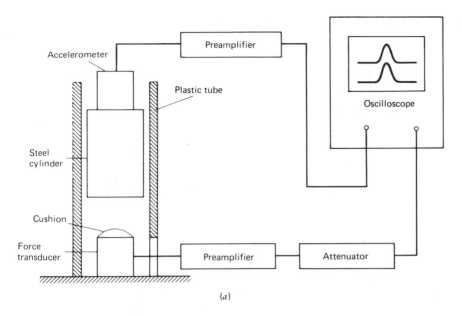

*Figure 7.63* Gravimetric calibration system for accelerometers. (*a*) Components of a gravimetric calibration system. (*b*) Gravimetric calibrator. (Courtesy of PCB Piezotronics.)

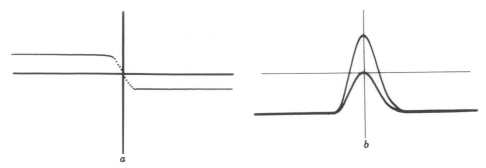

**Figure 7.64**  Typical voltage–time curves from a gravimetric accelerometer calibration. (*a*) Weight lifting. (*b*) Impacting. (Courtesy of PCB Piezotronics.)

in). The attenuator of Fig. 7.63 is used to vary the amplitude of the signal from the force transducer. If the attenuator is adjusted to give $E_f = E_a$ during impact, the voltage $E_{mg}$ obtained *after* the attenuator is set becomes a direct measure of the voltage sensitivity $S_a$ of the accelerometer. A typical set of superimposed force and acceleration signals with $E_f = E_a$ that illustrate the linearity of the system (no detectable difference in signal) is shown in Fig. 7.65*a*. The linearity is further illustrated in Fig. 7.65*b* where a plot of acceleration versus force for one of the pulses is shown. Typical gravimetric calibration results for an accelerometer are listed in Table 7.8.

The data of Table 7.8 indicate that the gravimetric calibration method is capable of providing an accuracy that is equivalent to that provided by the back-to-back comparison method with sinusoidal motion. The two quantities that must be accurately known are the local acceleration of gravity $g$ since $S_a$, as defined by Eq. (7.101), is expressed in terms of local g's and the voltage $E_{mg}$ due to the weight of the test cylinder and accelerometer.

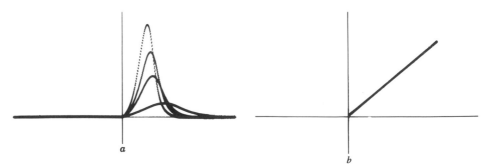

**Figure 7.65**  Linearity illustrations for gravimetric accelerometer calibration. (*a*) Superimposed acceleration and force versus time. (*b*) Acceleration versus force. (Courtesy of PCB Piezotronics.)

**TABLE 7.8**  Gravimetric Calibration of an Accelerometer[a]
PCB Model 302A, S/N 3116[b]

| | Test Number | | |
|---|---|---|---|
| | **1** | **2** | **3** |
| $E_{mg}$ (mV) | 38.8 | 38.8 | 38.8 |
| $E_f$ (mV) | 177 | 492 | 1016.5 |
| $F_a$ (mV) | 45.5 | 126.3 | 261.9 |
| $a$ (g's) | 4.56 | 12.7 | 26.2 |
| $S_a$ (mV/g) | 9.97 | 9.97 | 10.00 |
| Deviation (%) | $-0.2$ | $-0.2$ | $+0.1$ |

[a] Lally, R. W.: "Gravimetric Calibration of Accelerometers," available from PCB Piezotronics, Inc., Buffalo, NY.
[b] Reference sinusoidal calibration $S_a$ = 9.99 mV/g.

## Force-Transducer Calibration with a Constant Input

Static calibration of force transducers with electrical-resistance strain-gage sensors was discussed in Section 6.8. Typical procedures involve loading with a testing machine whose accuracy has been certified, loading with standard weights, or comparison with a standard force transducer. Similar calibrations are possible with force transducers having piezoelectric sensors; however, the loads must be applied and then removed very quickly due to the low-frequency response of piezoelectric sensors.

## Force-Transducer Calibration with a Sinusoidal Input

Dynamic calibration of a force transducer can be performed with a vibration generator by using the arrangement illustrated in Fig. 7.66. Here, known calibration masses $m_c$ are attached in sequence to the seismic mass $m$ of the force transducer, after which the base is subjected to a sinusoidal oscillation that is measured with the accelerometer. With this arrangement, the external force $F(t)$ applied to the force transducer is

$$F(t) = -m_c \ddot{y}$$

or in terms of the relative motion between the seismic mass and the base as given by Eq. (7.40),

$$F(t) = -m_c(\ddot{x} + \ddot{z}) \tag{7.102}$$

The differential equation governing behavior of the force transducer during

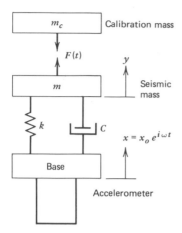

**Figure 7.66**   Calibration of a force transducer with a vibration generator.

calibration is obtained by substituting Eq. (7.102) into Eq. (7.41). Thus

$$(m + m_c)\ddot{z} + C\dot{z} + kz = -(m + m_c)\ddot{x} \qquad (7.103)$$

Equation (7.103) indicates that the natural frequency of the transducer during calibration $\omega_{nc}$ decreases as the calibration mass $m_c$ is increased. This behavior must be given careful consideration during calibration in order to avoid errors due to resonance effects. The natural frequency $\omega_{nc}$ is

$$\omega_{nc} = \sqrt{\frac{k}{m + m_c}} = \frac{\sqrt{k/m}}{\sqrt{1 + m_c/m}} = \frac{\omega_n}{\sqrt{1 + m_c/m}} \qquad (7.104)$$

The effective force during calibration is the sum of the inertia force $m_c\ddot{x}$ of the calibration mass and the inertia force $m\ddot{x}$ of the seismic mass of the transducer. Thus, the output voltages $E_f$ of the force transducer and $E_a$ of the accelerometer are

$$E_f = S_f(m + m_c)\ddot{x} \qquad (7.105)$$

$$E_a = S_a(\ddot{x}/g) \qquad (7.106)$$

From Eqs. (7.105) and (7.106) it is obvious that the voltage ratio $E_f/E_a$ is related to the sensitivity ratio $S_f/S_a$ by the expression

$$\frac{E_f}{E_a} = \frac{S_f}{S_a}(m + m_c)g = \frac{S_f}{S_a}(W + W_c) \qquad (7.107)$$

**where**   $W$ is the weight of the seismic mass.
   $W_c$ is the weight of the calibration mass.

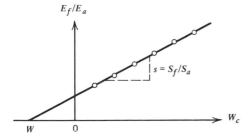

**Figure 7.67** Voltage ratio $E_f/E_a$ versus calibration weight $W_c$ from a sinusoidal calibration of a force transducer.

A plot of voltage ratio $E_f/E_a$ versus calibration weight $W_c$ from a typical sinusoidal calibration of a force transducer is shown in Fig. 7.67. It is evident from Eq. (7.107) that the slope of such a plot is the sensitivity ratio $S_f/S_a$. Either a peak- or RMS-voltage ratio can be used for the calibration. The intercept point on the horizontal axis of Fig. 7.67 should verify the weight $W$ of the seismic mass of the force transducer being calibrated. During a given calibration, plots of voltage ratio $E_f/E_a$ versus calibration weight $W_c$ should be obtained over a wide range of vibration amplitudes (to check linearity) and frequencies (to check resonance effects). Once the slope $s$ has been accurately established, the voltage sensitivity $S_f$ of the force transducer is simply

$$S_f = sS_a \qquad (7.108)$$

It is evident from Eq. (7.108) that the voltage sensitivity $S_a$ of the accelerometer must be accurately known. By using standard accelerometers and exercising care during the measurements, accuracies of $\pm 2$ percent are obtainable with this method of calibration for force transducers.

## Force Transducer Calibration for Impact Applications

A modern method of structural testing, which utilizes advanced frequency analysis techniques to construct a dynamic response model of the structure for vibration analyses, mode-shape determinations, mechanical-impedance determinations, and structural-integrity studies, requires striking of the structure with a hammer-type device and simultaneous recording of the impacting force together with accelerations at different points of the structure. A typical hammer with attached force transducer and impact head for use in such studies is shown in Fig. 7.68.

Calibration of the force transducer used in this impact application can be accomplished by using the pendulum system illustrated in Fig. 7.69. The differential equation describing motion of the hammer head and attached seismic mass is

$$m_h \ddot{z} + k_f z = F(t) - m_h \ddot{x}_b \qquad (7.109)$$

*a*

***Figure 7.68*** Impact hammer with attached force transducer and impact head. (Courtesy of PCB Piezotronics.)

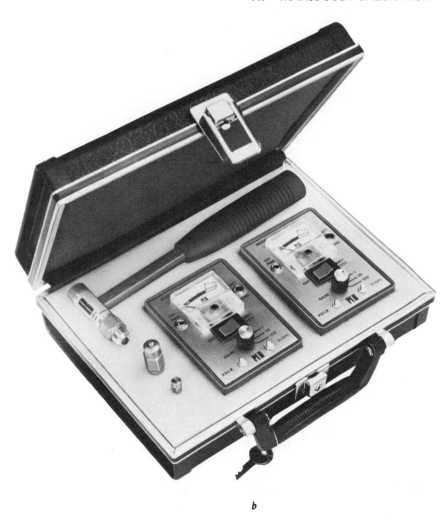

***Figure 7.68***   *(Continued).*

**where**   $m_h$ is the mass of the hammer head and the seismic mass of the transducer.

$m_b$ is the mass of the hammer body and transducer base.

$k_f$ is the spring constant of the piezoelectric sensor.

$z$  is the relative motion between the seismic mass and base of the
transducer.

When the impacting frequencies are well below the resonant frequency of the
force transducer, the dominant term on the left side of Eq. (7.109) is the spring

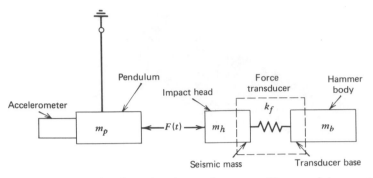

**Figure 7.69**  Calibration of an impact hammer with a pendulum system.

force $k_f z$. Relative motion $z$ is also very small in comparison to $x_b$ and $x_h$; therefore, it can be assumed that $x_b \approx x_h$. When the previous conditions are satisfied,

$$F(t) = m_h \ddot{x}_h + m_b \ddot{x}_b \approx (m_h + m_b)\ddot{x}_b$$

$$z \approx \frac{m_b}{m_h + m_b} \frac{F(t)}{k_f}$$

Output voltage $E_f$ from the force transducer can be expressed in terms of the relative motion $z$ as

$$E_f = S_z z = \frac{m_b}{m_h + m_b} \frac{S_z}{k_f} F(t) = S_f F(t) \qquad (7.110)$$

where $S_z$ is the voltage sensitivity of the force transducer to relative motion $z$. The quantities $S_z$ and $k_f$ of Eq. (7.110) are sensor properties that remain constant; however, the mass $m_b$ of the hammer and the mass $m_h$ of the impact head can be easily changed, and any such change will significantly affect the voltage sensitivity $S_f$ of the instrument to force $F(t)$.

During calibration of an impacting hammer with the system of Fig. 7.69, the force $F(t)$ applied by the pendulum is

$$F(t) = m_p a = m_p g(a/g) = W_p(a/g) \qquad (7.111)$$

**where**  $m_p$ is the combined mass of the pendulum and accelerometer.
$W_p$ is the combined weight of the pendulum and accelerometer.
$a$   is the acceleration of the pendulum during impact.

The output voltage $E_a$ from the accelerometer during impact is

$$E_a = S_a(a/g) \qquad (7.112)$$

Thus, from Eqs. (7.111) and (7.112),

$$\frac{E_f}{E_a} = \frac{S_f}{S_a}W_p = sW_p \tag{7.113}$$

Equation (7.113) is similar to Eq. (7.107); therefore, the calibration procedure with impact inputs is identical to that used for sinusoidal inputs. In the case of impact inputs, the voltage ratio $E_f/E_a$ is plotted as a function of pendulum weight $W_p$ to obtain the slope $s = S_f/S_a$. The voltage sensitivity $S_f$ of the impact hammer to force is then simply

$$S_f = sS_a \tag{7.114}$$

Again, by using standard accelerometers and exercising care during the measurements, accurate calibration ($\pm 2$ percent) can be achieved.

The calibration procedure outlined for impact hammers is adequate only for operating frequencies well below the natural frequency of the force transducer used in the hammer. At higher operating frequencies, where stress-wave-propagation effects become important, the force measurements made with such hammer devices may exhibit considerable error.

## Overall System Calibration

Electronic measurement systems usually contain a number of components, such as power supplies, signal conditioning circuits, amplifiers, and recording instruments, in addition to the specific transducer required for the measurement. It is possible to calibrate each component and thus determine the input-output relationship for the complete system; however, this procedure is time-consuming and subject to the calibration errors associated with each component of the system. A more precise and direct procedure establishes a single calibration constant for the complete system that relates readings from the recording instrument directly to the quantity being measured.

System calibration involves not only voltage sensitivity determinations but also determinations of such dynamic response characteristics as rise time, overshoot, undershoot, and time constant. A typical type of system calibration, which does not require that the transducer be disconnected from either the system or the object on which it is mounted, uses a voltage input on the transducer end of the system. For piezoelectric transducers commonly used for force and acceleration measurements, the voltage generator can be connected to the system through the calibration capacitor $C_{cal}$, which is provided on the input side of most charge amplifiers as shown in Fig. 7.48. This type of system calibration is based on the fact that a charge generator in parallel with a capacitor is equivalent to a voltage generator in series with a capacitor. This charge amplifier circuit, shown in Fig. 7.48, will be used for the discussion of system calibration that follows.

The calibration voltage $E_{cal}$ needed to simulate a calibration charge $q_{cal}$ is

$$E_{cal} = \frac{q_{cal}}{C_{cal}} \qquad (7.115)$$

Also, the calibration charge $q_{cal}$ is related to the quantity being simulated $a_{cal}$ by the expression

$$q_{cal} = S_q a_{cal} \qquad (7.116)$$

where $S_q$ is the charge sensitivity of the piezoelectric sensor being used in the transducer. The voltage sensitivity of the charge amplifier circuit of Fig. 7.48 is given by Eq. (7.84) as

$$S_v = \frac{S_q}{bC_f} \qquad (7.84)$$

The output voltage $E_{cal}$ from the charge amplifier circuit, which results from application of the calibration voltage $E_{cal}^*$, is

$$E_{cal} = S_v a_{cal} = \frac{S_q a_{cal}}{bC_f} = \frac{C_{cal} E_{cal}^*}{bC_f} \qquad (7.117)$$

Equation (7.117) shows that the potentiometer position $b$ can be adjusted to provide a standardized output for the circuit. Range is then established by selecting the proper feedback capacitance $C_f$.

The calibration voltages used to simulate transducer loadings may be sinusoidal, periodic, or transient. Precision voltage calibrators are available that provide either a 0- to $-10$-V full-scale step function or a 0- to $-10$-V full-scale 100-Hz square wave. The voltage is adjustable over the full range with a resolution of 0.02 percent of full scale and a linearity of $\pm 0.25$ percent of full scale. Precision calibration capacitors are available for use with charge amplifiers that do not have built-in calibration capacitors.

The calibration and check-out of measuring systems that have piezoelectric transducers and high-input-impedance amplifiers as components can also be accomplished by applying a calibration voltage $E_{cal}$ across a series resistance $R_s$ in the ground side of the circuit[7] as shown in Fig. 7.70. The series resistance $R_s$ is usually of the order of 10 to 100 $\Omega$, while the input resistance $R_a$ of the

---

[7] "In-place Calibration of Piezoelectric Crystal Accelerometer Amplifier Systems," by D. Pennington, ISA National Flight Test Instrument Symposium (San Diego, CA, May 1960). See also, Endevco Corp., Tech. Paper Nos. 211 and 216.

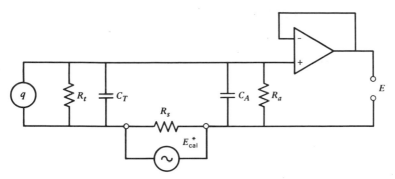

***Figure 7.70***  Voltage insertion method of system calibration.

amplifier is of the order of 100 MΩ; therefore, the presence of $R_s$ exerts only an insignificant effect on the circuit during normal operation and calibration. Special adaptors that can be inserted into the transducer connecting cable are commercially available.[8] Such adaptors contain the series resistor and provide connections for the voltage generator used for calibration.

The capacitance $C_T$ of Fig. 7.70 includes the transducer capacitance $C_t$, the cable capacitance from the transducer to the adaptor $C_{c1}$, and the adaptor connector capacitance $C_{r1}$. Similarly, the capacitance $C_A$ of Fig. 7.70 includes the adaptor connector capacitance $C_{r2}$, the cable capacitance from the adaptor to the amplifier $C_{c2}$, the amplifier input capacitance $C_a$, and any standardization capacitance $C_s$ used to yield a standard voltage sensitivity. For the purpose of the analysis that follows, the amplifier (cathode follower or charge amplifier) is assumed to have a gain $G$ that is constant over a broad range of frequencies.

The sinusoidal input-output relationship for the charge generator circuit of Fig. 7.70 during normal operation (no calibration voltage inserted) can be obtained from Eq. (7.74). Thus, with the terms used in this discussion

$$E = \frac{iR_a C\omega}{1 + iR_a C\omega} \frac{GS_q}{C} a(t) \tag{7.118}$$

where

$$C = C_T + C_A$$

When the transducer is inactive and the calibration voltage $E_{cal}^*$ is inserted into the circuit, the equivalent circuit shown in Fig. 7.71 can be used for analysis.

---

[8] Such as Bruel and Kjaer Adaptor UA 0322 ($R_s = 10\ \Omega$).

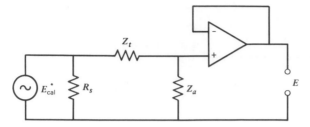

**Figure 7.71**   Equivalent circuit for analysis with voltage inserted.

In this circuit,

$$Z_t = \frac{1}{iC_T\omega} \qquad \text{since } R_t \to \infty$$

$$Z_a = \frac{R_a}{1 + iR_aC_A\omega} \tag{7.119}$$

The output voltage $E_{cal}$ from the circuit of Fig. 7.71 is

$$E_{cal} = \frac{GZ_a}{Z_t + Z_a}E_{cal}^* \tag{7.120}$$

Substituting Eqs. (7.119) into Eq. (7.120) yields

$$E_{cal} = \frac{iR_aC\omega}{1 + iR_aC\omega}\frac{GC_T}{C}E_{cal}^* \tag{7.121}$$

If it is assumed that an input $a(t) = a_{cal}$ is applied to the circuit of Fig. 7.70 during normal operation, then Eq. (7.118) indicates that the output voltage $E_{cal}$ would be

$$E_{cal} = \frac{iR_aC\omega}{1 + iR_aC\omega}\frac{GS_q}{C}a_{cal} \tag{7.122}$$

From Eqs. (7.121) and (7.122) it is evident that

$$E_{cal}^* = \frac{S_q}{C_T}a_{cal} = S_{oc}a_{cal}$$

The quantity $S_{oc} = S_q/C_t$ is the open-circuit voltage sensitivity[9] of the transducer. When the quantity $R_a C \omega \gg 1$, Eq. (7.121) becomes

$$E_{cal} = \frac{GC_T}{C} E_{cal}^* = \frac{GC_T}{C} S_{oc} a_{cal} = S_a a_{cal} \qquad (7.123)$$

where $S_a$ is the effective voltage sensitivity of the system.

The voltage insertion method of calibration provides a convenient means for establishing all "system" parameters. The time constant $T = R_a C$ for the complete system can be obtained from the 3-dB point where $R_a C \omega = 1.0$. The ratio $GC_t/C$ can be obtained from the voltage ratio $E_{cal}/E_{cal}^*$ when $R_a C \omega \gg 1.0$. The values of $G$ and $R_a$ for the amplifier are usually known and the capacitance $C_t$ of the transducer is provided by the manufacturer or can easily be measured. Thus, all of the electrical parameters of the system can be determined. The integrity of the transducer, cables, and connections are checked, since failure of any one of these units result in a zero output voltage. This "Go"–"No Go" check is very useful in large multichannel systems, since a simple "tap" on the transducer establishes integrity of the circuit.

A step voltage input $E_s$ is commonly used for the voltage insertion type of calibration, since it thoroughly tests the fidelity of the measuring system. For the circuit of Fig. 7.70, the governing differential equation is

$$\dot{E} + \frac{E}{R_a C} = \frac{GC_T}{C} S_{oc} \dot{a} = \frac{GC_T}{C} \dot{E}_{cal} \qquad (7.124)$$

which has a step-voltage response of

$$E = \frac{GC_T}{C} E_s e^{-t/R_a C}$$

$$= S_a a_{cal} e^{-t/R_a C} \qquad (7.125)$$

This response illustrates the exponential decay characteristics of the circuit. A typical recording of the response of a system to a step input is shown in Fig. 7.72. This response establishes the rise time, overshoot, and exponential decay characteristics of the measuring system. Each of these quantities impose limitations on the use of the system. Unfortunately, the transducer's natural frequency is not excited by this calibration procedure. Also, the natural frequency

---

[9] Some manufacturers use $S_{oc} = S_q/C_t$ for the open-circuit voltage sensitivity, while others use $S_{oc} = S_q/(C_t + 300)$, since many 3-m-long cables have approximately 300 pF of capacitance. Note that $S_q$ (the charge sensitivity) does not depend on capacitance.

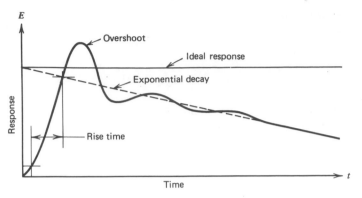

*Figure 7.72*   Typical response of a measurement system to a step input.

of the recording system is not always provided since the oscillations of the signal about the decay line are sometimes difficult to interpret.

Calibration of transducers and complete measurement systems must be performed periodically to ensure satisfactory performance. As the number of channels in a measurement system increases, the need for quick, accurate, and efficient calibration procedures becomes more and more pressing. Calibration must always be regarded as one of the most important steps in the measurement process.

## 7.10   SUMMARY

A broad range of topics associated with motion measurement, the influence of motion on force and pressure measurements, and the characteristic response of various measurement systems to sinusoidal and transient events was considered in this chapter. A brief review of displacement, velocity, and acceleration relationships in several different coordinate systems was presented at the beginning of the chapter. Simple-harmonic vibratory motion was then reviewed and a phasor representation (with real and imaginary components in the complex plane) of sinusoidal vibratory motion which provides a convenient means for relating displacement, velocity, and acceleration without extensive trigonometric identity manipulation and which provides the final expressions in a simple form that contains both magnitude and phase information was developed.

Common dimensional measurement methods involving comparisons with secondary standards of length such as metal or wood scales, metal tapes, vernier calipers, micrometers, dial indicators, and fixed-scale, traveling, and traveling-stage microscopes are mentioned but not discussed in detail. A more extended discussion of the less familiar methods of optical interferometry and pneumatic gaging is provided. Finally, the use of gage blocks as a precise length reference and for calibration of secondary standards of length is discussed.

Displacement and velocity measurements with respect to a fixed reference frame were considered next. Included in the discussions are variable-resistance and variable-capacitance sensors, which are widely used for small static and dynamic displacement measurements, differential transformers (variable-inductance sensors), which are used when a larger range is needed, and resistance potentiometers, which are used when less accuracy but still greater range is required for the measurement. Photosensing transducers and microswitch position indicators are described in detail since they provide a convenient means for introducing several circuits that have been designed and perfected for displacement measurements. Direct-reading linear- and angular-velocity transducers (based on the principle of electromagnetic induction) are also introduced and described in considerable detail. Finally, moire methods, which provide full-field displacement information, are briefly introduced.

Motion measurement without a fixed reference requires use of a seismic-type instrument. Basically, such instruments detect relative motion between a base, which is attached to the structure of interest, and a seismic mass, which, due to inertia, tends to resist any changes in velocity. The theory of operation of such instruments is presented and equations are developed which describe their response to displacement, velocity, acceleration, force, and pressure inputs. Results of this analysis shows that seismic displacement and velocity transducers must be constructed with soft springs in order to have the desired response characteristics. As a result, such instruments exhibit large static deflections and low natural frequencies. Commonly used sensing elements for displacement transducers include linear variable differential transformers (LVDTs) and electrical resistance strain gages on flexible elastic members. Acceleration transducers, on the other hand, require a very stiff spring and a very small seismic mass. As a result, accelerometers exhibit small deflections and high natural frequencies. The sensing elements most widely used in accelerometers (and in dynamic force and pressure transducers) are piezoelectric crystals. Since this type of dynamic force or pressure transducer responds to acceleration in addition to the quantity being measured (force or pressure), extreme care must be exercised in controlling the acceleration of the base of the force or pressure instrument.

The three common circuits used with transducers having piezoelectric crystals as sensing elements are the charge-amplifier circuit, the cathode-follower circuit, and the built-in-amplifier circuit. The characteristics of each of these circuits was examined in detail. The charge-amplifier circuit is the most versatile but also the most costly. The built-in-amplifier circuit is the least costly and also the easiest to use.

Transient-signal analysis shows that both mechanical and electrical characteristics of a system can significantly alter the response. Transducers require time to respond to any transient mechanical event. In order to prevent serious transducer ringing, at least five complete natural oscillations of the transducer should occur during the rise time of the mechanical event. Similarly, the electrical time constant of the system must be sufficient to prevent serious signal decay during the transient event. The amount of signal undershoot at the end of a transient event can be used to judge the adequacy of the time constant of a system for a particular type of measurement.

Methods that utilize constant, sinusoidal, and transient mechanical inputs for accelerometer, force transducer, and pressure transducer calibration were described. Use of constant, sinusoidal, and transient voltage inputs for overall system calibration (the voltage insertion method of system calibration) was also discussed. Periodic calibration of individual transducers and complete systems must be performed to ensure continued satisfactory performance. Calibration must always be regarded as one of the most important steps in the measurement.

## REFERENCES

1. Broch, Jens Trampe: "Mechanical Vibration and Shock Measurements," Available from Bruel and Kjaer Instruments, Inc., Marlborough, Massachusetts, October 1980.

2. Bruel and Kjaer Instruments: "Piezoelectric Accelerometer and Vibration Preamplifier Handbook," Available from Bruel and Kjaer Instruments, Inc., Marlborough, Massachusetts, March 1978.

3. Bruel and Kjaer Instruments: *Technical Review,* A quarterly publication available from Bruel and Kjaer Instruments, Inc., Marlborough, Massachusetts.
    (a) *Vibration Testing of Components,* no. 2, 1958.
    (b) *Measurement and Description of Shock,* no. 3, 1966.
    (c) *Mechanical Failure Forecast by Vibration Analysis,* no. 3, 1966.
    (d) *Vibration Testing,* no. 3, 1967.
    (e) *Shock and Vibration Isolation of a Punch Press,* no. 1, 1971.
    (f) *Vibration Measurement by Laser Interferometer,* no. 1, 1971.
    (g) *A Portable Calibrator for Accelerometers,* no. 1, 1971.
    (h) *High Frequency Response of Force Transducers,* no. 3, 1972.
    (i) *Measurement of Low Level Vibrations in Buildings,* no. 3, 1972.
    (j) *On the Measurement of Frequency Response Functions,* no. 4, 1975.
    (k) *Fundamentals of Industrial Balancing Machines and Their Applications,* no. 1, 1981.
    (l) *Human Body Vibration Exposure and Its Measurement,* no. 1, 1982.

4. Dove, R. C., and P. H. Adams: *Experimental Stress Analysis and Motion Measurement,* Merrill, Columbus, Ohio, 1964.

5. Pennington, D.: "Piezoelectric Accelerometer Manual," Endevco Corporation, Pasadena, California, 1965.

6. Peterson, A. P. G., and E. E. Gross, Jr.: *Handbook of Noise Measurement,* Available from General Radio Company, Concord, Massachusetts, 1972.

7. Jones, E., S. Edelman, and K. S. Sizemore: Calibration of Vibration Pick-

ups at Large Amplitudes, *Journal of the Acoustical Society of America,* vol. 33, no. 11, November 1961, pp. 1462–1466.

8. Perls, T. A., and C. W. Kissinger: High Range Accelerometer Calibrations, Report No. 3299, National Bureau of Standards, June 1954.

9. Jones, E., D. Lee, and S. Edelman: Improved Transfer Standard for Vibration Pickups, *Journal of the Acoustical Society of America,* vol. 41, no. 2, February 1967, pp. 354–357.

10. Schmidt, V. A., S. Edelman, E. R. Smith, and E. T. Pierce: Modulated Photoelectric Measurement of Vibration, *Journal of the Acoustical Society of America,* vol. 34, no. 4, April 1966, pp. 455–458.

11. Schmidt, V. A., S. Edelman, E. R. Smith, and E. Jones: Optical Calibration of Vibration Pickups at Small Amplitudes, *Journal of the Acoustical Society of America,* vol. 33, no. 6, June 1961, pp. 748–751.

12. Crosswy, F. L., and H. T. Kalb: Dynamic Force Measurement Techniques, Part I: Dynamic Compensation, *Instrument and Control Systems,* February 1970, pp. 81–83.

13. Crosswy, F. L., and H. T. Kalb: Dynamic Force Measurement Techniques, Part II: Experimental Verification, *Instrument and Control Systems,* March 1970, pp. 117–121.

14. Bell, R. L.: Development of 100,000 g Test Facility, *Shock and Vibration Bulletin,* Bulletin 40, Part 2, December 1969, pp. 205–214.

15. Graham, R. A., and R. P. Reed (Ed.): *Selected Papers on Piezoelectricity and Impulsive Pressure Measurements,* Sandia Laboratories Report, SAND78-1911, December 1978.

16. Otts, J. V.: Force-Controlled Vibration Testing, Sandia Laboratories Technical Memorandum, SC-TM-65-31, February 1965.

17. Hunter, Jr., N. F., and J. V. Otts: Electronic Simulation of Apparent Weight in Force-Controlled Vibration Tests, Preprint No. P15-1-PHYM-MID-67, Presented at the 22nd Annual ISA Conference (Chicago, Illinois, September 11–14, 1967).

18. Change, N. D.: "General Guide to ICP Instrumentation," Available from PCB Piezotronics, Inc., Depew, New York.

19. Graneek, M., and J. C. Evans: A Pneumatic Calibrator of High Sensitivity, *Engineer,* July 13, 1951, p. 62.

20. Kistler, W. P.: Precision Calibration of Accelerometers for Shock and Vibration, *Test Engineer,* May 1966, p. 16.

---

## EXERCISES

**7.1** The earth circles the sun every 365 days at an average radius of 92,600,000 miles. The mean radius of the earth is 3,960 miles. Show that the average acceleration of the earth's center is 0.0006 g toward the sun and that the maximum acceleration on the surface of the earth at the equator relative to the earth's center is approximately 0.0034 g toward the center of the earth.

**7.2**   The equation of the spiral of Fig. E7.2 can be expressed as $r = 2\sqrt{\theta}$, where $r$ is in inches when $\theta$ is in radians. If angle $\theta$ can be expressed as a function of time in seconds as $\theta = 9t$, determine the velocity and acceleration of point $P$ on the curve when $t = \pi/54$ s. Express your answers in both rectangular and polar coordinate form.

**7.3**   The expression $x = 4 \cos 9t + 3 \sin 9t$ is a harmonic function. Show that $x$ can be written as either $x = 5 \cos(9t + \theta_1)$ or $x = 5 \sin(9t + \theta_2)$. Determine the phase angles $\theta_1$ and $\theta_2$. Which angle is leading and which angle is lagging the reference vector? What is the reference vector for each of the expressions for $x$?

**7.4**   A simple harmonic motion has an amplitude of 0.001 in. and a frequency of 100 Hz. Determine the maximum velocity (in inches per second) and maximum acceleration (in gravitational acceleration) associated with the motion. If the amplitude of motion doubles and the frequency remains the same, what is the effect on maximum velocity and maximum acceleration? If the frequency doubles and the amplitude of motion remains the same, what is the effect on maximum velocity and maximum acceleration?

**7.5**   Express the following complex numbers in exponential $(Ae^{i\theta})$ form:

(a)   $1 - i\sqrt{3}$
(b)   $-4$
(c)   $7i$
(d)   $6/(\sqrt{3} - i)$
(e)   $(\sqrt{3} + i)/(3 + 4i)$
(f)   $(\sqrt{3} - i)/(4 + 3i)$

**7.6**   Light having a wavelength of 546.1 nm is directed at normal incidence (see Fig. 7.11) onto a thin film of transparent material having an index of refraction of 1.59. Ten dark and nine bright fringes are observed

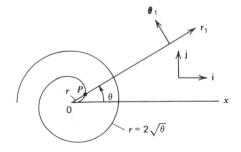

*Figure E7.2*

over a 50-mm length of the film. Determine the thickness variation over this length.

**7.7** An optical flat, shown schematically in Fig. 7.11 as material two, is being used to evaluate the finish of a ground and polished machine surface (specimen in Fig. 7.11). The optical flat rests on several high points of the specimen surface, but does not make contact over a large part of the surface. When a beam of sodium light having a wavelength of 589.0 nm is directed onto the surface of the optical flat at essentially normal incidence, an interference pattern forms that indicates that the deepest valley on the surface produces an interference fringe of order 5. Determine the difference in elevation between the contact points and the valley floor.

**7.8** For the pneumatic displacement gage shown in Fig. 7.12, show that

$$\frac{h}{H} = \frac{1}{1 + (A_S/A_R)^2}$$

A linear relationship proposed by Granek and Evans[10] is often used to simplify measurements made with such devices. The linear relationship is

$$\frac{h}{H} = 1.10 - 0.5(A_S/A_R)$$

Compare the two equations by plotting $h/H$ versus $A_S/A_R$ over the range $0.4 < h/H < 0.9$ for which the linear equation is proposed to be valid.

**7.9** A pneumatic displacement device is constructed as shown in Fig. E7.9. Assume that the relationship $h/H = 1.10 - 0.5(A_S/A_R)$ for $0.4 < h/H < 0.9$ is valid. Show that the relationship between system parameters $d_R$, $d_S$, and $x$ and manometer readings $h/H$ is

$$\frac{h}{H} = 1.10 - 2.0(d_S/d_R^2)x$$

**7.10** How many Class A reference blocks are required to accurately establish a length of 1.2693 in.? What tolerance limits should be placed on this dimension due to block tolerance and oil film effects?

**7.11** A 200-ft stainless steel ($\alpha = 9.6 \times 10^{-6}/°F$) surveyor's tape is left lying along the edge of a highway. The tape temperature is approximately

---

[10] "A Pneumatic Calibrator of High Sensitivity," by M. Graneek and J. C. Evans, *Engineer,* July 13, 1951, p. 62.

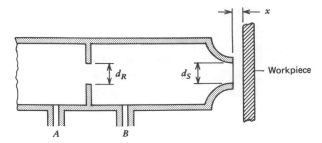

**Figure E7.9**

140°F when it is used to measure a distance of 186 ft. What is the actual distance measured if the tape is accurate at 60°F?

**7.12**  A 10-kΩ single-turn potentiometer has been incorporated into a displacement measuring system as shown in Fig. E7.12. The potentiometer consists of 0.005-in.-diameter wire wound onto a 1.25-in.-diameter ring, as shown in the figure. The pulley diameters are 2.00 in. Determine:

   (a)  The minimum load resistance that can be used if nonlinearity error must be limited to 0.25 percent
   (b)  The smallest motion $x$ that can be detected with this system

**7.13**  The potentiometer circuit shown in Fig. 7.16 is being used to measure the angular position of a shaft. The potentiometer being used can rotate 320 degrees, has a resistance $R_P$ of 2 kΩ, and is capable of dissipating 0.01 W of power in most environments. Show that

   (a)  The maximum voltage that can be applied to the potentiometer is 4.47 V
   (b)  The output sensitivity $S$ is given by the expression

$$S = \left(\frac{R_M}{R_M + R_L}\right)\left(\frac{R_P}{R_P + R_S}\right)\frac{E_i}{\theta_T}$$

where $\theta_T$ is the range of the potentiometer in degrees

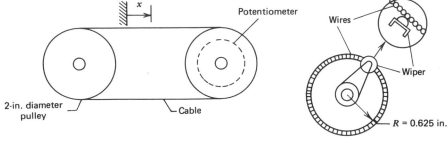

**Figure E7.12**

**7.14**  For the circuit of Exercise 7.13, a minimum sensitivity of 10 mV/degree is desired when 350 ft of AWG 28 copper wire (66.2 $\Omega$ per 1000 ft) is being used as lead wires. The op-amp, which is capable of driving an output circuit continuously at 20 mA, requires a minimum 2-V drop across resistor $R_s$ in order to operate satisfactorily.

(a)  What values of $R_M$ and $R_S$ would you select to achieve the desired sensitivity?

(b)  How would you calibrate and adjust the system? Keep in mind that 5-, 8-, 10-, 15-, and 18-Volt regulators are commercially available.

**7.15**  The multiple-resistor displacement-measuring system shown in Fig. 7.17 has $n$ equal resistors $R_n$. Each of these resistors has a value $k$ times that of the output resistor $R_o$ (i.e., $R_n = kR_o$).

(a)  Show that the normalized output voltage ratio $E_o/E_0$ is given by the expression

$$\frac{E_o}{E_0} = \frac{(n - p)(n + k)}{(n + k - p)n}$$

where $p$ is the number of resistors removed from the circuit by either a switch or a broken wire.

(b)  Plot $E_o/E_0$ as a function of $p$ for $n = 10$ and $k = 1, 10,$ and 100.

(c)  If linear output with $p$ is a desirable feature, what value of $k$ should be selected for the measurement system?

**7.16**  Show that the switch-debouncer circuit of Fig. 7.24 performs its intended function irrespective of the logic state present when the switch first makes momentary contact in the NO position. Which output ($Y_1$ or $Y_2$) is high when NO is high?

**7.17**  The moire fringe pattern of Fig. 7.28 was formed by rotating one grating through an angle $\theta$ with respect to the other grating. The fringes have formed in a direction that bisects the obtuse angle between the lines of the two gratings. If the angle of inclination $\phi$ of the moire fringes and the angle of rotation $\theta$ of one grating with respect to the other are both measured in the same direction and with respect to the lines of the fixed grating, show that

$$\theta = 2\phi - \pi$$

**7.18**  The voice coil of a small speaker can be used as a velocity sensor. The coil, shown schematically in Fig. E7.18, has a diameter of 15 mm and has 100 turns of wire wrapped onto the nonmagnetic core. During calibration, the sensor exhibited a sensitivity of 52 mV/(m/s) over a

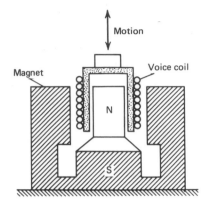

*Figure E7.18*

range of frequencies. Estimate the magnetic field strength $B$ in the gap occupied by the voice coil.

**7.19** Show that the voltage generated by $N$ windings of wire, similar to the one shown in Fig. 7.29b, rotating in a magnetic field of flux density $B$ can be expressed as

$$E = NBA\omega \sin \theta$$

where $A$ is the area of the coil ($A$ = length $l$ × width $d$), $\omega$ is the angular velocity of the coil, and $\theta$ is the angular position of the coil measured with respect to the position shown in Fig. 7.29b.

**7.20** The magnetic flux density $B$ (Webers per square meter) generated by an electric solenoid can be expressed as

$$B = 12.57(10^{-7})NI/l$$

where $N$ is the number of turns, $I$ is the current in amperes, and $l$ is the length of the solenoid in meters. The ring magnet, shown in Fig. E7.20, is being designed to have a field strength of 0.10 Wb/m². The

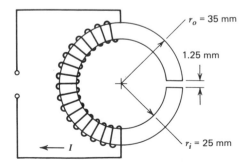

$r_o$ = 35 mm

1.25 mm

$r_i$ = 25 mm

$I$

*Figure E7.20*

solenoid will consist of 2000 turns of wire wrapped uniformly over half of the 10-mm-diameter steel ring. Determine the required current $I$ and estimate the voltage that would be induced in a single length of wire as it passed through the gap at a speed of 1.0 m/s.

**7.21** A linear-velocity transducer (LVT) has an inductance of 16.5 mH, a resistance of 6.2 $\Omega$, and a sensitivity of 30 mV/(in./s). Evaluate the performance of this sensor when it is used in conjunction with a recording instrument having an input resistance of (a) 100 $\Omega$ and (b) 1000 $\Omega$ in terms of frequency response and signal attenuation.

**7.22** The resistance of CPA 01 and CPA 02 crack-propagation gages in parallel with a 50-$\Omega$ shunt resistor is shown in Fig. 7.35. If the combined resistance is 34.5 $\Omega$ when 19 of the 20 strands are broken, determine the resistance of that single strand. Compare this single-strand resistance to the average strand resistance when all 20 strands are connected in parallel. Based on this information, design a circuit that will give a nearly linear voltage output as a function of the number of strands broken. Compare this circuit to Fig. 7.36.

**7.23** A velocity meter is being designed with a natural frequency of 5 Hz, damping of 10 percent, and a sensitivity of 8.3 mV/(in./s). The magnetic core weighs 0.25 lb, and is mounted on soft springs. Determine the required spring constant for the springs supporting the core. The velocity meter is mounted on a surface that is vibrating with a maximum velocity of 8.0 in./s. Determine the peak output voltage and phase angle if the frequency of vibration is (a) 8.0 Hz and (b) 20 Hz. Which measurement has the greatest error? Why?

**7.24** The force transducers listed in Table 7.2 have stiffnesses that range from $5(10^6)$ to $100(10^6)$ lb/in. Based on the information given in Table 7.2, estimate the weight and mass of the seismic mass for each model. What is the acceleration sensitivity compared to the force sensitivity for each transducer as constructed? (*Hint:* Use the ratios of relative motion $z$.) Which transducer is the least acceleration sensitive?

**7.25** An accelerometer is used to measure a periodic signal that can be expressed as

$$a(t) = a \sin(0.2\omega_n t) - 0.4a \sin(0.6\omega_n t)$$

where $\omega_n$ is the natural frequency of the accelerometer. The relative motion of the transducer can be expressed as

$$z(t) = b_i \sin(0.2\omega_n t - \phi_1) + b_3 \sin(0.6\omega_n t - \phi_3)$$

Determine the coefficients $b_1$ and $b_3$ and phase angles $\phi_1$ and $\phi_3$ when the transducer damping is (a) 5 percent and (b) 60 percent. Which damping condition gives the best modeling? Why? Compare $a(t)$ and $z(60$ percent).

**7.26** The Kistler model 912 force gage in Table 7.2 is connected to a structure with a $\frac{3}{8}$-in.-diameter by 1-in.-long coupling that weighs 0.0313 lb. Show that the nominal natural frequency is reduced to 34,400 Hz and that one g of base acceleration gives a signal that is equivalent to 0.0413 lb of force.

**7.27** The PCB 208A03 general-purpose force transducer listed in Table 7.2 is being used to measure the driving force from an electrodynamic vibration exciter, as shown in Fig. E7.27. The base of the force transducer is connected to the structure under test, and the seismic mass is connected to the exciter through a 2.00-lb connector. The effective mass of the structure (including the force transducer) is $m_s = 0.250$ lb · s$^2$/in., the effective spring constant is $k_s - 15,620$ lb/in., and the effective damping is 6 percent. Show that the natural frequency of the transducer is lowered to 6970 Hz when used this way, that the natural frequency of the structure is 39.8 Hz, and that the driving force measured by the transducer is related to the force generated between the exciter and connector by the expression

$$F(t) = R_o\left[1 + \frac{(m + m_c)\omega^2 e^{-i\phi}}{\sqrt{(k_s - m_s\omega^2)^2 + (C_s\omega)^2}}\right]e^{i\omega t}$$

where

$$\tan \phi = \frac{C_s\omega}{k_s - m_s\omega^2}$$

Show that the measurement is 2.66 percent too high when the driving frequency is 30 Hz, and 17.6 percent too high at the resonant frequency of the structure. What would happen if the force between the connector

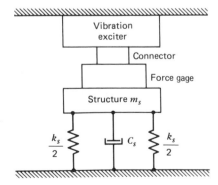

*Figure E7.27*

and the force gage were to be measured instead of the force $R_o$ between the connector and the exciter?

**7.28** The charge sensitivity, voltage sensitivity, and capacitance for an Endevco 2501-2000 pressure transducer are listed in Table 7.3. Typical coaxial transducer cable has a nominal capacitance of 30 pF/ft. Show that approximately 40 in. of cable must have been connected to the transducer to give the voltage sensitivity listed in the table. Show that the 3-dB cutoff frequency is 4.0 Hz when the transducer is connected to a voltage follower with an input resistance of 100 MΩ. Show that the voltage sensitivity drops to 18.67 mV/psi when the transducer is connected to the voltage follower with 15 ft of transducer cable. At what frequency (in hertz) will the attenuation be 10 percent with this longer cable?

**7.29** From the data in Table 7.5, estimate the minimum open-loop gain for the first op-amp in the charge amplifier. If $C_{cal}$ (1000 pF) in Fig. 7.48 is connected to ground, how is this open-loop gain $G$ changed? Consider only the cases when $C_f = 10$ pF and $C_f = 1000$ pF.

**7.30** A force transducer with a charge sensitivity of 48.6 pC/lb is connected to a charge amplifier. The transducer sensitivity dial is set at 46.8 pC/lb instead of 48.6 pC/lb. The range capacitor is set at the $C_f = 10,000$ pF position. The user initially thinks the sensitivity is 100 lb/V, as listed in Table 7.5; however, he later recognizes his error. What is the actual sensitivity? How can the results obtained by using a sensitivity of 100 lb/V be corrected to yield the true values?

**7.31** The Endevco model 2215E accelerometer listed in Table 7.1 will be used to measure some low-level acceleration signals. The transducer cable is 30 ft long and has a nominal capacitance of 30 pF/ft. Assume that the minimum op-amp gain $G_1 \geqslant 20,000$. What minimum feedback capacitance can be used if error due to source capacitance must be limited to 0.5 percent? What is the unit sensitivity (gravitational acceleration per volt) if $C_f = 200$ pF and the transducer sensitivity dial is set at (a) b = 0.085, (b) b = 0.170, and (c) b = 0.340? Which transducer dial setting gives the largest voltage signal for a given level of acceleration?

**7.32** A Kistler model 606L pressure transducer is being used in conjunction with a voltage follower, as shown in Fig. E7.32. The connecting cable is 10 ft long and has a nominal capacitance of 30 pF/ft. Determine:

(a)  The minimum blocking capacitance $C_b$ required if error due to $C_b$ must be less than 1 percent
(b)  The gain required to obtain a voltage sensitivity of 50 mV/psig.
(c)  The time constant if $R = 100$ MΩ

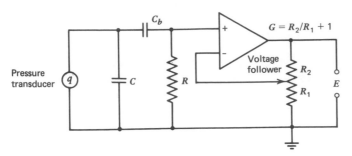

*Figure E7.32*

**7.33** The PCB model 302A accelerometer listed in Table 7.1 is connected to a power supply that has a 0.05-$\mu$F blocking capacitor and to an oscilloscope that has a 1.0-M$\Omega$ input impedance.

(a) Show that the 3-dB down point in Fig. 7.51 occurs when $\omega T_1 =$ 1.0.

(b) Show that the transfer function plots the same as the curve for $T_1 = 10T$ in Fig. 7.51.

(c) At what value of $\omega T_1$ is the signal attenuated 5 percent?

(d) What is the effect of increasing the blocking capacitance to 5 $\mu$F?

**7.34** For the accelerometer of Exercise 7.33, determine the phase shift when $C_1 = 0.05$ $\mu$F and (a) $\omega T_1 = 0.1$ and (b) $\omega T_1 = 1.0$. Compare your results to Fig. 7.51$b$.

**7.35** The "pop test" illustrated in Fig. E7.35 provides a convenient means for testing the overall performance of a pressure transducer. The pressure variation resulting from rupture of the diaphragm approaches a step input; therefore, typical transducer response is as shown in Fig. 7.54. During a specific test, the first three peaks of Fig. 7.54 were 8.51, 7.19, and 6.37 mV, while the first three valleys were 0, 1.879, and 3.05 mV for an initial chamber pressure of 3.0 psig. The final steady-state response was 5.0 mV. The three peaks occurred at 0.375, 0.875, and 1.375 ms, while the valleys occurred at 0, 0.50, and 1.00 ms. Show that

(a) The natural frequency of the transducer is 2000 Hz

(b) The damping is 7.5 percent

(c) The transducer sensitivity is 1.667 mV/psi

**7.36** Compare the rise times calculated from Eq. (7.89) with those obtained from the expression $(1 - \cos \omega_n t)$ if (a) $d = 5$ percent and (b) $d = 15$ percent.

**7.37** Show that Eq. (7.94) represents the electrical response of the cathode-follower circuit to the rectangular input pulse of Fig. 7.57.

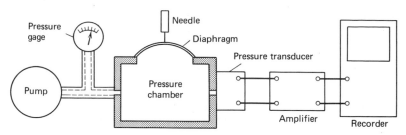

**Figure E7.35**

**7.38** The triangular pulse in Table 7.7 has a form that can be expressed as

$$a(t) = \frac{a_o}{t_1}t \qquad \text{for } 0 \leq t \leq t_1$$

Show that the transient solution of Eq. (7.70) for this input is

$$E_o = \frac{S_q a_o}{C}\frac{RC}{t_1}(1 - e^{-t/RC})$$

Show that the error in Table 7.7 can be estimated from the expression

$$\mathscr{E}_{max} = \frac{1}{2}\frac{t_1}{RC}$$

**7.39** Use Eq. (7.97) to show that the effective time constant of Eq. (7.98) can be obtained by equating the intitial slope of the actual response curve to that for an equivalent circuit that can be expressed as

$$E_{oe} = S_v a_o e^{-t/T_e}$$

Compare results when (a) $T_1 = T$, (b) $T_1 = 10T$, and (c) $T = 100T_1$.

**7.40** A centrifuge is being designed to obtain a 60,000 g acceleration field at a radius of 10.0 in.

  (a) Determine the required speed of the centrifuge (rpm).
  (b) Determine the bolt load for an accelerometer weighing 26 grams.
  (c) Will a $\frac{3}{16}$-in.-diameter brass bolt keep the accelerometer in place?

**7.41** A traveling microscope with a least count of 0.0001 in. is used to measure the peak-to-peak displacement during a sinusoidal calibration of an accelerometer. The microscope is focused on an object having a diameter of 0.0011 in., as shown in Fig. E7.41. For the readings shown, determine the acceleration (in g's) if the frequency of oscillation is

(a)  50 Hz
(b)  100 Hz
(c)  500 Hz
(d)  1000 Hz
(e)  What is your estimate of the perent error in these measurements?
(f)  Would any of these acceleration levels be difficult to obtain if the moving mass weighs 0.35 lb and the vibration exciter can deliver a maximum force of 100 lb?

**7.42**  Several calibration methods require accurate knowledge of the local acceleration of gravity. A common method for obtaining this quantity utilizes a simple pendulum (see Section 3.10). If a pendulum having a length of 25 in. is used with a stopwatch that can be started and stopped consistently within 0.15 s, which variable—pendulum length or period of oscillation—must be measured most accurately? How can the 0.15-s uncertainty in the period measurement be overcome? What shape should the pendulum mass have so that its center of mass can most easily be located accurately? Would increasing the pendulum length or using multiple period measurements increase the accuracy of the measurement of g?

**7.43**  A Michelson interferometer with a helium–neon ($\lambda = 632.8$ nm) laser light source is being used for the absolute calibration of an accelerometer. The number of changes in intensity per vibration cycle was recorded as 4008 at a frequency of oscillation of 750 Hz. The peak-to-peak output voltage was 6370 mV. Determine:

(a)  The acceleration (peak-to-peak)
(b)  The voltage sensitivity of the transducer

**7.44**  Data obtained from four gravimetric calibration tests are listed below:

|                | Test Number |       |       |       |
|----------------|-------------|-------|-------|-------|
|                | 1           | 2     | 3     | 4     |
| $E_{mg}$ (mV)  | 46.6        | 46.5  | 46.7  | 46.5  |
| $E_f$ (mV)     | 19.48       | 78.1  | 302   | 1162  |
| $E_a$ (mV)     | 4.23        | 16.83 | 64.5  | 254.0 |

Determine the sensitivity $S_a$ of the accelerometer. After the tests were conducted it was determined that the local acceleration of gravity was 31.30 ft/s² instead of the nominal 32.17 ft/s². What effect will this change have on the previously determined values of $S_a$?

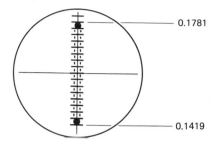

0.1781

0.1419

*Figure E7.41*

**7.45** The RMS voltage ratio $E_f/E_a$ versus calibration weight $W_c$ curve from a typical sinusoidal calibration of a dynamic force gage is shown in Fig. 7.67. If a similar curve for a dynamic force gage being calibrated has a slope of 6.08 g's/lb and intersects the vertical axis at a value of 0.304 V/V, determine:

(a)  The sensitivity of the force gage if the sensitivity of the accelerometer is 7.65 mV/g

(b)  The weight of the seismic mass

**7.46** In a dynamic force gage calibration test, the transducer sensitivity dial of the charge amplifier being used with the accelerometer was set at $b$ = 1.00 and the transducer sensitivity dial of the charge amplifier being used with the force gage was set at $b$ = 0.276. The charge sensitivity of the accelerometer is 2.76 pC/g. The feedback capacitor of the charge amplifier being used with the accelerometer was set at 100 pF, while that of the force transducer was set at 2000 pF. The slope of the $E_f/E_a$ versus $W_c$ curve is 7.82 g's/lb. Determine the charge sensitivity of the force gage.

**7.47** An impact hammer with attached force transducer is to be calibrated by impacting a mass suspended as a pendulum, as shown in Fig. 7.69. The sensitivity of the accelerometer is 3.65 mV/g. The peak voltage ratio $E_f/E_a$ versus pendulum weight $W_p$ curve has a slope of 14.38 g's/lb. The hammer head weighed 0.165 lb and the hammer body weighed 0.769 lb during the calibration tests. Show that the voltage sensitivity of the force gage is 52.5 mV/lb. At a later date, the weight of the hammer head was increased to 0.611 lb and the weight of the body was increased to 1.278 lb in order to obtain some desired impact force characteristics. Show that the voltage sensitivity of the force gage changes to 43.2 mV/lb.

**7.48** A pressure transducer having a charge sensitivity of 1.46 pC/psi is to be used with a charge amplifier to measure hydraulic pump pressures that range from 100 to 1000 psi. The charge amplifier has a calibration capacitor ($C_{cal}$ = 1000 pF), as shown in Fig. 7.48. The required voltage sensitivity for this application is 5 mV/psi. Specify:

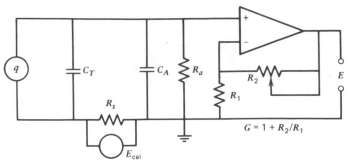

$$G = 1 + R_2/R_1$$

*Figure E7.49*

(a) The transducer sensitivity setting $b$
(b) The feedback capacitor setting $C_f$ (see Table 7.5 for standard values)
(c) The peak calibration charges to simulate 100 psi and 1000 psi
(d) The peak calibration voltages required to simulate 100 psi and 1000 psi
(e) The anticipated calibration output voltages corresponding to the two pressures

**7.49** A Kistler model 912 force transducer is connected to the voltage insertion adaptor with a cable having a capacitance of 107 pF. The voltage insertion adaptor is connected to the amplifier with a cable having a capacitance of 356 pF. The amplifier has an input capacitance of 15 pF and an effective resistance of 700 MΩ. The amplifier gain $G$ can be adjusted from 1 to 10 by using the method illustrated in Fig. E7.49. The insertion adapter has a resistance $R_s$ of 16.2 Ω. Determine:

(a) The calibration voltage requried to simulate a 55.0-lb load
(b) The minimum test frequency for which there is less than a 0.5 percent low-frequency attenuation
(c) The gain $G$ required to yield a voltage sensitivity of 200 mV/lb

# EIGHT
# TEMPERATURE MEASUREMENTS

## 8.1 INTRODUCTION

*Temperature,* unlike other quantities such as length, time, or mass, is an abstract quantity that must be defined in terms of the behavior of materials as the temperature changes. Some examples of material behavior that have been used in the measurement of temperature include change in volume of a liquid, change in length of a bar, change in electrical resistance of a wire, change in pressure of a gas at constant volume, and change in color of a lamp filament.

Temperature is related to the kinetic energy of the molecules at a localized region in a body; however, these kinetic energies cannot be measured directly and the temperature inferred. To circumvent this difficulty, the International Practical Temperature Scale has been defined in terms of the behavior of a number of materials at thermodynamic fixed points.

The International Practical Temperature Scale is based on six fixed points that cover the temperature range from $-183°C$ ($-297°F$) to $1065°C$ ($1949°F$). These six points, each of which correspond to an equilibrium state during a phase transformation of a particular material, are listed in Table 8.1.

**TABLE 8.1** The Six Primary Points of the International Practical Temperature Scale

| Material | Equilibrium State Phase Transformation | Temperature °C | Temperature °F |
|---|---|---|---|
| Oxygen | Liquid–vapor | − 182.962 | − 297.33 |
| Water[a] | Solid–liquid–vapor | 0.01 | 32.02 |
| Water | Liquid–vapor | 100.00 | 212.00 |
| Zinc | Solid–liquid | 419.58 | 787.24 |
| Silver | Solid–liquid | 961.93 | 1763.47 |
| Gold | Solid–liquid | 1064.43 | 1947.97 |

[a] Triple point of water.

All of the primary fixed points with the exception of the triple point of water are at a pressure of one standard atmosphere. The temperature, associated with each of these thermodynamic states, is specified in degrees Celsius and converted to degrees Fahrenheit when required for engineering applications. Other secondary fixed points are listed in Table 8.2.

Between the fixed points (both primary and secondary), the temperature is defined by the response of specified temperature sensors and interpolation equations. The scale is divided into four ranges with the sensors, fixed points, and temperature span prescribed as indicated in Table 8.3.

The International Practical Temperature Scale is an empirical reference that is used to substitute for the direct measurement of the kinetic energy of a molecule. The scale will change with time as scientists improve sensors and other apparatus. However, for engineering purposes, the accuracy of the existing scale (1968) is more than adequate. Of more importance to engineering applications

**TABLE 8.2** Other Secondary Points of the International Practical Temperature Scale

| Material | Equilibrium State Phase Transformation | Temperature °C | Temperature °F |
|---|---|---|---|
| Hydrogen | Solid–liquid–vapor | − 259.34 | − 434.81 |
| Hydrogen | Liquid–vapor | − 252.87 | − 423.17 |
| Neon | Liquid–vapor | − 218.789 | − 361.820 |
| Water | Liquid–solid | 0 | 32 |
| Tin | Liquid–solid | 231.9681 | 449.5426 |
| Lead | Liquid–solid | 327.502 | 621.504 |
| Sulfur | Liquid–vapor | 444.6 | 832.3 |
| Antimony | Liquid–solid | 630.74 | 1167.33 |
| Aluminum | Liquid–solid | 660.37 | 1220.67 |

**TABLE 8.3** Temperature Range, Sensors, and Interpolation Equations for the International Practical Temperature Scale

| Temperature Range (°C) | Sensor | Fixed Point | Equation |
|---|---|---|---|
| − 190 to 0 | Platinum thermometer | Oxygen, ice, steam, sulfur | Reference equation |
| 0 to 660 | Platinum thermometer | Ice, steam, sulfur | Parabola |
| 660 to 1063 | 10% rhodium platinum thermocouple | Antimony, silver, gold | Parabola |
| Above 1063 | Optical pyrometer | — | Planck's Law |

is the selection of the temperature sensor, its installation, the instrumentation system for recording and displaying the output signal, and the use of the temperature data in product design or process control. This chapter deals with the issues of importance to the measurement of temperature for engineering applications.

## 8.2 RESISTANCE THERMOMETERS

Resistance thermometers consist of a sensor element that exhibits a change in resistance with any change in temperature, a signal conditioning circuit that converts the resistance change to an output voltage, and appropriate instrumentation to record and display the output voltage. Two different types of sensors are normally employed: resistance temperature detectors (RTDs) and thermistors.

Resistance temperature detectors are simple resistive elements formed of such materials as platinum, nickel, or a nickel–copper alloy known commercially as Balco.[1] These materials exhibit a positive coefficient of resistivity and are used in RTDs because they are stable and provide a reproducible response to temperature over long periods of time.

Thermistors are fabricated from semiconducting materials such as oxides of manganese, nickel, or cobalt. These semiconducting materials, which are formed into the shape of a small bead by sintering, exhibit a high negative coefficient of resistivity. In some special applications, where very high accuracy is required, doped silicon or germanium is used as the thermistor material.

The equations governing the response of RTDs and thermistors to a temperature change and the circuits used to condition their outputs are different; therefore, they will be treated separately in the following subsections.

---

[1] Balco is a trade name for a product of the W. B. Driver Co.

### Resistance Temperature Detectors (RTDs)

A typical RTD consists of a wire coil for a sensor with a framework for support and a sheath for protection, a linearizing circuit, a Wheatstone bridge, and a voltage display instrument. The sensor is a resistive element that exhibits a resistance–temperature relationship given by the expression

$$R = R_o(1 + \gamma_1 T + \gamma_2 T^2 + \cdots + \gamma_n T^n) \tag{8.1}$$

**where**   $\gamma_1, \gamma_2, \ldots, \gamma_n$ are temperature coefficients of resistivity.
$R_o$ is the resistance of the sensor at a reference temperature $T_o$.

The reference temperature is usually specified as $T_o = 0°C$.

The number of terms retained in Eq. (8.1) for any application depends upon the material used in the sensor, the range of temperature, and the accuracy required in the measurement. Resistance temperature curves for platinum, nickel, and copper, which illustrate the nonlinearity in resistance $R$ with temperature $T$ for each of these materials, are shown in Fig. 8.1. For a limited range of temperature, the linear form of Eq. (8.1) is often used to relate resistance change to temperature change. Equation (8.1) is then expressed as

$$\frac{\Delta R}{R_o} = \gamma_1(T - T_o) \tag{8.2}$$

When error due to the neglect of nonlinear terms becomes excessive, either linearizing circuits can be used to compensate for the nonlinearities, or additional terms can be retained from Eq. (8.1) to relate the measured $\Delta R$ to the unknown temperature $T$. Retaining the temperature coefficients $\gamma_1$ and $\gamma_2$ from Eq. (8.1)

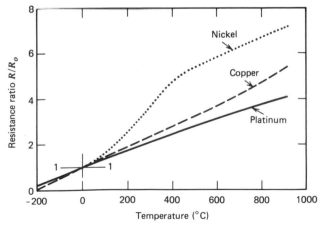

***Figure 8.1***   Resistance–temperature curves for nickel, copper, and platinum.

yields the second-order relationship

$$\frac{\Delta R}{R_o} = \gamma_1(T - T_o) + \gamma_2(T - T_o)^2 \tag{8.3}$$

Equation (8.3) is more cumbersome to employ in practice but provides accurate results over a wider temperature range.

Sensing elements are available in a wide variety of forms. The different RTD sensors shown in Fig. 8.2 illustrate the wide range of commercially available products. One widely used sensor consists of a high-purity platinum wire wound on a ceramic core. The element is stress relieved after winding, immobilized against strain, and artificially aged during fabrication to provide for long-term stability. Drift is usually less than 0.1°C when such a sensor is used at its upper temperature limit.

The sensing element is usually protected by a sheath fabricated from stainless steel, glass, or a ceramic. Such sheaths are made pressure tight to protect the sensing element from the corrosive effects of both moisture and the process medium.

Lead wires from the sensor exit from the sheath through a specially designed

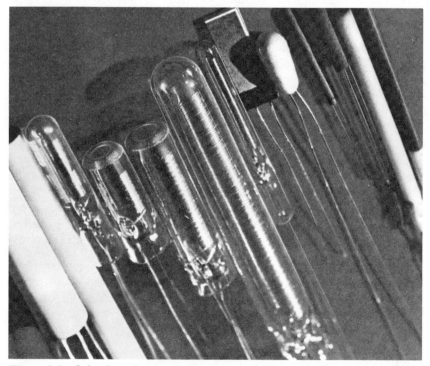

*Figure 8.2*   Selection of resistance temperature detectors. (Courtesy of Omega Engineering, Inc.)

seal. The method of sealing between the sheath and the lead wires depends on the upper temperature limit of the sensor. Epoxy cements are used for the low-temperature range (<260°C), while glass and ceramic cements are used for the high-temperature range (>260°C). Since the temperature at the sheath exit is usually much lower than the process temperature being monitored, lead wires insulated with Teflon or impregnated fiberglass are often suitable for use with process temperatures as high as 750°C (1380°F).

The sensors shown in Fig. 8.2 are immersion type transducers that are inserted in the medium to measure fluid temperatures. The response time in this application is relatively long (between 1 and 5 s is required to approach 100 percent response). This relatively long response time for immersion thermometers is not usually a serious concern, since the rate of change of liquid temperatures in most processes is relatively slow.

Resistance temperature detectors can also be employed to measure temperatures on the surface of an object by using a different type of sensor. Resistance temperature detectors for surface temperature measurements utilize either a thin-wire element, such as the one shown in Fig. 8.3, or a thin-film element that resembles an electrical resistance strain gage and is fabricated by using the photoetching process developed in recent years to produce high-quality strain gages. The foil sensors are available on either polyimide or glass-fiber-reinforced epoxy resin carriers. The wire models are available with either teflon or phenolic-glass carriers or as free filaments. The sensors with carriers are bonded to the surface with an adhesive suitable for the temperature range to be encountered. The free filaments are normally mounted by flame spraying. The response time of a thin-film sensor compares favorably with a small thermocouple; therefore, measurement of rapidly changing surface temperatures is possible.

An example of a bondable type of dual-grid resistance temperature detector is shown in Fig. 8.4. This construction detail shows two thin-foil sensing elements connected in series and laminated in a glass-fiber reinforced epoxy resin matrix. One of the two sensing elements is fabricated from nickel and the other from manganin. These two materials were selected since they exhibit equal but opposite nonlinearities in their resistance–temperature characteristics over a significant temperature range. By connecting the nickel and manganin in series, the nonlinear effects cancel and the composite sensor provides a linear response

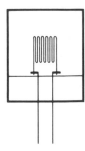

*Figure 8.3* Resistance temperature detector for surface temperature measurements. (Courtesy of BLH Electronics.)

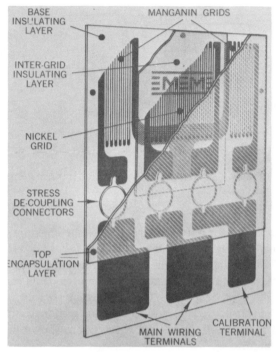

BASE
INSULATING
LAYER

MANGANIN GRIDS

INTER-GRID
INSULATING
LAYER

NICKEL
GRID

STRESS
DE-COUPLING
CONNECTORS

TOP
ENCAPSULATION
LAYER

MAIN WIRING
TERMINALS

CALIBRATION
TERMINAL

*Figure 8.4* Cryogenic linear temperature sensor. (Courtesy of Micro-Measurements.)

with respect to temperature over the temperature range from $-269°C$ ($-452°F$) to $24°C$ ($75°F$). The bondable RTD is fabricated with integral printed-circuit terminals to provide for easy attachment of the lead wires.

The output from a resistance temperature detector (RTD) is a resistance change $\Delta R/R$ that can be conveniently monitored with a Wheatstone bridge, as illustrated schematically in Fig. 8.5. The RTD is installed in one arm of the bridge, a decade resistance box is placed in an adjacent arm, and a matched pair of precsion resistors are inserted in the remaining arms to complete the bridge. Careful consideration must be given to the lead wires since any resistance change $\Delta R/R$ in the lead wires will produce an error in the readout. With the three-lead-wire arrangement shown in Fig. 8.5, any temperature-induced resistance change in the lead wires is canceled. The Wheatstone bridge shown in Fig. 8.5 can be balanced by adjusting the decade resistance box. In the null position, the reading on the box is exactly equal to the resistance of the RTD. The temperature is then determined from a table of resistance versus temperature for the specific RTD being used. Care must also be exercised in powering the bridge to avoid excessive currents in the sensor. If the excitation voltage is excessive, errors will occur due to self-heating of the sensor. This problem can be avoided by maintaining excitation at 0.25 V or less. Fortunately, the resistance change with temperature in the RTD is large; therefore, adequate resolution of temperature can be achieved with very low excitation voltages.

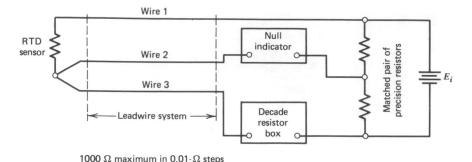

1000 Ω maximum in 0.01-Ω steps

*Figure 8.5* Wheatstone-bridge circuit with lead-wire compensation and manual reading of the output from a resistance temperature detector.

Another circuit that can be employed for automatic readout is the constant-current potentiometer circuit shown in Fig. 8.6. The output voltage $E = IR$ from this circuit can be monitored with a digital voltmeter. If a constant current of 1 mA is supplied to the sensor, the output of the digital voltmeter converts easily to resistance ($R = E/I$). The temperature is then determined from a resistance–temperature table for the sensor. Errors due to resistance changes in the lead wires are also eliminated by using the four-lead-wire system shown in Fig. 8.6 when $R_M$ of the DVM is high.

The circuits shown in Figs. 8.5 and 8.6 provide simple and accurate methods for measuring the sensor resistance. However, since the sensor resistance $R$ is a nonlinear function of temperature $T$, as shown in Fig. 8.7, tables must be used to relate the measured resistance to the temperature. This tabular conversion procedure is time-consuming and prevents direct display of the temperature on a strip-chart recorder or other instrument.

The nonlinear response of the bonded RTD can be significantly improved by utilizing the simple shunting circuit illustrated in Fig. 8.8. A shunt resistor, having a resistance value three times that of the sensor resistance, improves the linearity but reduces the output of the sensor, as shown in Fig. 8.7. Fortunately, this reduction in output is not a serious disadvantage because of the very high signal output from a typical RTD. Use of a shunt resistor in the circuit does not

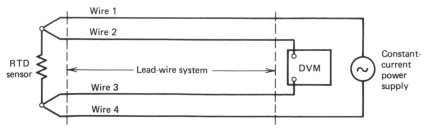

*Figure 8.6* Constant-current potentiometer circuit with lead-wire compensation and automatic reading of the output from an RTD sensor.

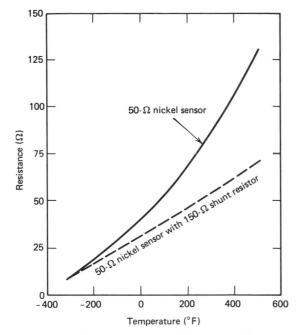

*Figure 8.7* Response of nickel foil resistance temperature detectors.

completely compensate for the nonlinear relationship between sensor resistance and temperature; however, the deviation from linearity is small and within acceptable limits for many applications. The deviation from linearity for a typical RTD in a shunt circuit over the temperature range from $-196°C$ ($-320°F$) to $260°C$ ($500°F$) is shown in Fig. 8.9.

Four common errors encountered during use of RTDs for temperature measurements result from lead-wire effects, stability, self-heating, and sensitivity of the RTD to strain. Lead-wire errors can be minimized by making the lead wires as short as possible. The total resistance of a two-wire system should always be less than 1 percent of the sensor resistance. The effect of lead-wire resistance is to increase the apparent resistance of the sensor and thus cause a zero shift (offset) and a reduction in sensitivity. The error due to temperature-induced resistance changes in the lead wires is usually small enough to be neglected. For example, a temperature change of $10°C$ ($18°F$) over the entire length of a two-

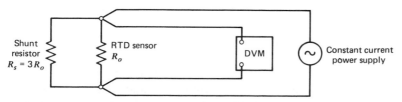

*Figure 8.8* Shunt method for improving the linearity of bonded RTD sensors.

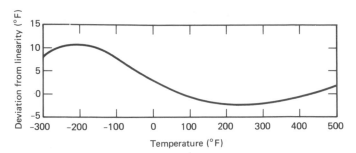

***Figure 8.9*** Deviation from linearity for a 50-Ω RTD sensor with a 150-Ω shunt resistor when the sensor is bonded to 1018 steel. (Courtesy of Micro-Measurements.)

wire circuit ($R_L$ = 0.5 Ω) with a 50-Ω nickel sensor will produce an apparent temperature offset of only 0.2°C (0.4°F). All error due to lead wires can be eliminated by using either the three- or four-wire circuits illustrated in Figs. 8.5 or 8.6.

Stability of the sensors is usually assured by aging of the elements during the manufacturing process. Stability may become a source of error when the upper temperature limit of the sensor is exceeded either by design or accident. Anytime the upper temperature limit of a sensor is exceeded, any new temperature measurements should be repeated until stable and reproducible readings are obtained. Stability can also be affected by the polymeric carrier used with bondable RTDs. These carriers have a finite life and lose their strength at temperatures in excess of 120°C (250°F).

Self-heating errors are produced when excitation voltages or currents are used in the signal conditioning circuits. Usually there is no reason for large excitation signals, since an RTD is a high-output sensor (a typical output is 0.9 mV/V · °C or 0.5 mV/V · °F). Self-heating errors can be minimized by limiting the power dissipation in the RTD to less than 2 mW. In those applications where small temperature changes are to be measured and very high sensitivity is required, sensors with large surface areas should be employed. These sensors with large surface areas can dissipate larger amounts of heat; therefore, higher excitation voltages can be used without introducing self-heating errors.

Bonded RTD sensors resemble strain gages and, in fact, they respond to strain. Fortunately, the strain sensitivity of the sensor is small in comparison to the temperature sensitivity. A bonded RTD with a nickel sensor exhibits an apparent temperature change of 1.7°C (3°F) when subjected to an axial tensile strain of 1000 μm/m along the filaments of the gage grid. The magnitude of the strain effect is such that it can be neglected in most applications.

## Thermistors

*Thermistors* are temperature-sensitive resistors fabricated from semiconducting materials, such as oxides of nickel, cobalt, or manganese and sulfides of iron, aluminum, or copper. Thermistors with improved stability are obtained when oxide systems of manganese–nickel, manganese–nickel–cobalt, or manganese–nickel–iron are used. Conduction is controlled by the concentration of oxygen in the oxide semiconductors. An excess or deficiency of oxygen from exact stoichiometric requirements results in lattice imperfections known as Schottky defects and Frankel defects. *N-type oxide semiconductors* are produced when the metal oxides are compounded with a deficiency of oxygen that results in excess ionized metal atoms in the lattice (Frankel defects). *P-type oxide semiconductors* are produced when there is an excess of oxygen that results in a deficiency of ionized metal atoms in the lattice (Schottky defects).

Semiconducting materials, unlike metals, exhibit a decrease in resistance with an increase in temperature. The resistance–temperature relationship for a thermistor can be expressed as

$$\ln (R/R_o) = \beta(1/T - 1/T_o)$$

or

$$R = R_o e^{\beta(1/T - 1/T_o)} \tag{8.4}$$

**where**    $R$  is the resistance of the thermistor at temperature $T$.
$R_o$ is the resistance of the thermistor at reference temperature $T_o$.
$\beta$  is a material constant that ranges from 3000 K to 5000 K.
$T$ and $T_o$ are absolute temperatures, K.

The sensitivity $S$ of a thermistor is obtained from Eq. (8.4) as

$$S = \frac{\Delta R/R}{\Delta T} = -\frac{\beta}{T^2} \tag{8.5}$$

For $\beta = 4000$ K and $T = 298$ K, the sensitivity $S$ equals $-0.045/$K, which is more than an order of magnitude higher than the sensitivity of a platinum resistance thermometer ($S = +0.0036/$K). The very high sensitivity of thermistors results in a large output signal and good accuracy and resolution in temperature measurements. For example, a typical thermistor with $R_o = 2000$ $\Omega$ and a sensitivity $S = -0.04/$K exhibits a response $\Delta R/\Delta T = 80$ $\Omega/$K. This very large resistance change can be converted to a voltage with a simple bridge circuit. The voltage change associated with a temperture change as small as 0.0005 K can be easily and accurately monitored.

Equation (8.4) indicates that the resistance $R$ of a thermistor decreases exponentially with an increase in temperature. Typical response curves for a family of thermistors is shown in Fig. 8.10. Since the output from the thermistor is nonlinear, precise determinations of temperature must be made by measuring the resistance $R$ and using a calibration table similar to the one presented in Appendix A (Table A.1). Linearity of the output can be improved by using modifying potentiometer and/or bridge circuits; however, these circuits reduce the sensitivity and the output of the thermistor.

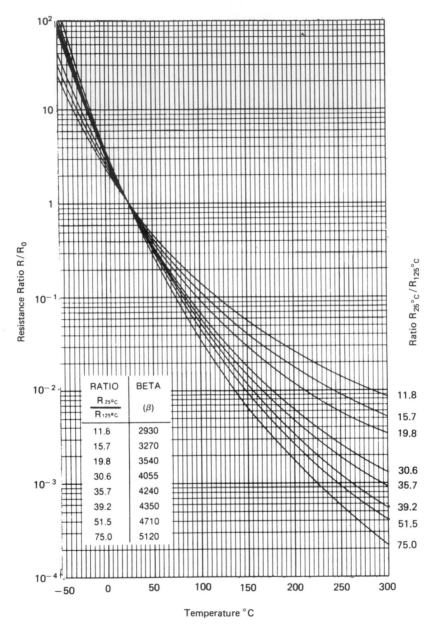

**Figure 8.10** Resistance as a function of temperature for different thermistors ($T_0 = 25°C$). (Courtesy of Thermometrics, Inc.)

Thermistors are produced by mixing two or more semiconducting oxide powders with a binder to form a slurry. Small drops (beads) of the slurry are formed over the lead wires, dried, and fired in a sintering furnace. During sintering, the metallic oxides shrink onto the lead wires and form an excellent electrical connection. The beads are then hermetically sealed by coating the beads with glass. The glass coating improves stability of the thermistor by eliminating water absorption into the metallic oxide. Thermistor beads, such as those shown in Fig. 8.11, are available in diameters that range from 0.005 in. (0.125 mm) to 0.060 in. (1.5 mm). Thermistors are also produced in the form of disks, wafers, flakes, rods, and washers to provide sensors of the size and shape required for a wide variety of applications.

A large variety of thermistors are commercially available with resistances at the reference temperature $T_o$ that vary from a few ohms to several megohms. When a thermistor is selected for a particular application, the minimum resistance at high temperature must be sufficient to avoid overloading of the readout device. Similarly, the maximum resistance at low temperature must not be so high that noise pickup becomes a serious problem. Thermistors having a resistance $R_o = 3000 \ \Omega$ that varies from a low of about $2000 \ \Omega$ to a high of $5000 \ \Omega$ over the temperature range are commonly employed.

Thermistors can be used to measure temperatures from a few degrees above absolute zero to about 315°C (600°F). They can be used at higher temperatures, but the stability begins to decrease significantly above this limit. The range of a thermistor is usually limited to about 100°C (180°F), particularly, if it is part of an instrumentation system with a readout device that has been compensated to provide nearly linear output. The accuracy of thermistors depends upon the techniques employed for the measurement of $\Delta R/R$ and for the calibration of the sensor. With proper techniques and glass-bead thermistors, temperatures of 125°C can be measured with an accuracy of 0.01°C. Long-term drift data indicate that stabilities better than 0.003°C/year, when cycled between 20 and 125°C, can be achieved.

The accuracy of the measurement of temperature with a thermistor depends on the instrumentation system employed and the method used to account for the nonlinear response. Both Wheatstone bridge and potentiometer circuits can be used to determine the resistance changes in a thermistor resulting from a change in temperature.

Use of a thermistor as the active element in a Wheatstone bridge is shown schematically in Fig. 8.12$a$. If the Wheatstone bridge is initially balanced ($R_T R_3 = R_2 R_4$) and if resistors $R_2$, $R_3$, and $R_4$ are fixed-value precision resistors, then the output voltage $\Delta E_o$ produced by a temperature-induced change in resistance $\Delta R_T$ in the thermistor is given by Eq. (4.19) as

$$\frac{\Delta E_o}{E_i} = \frac{\Delta R_T R_3}{(R_T + \Delta R_T + R_2)(R_3 + R_4)} \tag{8.6}$$

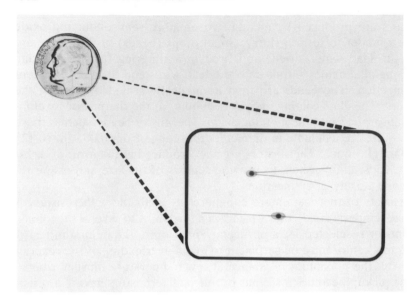

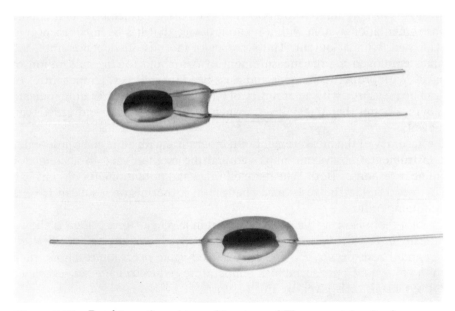

***Figure 8.11*** Bead-type thermistors. (Courtesy of Thermometrics, Inc.)

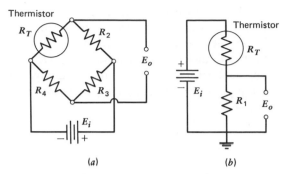

*(a)*                         *(b)*

**Figure 8.12**   Constant-voltage Wheatstone bridge and potentiometer circuits used with thermistors. (*a*) Wheatstone-bridge circuit. (*b*) Potentiometer circuit.

For the common case where $R_2 = R_3$ and $R_T = R_4$, Eq. (8.6) reduces to

$$\frac{\Delta E_o}{E_i} = \frac{\Delta R_T/R_T}{(1 + \Delta R_T/R_T + R_2/R_T)(1 + R_T/R_2)}$$

$$= \frac{\Delta R_T/R_T}{2 + R_T/R_2 + R_2/R_T + \Delta R_T/R_T + \Delta R_T/R_2} \qquad (8.7)$$

For thermistors, the terms $\Delta R_T/R_T$ and $\Delta R_T/R_2$ in the denominator of Eq. (8.7) are not small with respect to the other terms; therefore, they cannot be neglected to simplify the solution of the equation for $\Delta R_T/R_T$. Consequently, the output from the Wheatstone bridge is also a nonlinear function of the change in thermistor resistance, and the determination of $\Delta R_T/R_T$ from the bridge output voltage is not trivial. For the special case of an equal-arm bridge ($R_T = R_2 = R_3 = R_4$), Eq. (8.7) reduces to a simpler form and the change in thermistor resistance can be expressed in terms of the bridge output voltage $\Delta E_o$ as

$$\frac{\Delta R_T}{R_T} = \frac{4\Delta E_o/E_i}{1 - 2\Delta E_o/E_i} \qquad (8.8)$$

The thermistor resistance $R_T^*$ at any temperature $T$ is then given by the simple expression

$$R_T^* = R_T + \Delta R_T = R_T(1 + \Delta R_T/R_T) \qquad (8.9)$$

Substituting Eq. (8.8) into Eq. (8.9) yields

$$R_T^* = R_T \left( \frac{1 + 2\Delta E_o/E_i}{1 - 2\Delta E_o/E_i} \right) \qquad (8.10)$$

The value of $R_T^*$ obtained from Eq. (8.10) is converted to temperature by using tables that list $T$ as a function of $R_T^*$ for the specific thermistor being used. This procedure accounts for nonlinearities in both the bridge and the thermistor.

If the constant-voltage supply to the equal-arm Wheatstone bridge ($R_2$, $R_3$, and $R_4$ are fixed-value resistors) is replaced with a constant-current supply, the output voltage $\Delta E_o$ produced by a temperature-induced change in resistance $\Delta R_T$ in the thermistor is given by Eq. (4.33) as

$$\frac{\Delta E_o}{I} = \frac{R_T^2}{4R_T + \Delta R_T} \frac{\Delta R_T}{R_T} \tag{8.11}$$

Also, since the voltage drop $E_T$ across the thermistor equals $IR_T$, Eq. (8.11) can be expressed in terms of the voltage drop $E_T$ as

$$\frac{\Delta E_o}{E_T} = \frac{1}{4 + \Delta R_T/R_T} \frac{\Delta R_T}{R_T} \tag{8.12}$$

Solving Eq. (8.12) for $\Delta R_T/R_T$ yields

$$\frac{\Delta R_T}{R_T} = \frac{4\Delta E_o/E_T}{1 - \Delta E_o/E_T} \tag{8.13}$$

Substituting Eq. (8.13) into Eq. (8.9) gives the resistance $R_T^*$ as

$$R_T^* = R_T \left( \frac{1 + 3\Delta E_o/E_T}{1 - \Delta E_o/E_T} \right) \tag{8.14}$$

A comparison of Eqs. (8.10) and (8.14) shows that the nonlinearity of the Wheatstone bridge has been improved by using a constant-current supply; however, the nonlinearity remains significant and the resistance $R_T^*$ must be converted to temperature by using the appropriate conversion table for the thermistor.

Potentiometer circuits can also be employed to convert the resistance change $\Delta R_T$ of the thermistor to a voltage change $\Delta E_o$. If the thermistor is placed in position $R_2$ of the potentiometer circuit, as shown in Fig. 8.12b, Eq. (4.2) indicates that

$$\frac{\Delta E_o}{E_i} = - \frac{\dfrac{r}{(1 + r)^2} (\Delta R_T/R_T)}{1 + \dfrac{r}{1 + r} (\Delta R_T/R_T)}$$

$$= - \frac{\dfrac{r}{1 + r} (\Delta R_T/R_T)}{1 + r(1 + \Delta R_T/R_T)} \tag{8.15}$$

where $r = R_T/R_1$. Equation (8.15) again shows the presence of nonlinear terms that may be significant. For the special case of $r = 1$, Eq. (8.15) reduces to

$$\frac{\Delta R_T}{R_T} = -\frac{4\Delta E_o/E_i}{1 + 2\Delta E_o/E_i} \tag{8.16}$$

The resistance of the thermistor $R_T^*$ is obtained by substituting Eq. (8.16) into Eq. (8.9). Thus

$$R_T^* = R_T \left(\frac{1 - 2\Delta E_o/E_i}{1 + 2\Delta E_o/E_i}\right) \tag{8.17}$$

A comparison of Eq. (8.17) for the potentiometer circuit with Eq. (8.10) for the Wheatstone-bridge circuit shows that the equations are identical except for the signs of the $\Delta E_o/E_i$ terms. Since $\Delta E_o/E_i$ can be either positive or negative, depending on the direction (increase or decrease) of the temperature change, the two equations are identical in form; therefore, one circuit has no advantage over the other.

A simple circuit for determining thermistor resistance $R_T^*$ is shown in Fig. 8.13. This circuit employs a constant-current power supply directly across the thermistor. Since the output voltage $E_o$ equals $IR_T$, the voltage change $\Delta E_o/E_o$ is given by the simple expression

$$\frac{\Delta E_o}{E_o} = \frac{\Delta R_T}{R_T} \tag{8.18}$$

Substituting Eq. (8.18) into Eq. (8.9) yields

$$R_T^* = R_T(1 + \Delta E_o/E_o) \tag{8.19}$$

Thus, the simple circuit shown in Fig. 8.13 exhibits a linear relationship between output voltage and sensor resistance; therefore, the simple circuit is superior to the more traditional Wheatstone-bridge and potentiometer circuits for temperature measurements with thermistors if the accuracy obtained with the DVM is sufficient and if regulation of the constant-current power supply is adequate. Many modified bridge and potentiometer circuits have been developed to linearize the output of the thermistor. Some of these circuits are covered in the exercises at the end of the chapter.

When thermistors are used to measure temperature, errors resulting from lead-wire effect are usually small enough to be neglected even for relatively long

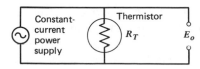

**Figure 8.13** Constant-current potentiometer circuit used with thermistors.

lead wires. The sensitivity of a thermistor is high; therefore, the change in resistance $\Delta R_T$ resulting from a temperature change is much greater than the small change in resistance of the lead wires due to the temperature variation. Also, the resistance of the thermistor is very large relative to the resistance of the lead wires ($R_T/R_L \approx 1000$); consequently, any reduction in sensitivity of the sensor due to lead-wire resistance is negligible.

Errors may occur as a result of self-heating since the power ($P = I^2R_T$) dissipated in the thermistor will heat it above its ambient temperature. Recommended practice limits the current flow through the thermistor to a value such that the temperature rise due to the $I^2R_T$ power dissipation is smaller than the precision to which the temperature is to be measured. A typical thermistor with $R_T = 5000 \ \Omega$ is capable of dissipating 1 mW/°C above the ambient temperature. Thus, if the temperature is to be determined with an accuracy of 0.5°C, the power to be dissipated should be limited to less than 0.5 mW. This limitation establishes a maximum value for the current $I$ at

$$I = \sqrt{P/R_T} = \sqrt{0.0005/5000} = 316(10^{-6})A = 316 \ \mu A$$

In this example, it would be prudent to limit the current $I$ to approximately 100 $\mu A$. Adequate response can be obtained, even at these low currents, because the sensitivity of a thermistor is so very high. Precise measurements of $\Delta E_o$ can be made easily with a digital millivoltmeter.

## 8.3  EXPANSION THERMOMETERS

The expansion (or contraction) per unit length $\Delta l/l$ of a material experiencing an increase (or decrease) in temperature $\Delta T$ is given by the expression

$$\Delta l/l = \alpha \Delta T \tag{8.20}$$

where $\alpha$ is the thermal coefficient of expansion of the material. Since $\alpha$ is very small for metals (it ranges from 1 to 26 $\mu m/m \cdot °C$), direct measurement of $\Delta l$ to infer $\Delta T$ is difficult. In order to circumvent this sensitivity problem, devices that utilize the differential expansion between two different materials have been developed to measure temperature change $\Delta T$. Devices of this type include the familiar liquid-in-glass thermometer, bonded bimetallic strips of two metals, and pressure thermometers. Each of these devices is described in the following subsections.

### Liquid-in-Glass Thermometers

The well-known and widely used glass thermometer provides a simple, convenient, and inexpensive means for measuring temperature in many applications. The thermometer consists of an indexed glass capillary tube with a bulb at one end to hold a supply of fluid. The fluids commonly used are mercury and alcohol. Mercury can be used for temperatures between $-39°C$ ($-38°F$) and $538°C$ ($1000°F$). When a lower temperature limit is needed, alcohol permits measurements at temperatures as low as $-62°C$ ($-80°F$); pentane can be used for measurements as low as $-218°C$ ($-360°F$).

Glass thermometers are designed for either partial or full immersion. As the name implies, full-immersion thermometers are calibrated to read correctly when the thermometer is completely immersed in the fluid whose temperature is being measured. Partial-immersion types are marked and should be immersed only to the depth indicated by the immersion mark.

The accuracy that can be achieved with a glass thermometer depends upon the quality and range of the particular thermometer being used. Also, strict attention must be paid to immersion requirements, since corrections must be made when these requirements are not satisfied. With a good-quality, full-immersion thermometer having a range from $0°C$ to $100°C$, the temperature can be determined to within $\pm 0.1°C$.

Glass thermometers provide a low-cost means for measuring temperatures with reasonable accuracy over the range from about $-200°C$ to $500°C$. Since the readout is visual, they are not used in automatic data systems or in automatically controlled processes in industry.

## Bimetallic Thermometers

The sensing element in a bimetallic thermometer consists of a bonded composite of two materials, as illustrated in Fig. 8.14. Material $A$ is usually a copper-based alloy with a large coefficient of thermal expansion, while material $B$ is usually Invar (a nickel steel), which has a very small coefficient of thermal expansion. When the bonded bimetallic strip is subjected to a temperature change, the differential expansion causes it to bend into a circular arc. The radius of curvature $r_a$ of the arc is given by the expression

$$r_a = \frac{[3(1 + \theta)^2 + (1 + \theta e)(\theta^2 + 1/\theta e)]t}{6(\alpha_A - \alpha_B)(1 + \theta)^2 \Delta T} \tag{8.21}$$

**where**   $\theta = t_B/t_A$   is the thickness ratio.
         $e = E_B/E_A$   is the modulus ratio.

Bimetallic elements in the form of cantilever beams, spirals, washers, and helixes are inexpensive and deform significantly with relatively small changes in temperature; therefore, they are used in a wide variety of temperature sensing and temperature control devices. In thermostats, they are used to control temper-

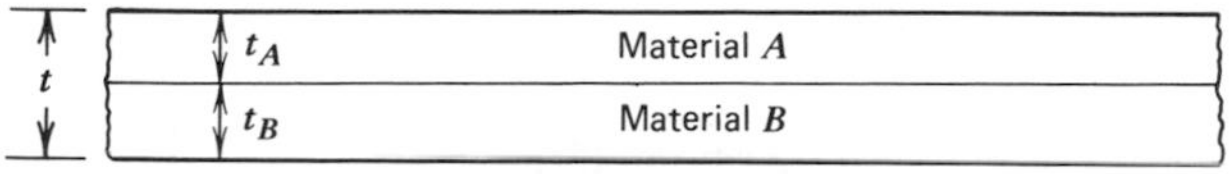

*Figure 8.14*   Beam-type bimetalic element.

ature by switching the heat source on and off. As overload switches in electrical equipment, they are activated by excessive current flows and turn off the equipment. Finally, they are often used in conjunction with a linear-displacement sensor such as a potentiometer or linear variable-differential transformer (LVDT) to provide a temperature indicating instrument.

The accuracy of bimetallic thermometers varies; therefore, they are usually used in control applications where low cost is more important than accuracy. For those applications where accuracy is important, high-quality bimetallic thermometers are available with guaranteed accuracies of about 1 percent.

## Pressure Thermometers

A typical pressure thermometer, illustrated schematically in Fig. 8.15, consists of a bulb filled with a liquid such as mercury or xylene, a capillary tube, and a pressure sensor. When the bulb is subjected to a temperature change, both the bulb and the fluid experience a volume change. The differential volume change $\Delta V_d$ is proportional to the temperature change $\Delta T$. In a closed system completely filled with liquid under an initial pressure, the pressure changes in response to the differential volume change. The pressure is transmitted through the capillary tube to a pressure measuring transducer, such as a bourdon tube, bellows, or diaphragm. Movement of the bourdon tube or bellows can be transmitted through a suitable linkage system to a pointer whose position relative to a calibrated scale gives an indication of the temperature. The bourdon tube or bellows can also be used with a potentiometer or linear variable-differential transformer (LVDT) to construct a temperature measuring and recording instrument. Similarly, an electrical resistance strain gage on a diaphragm provides the sensor for a temperature measuring and recording instrument.

The dynamic response of a pressure thermometer is poor because of the thermal lag associated with the mass of fluid in the bulb; therefore, such instruments cannot be used to measure temperatures in fluids undergoing rapid changes in temperature. The pressure thermometer can, however, provide a

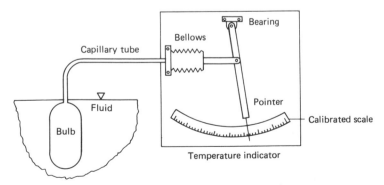

*Figure 8.15*   Schematic illustration of a pressure thermometer.

simple, low-cost, reliable, and trouble-free method of measuring temperature in systems undergoing relatively slow changes in temperature.

Pressure thermometers filled with mercury cover the range from $-39°C$ to $538°C$ ($-38°F$ to $1000°F$), while those filled with xylene are used for the range from $-100°C$ to $400°C$ ($-150°F$ to $750°F$). The response is linear over a large portion of the range. Capillary tubes as long as 60 m (200 ft) have been used sucessfully for remote measurements. Temperature variations along the capillary tube and at the pressure sensing device require compensation. A common compensation scheme utilizes an auxiliary pressure sensor and capillary tube. Bimetal elements can also be used to effect partial compensation. The accuracy of pressure thermometers under the best of conditions is approximately $\pm 0.5$ percent of the scale range.

## 8.4 THERMOCOUPLES

A thermocouple is a very simple temperature sensor, consisting essentially of two dissimilar wires in thermal contact, as indicated in Fig. 8.16a. The operation of a thermocouple is based on the Seebeck effect, which results in the generation of a thermoelectric potential when two dissimilar metals are joined together to form a junction. The thermoelectric effect is produced by diffusion of electrons across the interface between the two materials. The electric potential of the material accepting electrons becomes negative at the interface, while the potential of the material providing the electrons becomes positive. Thus, an electric field is established by the flow of electrons across the interface. When this electric field becomes sufficient to balance the diffusion forces, a state of equilibrium with respect to electron migration is established. Since the magnitude of the diffusion force is controlled by the temperature of the thermocouple junction, the electric potential developed at the junction provides a measure of the temperature.

The electric potential is usually measured by introducing a second junction in an electric circuit, as shown in Fig. 8.16b, and measuring the voltage $E_o$ across one leg with a suitable voltmeter. The voltage $E_o$ across terminals M-N can be

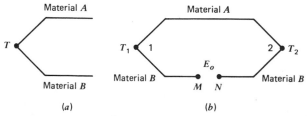

(a)                                    (b)

**Figure 8.16** Thermocouple sensor and circuit for measuring the temperature difference $T_1 - T_2$. (a) Single junction. (b) Dual junction.

represented approximately by an empirical equation having the form

$$E_o = C_1(T_1 - T_2) + C_2(T_1^2 - T_2^2) \tag{8.22}$$

**where**    $C_1$ and $C_2$ are thermoelectric constants that depend on the materials used to form the junctions.

$T_1$ and $T_2$ are junction temperatures.

In practice, junction 1 is used to sense an unknown temperature $T_1$, while junction 2 is maintained at a known reference temperature $T_2$. Since the reference temperature $T_2$ is known, it is possible to determine the unknown temperature $T_1$ by measuring the voltage $E_o$. It is clear from Eq. (8.22) that the response of a thermocouple is a nonlinear function of the temperature. Also, experience has shown that Eq. (8.22) is not a sufficiently accurate representation of the voltage–temperature relationship to be used with confidence when precise measurements of temperature are required. For this reason, thermocouples are calibrated over the complete range of temperature for which they are useful and tables are obtained which can be used to relate temperature $T_1$ to the thermoelectric voltage $E_o$. Thermocouple tables are presented in Appendix A for iron-constantan (Table A.2), Chromel-Alumel (Table A.3), Chromel-constantan (Table A.4), and copper-constantan (Table A.5) thermocouples. It is important to note that the reference temperature is $T_2 = 0°C$ (32°F) in these tables.

## Reference Junction Temperature

Since a thermocouple circuit responds to a temperature difference $(T_1 - T_2)$, it is essential that the reference junction be maintained at a constant and accurately known temperature $T_2$. Four common methods are used to maintain the reference temperature.

The simplest and most popular technique utilizes an ice and water bath, as illustrated in Fig. 8.17. The reference junction is immersed in a mixture of ice and water in a thermos bottle that is capped to prevent heat loss and temperature

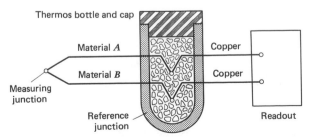

**Figure 8.17**    The ice bath method for maintaining a reference temperature at 0°C (32°F).

gradients. Water (sufficient only to fill the voids) must be removed and ice must be replaced periodically to maintain a constant reference temperture. Such an ice bath can maintain the water temperature (and thus the reference tempera-ture) to within 0.1°C (0.2°F) of the freezing point of water.

A very-high-quality reference temperature source is available that employs thermoelectric refrigeration (Peltier cooling effect). Thermocouple wells in this unit contain air-saturated water that is maintained at precisely 0°C (32°F). The outer walls of the wells are cooled by the thermoelectric refrigeration elements until freezing of the water in the wells begins. The increase in volume of the water as it begins to freeze on the walls of the wells expands a bellows that contacts a microswitch and deactivates the refrigeration elements. The cyclic freezing and thawing of the ice on the walls of the wells accurately maintains the temperature of the wells at 0°C (32°F). This automatic and precise control of temperature can be maintained over extended periods of time.

The electrical-bridge method, illustrated in Fig. 8.18, is usually used with potentiometric, strip-chart, recording devices to provide automatic compensa-tion for reference junctions that are free to follow ambient temperature con-ditions. This method incorporates a Wheatstone bridge with a resistance tem-perature detector (RTD) as the active element into the thermocouple circuit. The RTD and the reference junctions of the thermocouple are mounted on a reference block that is free to follow the ambient temperature. As the ambient temperature of the reference block varies, the RTD changes resistance. The bridge is designed to produce an output voltage that is equal but opposite to the voltage developed in the thermocouple circuit as a result of the changes in temperature $T_2$ from 32°F (0°C). Thus, the electrical-bridge method automati-cally compensates for changing ambient conditions. This method is widely used with potentiometric recording devices that are used to display one or more

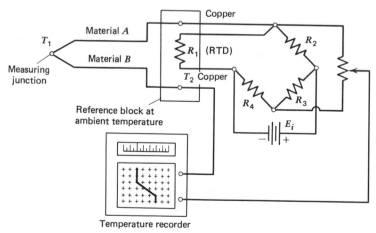

**Figure 8.18** The electrical-bridge method of compensation for changes in the reference temperature.

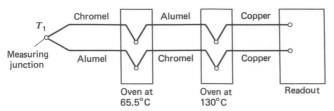

*Figure 8.19* Double oven method for reference junction control.

temperatures over long periods of time when it is obviously not practical to maintain the simple ice bath.

A different type of reference temperature control is obtained by using an oven that maintains a fixed temperature higher than any expected ambient temperature. This system is practical when heating is more easily obtained than cooling; however, the thermoelectric voltage–temperature tables must be corrected for the higher reference junction temperature. A popular technique, which eliminates the need for reference junction temperature corrections, employs two ovens at different temperatures, as illustrated in Fig. 8.19, to simulate a reference temperature of 0°C (32°F). In the example shown in Fig. 8.19, each of the junctions (one Chromel–Alumel and one Alumel–Chromel) in the first oven produces a voltage of 2.66 mV at an oven temperature of 65.5°C (150°F). This total voltage of 5.32 mV is canceled by the double junction of Alumel–copper and copper–Chromel in the second oven at a temperature of 130°C (266°F). The net effect of the four junctions in the two ovens is to produce the thermoelectric equivalent of a single reference junction at a temperature of 0°C (32°F).

## Thermoelectric Behavior

The thermoelectric behavior of a thermocouple is based on a combination of the Seebeck effect, the Thompson effect, and the Peltier effect. A complete explanation of the contribution of each of these effects to thermocouple behavior requires a thorough understanding of thermodynamics and several aspects of physics. No attempt is made here to describe in detail why a thermocouple circuit produces a given voltage. Instead, emphasis is placed on the practical aspects of employing thermocouples to measure a wide range of temperatures with good accuracy and at low cost.

The practical use of thermocouples is based on six operating principles that are stated below and illustrated in Fig. 8.20.

## PRINCIPLES OF THERMOCOUPLE BEHAVIOR

1. A thermocouple circuit must contain at least two dissimilar materials and at least two junctions, as illustrated in Fig. 8.20*a*.

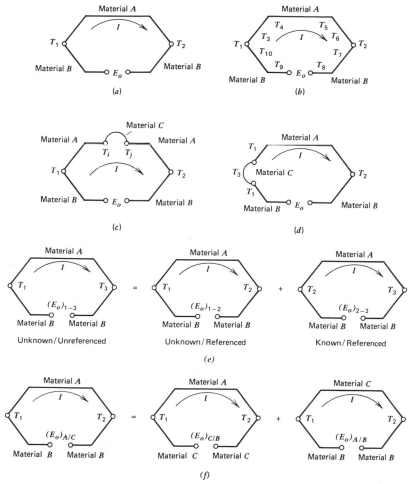

***Figure 8.20*** Typical situations encountered during use of thermocouples. (*a*) Basic thermocouple circuit. (*b*) Output depends on ($T_1 - T_2$) only. (*c*) Intermediate metal in circuit. (*d*) Intermediate metal in junction. (*e*) Voltage addition from identical thermocouples at different temperatures. (*f*) Voltage addition from different thermocouples at identical temperatures.

2.  The output voltage $E_o$ of a thermocouple circuit depends only on the difference between junction temperatures ($T_1 - T_2$) and is independent of the temperatures elsewhere in the circuit (see Fig. 8.20*b*).

3.  If a third metal $C$ is inserted into either leg (*A* or *B*) of a thermocouple circuit, the output voltage $E_o$ is not affected, provided the two new junctions $A/C$ and $C/A$, as shown in Fig. 8.20*c*, are maintained at the same temperature (for example, temperature $T_3$).

4.  The insertion of an intermediate metal $C$ into junction 1 does not affect the output voltage $E_o$, provided the two junctions formed by the insertion ($A/C$ and $C/B$) are maintained at the same temperature $T_1$. (See Fig. 8.20$d$.)

5.  A thermocouple circuit with temperatures $T_1$ and $T_2$ produces an output voltage $(E_o)_{1-2} = f(T_1 - T_2)$, and one exposed to temperatures $T_2$ and $T_3$ produces an output voltage $(E_o)_{2-3} = f(T_2 - T_3)$. If the same circuit is exposed to temperatures $T_1$ and $T_3$, the output voltage $(E_o)_{1-3} = f(T_1 - T_3) = (E_o)_{1-2} + (E_o)_{2-3}$. (See Fig. 8.20$e$.)

6.  A thermocouple circuit fabricated from materials $A$ and $C$ generates an output voltage $(E_o)_{A/C}$ when exposed to temperatures $T_1$ and $T_2$, while a similar circuit fabricated from materials $C$ and $B$ generates an output voltage $(E_o)_{C/B}$. Furthermore, a thermocouple fabricated from materials $A$ and $B$ generates an output voltage $(E_o)_{A/B} = (E_o)_{A/C} + (E_o)_{C/B}$ (See Fig. 8.20$f$.)

The six principles of thermoelectric behavior are important since they provide the basis for the design, circuitry, and application of thermocouples to temperature measurements.

The first principle formalizes the experimental observation that a thermocouple circuit must be fabricated with two different materials in such a way that two junctions are formed. The output voltage $E_o$ (see Fig. 8.20$a$) has been observed to be a nonlinear function of the difference in temperature $(T_1 - T_2)$ at these two junctions. For clockwise current flow as illustrated in Fig. 8.20$a$, the output voltage $E_o$ can be expressed as

$$E_o = e_{B/A} T_1 + e_{A/B} T_2 \qquad \text{(a)}$$

**where**   $e_{B/A}$ is the junction potential per unit temperature at a junction as the current flows from material $B$ to material $A$.

$e_{A/B}$ is the junction potential per unit temperature at a junction as the current flows from material $A$ to material $B$.

Since $e_{B/A} = -e_{A/B}$, Eq. (a) can be written in its more familiar form as

$$E_o = e_{B/A}(T_1 - T_2) \qquad \text{(b)}$$

Experiments indicate that the relationship between $E_o$ and temperature difference $(T_1 - T_2)$, as expressed by Eq. (b), is nonlinear and dependent upon the two materials used to fabricate the thermocouple.

Thermocouple calibration tables such as those presented in Tables 8.4 to 8.7 are used to relate temperature difference $(T_1 - T_2)$ to a measured output voltage $E_o$. Since an unknown temperature $T_1$ is being measured, the reference temperature $T_2$ must be known. The calibration information presented in Tables 8.4 to 8.7 is based on a reference temperature $T_2 = 0°C$ ($32°F$). If the reference temperature $T_2$ is not 0°C (but rather some other known value, such as 100°C), it is still possible to determine $T_1$, but the procedure involves application of the fifth principle of thermoelectric behavior.

The second principle indicates that the voltage output $E_o$ from a thermocouple circuit is not influenced by the temperature distribution along the conductors, except at points where connections are made to form junctions (see Fig. 8.20b). This principle provides assurance that the output voltage $E_o$ of the thermocouple circuit will be independent of the length of the lead wires and the temperature distribution along their length.

The third principle deals with insertion of an intermediate conductor (such as copper lead wires or a voltage measuring instrument) into one of the legs of a thermocouple circuit (see Fig. 8.20c). The effect of this insertion of material $C$ into the $A$-$B$-type thermocouple can be determined by writing the equation for the output voltage $E_o$ as

$$E_o = e_{B/A}T_1 + e_{A/C}T_i + e_{C/A}T_j + e_{A/B}T_2 \tag{c}$$

Since $e_{B/A} = -e_{A/B}$ and $e_{A/C} = -e_{C/A}$, Eq. (c) can be written as

$$E_o = e_{B/A}(T_1 - T_2) + e_{A/C}(T_i - T_j) \tag{d}$$

Equation (d) indicates that the effect of the $A/C$ junctions can be reduced to zero if $T_i = T_j$. A similar analysis will show that the effect of $B/C$ junctions can be reduced to zero if $T_i = T_j$ when the third metal $C$ is inserted in the $B$ leg of the thermocouple. This principle verifies that insertion of a third material $C$ into the circuit will have no effect on the output voltage $E_o$, provided the junctions formed in either leg $A$ or leg $B$ are maintained at the same temperature $T_i = T_j = T_3$.

The fourth principle deals with insertion of an intermediate metal into a junction during fabrication or utilization of a thermocouple. Such a situation can occur when junctions are formed by twisting the two thermocouple materials $A$ and $B$ together and soldering or brazing the connection with an intermediate metal $C$ (see Fig. 8.20d) or when the thermocouple junction is attached to the surface or embedded into a specimen. The influence of the presence of the intermediate metal in the junction can be evaluated by considering the expression for output voltage $E_o$ which can be written as

$$E_o = e_{B/C}T_1 + e_{C/A}T_1 + e_{A/B}T_2 \tag{e}$$

Since $e_{C/A} = e_{C/B} + e_{B/A}$, Eq. (e) reduces to

$$E_o = e_{B/A}(T_1 - T_2) \tag{f}$$

Equation (f) verifies that the output voltage $E_o$ is not affected by the presence of a third material $C$, which may be inserted during fabrication of the thermo-couple, if the two junctions $B/C$ and $C/A$ are at the same temperature.

The fifth principle deals with the relationship between output voltage $E_o$ and the reference junction temperature. As mentioned previously, Tables A.2 to A.5 are based on a reference temperature of 0°C (32°F). In some instances it may be more convenient to use a different reference temperature (say boiling water at 100°C). The effect of this different reference temperature can be accounted for by using the equivalent thermocouple system illustrated in Fig. 8.20e. The outputs from such a system can be expressed in equation form as

$$(E_o)_{1-3} = f(T_1 - T_3) = (E_o)_{1-2} + (E_o)_{2-3} \tag{8.23}$$

Use of Eq. (8.23) for the case of an arbitrary reference temperature $T_3$ can be illustrated by considering the example of an iron–constantan thermocouple exposed to an unknown temperature $T_1$ while the reference temperature $T_3$ is maintained at 100°C. Assume that an output voltage $(E_o)_{1-3} = 28.388$ mV is recorded under these conditions. The voltage $(E_o)_{2-3}$ of Eq. (8.23) can be determined from Table A.2 since it is known that $T_2 = 0°C$ and $T_3 = 100°C$. Thus, $(E_o)_{2-3} = -(E_o)_{3-2} = -5.268$ mV. Solving Eq. (8.23) for $(E_o)_{1-2}$ yields

$$(E_o)_{1-2} = (E_o)_{1-3} - (E_o)_{2-3}$$

$$= 27.388 - (-5.268) = 32.656 \text{ mV}$$

Table A.2 indicates that an output voltage $(e_o)_{1-2} = 32.656$ mV would be produced by a temperature $T_1 = 592.5°C$. The same procedure can be used to correct for any arbitrary (but known) reference temperature.

The sixth principle pertains to voltage addition for thermocouple circuits fabricated from different materials as illustrated in Fig. 8.20f. Output voltage $(E_o)_{A/B}$ can be expressed in equation form as

$$(E_o)_{A/B} = (E_o)_{A/C} + (E_o)_{C/B} \tag{8.24}$$

As a result of this principle, calibration tables can be constructed for any pair of materials by knowing the calibrations for the individual materials when they are paired with a standard thermocouple material such as platinum. For example, materials $A$ and $B$, when paired with the standard material $C$, would provide $(E_o)_{A/C}$ and $(E_o)_{C/B} = -(E_o)_{B/C}$. The calibration for a junction formed by using materials $A$ and $B$ could then be determined by using Eq. (8.24). Use of this principle eliminates the need to calibrate all possible combinations of materials to establish their usefulness.

## Thermoelectric Materials

The thermoelectric effect occurs whenever a thermocouple circuit is fabricated from any two dissimilar metals; therefore, the number of materials suitable for use in thermocouples is very large. In most cases, materials are selected to

1. Maximize sensitivity over the range of operation.
2. Provide long-term stability at the upper temperature levels.
3. Ensure compatibility with available instrumentation.
4. Minimize cost.

The sensitivity of a number of materials in combination with platinum is presented in Table 8.4.

The results from Table 8.4 can be used to determine the sensitivity $S$ at 0°C (32°F) of a thermocouple fabricated from any two materials listed in the table. For instance, the sensitivity of a Chromel–Alumel thermocouple is

$$S_{\text{Chromel/Alumel}} = 25.8 - (-13.6) = 39.4 \ \mu V/°C$$

It is important to recall that the sensitivity $S$ of a thermocouple is not constant, since the output voltage $E_o$ is a nonlinear function of the difference in junction temperatures $(T_1 - T_2)$. Sensitivity $S$ as a function of temperature for the six most frequently used material pairs is listed in Table 8.5.

**TABLE 8.4**  Thermoelectric Sensitivity $S$ of Several Materials in Combination with Platinum at 0°C (32°F)

| Material | Sensitivity $S$ | | Material | Sensitivity $S$ | |
|---|---|---|---|---|---|
|  | $\mu V/°C$ | $\mu V/°F$ |  | $\mu V/°C$ | $\mu V/°F$ |
| Bismuth | − 72 | − 40 | Copper | +  6.5 | +  3.6 |
| Constantan | − 35 | − 19.4 | Gold | +  6.5 | +  3.6 |
| Nickel | − 15 | −  8.3 | Tungsten | +  7.5 | +  4.2 |
| Alumel | − 13.6 | −  7.6 | Iron | + 18.5 | + 10.3 |
| Platinum | 0 | 0 | Chromel | + 25.8 | + 14.3 |
| Mercury | + 0.6 | + 0.3 | Germanium | +300 | +167 |
| Carbon | + 3 | + 1.7 | Silicon | +440 | +244 |
| Aluminum | + 3.5 | + 1.9 | Tellurium | +500 | +278 |
| Lead | + 4 | + 2.2 | Selenium | +900 | +500 |
| Silver | + 6.5 | + 3.6 |  |  |  |

**TABLE 8.5**[a]  Sensitivity as a Function of Temperature for Six Different Types of Thermocouples ($\mu$V/°C)

| Temperature (°C) | E[b] | J[c] | K[d] | R[e] | S[f] | T[g] |
|---|---|---|---|---|---|---|
| −200 | 25.1 | 21.9 | 15.3 | — | — | 15.7 |
| −100 | 45.2 | 41.1 | 30.5 | — | — | 28.4 |
| 0 | 58.7 | 50.4 | 39.5 | 5.3 | 5.4 | 38.7 |
| 100 | 67.5 | 54.3 | 41.4 | 7.5 | 7.3 | 46.8 |
| 200 | 74.0 | 55.5 | 40.0 | 8.8 | 8.5 | 53.1 |
| 300 | 77.9 | 55.4 | 41.4 | 9.7 | 9.1 | 58.1 |
| 400 | 80.0 | 55.1 | 42.2 | 10.4 | 9.6 | 61.8 |
| 500 | 80.9 | 56.0 | 42.6 | 10.9 | 9.9 | — |
| 600 | 80.7 | 58.5 | 42.5 | 11.3 | 10.2 | — |
| 700 | 79.8 | 62.2 | 41.9 | 11.8 | 10.5 | — |
| 800 | 78.4 | — | 41.0 | 12.3 | 10.9 | — |
| 900 | 76.7 | — | 40.0 | 12.8 | 11.2 | — |
| 1000 | 74.9 | — | 38.9 | 13.2 | 11.5 | — |

[a] From NBS Monograph 125, March 1974.
[b] Chromel–constantan thermocouple (E type).
[c] Iron–constantan thermocouple (J type).
[d] Chromel–Alumel thermocouple (K type).
[e] Platinum 13 percent rhodium–platinum thermocouple (R type).
[f] Platinum 10 percent rhodium–platinum thermocouple (S type).
[g] Copper–constantan thermocouple (T type).

The voltage output $E_o$ as a function of temperature for several popular types of thermocouples is shown in Fig. 8.21. This graphical display shows that the E-type (Chromel–constantan) thermocouple generates the largest output voltage at a given temperature; unfortunately, it has an upper temperature limit of only 1000°C (1832°F). The upper limit of the temperature range is increased (but the sensitivity is decreased) to 1260°C (2300°F) with a K-type (Chromel–Alumel) thermocouple, to 1538°C (2800°F) with an S-type (platinum 10 percent rhodium–platinum) thermocouple, and to 2800°C (5072°F) with a G-type (tungsten–tungsten 26 percent rhenium) thermocouple. The operating range of temperature, together with the span of output voltages for most of the popular types of thermocouples, are listed in Table 8.6.

## Fabrication and Installation Procedures

Proper installation of a thermocouple may involve fabrication of the junction, selection of the lead wires (diameter and insulation), and placement of the thermocouple on the surface of the component or in the fluid at the point where

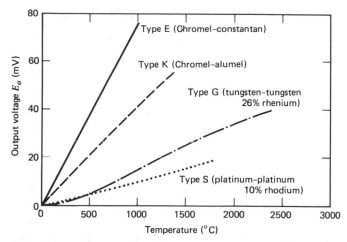

***Figure 8.21*** Output voltage $E_o$ versus temperature $T_1$ with a reference temperature $T_2$ = 0°C for several types of thermocouples.

**TABLE 8.6**  Operating Range and Voltage Span for Several Different Types of Thermocouples

| Type of Thermocouple | Temperature Range | | Voltage Span (mV) |
|---|---|---|---|
| | °C | °F | |
| Copper–constantan | − 185 to 400 | − 300 to 750 | − 5.284 to 20.805 |
| Iron–constantan | − 185 to 870 | − 300 to 1600 | − 7.52  to 50.05 |
| Chromel–Alumel | − 185 to 1260 | − 300 to 2300 | − 5.51  to 51.05 |
| Chromel–constantan | 0 to 980 | 32 to 1800 | 0      to 75.12 |
| Platinum 10% rhodium–platinum | 0 to 1535 | 32 to 2800 | 0      to 15.979 |
| Platinum 13% rhodium–platinum | 0 to 1590 | 32 to 2900 | 0      to 18.636 |
| Platinum 30% rhodium–platinum 6% rhodium | 38 to 1800 | 100 to 3270 | 0.007 to 13.499 |
| Platinel 1813–Platinel 1503 | 0 to 1300 | 32 to 2372 | 0      to 51.1 |
| Iridium 60% rhodium 40% iridium | 1400 to 1830 | 2552 to 3326 | 7.30  to  9.55 |
| Tungsten 3% rhenium–tungsten 25% rhenium | 10 to 2200 | 50 to 4000 | 0.064 to 29.47 |
| Tungsten–tungsten 26% rhenium | 16 to 2800 | 60 to 5072 | 0.042 to 43.25 |
| Tungsten 5% rhenium–tungsten 26% rhenium | 0 to 2760 | 32 to 5000 | 0      to 38.45 |

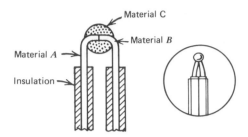

*Figure 8.22* Fabrication details for a thermocouple junction.

the temperature is to be measured. The recommended procedure for forming a thermocouple junction, illustrated in Fig. 8.22, consists of butting the two wires together and fastening by welding, brazing, or soldering to form a small bead of material around the junction. The wire diameter used in the fabrication of the thermocouple depends upon the dynamic response required of the thermocouple and the degree of hositility of the environment in which the thermocouple must operate. When temperature fluctuations are rapid, the wire diameter must be small and any protective sheathing must be thin so as to minimize thermal lag. Wire diameters as small as 0.125 mm (0.005 in.) are routinely employed when response time becomes an important factor in the temperature measurement. On the other hand, thermocouples are often required to operate over long periods of time at high temperatures in either reducing or oxidizing atmospheres. In these applications, heavy-gage wires are used with relatively large-diameter junctions so that part of the junction can be sacrificed to extend the period of stable operation.

The material used to provide insulation for the lead wires is determined by the maximum temperature to which the thermocouple will be subjected. Types of insulation, together with their temperature limits, are listed in Table 8.7. For higher-temperature applications, thermocouple wire is available with a ceramic insulation swaged into a metal sheath. For extremely high-temperature applications (2315°C or 4200°F), ceramic (Beryllia) tubes are often used to insulate the wires.

**TABLE 8.7**  Characteristics of Thermocouple-Wire Insulation

| Material | Abrasion Resistance | Flexibility | Temperature (°C) | | Temperature (°F) | |
|---|---|---|---|---|---|---|
| | | | Max. | Min. | Max. | Min. |
| Polyvinyl chloride | Good | Excellent | 105 | −55 | 220 | −65 |
| Polyethylene | Good | Excellent | 75 | −75 | 165 | −100 |
| Nylon | Excellent | Good | 150 | −55 | 300 | −65 |
| Teflon-FEP | Excellent | Good | 200 | −165 | 390 | −265 |
| Silicone rubber | Fair | Excellent | 200 | −75 | 390 | −100 |
| Asbestos | Good | Good | 540 | −75 | 1000 | −100 |
| Glass | Poor | Good | 540 | −75 | 1000 | −100 |
| Refrasil | Poor | Good | 980 | −75 | 1800 | −100 |

A.

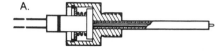

***Figure 8.23*** Details of a typical thermocouple probe. (Courtesy of BLH Electronics.)

Thermocouples are widely used to measure fluid temperatures in tanks, pipes, boilers, reactors, etc. The thermocouple is usually protected from the fluid by metal wells or probes and insulated with swaged and compacted ceramic powders. A wide variety of commercial products, similar to the one shown schematically in Fig. 8.23, are available to provide for easy installation and maintenance over extended periods of operation.

Surface installations are usually made by bonding, welding, or brazing the thermocouple to the surface. In some cases, the thermocouple is embedded in a shallow, small-diameter hole prior to welding. Care should be exercised in minimizing the weld material and in maintaining the geometry of the surface being instrumented.

## Recording Instruments for Thermocouples

In the previous subsections it was shown that thermocouples are voltage generators; however, the voltage generated by a thermocouple is quite small (see Table 8.5); therefore, small voltage losses due to current flow in the circuit ($\Delta E_o = IR$) can produce significant errors in the temperature measurements. Also, when a temperature gradient and a current flow are in the same direction, heat is generated (the Thompson effect). Similarly, when the directions are opposite, heat is absorbed. Obviously, the Thompson effect can also produce significant errors in temperature measurements when current flows in a thermocouple circuit. Thus, the output voltage $E_o$ from a thermocouple circuit must be measured and recorded with a recording instrument having a high input impedance so that current flow in the circuit is minimized.

The most common instruments used to measure and record the output voltage $E_o$ from a thermocouple circuit are the digital voltmeter (DVM), the strip-chart recorder, and the oscilloscope. All of these are high-input-impedance instruments ($R_M > 10^6 \ \Omega$).

The digital voltmeter is ideal for static and quasi-static measurements where it can be used in the manual mode of operation or as the voltage measuring component of a data logging system for automatic recording at a rate of approximately 20 points per second. A digital voltmeter with an input impedance of 10 M$\Omega$ will limit current flow in the thermocouple circuit to $10^{-10}$ A at a thermocouple voltage of 1 mV. Under such conditions, the $IR$ losses and junction heating and cooling due to the Thompson effect are negligible. A commercial DVM which has been adapted to provide a readout directly in terms of degrees Celsius for several different types of thermocouples is shown in Fig. 8.24.

The strip-chart recorder, with its servomotor-driven, null-balance potentiometric circuit (see Section 2.4), is an ideal instrument for quasi-static temperature

***Figure 8.24*** A digital thermometer. (Courtesy of Soltec Corp.)

measurements over long periods of time. Such strip-chart recorders are usually equipped with a bridge-compensation device that provides the reference junction required for thermocouple operation. In these instances, the scale of the recorder is usually calibrated to read temperature directly for a particular type of thermocouple.

High-frequency variations in temperature can be recorded with either an oscilloscope or an oscillograph. Use of an oscilloscope is straightforward since its input impedance and sensitivity are usually well matched to the thermocouple circuit. Use of an oscillograph, however, requires considerable caution if it exhibits a low input impedance. Any oscillograph used for temperature measurements with thermocouples must be equipped with a high-impedance input amplifier to limit current flow in the thermocouple circuit; otherwise, significant errors will occur as a result of *IR* losses and Thompson effects.

The use of dc voltmeters without preamplifiers is not recommended. While it is possible to use such instruments and correct for *IR* losses and Thompson effects, the probability for error is high. The difference in cost between a digital voltmeter and a dc voltmeter is modest; therefore, cost does not usually become an important consideration in the decision to avoid use of a dc voltmeter.

## 8.5 TWO-TERMINAL INTEGRATED-CIRCUIT TEMPERATURE TRANSDUCERS

The two-terminal integrated-circuit temperature transducer is a device that provides an output current $I$ that is proportional to absolute temperature $T_A$ when an input voltage $E_i$ (between 4 and 30 V) is applied to the terminals of the transducer. This type of temperature transducer is a high-impedance constant-current regulator over the temperature range from $-55°C$ ($-70°F$) to $150°C$ ($300°F$). It exhibits a nominal current sensitivity $S_I$ of 1 $\mu A/K$. The current sensitivity is controlled by an internal resistance that is laser trimmed at the factory to give an output of 298.2 $\mu A$ at a temperature of 298.2K (25C). Typical input voltage–current–temperature characteristics of a two-terminal integrated-circuit temperature transducer are shown in Fig. 8.25.

The two-terminal integrated-circuit temperature transducer is ideally suited for remote temperature measurements since it acts like a constant-current source, and as a result, lead wire resistance $R_L$ has no effect on the output voltage of the transducer circuit at the recording instrument. A well-insulated pair of twisted lead wires can be used for distances of several hundred feet. Also, many of the problems associated with use of thermocouples or RTD devices such as small output signals, need for precision amplifiers and linearization circuitry, cold junction compensation, and thermoelectric effects at connections are not encountered with the two-terminal temperature transducer.

The output voltage $E_o$ from the two-terminal temperature sensor circuit is controlled by series resistance $R_S$ as shown in Fig. 8.26. Since the temperature sensor serves as a current source, this output voltage can be expressed as

$$E_o = IR_S = S_I T_A R_S = S_T T_A \qquad (8.25)$$

**where** $S_I$ is the current sensitivity of the sensor.
$R_S$ is the series resistance across which the output voltage is measured.
$T_A$ is the absolute temperature.
$I$ is the current output at absolute temperature $T_A$.
$S_T$ is the voltage sensitivity of the circuit.

The output resistance $R_S$ often contains a trim potentiometer, as shown in Fig. 8.26, which is used to standardize the output voltage to a value such as 1

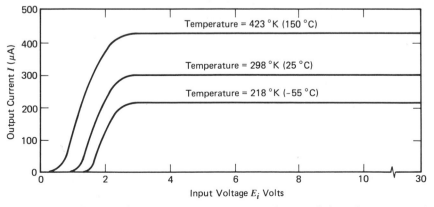

*Figure 8.25* Input voltage-current-temperature characteristics of a two-terminal integrated-circuit temperature transducer.

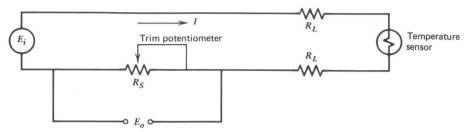

**Figure 8.26**   Two-terminal temperature sensor circuit with leadwire resistance and a series output resistance with trim potentiometer for standardizing sensitivity.

mV/K or 10 mV/K. This trim adjustment also permits the sensor's calibration error at a given temperature to be adjusted so as to improve accuracy over a given range of temperatures as shown in Fig. 8.27.

Unfortunately, the two-terminal integrated-circuit temperature transducer is limited to use in the range of temperatures from $-55°C$ to $150°C$. In this temperature range, it is an excellent temperature measuring device.

## 8.6   RADIATION METHODS (PYROMETRY)

As the temperature of a body increases it becomes increasingly difficult to measure the temperature with resistance temperature detectors (RTDs), thermistors, or thermocouples. The problems associated with measurement of high temperatures by means of these conventional methods (lack of stability, breakdown of insulation, etc.) provided the motivation for initial developments in *pyrometry* (inferring temperature from a measurement of the radiation emitted by the body). As the art of pyrometry developed, two other applications emerged. In certain measurements of temperature, the presence of the sensor affects the temperature; therefore, a noncontact method of measurement, such as pyrom-

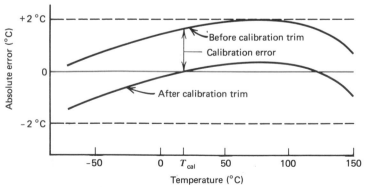

**Figure 8.27**   Typical nonlinearity and calibration error for a two-terminal integrated-circuit temperature transducer.

etry, is desirable. Another application that has developed is the measurement of complete temperature fields. Thermocouples, thermistors, and RTDs provide data only at a point. Temperature distributions over the entire body are often needed, and the radiation emitted from a body provides this information if it is properly recorded and interpreted. Thus, pyrometry is a useful method for the measurement of very high temperatures, for providing a noncontacting method for measurement, and for obtaining full-field temperature distributions.

## Principles of Radiation Measurements of Temperature

When a body is heated it radiates energy that can be detected and related to the temperature of the body. The relationship between intensity of radiation, wavelength of the radiation, and temperature is known as Planck's Law and can be expressed as

$$W_\lambda = \frac{2\pi c^2 h}{\lambda^5 (e^{hc/k\lambda T} - 1)} = \frac{C_1}{\lambda^5 (e^{C_2/\lambda T} - 1)} \tag{8.26}$$

**where**   $W_\lambda$ is the spectral radiation intensity for a black body (W/m³).
   $\lambda$   is the wavelength of the radiation (m).
   $T$   is the absolute temperature (K).
   $h$   is Planck's constant $= 6.626 \ (10^{-34}) \ (\text{J} \cdot \text{s})$.
   $c$   is the velocty of light $= 299.8 \ (10^6) \ (\text{m/s})$.
   $k$   is Boltzmann's constant $= 1.381 \ (10^{-23}) \ (\text{J/K})$.

$C_1$ and $C_2$ are constants.

From Eq. (8.26) it is evident that

$$C_1 = 2\pi c^2 h = 3.75(10^{-16}) \ \text{W} \cdot \text{m}^2$$

$$C_2 = hc/k = 1.44 \ (10^{-2}) \ \text{m} \cdot \text{K}$$

The spectral radiation intensity $W_\lambda$ is the amount of energy emitted by radiation of wavelength $\lambda$ from a flat surface at temperature $T$ into a hemisphere. It is evident from Eq. (8.26) that the spectral radiation intensity $W_\lambda$ depends upon both wavelength $\lambda$ and temperature $T$. A plot of $W_\lambda$ versus $\lambda$ for several different temperatures is shown in Fig. 8.28. Note that $W_\lambda$ peaks at a specfic wavelength, which depends on temperature, and that the wavelength associated with the peak $W_\lambda$ increases as the temperature decreases. The wavelength $\lambda_p$ associated with the peak in $W_\lambda$ can be expressed as

$$\lambda_p = 2891(10^{-6})/T \tag{8.27}$$

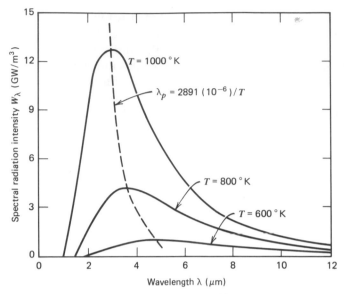

*Figure 8.28* Blackbody radiation at different temperatures.

The area under each of the curves in Fig. 8.28 is the total power $W$ emitted at the particular temperature $T$. Thus

$$W = \int_\lambda W_\lambda d\lambda = 5.67(10^{-8})T^4 \text{ W/m}^2 \qquad (8.28)$$

Equation (8.28) is the Stefan–Boltzmann Law with the emissivity $\varepsilon$ equal to unity ($\varepsilon = 1$). From the previous discussion it is evident that

1. The total power $W$ increases as a function of the fourth power of the temperature.
2. The peak value of spectral radiation intensity $W_\lambda$ occurs at shorter wavelengths as the temperature increases.

Both of these physical principles are used as the basis for a measurement of temperature.

### The Optical Pyrometer

The optical pyrometer, illustrated schematically in Fig. 8.29a, is used to measure temperature over the range from 700°C to 4000°C (1300°F to 7200°F). The radiant energy emitted by the body is collected with an objective lens and focused onto a calibrated pyrometer lamp. An absorption filter is inserted into the optical system between the objective lens and the pyrometer lamp when the temperature of the body exceeds 1300°C (2370°F). The radiant energy from both the hot body and the filament of the pyrometer lamp is then passed through a red filter with a sharp cutoff below $\lambda = 0.63$ μm. The light transmitted through the filter is then collected by an objective lens and focussed for viewing with an ocular lens. The image observed through the eyepiece of the pyrometer is that of the lamp filament superimposed on a background intensity due to the hot body. The current to the filament of the pyrometer lamp is adjusted until the brightness of the filament matches that of the background. Under a matched condition, the filament disappears (hence the commonly used name—disappearing-filament optical pyrometer), as illustrated in Fig. 8.29*b*. The current required to produce the brightness match is measured and used to establish the temperature of the hot body. Pyrometers are calibrated by visually comparing the brightness of the tungsten filament with a blackbody source of known temperature ($\varepsilon = 1$).

When the brightness of the background and the filament are matched, it is evident from Eq. (8.26) that

$$\frac{\varepsilon}{e^{C_2/\lambda_r T} - 1} = \frac{1}{e^{C_2/\lambda_r T_f} - 1} \tag{8.29}$$

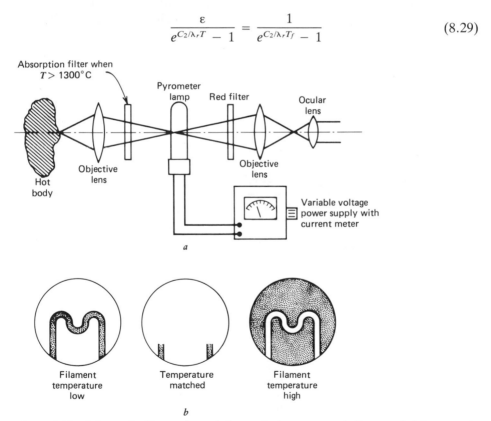

*Figure 8.29*   Schematic illustration of the optical system and filament brightness adjustment in an optical pyrometer. (*a*) Schematic illustration of an optical pyrometer. (*b*) Filament brightness adjustment in an optical pyrometer.

**where** $\lambda_r$ is the wavelength of the red filter $\lambda_r = 0.63$ μm.
$\varepsilon$ is the emissivity of the surface of the hot body at $\lambda = 0.63$ μm.
$T_f$ is the temperature of the filament.
$T$ is the unknown surface temperature.

When $T < 4000°C$ (7200°F), the term $e^{C_2/\lambda T} \gg 1$ and Eq. (8.29) reduces to

$$T = \frac{1}{\lambda_r (\ln \varepsilon)/C_2 + 1/T_f} \tag{8.30}$$

It is obvious from Eq. (8.30) that $T = T_f$ only when $\varepsilon = 1$. If $\varepsilon \neq 1$, then $T \neq T_f$ and Eq. (8.30) must be used to determine the temperature $T$ from the temperature $T_f$ indicated by the pyrometer. The emissivity of a number of materials (oxidation-free surface) are listed in Table 8.8.

If the emissivity of a surface is not known precisely, then an error will occur when Eq. (8.30) is used to determine the temperature $T$. The change in temperature as a function of change in emissivity is obtained from Eq. (8.30) as

$$\frac{dT}{T} = -\frac{\lambda T}{C_2} \frac{d\varepsilon}{\varepsilon} \tag{8.31}$$

Since $\lambda T/C_2 < 0.1$ for $T < 2000°C$ (3630°F), errors in temperature determinations are mitigated considerably with respect to errors in emissivity. For example, at

**TABLE 8.8**[a] Emissivity $\varepsilon$ of Engineering Materials at $\lambda = 0.65$ μm

| Material | Solid | Liquid | Material | Solid | Liquid |
|---|---|---|---|---|---|
| Beryllium | 0.61 | 0.61 | Thorium | 0.36 | 0.40 |
| Carbon | 0.80–0.93 | — | Titanium | 0.63 | 0.65 |
| Chromium | 0.34 | 0.39 | Tungsten | 0.43 | — |
| Cobalt | 0.36 | 0.37 | Uranium | 0.54 | 0.34 |
| Columbium | 0.37 | 0.40 | Vanadium | 0.35 | 0.32 |
| Copper | 0.10 | 0.15 | Zirconium | 0.32 | 0.30 |
| Iron | 0.35 | 0.37 | Steel | 0.35 | 0.37 |
| Manganese | 0.59 | 0.59 | Cast Iron | 0.37 | 0.40 |
| Molybdenum | 0.37 | 0.40 | Constantan | 0.35 | — |
| Nickel | 0.36 | 0.37 | Monel | 0.37 | — |
| Platinum | 0.30 | 0.38 | 90 Ni–10 Cr | 0.35 | — |
| Rhodium | 0.24 | 0.30 | 80 Ni–20 Cr | 0.35 | — |
| Silver | 0.07 | 0.07 | 60 Ni–24 Fe– | 0.36 | — |
| Tantalum | 0.49 | — | 16 Cr | | |

[a] From ASME Performance Test Codes PTC 19.3, 1974.

a temperature of 1500°K, a 20 percent error in emissivity produces only a 1.3 percent error in temperature $T$.

The disappearing-filament optical pyrometer is the most accurate of all radiation-type temperature measuring devices and as a consequence it is used to establish the International Practical Temperature Scale above 1063°C (1945°F). If the emissivity of the hot body is accurately known, the error in a temperature measurement is usually less than 1 percent.

## Photon Detector Temperature Instruments

There are many applications where temperature must be measured without contacting the body. Examples are temperature measurements involving thin films or foils where the presence of a sensor would markedly affect the measurement and, at the other extreme, scanning of buildings, piping systems or other large objects where a requirement exists for measuring the temperature distribution over a large area. Both of these types of measurements can be made with a temperature measuring instrument that incorporates a photon detector as the sensor.

A *photon detector* is a sensor that responds by generating a voltage that is proportional to the photon flux density $\phi$ impinging on the sensor. A schematic diagram of a photon detector system for measuring temperature is shown in Fig. 8.30. The photons emitted from a small area $A_s$ of a surface (not necessarily hot) are collected by a lens and focused on a photon detector of area $A_d$. The photon flux density $\phi$ at the detector, when the optical system is focused, can be expressed as

$$\phi = \frac{kD^2}{4f^2} \varepsilon g(T) \tag{8.32}$$

**where**   $k$   is the transmission coefficient of the lens (and filter if one is used).
  $D$   is the diameter of the lens.
  $f$   is the focal length of the lens.
  $g(T)$ is a known function of the temperature of the surface.
  $\varepsilon$   is the emissivity of the surface.

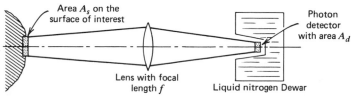

*Figure 8.30*   Schematic illustration of a temperature measuring instrument with a photon detector type of sensor.

The output voltage $E_o$ from the detector as a result of the flux density $\phi$ is

$$E_o = k_t \frac{D^2}{4f^2} \varepsilon g(T) \tag{8.33}$$

where $k_t$ is the system sensitivity, and includes the transmission coefficient of the lens, the amplifier voltage gain, and the detector sensitivity. The system sensitivity $k_t$ is essentially a constant; however, the lens in a typical instrument (see Fig. 8.31) can be changed to provide for different fields of view where the solid angle may range from 3.5 to 40 degrees. The term $\varepsilon g(T)$ depends only on the temperature of the surface and its emissivity. A typical response curve, shown in Fig. 8.32, indicates that the output voltage varies as a function of the cube of the temperature. Thus, Eq. (8.33) can be simplified to

$$E_o - K\varepsilon T^3 \tag{8.34}$$

where $K$ is a calibration constant for the instrument. In practice, $K$ is determined by calibrating the instrument with a blackbody source ($\varepsilon = 1$) over an appropriate range of temperatures. When the instrument is used for temperature measurements, the emissivity $\varepsilon$ of the surface must be considered, since it may differ significantly (see Table 8.12) from the calibration value. Any correction required is easily made by substituting the correct value of the emissivity into Eq. (8.34) and solving for the required temperature $T$. Thus

$$T = \left(\frac{E_o}{K\varepsilon}\right)^{1/3} \tag{8.35}$$

Errors in temperature due to inaccuracies in emissivity are mitigated by a factor of 0.333 since differentiation of Eq. (8.35) gives $dT/T = -(1/3)(d\varepsilon/\varepsilon)$.

Many different commercial instruments employ the photon detector; therefore, it is difficult to list specifications that cover the full range of products. Specifications for the scanner shown in Fig. 8.31 indicate that it can be used to

*Figure 8.31*  Photon detector-based temperature scanner. (Courtesy of AGA Corporation.)

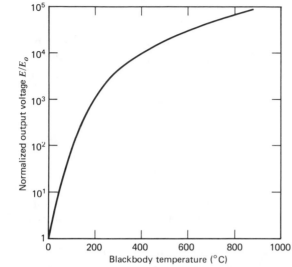

**Figure 8.32** Typical response curve for an indium antimonide photon detector.

measure temperatures in the range from $-20°C$ to $1600°C$ ($0°F$ to $2900°F$). Its sensitivity at $30°C$ ($86°F$) is $0.1°C$ ($0.2°F$).

A recent innovation with this type of instrument permits determination of temperature distributions over extended regions of a body. This improved capability is accomplished by inserting two mechanically driven cylindrical lenses into the optical path. As the two lenses are oscillated, a region of the surface of the body is scanned. At any instant, a relatively small target area is focused on the photon detector, and the temperature of this small target area is determined. Since the entire surface of the body is scanned in a short period of time, a full-field photograph of the temperature distribution representing an *x-y* array of the many small target areas can be obtained. A single frame typically contains 28,000 individual temperature measurements (280 lines with 100 elements per line). The voltage output can be displayed on a TV monitor in either gray scale or color. Photon-detector-type instruments can complete a scan of a field in about 40 ms. If a video recorder is used to store the images, the system can be used to study full-field dynamic temperature distributions.

## 8.7 CALIBRATION METHODS

Calibration of temperature sensors is usually accomplished by using the freezing-point (or boiling-point) method, the melting-wire method, or the comparison method.

The *freezing-point method* is the easiest and most frequently employed calibration technique. With this approach, the temperature sensor is immersed in a melt of pure metal that has been heated in a furnace to a temperature above the melting point. The temperature of the melt is slowly reduced while a temperature–time record, similar to the one shown in Fig. 8.33, is recorded. As the metal changes state from liquid to solid, the temperature remains constant at the melting-point temperature $T_M$ and provides an accurate reference temperature for calibration. The particular metal selected for the bath is determined by the temperature required for the calibration. Usually, a sensor should be calibrated at three points within its temperature range (preferably the minimum, the midpoint, and the maximum). The melting points of a number of metals are listed in Tables 8.1 and 8.2. These data indicate that the melting-point approach can be used to provide calibration temperatures in the range from 232°C (450°F) with tin to 1064°C (1948°F) with gold. The metals must be pure since small quantities of an impurity can significantly affect the melting point and thus affect the calibration. Melting point standards are commercially available for temperatures ranging from 125°F to 600°F in 25°F increments. These standards are accurate to ±1°F.

The lower range of the temperature scale is usually calibrated by using the *boiling phenomenon*. The temperature sensor is immersed in a liquid bath and heat is added slowly until the fluid begins to boil and a stable calibration temperature is achieved. Atmospheric pressure must be considered in ascertaining the boiling point of any liquid since pressure variations can significantly affect the calibration results. For example, reducing the atmospheric pressure from 29.922 in. of Hg to 26.531 in. of Hg results in a decrease in the saturation (boiling) temperature of water from 212°F to 206°F.

The *melting-wire method* of calibration is used with thermocouples. With this approach, the hot junction of the thermocouple is effected by connecting the two dissimilar wires with a pure third metal, such as silver or tin. As the hot junction is heated, the output voltage $E_o$ is recorded continuously. When the connecting material melts, the thermocouple junction is broken and the output voltage $E_o$ drops to zero. The output voltage $E_o$ just prior to the drop is associated with the melting point (calibration temperature $T_M$) of the specific material used for the joint.

The *comparison method* utilizes two temperature sensors: one of unknown quality and one of reference or standard quality. Both are immersed in a liquid

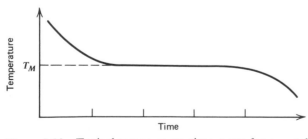

*Figure 8.33*   Typical temperature–time curve for a metal during solidification.

bath that is temperature cycled over the range of interest. The response of the "standard" sensor gives the temperature of the bath at any time that can be used as the calibration temperature for the unknown sensor. The "standard" temperature sensor must be calibrated periodically to ensure its accuracy.

## 8.8 DYNAMIC RESPONSE OF TEMPERATURE SENSORS

Temperature sensors are classified as first-order systems, since their dynamic response is controlled by a first-order differential equation that describes the rate of heat transfer between the sensor and the surrounding medium. Thus,

$$q = hA(T_m - T) = mc \frac{dT}{dt} \tag{8.36}$$

where   $q$   is the rate of heat transfer to the sensor by convection.
        $h$   is the convective heat-transfer coefficient.
        $A$   is the surface area of the sensor through which heat passes.
        $T_m$ is the temperature of the surrounding medium at time $t$.
        $T$   is the temperature of the sensor at time $t$.
        $m$   is the mass of the sensor.
        $c$   is the specific heat capacity of the sensor.

Equation (8.36) can also be expressed as

$$\frac{dT}{dt} + \frac{hA}{mc}T = \frac{hA}{mc}T_m \tag{8.37}$$

Solving Eq. (8.37) for the homogeneous part yields

$$T = C_1 e^{-t/\beta} \tag{8.38}$$

where   $C_1$        is a constant of integration.

$\beta = \dfrac{mc}{hA}$ is the time constant for the sensor.   (8.39)

A complete solution of Eq. (8.37) requires specification of the medium temperature as a function of time and the initial conditions. Two cases provide valuable insight into the behavior of temperature sensors; namely, response to step-function inputs and response to ramp-function inputs.

Consider first the response of a temperature sensor to a step-function input (sensor is suddenly immersed in a medium maintained at constant temperature $T_m$). In this case, the particular solution of Eq. (8.37) is $T = T_m$; therefore, the general solution is

$$T = C_1 e^{-t/\beta} + T_m \tag{8.40}$$

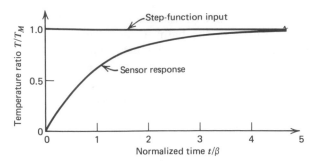

**Figure 8.34**    Response of a temperature sensor to a step-function input.

For the initial condition $T(0) = 0$, the integration constant $C_1$ in Eq. (8.40) equals $-T_m$; thus, the final expression for temperature $T$ as a funciton of time $t$ for the step-function input is

$$T = T_m(1 - e^{-t/\beta}) \qquad (8.41)$$

The results of Eq. (8.41), shown in nondimensional form in Fig. 8.34, indicate that a temperature sensor requires considerable time before it begins to approach the temperature of the surrounding medium $T_m$. The temperature of the sensor is within 5 percent of $T_m$ at $t = 3\beta$ and within 2 percent of $T_m$ at $t = 3.91\beta$. The response time can be improved by reducing $\beta$. This can be accomplished by designing a sensor with a small mass, a low specific heat capacity, and a large surface area. An example of a rapid-response thermocouple is shown in Fig. 8.35. This type of thermocouple is fabricated from a thin sheet of foil having a thickness of approximately 0.012 mm (0.0005 in.). The foil elements are mounted on a thin polymeric carrier to facilitate bonding to the component. The time constant $\beta$ for this type of thermocouple ranges from 2

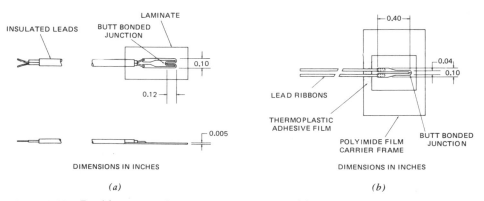

**Figure 8.35**    Rapid-response foil-type thermocouples. (*a*) Encapsulated foil element. (*b*) Free-filament foil element. (Courtesy of Omega Engineering, Inc.)

to 5 ms, depending primarily on the convective heat-transfer coefficient $h$ that exists between the sensor and the medium.

The second case of importance involves response of a temperature sensor to a ramp-function input, such as the one illustrated in Fig. 8.36. The sensor and the surrounding medium are initially at the same temperature; thereafter, the temperature of the medium increases linearly with time. Solving Eq. (8.37) for the particular solution yields

$$T = b(t - \beta)$$

where $b$ is the slope of the temperature–time ramp function, as illustrated in Fig. 8.37. Thus, the general solution of Eq. (8.37) for the ramp-function input is

$$T = C_1 e^{-t/\beta} + b(t - \beta) \tag{8.42}$$

For the initial condition $T(0) = T_m(0) = 0$, the integration constant $C_1$ in Eq. (8.42) equals $b\beta$; therefore, the response of a temperature sensor to a ramp-function input can be expressed as

$$T = b\beta e^{-t/\beta} + b(t - \beta) \tag{8.43}$$

The results of Eq. (8.43) are shown in Fig. 8.36. This data indicate that the initial response of the sensor is sluggish; however, after a short initial interval, the sensor tracks the rise in temperature of the medium surrounding the sensor with the correct slope, but with a time lag $\beta$. This behavior is evident in Eq. (8.43) where the first term is important during the initial response and the second term dominates the long-term response. The first term decreases rapidly with time and becomes negligible when $t > 3\beta$. Since the lag time $\beta$ can be determined from a simple experiment, accurate temperature measurements can be made for time greater than $3\beta$ by correcting for the time lag. Sensors with small time constants should be used to reduce the time lag and the transient period so that they are consistent with the time requirements of the process being monitored or controlled.

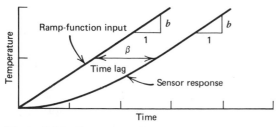

*Figure 8.36*  Response of a temperature sensor to a ramp-function input.

## 8.9 SUMMARY

Temperature is an abstract quantity and as such must be defined in terms of the behavior of materials as the temperature changes. This is accomplished by defining the temperature associated with phase transformations in several different materials over the temperature range from $-183°C$ to $1064°C$ ($-297°F$ to $\iota948°F$).

The different sensors available for temperature measurement include resistance temperature detectors (RTDs), thermistors, expansion thermometers, thermocouples, and pyrometers. Each type of sensor has advantages and disadvantages; selection of the proper sensor for a particular application is usually based on considerations of temperature range, accuracy requirements, environment, dynamic response requirements, and available instrumentation.

The most frequently used temperature measuring sensor is the thermocouple, since it is a low-cost transducer that is easy to fabricate and install. The signal output is relatively low and must be measured and recorded with an instrument having a high input impedance so that current flow in the circuit is minimized; otherwise, significant errors can be introduced. The nonlinear output of the thermocouple is clearly a disadvantage; however, modern instruments used to record the output voltage incorporate linearizing circuits that permit readout in terms of temperature directly. The range of temperatures that can be measured with thermocouples is very large ($-185°C$ to $2800°C$ or $-300°F$ to $5072°F$).

Resistance-based temperature sensors (RTDs and thermistors) are usually employed when a high sensitivity is required. Because of the high voltage output, higher accuracies can be achieved. The RTD-type sensor is available in coil form for fluid temperature measurements and in a bondable grid form for surface temperature measurements. These sensors are easy to install and the instrumentation used to monitor the output signal is inexpensive and easy to employ. Thermistors are used in many commercial temperature recorders because their high voltage output permits a reduction in complexity and cost of the readout system. The range of thermistors is limited and the output is extremely nonlinear.

Bimetallic thermometers are used primarily in control applications where long-term stability and low cost are important considerations. They are often used to activate switches in on–off temperature control situations where precision is not a stringent requirement.

Pressure thermometers are also employed primarily in control applications. The accuracy obtainable with a pressure thermometer is usually better than that of a bimetallic thermometer, but the costs are higher. Both bimetallic and pressure thermometers exhibit long-term stability and minimize need for electronic components in their control circuits.

Pyrometers are used primarily to monitor the extremely high temperatures associated with metallurgical processes. The disappearing-filament optical pyrometer has been a reliable instrument in many industrial applications for several decades. In more recent years, with the development of photon detectors, radiation methods of temperature measurement have been extended into the lower temperature range. Two significant advantages are offered by instruments that use photon detectors as sensors. First, they can be used to measure temperature without contacting the specimen in applications involving thin films, paper, or moving bodies. Second, they can be used to monitor temperature distributions over extended fields. The single disadvantage of the photon-detector-based instruments is their relatively high cost.

Temperature sensors are first-order systems that respond to a step change in temperature in a manner that is described by the equation

$$T = T_m(1 - e^{-t/\beta}) \qquad (8.41)$$

Errors due to the time required for heat transfer can be minimized by reducing the time constant $\beta$ for the sensor.

## REFERENCES

1. *The International Practical Temperature Scale of 1968,* A Committee Report, English version appeared in *Metrologia,* vol. 5, no. 2, April 1969.

2. American Society for Testing and Materials: *Evolution of the International Practical Temperature Scale of 1968,* ASTM STP 565, 1974.

3. *The International Practical Temperature Scale of 1968—Amended Edition of 1975,* A Committee Report, English version appeared in *Metrologia,* vol. 12, 1976, pp. 7–17.

4. Preston-Thomas, H.: The Origin and Present Status of the IPTS-68, *Temperature,* vol. 4, Instrument Society of America, Pittsburgh, 1972.

5. Curtis, D. J., and G. J. Thomas: Long Term Stability and Performance of Platinum Resistance Thermometers for Use to 1063 C, *Meterologia,* vol. 4, no. 4, October 1968, pp. 184–190.

6. American Society of Mechanical Engineers: Resistance Thermometers, *Temperature Measurement,* Chapter 4, Supplement to ASME Performance Test Codes, PTC 19.3, 1974.

7. Benedict, R. P., and R. J. Russo: Calibration and Application Techniques for Platinum Resistance Thermometers, *Transactions of the American Society of Mechanical Engineers, Journal of Basic Engineering,* June 1972, pp. 381–386.

8. Becker, J. A., C. B. Green, and G. L. Pearson: Properties and Uses of Thermistors, *Transactions of the American Institute of Electrical Engineers,* vol. 65, November 1946, pp. 711–725.

9. National Bureau of Standards: *Liquid-in-Glass Thermometry,* NBS Monograph 150, 1975.

10. American Society of Mechanical Engineers: Liquid-in-Glass Thermometers, *Temperature Measurement,* Chapter 5, Supplement to ASME Performance Test Codes, PTC 19.3, 1974.

11. Seebeck, T. J.: *Evidence of the Thermal Current of the Combination Bi-Cu by its Action on Magnetic Needle,* Royal Academy of Science, Berlin, 1822–1823, p. 265.

12. Peltier, J. C. A.: Investigation of the Heat Developed by Electric Currents in Homogeneous Materials and at the Junction of Two Different Conductors, *Annales de Chimie et Physique,* vol. 56 (2nd Ser.), 1834, p. 371.

13. Thompson, W.: On the Thermal Effects of Electric Currents in Unequal Heated Conductors, *Proceedings of the Royal Society,* vol. 7, May 1854.

14. Finch, D. I.: General Principles of Thermoelectric Thermometry, *Temperature,* vol. 3, part 2, Reinhold, New York, 1962.

15. Powell, R. L., W. J. Hall, C. H. Hyink, Jr., L. L. Sparks, G. W. Burns, M. G. Scroger, and H. H. Plumb: *Thermocouple Reference Tables Based on IPTS-68,* NBS Monograph 125, March 1974.

16. American Society for Testing and Materials: *Manual on the Use of Thermocouples in Temperature Measurement,* ASTM STP 470A, March 1974.

17. Nutter, G. D.: Radiation Thermometry, *Mechanical Engineering,* part 1, June 1972, p. 16, part 2, July 1972, p. 12.

18. American Society of Mechanical Engineers: Optical Pyrometers, *Temperature Measurement,* Chapter 7, Instruments and Apparatus Supplement to ASME Performance Test Codes, PTC 19.3, 1974.

19. Benedict, R. P.: *Fundamentals of Temperature, Pressure, and Flow Measurements,* 2nd ed., Wiley, New York, 1977.

20. Doebelin, E. O.: *Measurement Systems,* McGraw-Hill, New York, 1975.

21. Omega Engineering: *Temperature Measurement Handbook,* Omega Engineering, Stamford, Connecticut, 1981.

22. Cook, N. H., and E. Rabinowicz: *Physical Measurement and Analysis,* Addison-Wesley, Reading, Massachusetts, 1963.

## EXERCISES

**8.1** Why are the primary and secondary points on the International Temperature Scale important in the measurement of temperature?

**8.2** From the results shown in Fig. 8.1, determine the temperature coefficients of resistivity $\gamma_1$, $\gamma_2$, and $\gamma_3$ in Eq. (8.1) for a resistance temperature detector (RTD) fabricated from platinum for the temperature range from $-200°C$ to $1000°C$.

**8.3** From the results shown in Fig. 8.1, determine the temperature coefficients of resistivity $\gamma_1$, $\gamma_2$, and $\gamma_3$ in Eq. (8.1) for a resistance temperature detector (RTD) fabricated from cooper for the temperature range from $-200°C$ to $1000°C$.

**8.4** From the results shown in Fig. 8.1, determine the temperature coefficients of resistivity $\gamma_1$, $\gamma_2$, and $\gamma_3$ in Eq. (8.1) for a resistance temperature detector (RTD) fabricated from nickel for the temperature range from $0°C$ to $1000°C$.

**8.5** Repeat Exercise 8.2 for a temperature range from $0°C$ to $600°C$.

**8.6** Repeat Exercise 8.3 for a temperature range from $0°C$ to $600°C$.

**8.7** Repeat Exercise 8.4 for a temperature range from $0°C$ to $600°C$.

**8.8** From the results shown in Fig. 8.1, determine the temperature coefficients of resistivity $\gamma_1$ and $\gamma_2$ in Eq. (8.3) for a resistance temperature detector (RTD) fabricated from platinum for the temperature range from $100°C$ to $400°C$.

**8.9** Repeat Exercise 8.8 for a copper RTD.

**8.10** Repeat Exercise 8.8 for a nickel RTD.

**8.11** A resistance temperature detector (RTD) fabricated from platinum exhibits a temperature coefficient of resistivity $\gamma_1 = 0.003902/°C$. If the resistance of the sensor is $100 \ \Omega$ at $0°C$, find the resistance at

| | | | |
|---|---|---|---|
| (a) | $-270°C$ | (d) | $+300°C$ |
| (b) | $-100°C$ | (e) | $+500°C$ |
| (c) | $+100°C$ | (f) | $+660°C$ |

Assume $\gamma_2$ is negligible.

**8.12** Find $\gamma_2$ for the RTD of Exercise 8.11 by using the data of Fig. 8.1 for the temperature range from $-200°C$ to $600°C$. Determine the percent error introduced in the resistance determinations at each temperature level of Exercise 8.11 by neglecting $\gamma_2$.

**8.13** Show that lead-wire effects are completely eliminated by using the three-wire system illustrated in Fig. 8.5 to connect a resistance temperature detector (RTD) into a Wheatstone bridge.

**8.14** Show that lead-wire effects are completely eliminated by using the four-wire system illustrated in Fig. 8.6 to connect a resistance temperature detector (RTD) into a constant-current potentiometer circuit.

**8.15** A platinum RTD with a resistance of $100 \ \Omega$ at $0°C$ is used in the constant-current potentiometer circuit shown in Fig. 8.6. If the current $I$ equals 5 mA, determine the output voltage $E_o$ at the following temperatures:

| | | | |
|---|---|---|---|
| (a) | $-270°C$ | (d) | $+300°C$ |
| (b) | $-100°C$ | (e) | $+500°C$ |
| (c) | $+100°C$ | (f) | $+660°C$ |

**8.16**   Verify Eq. (8.5).

**8.17**   If $\beta$ = 4000 K and $R_o$ = 3000 $\Omega$ at $T_o$ = 298 K, determine the resistance of a thermistor at

(a)   $-80°C$          (d)   50°C
(b)   $-40°C$          (e)   75°C
(c)    0°C             (f)   150°C

**8.18**   Verify Eq. (8.7).

**8.19**   Verify Eq. (8.10) for a constant-voltage, equal-arm Wheatstone bridge.

**8.20**   Verify Eq. (8.14) for a constant-current, equal-arm Wheatstone bridge.

**8.21**   Verify Eq. (8.17) for the potentiometer circuit.

**8.22**   The thermistor described in Exercise 8.17 is connected into a constant-voltage Wheatstone bridge as illustrated in Fig. 8.12a. If the bridge is initially balanced at $T_o$ = 25°C, determine the output voltage ratio $\Delta E_o/E_i$ for the temperatures listed below. Assume $R_1 = R_2 = R_3 = R_4 = R_t$.

(a)   $-80°C$          (d)   50°C
(b)   $-40°C$          (e)   75°C
(c)    0°C             (f)   150°C

**8.23**   The thermistor-bridge combination described in Exercise 8.22 provided the following output voltage ratios $\Delta E_o/E_i$ in a series of tests. Determine the temperatures by using the attached table that was provided by the manufacturer for the thermistor (Table A.1).

(a)    0.1            (e)    0.3
(b)   $-0.1$          (f)   $-0.3$
(c)    0.2            (g)    0.4
(d)   $-0.2$          (h)   $-0.4$

**8.24**   The thermistor described in Exercise 8.17 is connected into a constant-current Wheatstone bridge. If the bridge is initially balanced at $T_o$ = 25°C, determine the output voltage ratio $\Delta E_o/E_T$ for the temperatures listed below. Assume $R_1 = R_2 = R_3 = R_4 = R_T$.

(a)   $-70°C$          (d)   60°C
(b)   $-35°C$          (e)   100°C
(c)    0°C             (f)   140°C

**8.25**   The thermistor-bridge combination described in Exercise 8.24 provided the following output voltage ratios $\Delta E_o/E_T$ in a series of tests. Determine the temperatures by using Table A.1.

(a)    0.1            (c)    0.2
(b)   $-0.1$          (d)   $-0.2$

(e)    0.3                               (g)    0.5
(f)    −0.3                              (h)    0.9

**8.26**  The thermistor described in Exercise 8.17 is connected into the simple circuit illustrated in Fig. 8.13. If $I = 5$ mA, determine the voltage ratio $\Delta E_o/E_o$ for the temperatures listed below. $T_o = 25°C$.

(a)    −80°C                            (c)    75°C
(b)      0°C                            (d)    150°C

**8.27**  The thermistor-circuit combination described in Exercise 8.26 provided the following output voltage ratios $\Delta E_o/E_o$ in a series of tests. Determine the temperatures by using Table A.1.

(a)    500                              (f)    0.1
(b)    100                              (g)    −0.1
(c)     50                              (h)    −0.5
(d)     10                              (i)    −0.90
(e)    1.0                              (j)    −0.98

**8.28**  The modified potentiometer circuit shown in Fig. E8.28 is often used with thermistors to improve the linearity of output voltage $E_o$ with temperature $T$. Determine the relationship between output voltage $E_o$ and temperature $T$. Prepare a graph showing $E_o$ versus $T$ for the range of temperature from 0°C to 100°C if $\beta = 4000$ K, $E_i = 1.0$ V, and $R_1 = R_2 = R_T = 3000$ Ω.

**8.29**  Repeat Exercise 8.28 for the circuit shown in Fig. E8.29.

**8.30**  Repeat Exercise 8.28 for the cirucit shown in Fig. E8.30.

**8.31**  Repeat Exercise 8.28 for the circuit shown in Fig. E8.31.

**8.32**  List the primary advantages and disadvantages of

(a)    Liquid-in-glass thermometers
(b)    Bimetallic thermometers
(c)    Pressure thermometers

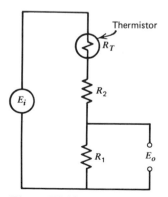

*Figure E8.28*

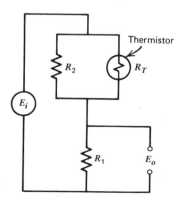

*Figure E8.29*

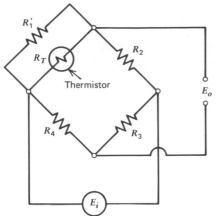

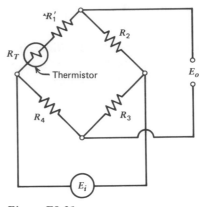

Figure E8.31

*Figure E8.30*

**8.33**   A digital voltmeter (DVM) is being used to measure the output voltage $E_o$ from a copper–constantan thermocouple, as shown in Fig. E8.33.

   (a)   Determine the output voltage $E_o$ indicated by the DVM.
   (b)   If the DVM reading changes to 2.078 mV, what is the new temperature $T_1$?
   (c)   Does temperatures $T_2$ or $T_3$ influence the measurement? Why?

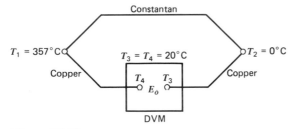

*Figure E8.33*

**8.34**   A digital voltmeter (DVM) is being used to measure the output voltage $E_o$ from a copper–constantan thermocouple, as shown in Fig. E8.34. Determine the output voltage indicated by the DVM.

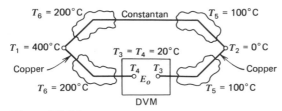

*Figure E8.34*

**8.35** A digital voltmeter (DVM) is being used to measure the output voltage $E_o$ from a Chromel–Alumel thermocouple, as shown in Fig. E8.35.

(a) Determine the output voltage $E_o$ indicated by the DVM.
(b) If the DVM reading changes to 20.470 mV, what is the new temperature $T_1$?
(c) Does the copper-Alumel junction at the DVM influence the reading on either (a) or (b)?

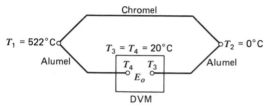

*Figure E8.35*

**8.36** A digital voltmeter (DVM) is being used to measure the output voltage $E_o$ from an iron–constantan thermocouple, as shown in Fig. E8.36.

(a) Determine the temperature $T_1$ associated with a DVM reading of 13.220 mV.
(b) Does the separation at junction 1 influence the measurement of $T_1$? List any assumptions made in reaching your answer.
(c) How far can the junctions be separated before errors will develop? Explain.

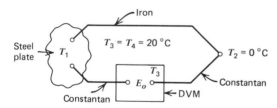

*Figure E8.36*

**8.37** A digital voltmeter (DVM) is being used to measure the output voltage $E_o$ from an iron–constantan thermocouple, as shown in Fig. E8.37.

(a) Determine the output voltage $E_o$ indicated by the DVM.
(b) If the DVM reading changes to 25.000 mV, what is the new temperature $T_1$?

**8.38** Use Eq. (8.24) to prepare a table for an iron–constantan thermocouple showing output voltage $E_o$ versus temperature $T_1$ in steps of 10°C over the temperature range from 0°C to 100°C.

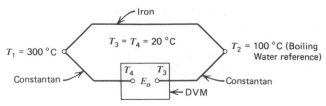

*Figure E8.37*

**8.39** A digital voltmeter (DVM) having an input impedance of 10 M$\Omega$ is being used to measure the output voltage $E_o$ of the iron–constantan thermocouple shown in Fig. E8.39. The thermocouple is fabricated from AWG No. 20 wire having a resistance of 0.357 $\Omega$ per double foot (1 ft of iron plus 1 ft of constantan). The distance between junctions 1 and 2 is 40 ft. Determine:

(a)  The *IR* drop due to the long lead wires
(b)  The output voltage $E_o$ indicated by the DVM
(c)  The temperature associated with the output voltage indicated by the DVM

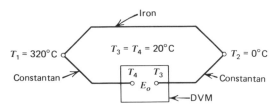

*Figure E8.39*

**8.40** A chromel–constantan thermocouple is accidentally grounded at both the active and reference junctions, as shown in Fig. E8.40. If the resistance of the thermocouple is 3 $\Omega$ and the resistance of the ground loop is 0.2 $\Omega$, determine the error introduced into the measurement of temperature $T_1$.

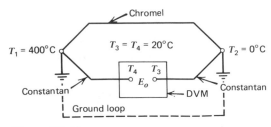

*Figure E8.40*

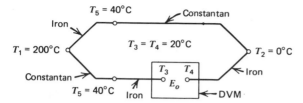

**Figure E8.41**

**8.41**   The extension wires of an iron–constantan thermocouple were improperly wired to produce the situation illustrated in Fig. E8.41. Determine the error introduced into the measurement of temperature $T_1$.

**8.42**   Graph the results of Planck's Law (Eq. 8.26) for temperatures of 500°C, 1000°C, 1500°C, and 2000°C.

**8.43**   Use Eq. (8.30) to prepare a graph showing the relationship between $T$ and $T_f$ if the emissivity $\varepsilon$ is

(a)   0.1          (d)   0.4
(b)   0.2          (e)   0.6
(c)   0.3          (f)   0.8

**8.44**   Use Eq. (8.31) to prepare a graph of error $dT/T$ versus temperature $T$ over the range from 1000°C to 4000°C for an optical pyrometer if the emissivity $\varepsilon$ of the surface is in error by

(a)   $d\varepsilon/\varepsilon = 0.1$
(b)   $d\varepsilon/\varepsilon = 0.2$

**8.45**   Use Eq. (8.34) to prepare a graph of the output ratio $E_o/K\varepsilon$ as a function of temperature $T$ for a photon detector over the temperature range from 500°C to 3000°C.

**8.46**   Derive an expression for the change in temperature $dT/T$ as a function of the change in emissivity $d\varepsilon/\varepsilon$ for a photon detector.

**8.47**   Use the results of Exercise 8.46 to prepare a graph of error in the temperature measurement $dT/T$ as a function of temperature $T$ for a photon detector if the estimate of the emissivity of a surface is in error by

(a)   $d\varepsilon/\varepsilon = 0.1$
(b)   $d\varepsilon/\varepsilon = 0.2$

**8.48**   Outline the procedure you would follow to measure the time constant $\beta$ for a thermocouple. See Eq. (8.41).

**8.49**   If $T_m = A \sin \omega t$ (a medium undergoing a cyclic variation in temperature), develop an expression for the temperature $T$ recorded by a temperature sensor as a function of time $t$. Interpret the results in terms of the response time of the sensor.

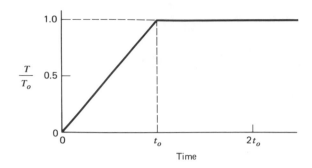

**8.50** Develop an expression for the response of a temperature sensor to the truncated-ramp type of input function shown in Fig. E8.50 for

(a) $0 < t \leqslant t_o$
(b) $t > t_o$

**8.51** Outline the procedure you would follow to determine the lag time associated with thermocouple response to a ramp-function type of input.

# NINE
# FLUID FLOW
# MEASUREMENTS

## 9.1 INTRODUCTION

Fluid flow measurements, expressed in terms of either volume flow rate or mass flow rate, are used in numerous applications, such as industrial process control, city water systems, petroleum or natural-gas pipeline systems, etc. The fluid may be a liquid, a gas, or a mixture of the two (mixed-phase flow). The flow can be confined or closed (as in a pipe or conduit), semiconfined (as in a river or open channel), or unconfined (as in the wake behind a fan or jet). In each case, several methods of flow measurement can be used to determine the required flow rates. Several of the more common measurement techniques are discussed in this chapter.

The concept of mass flow rate can be visualized by considering confined flow in a circular pipe as shown in Fig. 9.1. The mass flow rate $dm/dt$ (or $d\dot{m}$) through area $dA$ surrounding point $P$ (see Fig. 9.1$a$) can be expressed as

$$d\dot{m} = \rho V \, dA$$

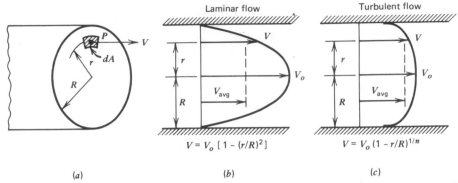

**Figure 9.1**    Mass flow rate in a closed conduit—general concept and velocity profiles. (a) General concept; (b) velocity profile; (c) velocity profile.

**where**    $\rho$ is the density of the fluid at point $P$.

   $V$ is the velocity of the fluid at point $P$ in a direction normal to area $dA$.

The total mass flow rate through the cross section of pipe containing point $P$ is

$$\dot{m} = \int_A d\dot{m} = \int_A \rho V \, dA \qquad (9.1)$$

Equation (9.1) is valid for any plane area and both fluid density and fluid velocity can vary over the cross section. When the fluid is either a liquid or a gas, the density is usually constant; therefore, the density term can be factored out of the second integral in Eq. (9.1) to yield an integral involving a volume per unit time (volume flow rate).

   For steady laminar flow, a parabolic velocity profile exists, as shown in Fig. 9.1b, which can be expressed as

$$V = V_o \left( 1 - \frac{r^2}{R^2} \right) \qquad (9.2)$$

**where**    $V_o$ is the centerline velocity.

   $R$   is the inside radius of the pipe.

   $r$   is a position parameter.

In those instances when the density of the fluid is a constant, Eq. (9.1) can be used to define an "average" velocity $V_{avg}$. Thus

$$\dot{m} = \int_A \rho V \, dA = \rho V_{avg} A \qquad (9.3)$$

For the case of laminar flow in a circular pipe, Eq. (9.3) gives

$$\dot{m} = \rho V_o \int_0^{2\pi} \int_0^R \left(1 - \frac{r^2}{R^2}\right) r \, dr \, d\theta$$

$$= \rho \pi R^2 \frac{V_o}{2} = \rho A V_{avg}$$

Since the cross-sectional area of the pipe $A$ equals $\pi R^2$,

$$V_{avg} = \frac{V_o}{2} \tag{9.4}$$

In fully developed turbulent flow in a smooth pipe, as shown in Fig. 9.1c, the velocity profile has been established by careful measurements to be of the form

$$V = V_o \left(1 - \frac{r}{R}\right)^{1/n} \tag{9.5}$$

The exponent $n$ depends upon Reynolds number, which for the circular pipe can be expressed as

$$R_e = \frac{\rho V_{avg} D}{\mu}$$

where $\mu$ is the absolute viscosity of the fluid and $D$ is the diameter of the pipe. For the case of the circular pipe it can be shown that the average velocity $V_{avg}$ is related to the centerline velocity $V_o$ by the expression

$$V_{avg} = \frac{2n^2}{(n+1)(2n+1)} V_o \tag{9.6}$$

Values of $n$ as a function of Reynolds number $R_e$ are shown in Fig. 9.2. Also plotted in this figure are velocity ratios $V_{avg}/V_o$ associated with the various values of $n$. From the laminar-flow results of Eq. (9.4) and the turbulent-flow results of Eq. (9.6), it is evident that accurate mass flow rates can be established from velocity measurements at a point only if the velocity profile is accurately known.

An approximation for the integral of Eq. (9.3), which can be applied to any flow problem and which is widely used in practice to establish mass flow rates, is the finite sum representation

$$\dot{m} = \sum_{i=1}^{n} \rho_i A_i V_i = \rho V_{avg} A \tag{9.7}$$

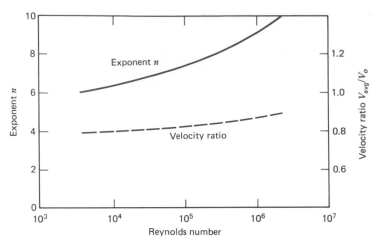

**Figure 9.2**  Exponent $n$ and velocity ratio $V_{avg}/V_o$ as a function of Reynolds number.

As an example, consider the open-channel-flow problem illustrated in Fig. 9.3. Here, the mass flow rate through each area $A_i$ is summed to obtain the total mass flow rate.

In the sections that follow, different flow measurement methods are presented that utilize either velocity measurements at a point or average velocity at a cross section. Usually, devices inserted into the flow (insertion-type transducers) provide velocity information at a point. Other devices, such as orifices, nozzles, and weirs, alter the basic flow in such a way that changes in pressure can be related to the average flow rate.

## 9.2  FLOW VELOCITY (Insertion-Type Transducers)

In this section, transducers designed to measure velocity at a point in the flow are discussed. The types of transducers considered are pitot tubes, hot-wire and hot-film anemometers, drag-force transducers, turbine meters, and vortex shedding devices.

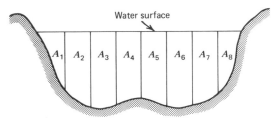

**Figure 9.3**  Measurement of mass flow rate in an open channel.

## Pitot Tube (Incompressible Flow)

The use of a pitot tube to measure velocity at a point in a fluid is illustrated schematically in Fig. 9.4. The velocity of the fluid at point $O$ is measured by inserting the open-ended pitot tube just downstream from point $O$. As the fluid particles move from point $O$ to point $S$ (the stagnation point at the center of the pitot-tube opening), their velocity decreases to zero as a result of the increased pressure at point $S$. The piezometer tube above point $O$ measures the static pressure at point $O$, while the pitot tube measures the total pressure at the stagnation point $S$. The dynamic pressure is the difference between the total pressure and the static pressure. For the streamline from point $O$ to point $S$, the Bernoulli equation gives

$$\frac{p_o}{\gamma} + \frac{V_o^2}{2g} = \frac{p_s}{\gamma} + \frac{V_s^2}{2g} \tag{9.8}$$

**where**   $V_o$ is the velocity of the fluid at point $O$.
$V_s$ is the velocity of the fluid at point $S$.
$g$  is the local acceleration due to gravity.
$\gamma$  is the weight per unit volume of the fluid.
$p_o$ is the static pressure at point $O$.
$p_s$ is the stagnation pressure at point $S$.

Since $V_s = 0$, the dynamic pressure ($p_s - p_o$) is

$$p_s - p_o = \frac{\gamma V_o^2}{2g} = \gamma h \tag{9.9}$$

The velocity at point $O$, as obtained from Eq. (9.9), is

$$V_o = \sqrt{2g \left( \frac{p_s - p_o}{\gamma} \right)} = \sqrt{2gh} \tag{9.10}$$

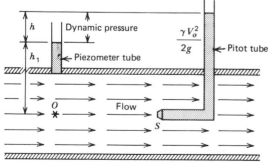

*Figure 9.4*   Velocity measurement with a pitot tube.

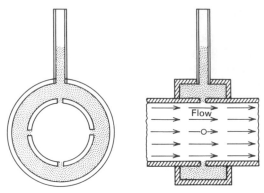

***Figure 9.5*** Piezometer ring for making static pressure measurements.

Velocity measurements made with a pitot tube require accurate static pressure measurements. Slight geometric errors in the pressure tap, such as a rounded corner or a machining burr, can lead to significant errors in the static pressure measurement. To minimize such errors, static pressure measurements are often made with a piezometer ring, as illustrated in Fig. 9.5. The use of multiple pressure taps around the periphery of the tube tends to minimize the static pressure errors.

*Pitot–static tubes* are compact, efficient, velocity measuring instruments that combine static-pressure measurements and stagnation-pressure measurements into a single unit, as illustrated schematically in Fig. 9.6. The static pressure recorded by a pitot–static tube is usually lower than the true static pressure because of the increase in velocity of the fluid near the tube. This difference between indicated and true static pressure can be accounted for by inserting an

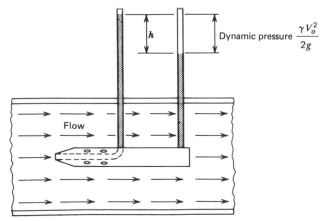

Dynamic pressure $\dfrac{\gamma V_o^2}{2g}$

***Figure 9.6*** Velocity measurement with a pitot–static tube.

instrument coefficient $C_I$ into Eq. (9.10). Thus

$$V_o = C_I \sqrt{2g\left(\frac{p_s - p_o'}{\gamma}\right)} \tag{9.11}$$

**where** $C_I$ is an experimentally determined calibration constant for the tube.
$p_o'$ is the indicated static pressure.

In a pitot–static tube designed by Prandtl, the static pressure tap is placed at a location where the drop in static pressure due to the increase in velocity of the fluid near the tube is exactly equal to the increase in static pressure due to fluid stagnation along the leading edge of the support stem. Thus, for the Prandtl pitot–static tube, the instrument coefficient $C_I$ is unity. For other pitot–static tubes, the instrument coefficient $C_I$ must be determined by calibration.

The stagnation pressure $p_s$ can be measured easily and accurately for most flow conditions. Four factors that affect the accuracy of the stagnation-pressure reading are geometry of the pitot tube, misalignment (yaw) of the pitot tube with the flow direction, viscous effects at low Reynolds numbers, and transverse-pressure gradients in flows with high-velocity gradients.

Geometric and misalignment effects are illustrated in Fig. 9.7, where dynamic pressure measurement error $\mathcal{E}$ is plotted as a function of pitot-tube orientation (angle of attack of the pitot tube with respect to the flow direction). In Fig. 9.7,

$$\mathcal{E} = \frac{\Delta p_d}{p_d} = \frac{p_d - p_d'}{p_d}$$

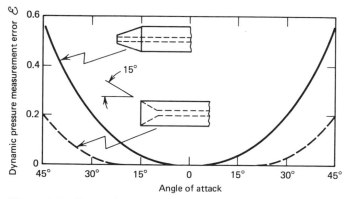

**Figure 9.7** Dynamic pressure measurement error as a function of pitot-tube orientation. (See NACA TN 2331, April 1951, for complete details.)

**where**   $p'_d$ is the measured dynamic pressure.

$p_d$ is the true dynamic pressure ($p_d = p_s - p_o = \gamma V_o^2/2g$).

From Fig. 9.7 it is evident that a square-end pitot tube with a 15-degree internal bevel angle is capable of providing dynamic pressure data with errors of less than 1 percent, provided the yaw angle is less than 25 degrees. Thus, with exercise of reasonable care and with normal flow conditions, small errors in pitot-tube orientation does not produce serious error in the dynamic pressure measurement.

A pitot-tube coefficient $C_P$ can be defined that provides a measure of viscous effects due to flow around the pitot tube itself. Thus,

$$C_P = \frac{p'_s - p_o}{p_d} = \frac{p'_s - p_o}{\gamma V_o^2/2g}$$

where $p'_s$ is the measured stagnation pressure in the presence of viscous effects. Experimental data showing pitot-tube coefficient $C_P$ as a function of Reynolds number $R_e$ is presented in Fig. 9.8. From these results it can be seen that for Reynolds numbers greater than 1000, there is no effect of viscosity, and $C_P$ is equal to unity. In the range $50 \leqslant R_e \leqslant 1000$, $C_P$ is slightly less than unity (has a minimum value of 0.99). For values of $R_e$ less than 50, $C_P$ is always greater than unity. For $R_e < 1$, the coefficient $C_P$ is given by the approximate expression $C_P \approx 5.6/R_e$. The data of Fig. 9.8 clearly show that viscous effects are important only for Reynolds numbers less than 50.

When a pitot tube is placed in a flow with a large velocity gradient, the flow around the tip of the tube can be significantly altered with respect to the uniform flow situaion away from the tube. This altered flow causes the sensed total pressure to be greater than normal, since the effective center of the pitot tube is shifted from the geometric center of the tube toward the region of higher velocity. Since the amount of center shift has been shown to be limited to

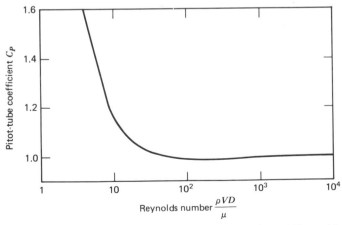

**Figure 9.8**   Pitot-tube coefficient $C_P$ as a function of Reynolds number.

approximately $0.2D$ for square-end pitot tubes (where $D$ is the diameter of the pitot tube), very-small-diameter tubes should be used for measurements in flows with high-velocity gradients.

In Figs. 9.4 and 9.6, the pressures $p_s$ and $p_o$ are illustrated with manometers. In practice, the pressures can be measured with manometers, pressure gages, or one of the pressure transducers discussed in Section 6.4. Differential pressure transducers are also commercially available for measuring the pressure difference $(p_s - p_o)$ directly. One type of differential pressure transducer utilizes a thin diaphragm as the elastic element and electrical resistance strain gages as the sensor. This differential pressure transducer, when used in conjunction with a Wheatstone bridge, provides for continuous monitoring of the pressure difference $(p_s - p_o)$, and thus for continuous monitoring of velocity $V_o$ and mass flow rate $\dot{m}$ by using Eqs. (9.10) and (9.7).

## Pitot Tube (Compressible Flow)

In previous discussions, the fluid was assumed to be incompressible. For compressible flow, the situation is much more complicated and requires use of the equation of motion for a compressible fluid, energy considerations, and a description of the flow process.

The equation of motion (Euler equation) for one-dimensional flow of a compressible ideal (nonviscous or frictionless) fluid can be expressed as:

$$\frac{dp}{\rho} + V\,dV = 0 \qquad \text{or} \qquad \frac{dp}{\gamma} + \frac{V\,dV}{g} = 0 \qquad (9.12)$$

**where** $\rho$ is the density of the fluid.

$\gamma$ is the specific weight of the fluid.

Equations (9.12) are based on the assumption that compressible-flow problems are usually concerned with gases of light weight and with flows in which changes of pressure and velocity predominate and changes of elevation are negligible. When there is no heat transfer and no work done by pumps or turbines in the flow of an ideal fluid, the flow is isentropic and the steady flow energy equation along a streamline reduces to Eq. (9.12). Thus, the energy and Euler equations are identical for isentropic flow of an ideal fluid.

When Eq. (9.12) is integrated along a streamline for isentropic flow of a perfect gas $(p_1/\rho_1^k = p_2/\rho_2^k)$, it becomes

$$\begin{aligned}
\frac{V_2^2 - V_1^2}{2} &= \frac{p_1}{\rho_1}\frac{k}{k-1}\left[1 - \left(\frac{p_2}{p_1}\right)^{(k-1)/k}\right] \\
&= \frac{p_2}{\rho_2}\frac{k}{k-1}\left[\left(\frac{p_1}{p_2}\right)^{(k-1)/k} - 1\right]
\end{aligned} \qquad (9.13)$$

where $k$ is the *adiabatic exponent* (ratio of specific heat at constant pressure to specific heat at constant volume for the gas) $k = c_p/c_v$.

In gas dynamics, Eq. (9.13) is usually expressed in terms of *Mach number* $M$ (ratio of velocity $V$ to sonic velocity $c$). Since the sonic velocity $c$ is given by the expression $c = \sqrt{kp/\rho}$, Eq. (9.13) can be written as

$$\frac{V_2^2}{c_1^2} = M_1^2 + \frac{2}{k-1}\left[1 - \left(\frac{p_2}{p_1}\right)^{(k-1)/k}\right]\tag{9.14}$$

Application of Eq. (9.14) to a stagnation point in a compressible flow ($V_2 = 0$ and $p_2 = p_s$) together with the free-stream conditions ($p_1 = p_o$ and $M_1 = M_o$) yields

$$\frac{p_s}{p_o} = \left[1 + \frac{k-1}{2}M_o^2\right]^{k/(k-1)}\tag{9.15}$$

Equation (9.15) indicates that measurements of stagnation pressure $p_s$ and static pressure $p_o$ provide sufficient data for the determination of free-stream Mach number of the undisturbed flow. Determination of the velocity $V_o$, however, requires measurement of a temperature in addition to the static and stagnation pressures. In practice, the stagnation temperature $T_s$ is easiest to measure since a thermocouple can be placed near the stagnation point on the pitot tube. Once $T_s$, $p_s$, and $p_o$ are known, the velocity $V_o$ can be determined by using the expression

$$\frac{V_o^2}{2} = c_p T_s\left[1 - \left(\frac{p_o}{p_s}\right)^{(k-1)/k}\right]\tag{9.16}$$

A comparison of velocities predicted by Eq. (9.10) with those given by Eq. (9.16) shows agreement within 1 percent for pressure differences ($p_s - p_o$) less than 0.83 psi or Mach numbers less than 0.28. For larger pressure differences or Mach numbers, the agreement becomes less satisfactory and Eq. (9.16) should be used for velocity determinations. At room temperature, the above values correspond to a velocity of 320 ft/s in air.

For supersonic flow ($M_o > 1$), a shock wave forms in front of the pitot tube upstream from the stagnation point. Velocity calculations for this complicated case are beyond the scope of this elementary text.

## Hot-Wire and Hot-Film Anemometers

Hot-wire and hot-film anemometers are devices that can be used to measure either velocity or velocity fluctuations (at frequencies up to 500 kHz) at a point in a fluid (liquid or gas) flow. Typical sensing elements (hot-wire and hot-film)

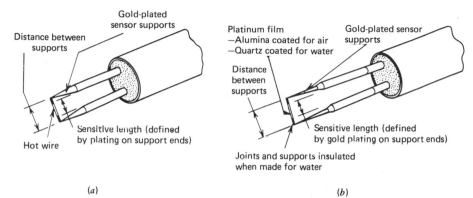

**Figure 9.9** Typical sensing elements (hot-wire and hot-film) and their supports. (a) Hot-wire sensor. (b) Hot-film sensor. (Courtesy of TSI Incorporated.)

and their supports are shown in Fig. 9.9. Hot-wire sensors are fabricated from platinum, platinum-coated tungsten, or a platinum–iridium alloy. Since the wire sensor is extremely fragile, hot-wire anemometers are usually used only for clean air or gas applications. Hot-film sensors, on the other hand, are extremely rugged; therefore, they can be used in both liquids and contaminated gas environments. In the hot-film sensor, the high-purity platinum film is bonded to a high-strength, fused-quartz rod. After the platinum film is bonded to the rod, the thin film is protected by using a thin coating of alumina if the sensor will be used in a gas, or a thin coating of quartz if the sensor will be used in a liquid. The alumina coatings have a high abrasion resistance and a high thermal conductivity. Quartz coatings are less porous and can be used in heavier layers for electrical insulation. Other hot-film anemometer shapes for special-purpose applications include conical, wedge, and hemispheric probes.

Hot-wire and hot-film anemometers measure velocity indirectly by relating power supplied to the sensor (rate of heat transfer from the sensor to the surrounding cooler fluid) to the velocity of the fluid in a direction normal to the sensor. Heat transfer from a heated wire placed in a cooler flowing fluid was studied by King,[1] who established that the heat transfer rate $dQ/dt$ is given by the expression

$$\frac{dQ}{dt} = (A + B\sqrt{\rho V})(T_a - T_f) = P = I_a^2 R_a \qquad (9.17)$$

**where**

      $A$ and $B$ are calibration constants.

      $\rho$   is the density of the fluid.

---

[1] "On the Convection of Heat from Small Cylinders in a Stream of Fluid, with Applications to Hot-Wire Anemometry," by L. V. King, *Philosophical Transactions of the Royal Society of London*, vol. 214, no. 14, 1914, pp. 373–432.

$V$ is the free-stream velocity of the fluid.

$T_a$ is the absolute temperature of the anemometer (hot wire or hot film).

$T_f$ is the absolute temperature of the fluid.

$I_A$ is the current passing through the wire (or film) sensor.

$R_a$ is the resistance of the wire (or foil) sensor.

The quantity $(T_a - T_f)$ is typically maintained at approximately 450°F in air and 80°F in water.

Materials used for hot-wire and hot-film sensors exhibit a change in resistance with temperature change. The resistance–temperature effect can be represented with sufficient accuracy for thermal anemometer applications by the linear expression

$$R_a = R_r[1 + \alpha(T_a - T_r)] \tag{9.18}$$

**where**    $R_r$ is the resistance of the sensor at reference temperature $T_r$.

$\alpha$ is the temperature coefficient of resistance of the wire or foil.

Equation (9.17) indicates that an indirect measurement of fluid velocity $V$ can be made through a measurement of either the current $I_a$ or the resistance $R_a$. In practice, the velocity measurement is made by using a hot-wire or a hot-film anemometer as the active element in a Wheatstone bridge. In one bridge configuration, sensor current $I_a$ is maintained at a constant value (the constant-current circuit) and sensor resistance $R_a$ changes with fluid flow and produces an output voltage $E_o$ from the Wheatstone bridge that can be related to the velocity of the fluid. In the second Wheatstone-bridge configuration (the constant-temperature circuit), the sensor resistance $R_u$ (and thus the sensor temperature $T_a$) is held at a constant value by varying the current passing through the sensor as the fluid velocity changes. In this circuit, the current $I_a$ is used to provide a measure of the velocity. A complete description of the characteristics of each of these circuits follows.

### Constant-Current Anemometer Circuit

A constant-current Wheatstone bridge with a hot-wire (or film) anemometer as the sensor is shown schematically in Fig. 9.10. In this bridge arrangement, resistance $R_2$ is much larger than sensor resistance $R_a$; therefore, current $I_a$ is for all practical purposes independent of changes in sensor resistance $R_a$. Variable resistor $R_4$ is used to balance the bridge under conditions of zero velocity. Any flow past the sensor cools the hot wire (or film), decreases its resistance as indicated by Eq. (9.18), and thereby unbalances the bridge. The unbalancing of the bridge produces an output voltage $E_o$, as given by Eq. (4.34), which is related to the fluid velocity $V$. The output voltage $E_o$ from the bridge is small;

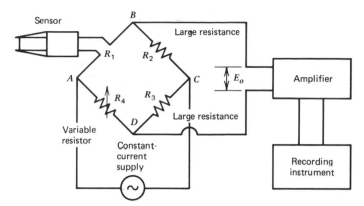

*Figure 9.10* Constant-current anemometer circuit.

therefore, considerable amplification is needed before it can be used to drive most voltage measuring instruments.

An important application of hot-wire or hot-film anemometry in addition to steady-flow velocity measurements is the measurement of turbulence. The voltage fluctuations between points $B$ and $D$ of the constant-current Wheatstone bridge (see Fig. 9.10) are a direct indication of velocity fluctuations (intensity of turbulence) in the flow. Thermal lag (the inability of the wire or film to transfer heat fast enough to the surrounding fluid) limits the resolution of these velocity fluctuations to several hundred hertz. Above a cutoff frequency $f_c$ (dependent upon sensor material, coating, and diameter), the wire or film acts as an integrator and attenuates the fluctuations at 6 dB/octave, as shown in Fig. 9.11. This attenuation problem can be solved electronically by passing the bridge output through a low-noise, high-gain amplifier to obtain a high-level signal that can be differentiated for frequencies above the cutoff frequency to obtain a flat (compensated) frequency response up to approximately 120 kHz. The steady flow and low-frequency components of the velocity (below 1 Hz) are attenuated

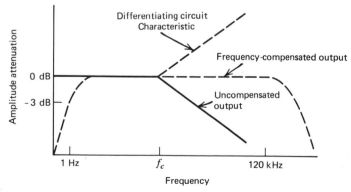

*Figure 9.11* Frequency response of a constant-current anemometer circuit.

as shown in Fig. 9.11. Since the cutoff frequency is different for each hot-wire (or film) probe, provision must be provided in the associated electronic system for making a cutoff frequency adjustment.

Two outstanding features of a constant-current anemometer system are its low noise level and its excellent sensitivity. Turbulence levels less than 0.005 percent of the mean velocity can be resolved in the 10-kHz frequency band. The range of frequencies for a typical anemometer is from 1 to 120,000 Hz. Also, the constant-current anemometer is very sensitive to small changes in velocity at low velocities (the calibration curve has a steep slope at zero velocity); therefore, it is an excellent instrument for low-velocity measurements.

The constant-current anemometer system has two disadvantages that have motivated development of the constant-temperature anemometer system. First, the frequency response of the constant-current anemometer is separated into two bands: the uncompensated low-frequency band and the compensated high-frequency band. Second, the compensated output is distorted when small high-frequency fluctuations are measured in the presence of large-amplitude, low-frequency oscillations.

### Constant-Temperature Anemometer Circuits

A simple constant-temperature circuit incorporating a hot-wire or hot-film anemometer as the sensor, which is ideal for steady-flow measurements, is shown schematically in Fig. 9.12. The hot-wire (or film) sensor is used as the active element in a Wheatstone bridge that is balanced under no-flow conditions by using variable resistor $R_4$. As flow past the sensor cools the hot wire (or film), sensor resistance $R_a$ decreases and the bridge becomes unbalanced. The unbalance is sensed with a galvanometer placed across points $B$ and $D$ of the bridge,

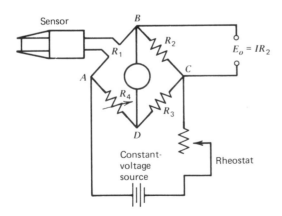

*Figure 9.12* Constant-temperature anemometer circuit.

as shown in Fig. 9.12. The balance condition can be restored by adjusting the rheostat to increase the input voltage $E_i$ to the bridge. The increase in bridge voltage increases the current flowing through the sensor and increases both sensor temperature $T_a$ and sensor resistance $R_a$ back to their no-flow values. Under conditions of constant sensor temperature and resistance, Eq. (9.17) can be reduced to

$$V = C_0 \left[ \left( \frac{I}{I_0} \right)^2 - 1 \right]^2 \tag{9.19}$$

**where**  $C_0$ is a calibration constant for the particular hot-wire (or film) probe.
$I_0$ is the current at zero velocity that gives the desired sensor temperature.
$I$ is the sensor current at velocity $V$.

The user must determine the constants $C_0$ and $I_0$ for his application. With precision bridge resistors, a sensitive galvanometer, and constant fluid temperature, better than 1 percent repeatability in steady-flow velocity measurements can be made with this simple, constant-temperature anemometer system. Any significant changes in fluid temperature, however, require rebalance and recalibration of the system, since the quantity $(T_a - T_f)$ in Eq. (9.17) must remain constant for Eq. (9.19) to be valid.

The current $I$ or $I_0$ passing through the sensor is easily measured by recording the voltage drop across resistance $R_2$ in the Wheatstone bridge of Fig. 9.12. Since the sensor current is proportional to the input voltage $E_i$ for this constant resistance bridge, input voltage, also, can be used as a measure of sensor current. A typical calibration curve showing sensor current $I$ $(E_o = IR_2)$ as a function of velocity $V$ is shown in Fig. 9.13.

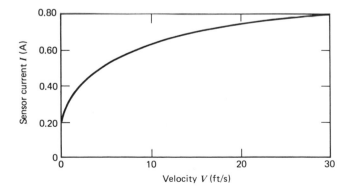

*Figure 9.13* Typical calibration curve for a hot-film anemometer in water.

A more sophisticated constant-temperature anemometer system is illustrated schematically in Fig. 9.14. Again, the hot-wire or hot-film sensor is used as the active element in a Wheatstone bridge that is balanced under no-flow conditions. As flow past the sensor cools the hot wire (or film), its resistance decreases and unbalances the bridge. A high-gain, differential amplifier is used to sense this unbalanced condition and changes the bridge voltage to increase the current flowing through the sensor. The increase in current flow restores sensor temperature $T_a$ and sensor resistance $R_a$ to no-flow values and balances the bridge. This constant-temperature anemometer system automatically maintains a balanced bridge and corrects for the thermal lag of the wire (or the film). In some systems, resistor $R_4$ of the bridge is mounted in the probe and is used to sense and correct for any change in fluid temperature.

Since the Wheatstone bridge is operated as a constant-resistance bridge in this system, the output from the amplifier (input voltage to the bridge) contains steady and fluctuating velocity information up to the upper limiting frequency of the system. Sensor current $I$, which can be easily measured by recording the voltage drop across resistance $R_2$, provides the same information. A typical constant-temperature anemometer system exhibits a frequency response from dc to approximately 500 kHz and a resolution of approximately 0.03 percent of the mean velocity in the 10-kHz band. The upper limiting frequency of this constant-temperature system is controlled by sensor size, amplifier gain, and other features of the components of the system.

An important feature of this constant-temperature anemometer system is the provision for linearized output that makes possible accurate measurements of large-amplitude, low-frequency velocity fluctuations. In addition, the con-

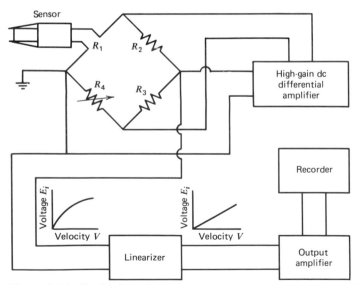

***Figure 9.14*** Sophisticated constant-temperature anemometer system.

stant-temperature type of anemometer system is preferred for studies involving steady-flow velocity profiles and flow-reversal measurement situations where it is impossible to distinguish the direction of flow. The constant-current type of anemometer system is preferred for low-turbulence and high-accuracy measurements. These general distinctions are gradually disappearing as probe design and amplifier technology progress to overcome specific limitations. Eventually, the constant-temperature system will be used exclusively due to its more desirable characteristics. Ultimately, the nonlinear problems will be overcome by using specially designed microprocessors for data reduction.

When hot-wire or hot-film sensors are used in liquids, other problems arise. First, liquids often carry dirt particles, lint, or organic matter. These materials can quickly coat the hot wire or film and cause significant reductions in heat transfer. Second, the presence of a current-carrying wire in a conducting medium can cause electrolysis of metals, shunting of the hot wire, and spurious changes in sensitivity. Third, the presence of the hot wire can cause formation of bubbles that significantly reduce the heat transfer. Bubbles can arise from entrained air or gas in the liquid, from electrolysis, or from boiling of the liquid. Successful use of anemometers in liquids usually require low wire temperatures, coatings on the hot wires, lower operating voltages, degasification of the liquid, and application of other bubble-removal techniques.

## Drag-Force Velocity Transducers

Drag-force velocity transducers operate on the principle that the drag force $F_D$ on a body in a uniform flow is related to the fluid velocity $V$, the fluid density $\rho$, and frontal area $A$ of the body normal to the flow direction by the expression

$$F_D = C_D \frac{\rho V^2 A}{2} \tag{9.20}$$

where $C_D$ is a nondimensional parameter known as the drag coefficient.

The drag coefficient for a specific transducer depends upon Reynolds number $R_e$ and the shape of the body. Drag coefficients for a sphere and for a circular disk are shown in Fig. 9.15. Since the drag coefficient for the circular disk is constant over a wide range of velocities ($C_D \approx 1.05$ for $R_e \geqslant 3000$) and since the magnitude of the drag coefficient for the disk is more than twice that for the sphere, the circular disk is the preferred sensing element for a drag-force transducer.

## Rotameter

The *rotameter* is a common flow measurement device whose operation is based on drag principles. The rotameter, shown schematically in Fig. 9.16, consists of a tapered tube and a solid float (bob) that is free to move vertically in

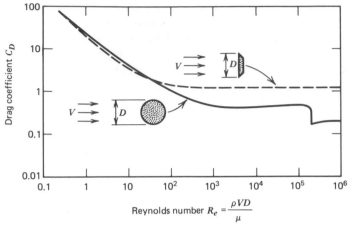

**Figure 9.15** Drag coefficients for a sphere and a circular disk as a function of Reynolds number.

the tube. At any flow rate within the range of the meter, fluid entering the bottom of the tube raises the float (thereby increasing the area between the float and the tube) until the drag and buoyancy forces are balanced by the weight of the float. This condition can be expressed by the equation

$$c_D \frac{\rho_f V^2 A_b}{2} + \rho_f V_b g = \rho_b V_b g \tag{9.21}$$

where $A_b$ is the frontal area of the float, $V_b$ is the volume of the float, $\rho_b$ is the density of the float, and $V$ is the mean velocity of the fluid in the annular space between the float and the tube. The first term in Eq. (9.21) is the drag force on the float, the second term is the buoyancy force on the float, and the third term is the weight of the float. The annular area $A$ can be expressed as

$$A = \frac{\pi}{4}[(D + ay)^2 - d^2] \approx \frac{\pi}{2} Day \tag{9.22}$$

where $a$ is a constant that describes the taper of the tube. The mass flow rate $\dot{m}$ is obtained from Eqs. (9.21) and (9.22) as

$$\dot{m} = \rho_f A V$$

$$= \frac{\pi}{2} Day \sqrt{\frac{2gV_b}{C_D A_b} (\rho_b - \rho_f)\rho_f} \tag{9.23}$$

$$= K\sqrt{(\rho_b - \rho_f)\rho_f}\, y$$

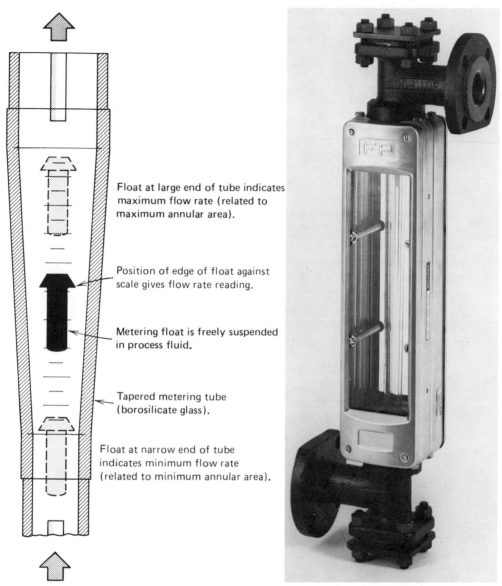

Float at large end of tube indicates maximum flow rate (related to maximum annular area).

Position of edge of float against scale gives flow rate reading.

Metering float is freely suspended in process fluid.

Tapered metering tube (borosilicate glass).

Float at narrow end of tube indicates minimum flow rate (related to minimum annular area).

*Figure 9.16*  Schematic illustration and photograph of a glass-tube industrial rotameter-type flowmeter. (*a*) Principle of operation. (*b*) Glass-tube rotameter. (Courtesy of Fischer & Porter.)

The drag coefficient can be made nearly independent of viscosity by using sharp edges on the float. The sensitivity to fluid density can be minimized by selecting $\rho_b = 2\rho_f$. Equation (9.23) then reduces to

$$\dot{m} = \frac{K\rho_b}{2} y \tag{9.24}$$

Thus, flow rate is indicated by the position of the float, which can be measured on a graduated scale or detected magnetically and transmitted to a remote location for recording.

## Current Meters

*Current meters* are mechanical devices that are widely used to measure water velocities in open rivers, channels, and streams. The rotational speed of the device is proportional to fluid velocity. A direct-reading, cup-type current meter with sensing unit, cable suspension, torpedo-shaped lead weight, and alignment fins is shown in Fig. 9.17. As the cup wheel rotates, a magnetically activated reed switch in the sensing unit produces a train of electrical pulses at a frequency proportional to the speed of the cup wheel. An electrical circuit in the indicating unit, which utilizes a self-contained battery as the power supply and a pulse rate integrator as the signal conditioning element, converts the train of pulses to a

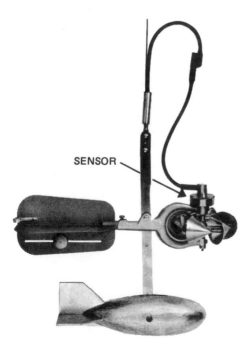

SENSOR

*Figure 9.17* Cup-type current meter with sensing unit, cable suspension, torpedo-shaped lead weight, and alignment fins. (Courtesy of Teledyne Gurley.)

direct display of the fluid velocity that is producing rotation of the wheel. Typically, the range of such instruments is 0 to 25 ft/s. Linearity of the unit shown in Fig. 9.17 is within ±5 percent over the full range of the unit. Error caused by temperature change is approximately 0.05 percent per degree Fahrenheit change from 75°F.

The U.S. Geological Survey (USGS) has established a typical velocity profile (by using data from thousands of measurements) for use in establishing flow rates (discharge rates) in large rivers, canals, and streams. This velocity profile, shown in Fig. 9.18, indicates that the velocity distribution in a typical large stream is parabolic with the maximum velocity occurring some distance (from $0.05h$ to $0.25h$, where $h$ is the depth of the stream) below the free surface. The free-surface velocity is typically 1.18 times the average velocity. The typical profile indicates that the mean velocity often occurs at a point slightly more than midway between the free surface and the bottom of the stream (at $0.6h$). An accurate estimate of the mean velocity can usually be obtained by averaging the velocities at points $0.2h$ and $0.8h$, as indicated in Fig. 9.18.

The flow rate in a stream is measured by using the procedure illustrated schematically in Fig. 9.19. First, a stretch of river with a fairly regular cross section is selected for the measurement. The cross section is then divided into vertical strips, as shown in Fig. 9.19. The mean velocity along each vertical line is determined by averaging velocity measurements at points $0.2h$ and $0.8h$ below the free surface. Thus,

$$V_i = \frac{V_{2i} + V_{8i}}{2} \tag{9.25}$$

The flow rate in a vertical strip is calculated by using the height of the strip, the width of the strip, and the mean velocity for the strip (averaged from the two

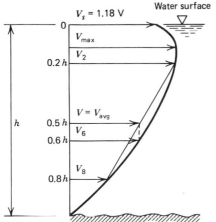

**Figure 9.18** Typical velocity profile in a large river.

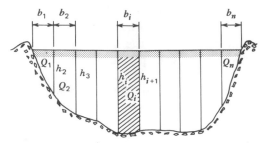

**Figure 9.19**    Flow-rate measurement procedure for a large river or stream.

vertical lines bounding the strip). Thus,

$$Q_i = b_i \left(\frac{h_i + h_{i+1}}{2}\right)\left(\frac{V_i + V_{i+1}}{2}\right) \tag{9.26}$$

The total volume flow rate for the stream is the sum of the flow rates for all of the strips:

$$Q = \sum_{i=1}^{n} Q_i \tag{9.27}$$

From the above it is obvious that many flow-rate data points are needed to establish accurate volume flow rates for rivers and streams.

## Turbine Flowmeters

Basically, a *turbine flowmeter* is a miniature propeller suspended in a pipe. This freely suspended axial turbine (see Fig. 9.20) is rotated by the flow of fluid (either liquid or gas) through the flowmeter. The rotational speed of the turbine is proportional to the velocity of the fluid. Since the flow passage is fixed, the turbine's rotational speed is also a true representation of the volume of fluid flowing through the flowmeter.

The only physical connection between the turbine and its housing is the turbine bearings. The rotation of the turbine is sensed by a pickoff coil in the flowmeter body that responds to the passage of each turbine blade past the coil. The output from the coil is a train of voltage pulses whose frequency is proportional to the volume flow rate. Once the pulses are transmitted to an appropriate recording system near the flowmeter or at a remote location, they can be amplified, counted, interfaced with a computer or microprocessor, and used to measure or control the fluid flow.

Flowmeters have been developed to an outstanding level of accuracy, linearity, durability, and reliability. Flowmeters are commercially available to measure fluid flow within the temperature range from $-430°F$ to $+750°F$. Accuracy

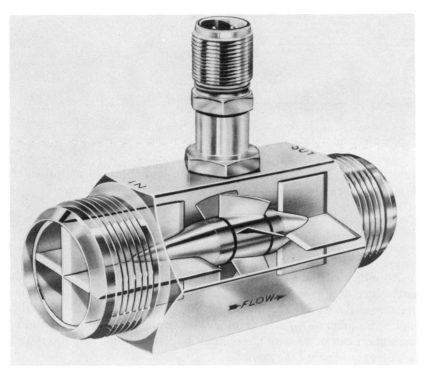

***Figure 9.20***   Cutaway view of a turbine flowmeter. (Courtesy of Flow Technology, Inc.)

within ±0.05 percent in liquids and ±0.5 percent in gases is easily obtained at flow rates from 0.03 to 20,000 gal/min. Turbine flowmeters are currently used to monitor and control critical flow rates in a number of different industrial processes.

## Vortex Shedding Transducers

When a circular cylinder is placed in a uniform flow with its axis perpendicular to the direction of the flow, eddies (called vortices) shed regularly (at a frequency $f_s$) from alternate sides of the cylinder in a pattern known as a Karman vortex street (see Fig. 9.21). The shedding frequency $f_s$, the diameter of the cylinder $D$, and the flow velocity $V_o$ are related by a dimensionless number known as the Strouhal number $S_N$. Thus,

$$S_N = \frac{f_s D}{V_o} \tag{9.28}$$

Experimental studies indicate that the Strouhal number is relatively constant ($0.20 \leqslant S_N \leqslant 0.21$) over the range of Reynolds numbers from 300 to 150,000.

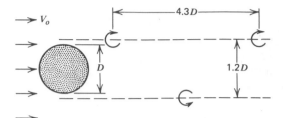

*Figure 9.21*    Schematic illustration of a Karman vortex street.

Since accurate frequency measurements are easy to make, Eq. (9.28) provides a means for making accurate velocity measurements in a flow. Thus,

$$V_o = \frac{f_s D}{S_N} \qquad (9.29)$$

Equation (9.29) indicates that small-diameter cylinders should be used for such measurements, since they give a higher frequency for a given flow velocity. The natural frequency of the mounted cylinder must be considerably higher than the vortex shedding frequency ($f_n/f_s > 3$); otherwise, significant nonlinear fluid–structure interaction can occur with consequent large-amplitude vibrations and large stresses that can destroy the device.

    A recent design of vortex shedding flowmeter uses a triangular wedge (bluff body) in a pipe with circular cross section, as shown in Fig. 9.22. In order to produce strong and consistent vortex shedding, the bluff body must extend across the full width of the pipe. Also, the height $h$ of the bluff body must be an appreciable fraction (approximately one-third) of the diameter of the pipe, and the length of the bluff body in the direction of flow should be approximately 1.3 times the height. The salient edges at the face of the bluff body force the location of the vortex shedding to be fixed, thus giving a consistent Strouhal number of $0.88 \pm 0.01$ over the range of Reynolds numbers from 10,000 to 1,000,000. Two sensor elements are mounted on the front face of the bluff body in such a way that their outputs are out of phase by 180 degrees. This arrangement gives a very good signal-to-noise ratio, since the vortex shedding signals add, while the common noise components cancel. Calibration studies have shown that air at 150, 600, and 1100 psi as well as water gives the same calibration value over a pipe Reynolds number range from 10,000 to 5,000,000. These results support the contention that calibration values for these probes are independent of pressure, temperature, and state of the fluid (liquid or gas). While Fig. 9.22 tends to imply that such devices can be used only in a closed pipe, such devices can also be used in a free-flow field by using a short length of pipe with the bluff body inside. The system, consisting of the bluff body and short pipe, can be calibrated and suspended in the flow in much the same manner as the current meter.

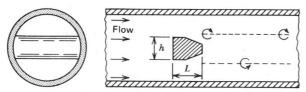

**Figure 9.22**  Use of a bluff body as a vortex shedding flowmeter.

## 9.3  FLOW RATES IN CLOSED SYSTEMS BY PRESSURE-VARIATION MEASUREMENTS

Common devices that measure average velocity or flow rate by means of either a constriction in the stream tube or a change in direction of the flow are considered in this section. The devices to be discussed are the venturi meter, the flow nozzle, the orifice meter, and the elbow meter. The operation of each of these devices is based on the fact that a change in geometry of the stream tube causes a corresponding change in velocity and pressure of the fluid within the tube.

The Bernoulli equation as applied to an ideal incompressible fluid ($\rho_1 = \rho_2$), as shown in Fig. 9.23 is

$$\frac{p_1}{\gamma} + \frac{V_1^2}{2g} + z_1 = \frac{p_2}{\gamma} + \frac{V_2^2}{2g} + z_2 \tag{9.30}$$

The velocity $V_1$ can be eliminated from Eq. (9.30) by using the continuity equation (conservation of mass) that requires that

$$Q = A_1 V_1 = A_2 V_2 \tag{9.31}$$

Thus, the ideal volume flow rate $Q_i$ (for an ideal frictionless fluid) can be expressed as

$$Q_i = A_2 V_2 \tag{9.32}$$

$$= \frac{A_2}{\sqrt{1 - (A_2/A_1)^2}} \sqrt{2g\left(\frac{p_1}{\gamma} + z_1 - \frac{p_2}{\gamma} - z_2\right)}$$

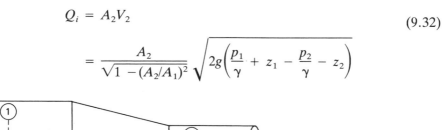

**Figure 9.23**  Illustration of conservation of mass at two locations in a closed system.

For real fluid flow and the same pressure drop term $(p_1/\gamma + z_1 - p_2/\gamma - z_2)$, the flow rate will be less than that predicted by Eq. (9.32) due to friction in the flow between the two pressure measuring points. This energy loss is usually accounted for by introducing an experimentally determined coefficient $C_V$ (coefficient of velocity) into Eq. (9.32). Actual flow rate $Q_a$ is then expressed as

$$Q_a = \frac{C_V A_2}{\sqrt{1 - (A_2/A_1)^2}} \sqrt{2g\left(\frac{p_1}{\gamma} + z_1 - \frac{p_2}{\gamma} - z_2\right)} \qquad (9.33)$$

The pressure difference term $(p_1/\gamma + z_1 - p_2/\gamma - z_2)$ in Eq. (9.33) can be measured with any of the standard pressure measuring devices, such as the differential manometer or the differential pressure transducer. Specific details are presented in the following subsections.

## Venturi Meter

A typical *venturi meter* consists of a cylindrical inlet section, a smooth entrance cone (acceleration cone) having an angle of approximately 21 degrees, a short cylindrical throat section, and a diffuser cone (deceleration cone) having an angle between 5 and 15 degrees. Recommended proportions and pressure tap locations for a venturi meter, as specified by the American Society of Mechanical Engineers (ASME), are shown in Fig. 9.24. Small diffuser angles tend to minimize head loss from pipe friction, flow separation, and increased turbulence. In order for the venturi meter to function properly, the flow must be well established as it enters the inlet pressure ring area. This can be accomplished by installing the meter downstream from a section of straight and uniform pipe having a length of approximately 50 pipe diameters. Straightening vanes can also be installed upstream of the venturi meter to reduce any rotational motion in the fluid. The pressures at the inlet section and at the throat section of the meter should be measured with piezometer rings, as shown in Fig. 9.24.

The coefficient of velocity $C_V$ for different size venturi meters having a pipe-to-throat diameter ratio $D/d$ of 2 is shown plotted as a function of Reynolds

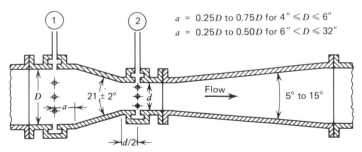

**Figure 9.24** Recommended proportions and pressure tap locations for a venturi meter.

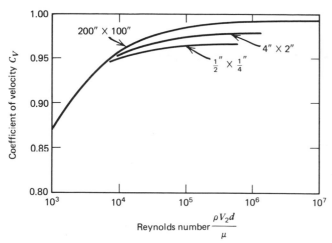

**Figure 9.25** Coefficient of velocity $C_V$ as a function of Reynolds number for venturi meters having a diameter ratio $D/d = 2$.

number at the throat $(R_e = \rho V_2 d/\mu)$ in Fig. 9.25. These data indicate that $C_V$ ranges from 0.97 to 0.99 over a wide range of sizes and Reynolds numbers. Experimental evidence at other diameter ratios indicates that $C_V$ decreases with increasing diameter ratios.

## Flow Nozzle

A *flow nozzle* is essentially a venturi meter with the diffuser cone removed. Since the diffuser cone exists primarily to minimize head loss caused by the presence of the meter in the system, it is obvious that larger head losses will occur in flow nozzles than in venturi meters. The flow nozzle is preferred over the venturi meter in many applications because of its lower initial cost and because of the fact that it can be easily installed between two flanges in any piping system. The recommended ASME geometry for a long-radius flow nozzle is shown in Fig. 9.26.

As shown in Fig. 9.26, the upstream pressure is measured through a port or piezometer ring in the pipe at a location one pipe diameter upstream from the inlet face of the nozzle, while the throat pressure is measured through a port or piezometer ring at a location one-half pipe diameter downstream from the inlet face of the nozzle. Errors associated with the measurement of throat pressure at this location can be corrected for, either in the velocity coefficient $C_V$ for the flow nozzle or in the discharge coefficient $C_D$ for the meter ($C_D = Q_a/Q_i$).

Extensive research on flow nozzles, sponsored by ASME and others, has produced a large number of reliable data on flow nozzle installation and velocity coefficients. Flow coefficients for the long radius ASME flow nozzle are shown in Fig. 9.27.

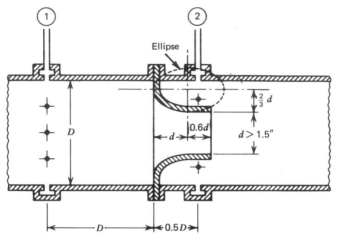

*Figure 9.26* Recommended geometery for an ASME long-radius flow nozzle.

## Orifice Meter

A restricted opening through which fluid flows is known as an orifice. An *orifice meter* consists of a plate with a sharp-edged circular hole (see Fig. 9.28) that is inserted between two flanges of a piping system for the purpose of establishing the flow rate from pressure-difference measurements across the orifice. The flow pattern associated with flow through a sharp-edged orifice plate is shown in Fig. 9.28. This flow pattern indicates that the streamlines tend to converge a short distance downstream from the plane of the orifice; therefore, the minimum-flow area is smaller than the area of the opening in the orifice plate. This minimum-flow area is known as the "contracted area of the jet" or

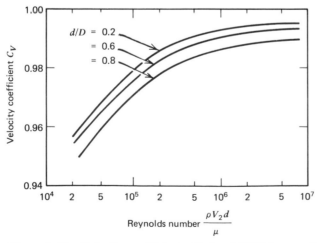

*Figure 9.27* Velocity coefficients $C_V$ for ASME long-radius flow nozzles.

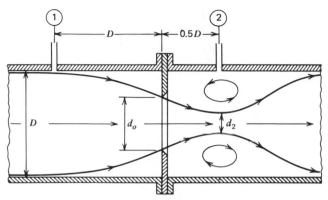

**Figure 9.28** Flow through a sharp-edged orifice plate.

the "vena contracta." The area at the vena contracta is introduced into Eq. (9.33) by using a contraction coefficient $C_C$ such that

$$A_2 = C_C A_0 \tag{9.34}$$

where $A_0$ is the area of the hole in the orifice plate. When Eq. (9.34) is substituted into Eq. (9.33), the flow rate through the orifice becomes

$$Q_a = \frac{C_V C_C A_0}{\sqrt{1 - (C_C A_0/A_1)^2}} \sqrt{2g\left(\frac{P_1}{\gamma} + z_1 - \frac{p_2}{\gamma} - z_2\right)}$$

$$= C A_0 \sqrt{2g\left(\frac{p_1}{\gamma} + z_1 - \frac{p_2}{\gamma} - z_2\right)} \tag{9.35}$$

where the orifice coefficient $C$ is defined as

$$C = \frac{C_V C_C}{\sqrt{1 - (C_C A_0/A_1)^2}} \tag{9.36}$$

Equation (9.36) shows that the value of the orifice coefficient $C$ depends upon the velocity coefficient $C_V$, the contraction coefficient $C_C$, and the area ratio $A_0/A_1$ of the installation. The orifice coefficient $C$ is also affected by the location of the pressure taps. Ideally, the pressure $p_2$ should be measured at the vena contracta; however, the location of the vena contracta changes with Reynolds number and area ratio. As a result, pressure taps are often placed one pipe diameter upstream and one-half pipe diameter downstream from the inlet face of the orifice plate. Typical variations of orifice coefficient $C$ as a function of the ratio of orifice diameter to pipe diameter are shown in Fig. 9.29 for different

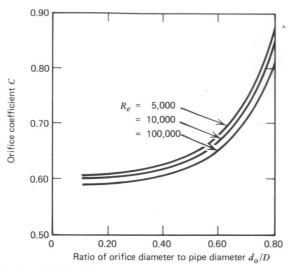

**Figure 9.29**  Orifice coefficient $C$ for different diameter ratios and Reynolds numbers.

Reynolds numbers based on pipe inlet conditions of velocity and diameter. For Reynolds numbers greater than 100,000, the value of $C$ remains essentially constant.

## Elbow Meter

The venturi meter, flow nozzle, and orifice meter, which are widely used devices for measuring flow rates in pipes, all contribute to the energy losses in the system. Elbow meters, on the other hand, do not introduce additional losses in the system, since they can simply replace an existing elbow in the system that is being used to change the direction of flow. The principle of operation of an elbow meter is illustrated in Fig. 9.30. Experimental studies indicate that the

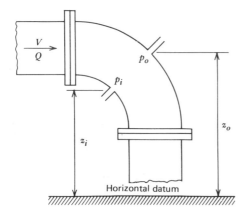

**Figure 9.30**  Principle of operation of an elbow meter.

velocity, pressure, and elevation for pressure taps on the inside and outside surfaces of a 90 degree elbow can be related by the expression

$$C_k \frac{V^2}{2g} = \frac{p_o}{\gamma} + z_o - \frac{p_i}{\gamma} - z_i \tag{9.37}$$

where $C_k$ is a coefficient that depends upon the size and shape of the elbow. Nominal values of $C_k$ range from 1.3 to 3.2. The volume flow rate $Q$ is obtained from pressure-difference measurements from the elbow meter by substituting Eq. (9.37) into Eq. (9.31). Thus

$$Q = AV = \frac{A}{\sqrt{C_k}} \sqrt{2g\left(\frac{p_o}{\gamma} + z_o - \frac{p_i}{\gamma} - z_i\right)}$$

$$= CA \sqrt{2g\left(\frac{p_o}{\gamma} + z_o - \frac{p_i}{\gamma} - z_i\right)} \tag{9.38}$$

where $C = 1/\sqrt{C_k}$ is known as the elbow meter coefficient. Typical values of $C$ range from 0.56 to 0.88.

The primary advantage of the elbow meter is the energy savings. The primary disadvantage is that each meter must be calibrated "in place" or in a calibration facility. The low operating cost can usually justify the calibration cost. The elbow meter, like the other meters in this section, requires a minimum of 10 to 30 pipe diameters of unobstructed upstream flow (to reduce large-scale turbulence and swirl) for satisfactory operation and accurate flow measurements; otherwise, flow straighteners must be used to stabilize the flow prior to entry into the flow metering device.

## 9.4 FLOW RATES IN PARTIALLY CLOSED SYSTEMS

Many variations of the orifice of the previous section are used in practice. For example, consider either the submerged orifice (*a*) or the free discharging orifice (*b*) of Fig. 9.31 that are often used to control fluid flow from one large reservoir to another. In these two cases, the area ratio $A_0/A_1$ is approximately zero; therefore, Eq. (9.35) becomes

$$Q_a = C_V C_C A_0 \sqrt{2g\left(\frac{p_1}{\gamma} + z_1 - \frac{p_2}{\gamma} - z_2\right)}$$

$$= CA_0 \sqrt{2g(h_1 - h_2)} \tag{9.39}$$

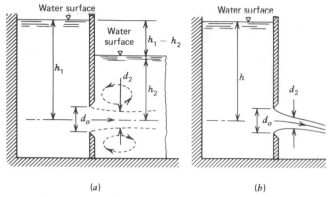

**Figure 9.31** Flow through an orifice between reservoirs. (*a*) Submerged orifice. (*b*) Free discharging orifice.

where $C = C_V C_C$ is the orifice or discharge coefficient, and $h_1$ and $h_2$ are the static heads. Orifice coefficient $C$ depends on the shape of the orifice as shown in Table 9.1.

The coefficients listed in Table 9.1 are nominal values for large-diameter ($d > 1$ in. or 25 mm) orifices operating under static heads ($h_1 - h_2$) in excess of 50 in. (1.25 m) of water. Above these limits of diameter and static head, the coefficients are essentially constant. For smaller diameter orifices and lower static heads, both viscous effects and surface tension effects begin to influence the discharge coefficient.

**TABLE 9.1** Orifices and Their Nominal Coefficients

|  | Sharp edged | Rounded | Short tube[a] | Borda |
|---|---|---|---|---|
| $C$ | 0.61 | 0.98 | 0.80 | 0.51 |
| $C_C$ | 0.62 | 1.00 | 1.00 | 0.52 |
| $C_V$ | 0.98 | 0.98 | 0.80 | 0.98 |

[a]$l \approx 2.5\, d.$

## 9.5 FLOW RATES IN OPEN CHANNELS FROM PRESSURE MEASUREMENTS

Accurate measurement of flow rates in open channels is important for navigation, flood control, irrigation, etc. The two broad classes of devices used for this type of measurement and control are the sluice gate and the weir. Both of these devices require placement of an obstruction (dam) in the flow channel, and thus cause a change in the flow. Pressure information is usually obtained by measuring free-surface elevations and deducing the pressures from these data.

### Sluice Gate

The *sluice gate* is an open-channel version of the orifice meter. As shown in Fig. 9.32, the flow through the gate exhibits jet contraction on the top surface, which produces a reduced area of flow or vena contracta just downstream from the gate. If it is assumed that there are no energy losses (ideal fluid) and that the pressure in the vena contracta is hydrostatic, the Bernoulli equation with respect to a reference at the floor of the channel can be written as

$$y_1 + \frac{V_1^2}{2g} = y_2 + \frac{V_2^2}{2g} \tag{9.40}$$

The velocity $V_1$ can be eliminated from Eq. (9.40) by using the continuity equation ($Q = A_1V_1 = A_2V_2$), and friction losses can be accounted for by introducing

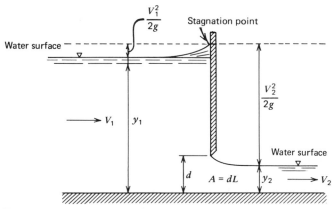

**Figure 9.32** Flow through a sluice gate.

a velocity coefficient $C_V$. With these substitutions, the actual flow rate $Q_a$ becomes

$$Q_a = C_V A_2 V_2$$

$$= \frac{C_V C_C A}{\sqrt{1 - (y_2/y_1)^2}} \sqrt{2g(y_1 - y_2)} \tag{9.41}$$

where    $A$   is the true area of the sluice gate opening.
         $C_C$ is a contraction coefficient that accounts for the reduced area at the vena contracta.

Equation (9.41) indicates that the flow rate or discharge through the sluice gate depends upon the coefficient of velocity $C_V$, the contraction coefficient $C_C$, the depth ratio $y_2/y_1$, and the difference in depths $(y_1 - y_2)$. Frequently, in practice, all of these effects are combined into a single discharge coefficient $C_D$, so that Eq (9.41) can be written simply as

$$Q_a = C_D A \sqrt{2gy_1} \tag{9.42}$$

Values for the discharge coefficient $C_D$ usually range between 0.55 and 0.60 so long as free flow is maintained downstream from the gate. When flow conditions downstream are such that submerged flow exists, the value of the discharge coefficient is significantly reduced.

With a constant upstream head $y_1$, it is obvious from Eq. (9.42) that the flow rate or discharge is controlled by the area of the sluice gate opening. Since the width of the gate is fixed, the height of the gate opening controls the flow rate. The position of the gate can be easily monitored with any displacement measuring transducer; therefore, the flow rate can be easily measured or controlled.

## Weirs

A *weir* can be simply defined as an obstruction in an open channel over which fluid flows. The flow rate or discharge over a weir is a function of the weir geometry and of the *weir head* (vertical distance between the weir crest and the liquid surface in the undisturbed region upstream from the weir). The basic discharge equation for a weir can be derived by considering a sharp-crested rectangular weir, as shown in Fig. 9.33, and applying the Bernoulli equation to a typical streamline with the weir crest as the reference. Thus

$$H + \frac{V_1^2}{2g} = (H - h) + \frac{V_2^2}{2g} \tag{9.43}$$

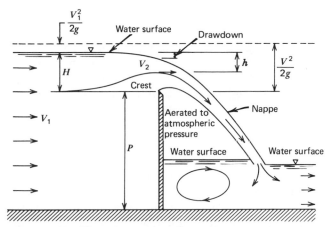

*Figure 9.33*  Flow over a sharp-crested rectangular weir.

**where**  $H$ is the head on the weir.

$h$ is a distance below the free surface where $V_2$ exists in the weir plane.

Solving Eq. (9.43) for $V_2$ yields

$$V_2 = \sqrt{2g\left(h + \frac{V_1^2}{2g}\right)} \tag{9.44}$$

In those instances when $V_1$ is small (usually the case), the velocity distribution in the flow plane above the crest of the weir can be expressed simply as

$$V_2 = \sqrt{2gh} \tag{9.45}$$

The ideal flow rate over the weir is obtained by integrating the quantity $V_2\, dA$ over the area of the flow plane above the weir. Thus,

$$Q = \int_A V_2\, dA = \int_0^H \sqrt{2gh}\, L\, dh$$

$$= \tfrac{2}{3}\sqrt{2g}\, LH^{3/2} \tag{9.46}$$

where $L$ is the width of the weir.

The actual flow rate $Q_a$ over a weir is less than the ideal flow rate $Q$ due to vertical drawdown contraction from the top, friction losses in the flow, and velocities being other than horizontal in the flow plane above the weir. These and other effects are accounted for by introducing a weir discharge coefficient $C_D$ such that

$$Q_a = C_D Q = \tfrac{2}{3}\sqrt{2g}\, C_D LH^{3/2} \tag{9.47}$$

Values of $C_D$ range from 0.62 to 0.75 as the ratio of weir head $H$ to weir height $P$ ranges from 0.1 to 2.0. The weir must be aerated and sharp for these coefficients to be valid. When the rectangular weir does not extend across the full width of the channel, additional end contractions occur so that the effective width of the weir is $(L - 0.1nH)$, where $n$ is the number of end contractions. Corrosion and algae often cause the weir to appear rough and rounded to the flowing fluid. This produces an increase in the weir coefficient due to a reduction in the edge contraction.

When flow rates are small, a triangular (V-notch) weir, such as the one illustrated in Fig. 9.34, is often used. This type of weir exhibits a higher degree of accuracy over a wider range of flow rates than does the rectangular weir. This type of weir also has the advantage that the average width of the flow section increases as the head increases. The basic discharge equation for the triangular weir can be derived in the same manner as the equation for the rectangular weir. The results are

$$Q_a = \tfrac{8}{15} \sqrt{2g} \ \tan(\theta/2) C_D H^{5/2} \tag{9.48}$$

Triangular weirs having vortex angles between 45 and 90 degrees have discharge coefficients $C_D$ that range between 0.58 and 0.60 provided the heads $H$ are in excess of 5 in. of water. At low heads, when the nappe clings to the weir plate, Eq. (9.48) is not applicable.

Equations (9.47) and (9.48) indicate that the flow rate depends on the head $H$ when weirs are being used as the measuring device. The weir head $H$ can be measured manually with a point gage or with a float-activated displacement transducer that serves as the sensor for a chart recorder or other data recording system.

## 9.6 COMPRESSIBLE FLOW EFFECTS IN CLOSED SYSTEMS

When a gas flows through a gradual contraction in a piping system or through a venturi type of flowmeter, compressibility effects occur and must be considered if accurate measurements are to be made (especially at Mach numbers greater than 0.3). By using the energy equation, the equation of state for a perfect gas,

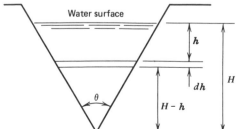

**Figure 9.34** A triangular or V-notch weir.

the continuity relationship for one-dimensional flow, and assuming the process to be isentropic, it can be shown that the mass flow rate $\dot{m}_C$ through a venturi-type contraction is given by the expression

$$\dot{m}_C = \frac{A_2}{\sqrt{1 - (p_2/p_1)^{2/k}(A_2/A_1)^2}} \sqrt{\frac{2k}{k-1}p_1\rho_1\left[\left(\frac{p_2}{p_1}\right)^{2/k} - \left(\frac{p_2}{p_1}\right)^{(k+1)/k}\right]} \tag{9.49}$$

The corresponding equation (simple Bernoulli equation) for an incompressible flow is

$$\dot{m}_B = \frac{A_2\rho_1}{\sqrt{1 - (A_2/A_1)^2}} \sqrt{2g\left(\frac{p_1 - p_2}{\gamma_1}\right)} \tag{9.50}$$

A comparison of Eqs. (9.49) and (9.50) indicates that an expansion factor $C_E$ can be incorporated into Eq. (9.50) so that the simple expression can be used for both compressible and incompressible flow. Energy losses are accounted for by introducing a velocity coefficient $C_V$, in the same manner as in Eq. (9.33), to correlate ideal and actual flow rates. With the introduction of these coefficients, the expression for mass flow rate $\dot{m}$ for both compressible and incompressible flow becomes

$$\dot{m} = \frac{C_V C_E A_2\rho_1}{\sqrt{1 - (A_2/A_1)^2}} \sqrt{2g\left(\frac{p_1 - p_2}{\gamma_1}\right)} \tag{9.51}$$

Values of $C_E$ for different pressure ratios $p_2/p_1$ are shown in Fig. 9.35. Equation (9.51) is limited to cases with subsonic ($M < 1$) flow velocities at the throat.

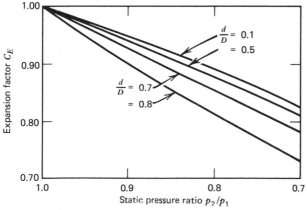

**Figure 9.35**   Expansion factor $C_E$ as a function of static pressure ratio for a venturi-type contraction.

The critical pressure ratio for a gas, above which the flow will be subsonic, is given by

$$\left(\frac{p_2}{p_1}\right)_{\text{critical}} = \left(\frac{2}{k+1}\right)^{k/(k-1)} \tag{9.52}$$

For air ($k = 1.4$), the critical pressure ratio is 0.528.

## 9.7 MISCELLANEOUS FLOW MEASUREMENT METHODS FOR CLOSED SYSTEMS

Several widely used, but totally different, flow-measurement methods for closed systems are discussed in this section. Included is the capillary flow meter for very small flow rates, positive-displacement flowmeters for use where high accuracy is needed under steady-flow conditions, hot-film mass flow transducers that are insensitive to temperature and pressure variations, and laser-Doppler anemometers for optical non-contacting flow measurements.

### Capillary Flowmeter

When very small flow rates must be measured, the capillary flowmeter shown schematically in Fig. 9.36 is very useful. Operation of this meter is based on the well-understood and experimentally verified conditions associated with laminar flow in a circular pipe. The Hagen–Poiseuille law, which governs flow under these conditions, can be writen as:

$$\frac{p_1 - p_2}{\gamma} = \frac{32\mu L}{\gamma D^2}V = \frac{\gamma_m - \gamma}{\gamma}h \tag{9.53}$$

Equation (9.53) is generally valid for Reynolds numbers ($R_e = \rho V D/\mu$) less than 2000. The velocity $V$ can be expressed in terms of the pressure difference ($p_1 - p_2$)

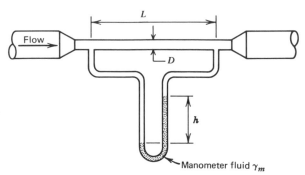

*Figure 9.36* Capillary-tube flowmeter.

or a differential manometer head $h$, as shown in Fig. 9.36. Thus

$$V = \frac{D^2}{32\mu L} (p_1 - p_2) = \frac{D^2}{32\mu L} (\gamma_m - \gamma)h \tag{9.54}$$

where $\gamma_m$ is the specific weight of the fluid in the differential manometer.
The volume flow rate, as obtained from Eq. (9.54), is

$$Q = AV = \frac{\pi D^4}{128\mu L} (p_1 - p_2) = \frac{\pi D^4}{128\mu L} (\gamma_m - \gamma)h$$

$$= K(p_1 - p_2) = K'h \tag{9.55}$$

The calibration constants $K$ and $K'$ are functions of temperature, since the viscosity and specific weight of the fluid and the diameter of the pipe are all temperature dependent. Once the constants have been established for the appropriate temperature, the flow rate $Q$ can be determined from a visual inspection or optical measurement of the differential manometer head $h$ or through a measurement of the pressure difference $(p_1 - p_2)$ with a differential pressure transducer and the appropriate automatic recording instruments.

## Positive-Displacement Flowmeters

Positive-displacement flowmeters are normally used where high accuracy is needed under steady-flow conditions (examples are home water meters and gasoline pump meters). Two common types of positive-displacement flowmeters are the nutating-disk meter (wobble meter) and the rotary-vane meter.

The nutating-disk device, shown schematically in Fig. 9.37, is widely used as the flow sensing unit in home water meters. In this type of meter, an inlet chamber is formed by the housing, the disk, and a partition between the inlet and outlet ports. Water is prevented from leaving the chamber by the disk that maintains line contact with the upper and lower conical surfaces of the housing.

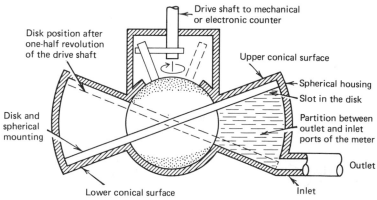

*Figure 9.37* Nutating-disk type of positive-displacement flowmeter.

When the pressure is reduced on the outlet side by a demand for water, the pressure difference causes the disk to wobble (but not rotate) about the vertical axis (axis of symmetry of the housing) and thus provide a passage for the flow around the partition. The wobble of the disk causes a small pin attached to the spherical mount for the disk to trace out a circular path about the vertical axis of the device. This motion of the pin is used to drive the recording mechanism. Since a fixed volume of water moves through the device during each revolution of the drive shaft, a simple mechanical or electronic counter can be used to monitor the flow rate. The nutating-disk flowmeter is accurate to within 1 percent when in good condition. When worn, the accuracy will be considerably less, especially for very small flow rates (such as a leaky faucet).

A second type of positive-displacement flowmeter is the rotary-vane type illustrated schematically in Fig. 9.38. This type of flowmeter consists of a cylindrical housing in which an eccentrically mounted drum with several spring-mounted vain pairs rotates. A fixed volume of fluid is transferred from the inlet port to the outlet port during each rotation of the drum. Thus, any type of counter can be used to monitor the flow rate. The rotary-vane flowmeter is generally more rugged and more accurate (about 1/2 percent) than the nutating-disk flowmeter.

### Hot-Film Mass Flow Transducers

The hot-film sensor, discussed in Section 9.2, provides the basis for a mass flow transducer that is relatively insensitive to gas temperature and gas pressure variations. The mass flow rate $\dot{m}$ ($\dot{m} = \rho V A$) depends on the cross-sectional area $A$ of the channel, the density $\rho$ of the fluid, and the flow velocity $V$. With a fixed channel area $A$, the momentum per unit area $\rho V$ provides a measure of the mass flow rate $\dot{m}$. Equation (9.17) as applied to mass flow transducer design can be written as

$$\frac{dQ}{dt} = [A + B(\rho V)^{1/n}](T_a - T_f) = I_a^2 R_a \tag{9.56}$$

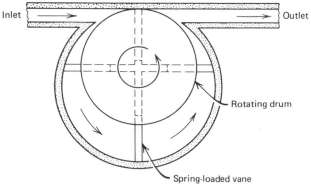

*Figure 9.38* Rotary-vane type of positive-displacement flowmeter.

Equation (9.56) indicates that it may be possible to measure the momentum per unit area $\rho V$ directly with a hot-film sensor.

A transducer that has been developed to measure $\rho V$ directly is shown schematically in Fig. 9.39. The hot-film probe $A$ in the center of the venturi throat measures $\rho V$, while the temperature compensator $B$ measures the fluid temperature. The inlet screens $C$ smooth the flow.

Initially, the platinum film sensor is heated by current from the anemometer control circuit to a temperature above that of the fluid. The fluid then carries heat away from the sensor in proportion to the flow rate. Two types of outputs are possible depending on the anemometer bridge circuit. A nonlinear signal comes directly from the anemometer bridge circuit. A linearized output can be obtained if a linearizer is incorporated into the anemometer circuit. Some typical calibration results from a circuit with a linearizer are shown in Fig. 9.40. These results show little deviation from the mean curve obtained under ambient conditions for temperatures ranging from 40°F to 100°F, or for pressures ranging from 15 psia to 30 psia. Similar results can be expected for temperature variations of $\pm 50°F$ and pressure variations of $\pm 50$ percent from calibration conditions. Flow ranges of 1000 to 1 with an accuracy of 0.5 percent, a repeatability of 0.05 percent, and a response time of 1 ms are possible.

## Laser–Doppler Anemometers

Introduction of the laser has made possible the development of the optical method of velocity measurement known as *laser–Doppler anemometry* or *laser–Doppler velocimetry*. In any form of wave propagation, frequency changes can occur as a result of movement of the source, receiver, propagating medium, or intervening reflector or scatterer. Such frequency changes are known as *Doppler shifts* and are named after the Austrian mathematician and physicist Christian Doppler (1803–1853), who first studied the phenomenon.

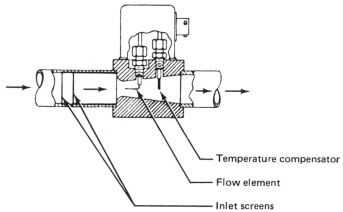

Temperature compensator

Flow element

Inlet screens

*Figure 9.39* Schematic diagram showing the components of a hot-film type of mass flow transducer. (Courtesy of TSI, Inc.)

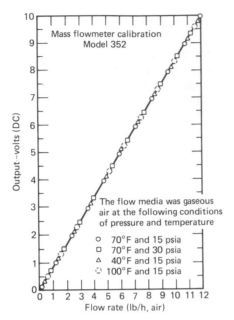

*Figure 9.40*   Calibration curve for a Model 352 mass flowmeter. (Courtesy of TSI, Inc.)

An example of the *Doppler effect* in the field of acoustics is the increase in pitch of a train whistle as the train approaches an observer, followed by a decrease in pitch as the train passes and moves away from the observer. The effect, illustrated in Fig. 9.41, is based on the fact that the observer perceives the number of sound waves arriving per second at his position as the frequency of the whistle.

The *Doppler shift* is also well known in the field of astronomy. The optical equivalent of the change in pitch of the train whistle occurs when light from a distant star is observed. If the distance between the star and the earth is decreasing, more light pulses are received in a given time interval and the color emitted from the star appears to be shifted toward the violet end of the spectrum. When the distance between the star and the earth is increasing, the light is shifted toward the red end of the sprectrum. The color shifts of remote galaxies has been accepted as evidence that the universe is expanding. Color shifts between approaching and receding sides of the sun with respect to the earth have been used to compute the period of rotation of the sun.

In laser–Doppler anemometry there is no relative movement between the source and the receiver. Instead, the shift is produced by the movement of particles (either natural or seeded) in the flow that scatter light from the source and thus permit it to reach the receiver. This same principle provides the basis for radar; however, in the case of radar, a much lower frequency part of the electromagnetic spectrum is used. The velocities commonly measured by using laser–Doppler anemometry methods are very small when compared to the velocity of light; therefore, the corresponding Doppler shifts are very small. With red light from a helium–neon laser ($\lambda$ = 632.8 nm, $f$ = 4.7 × $10^{14}$ Hz) and a

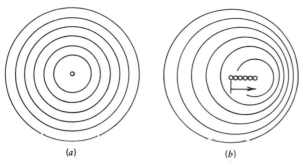

(a)                                              (b)

**Figure 9.41**   Schematic illustration of the Doppler effect. (*a*) Wave propagation from a stationary source. (*b*) Wave propagation from a moving source.

supersonic flow of 500 m/s, the Doppler shift in frequency is approximately 700 MHz. Since the resolution of a good-quality optical spectrometer is at best about 7 MHz, only velocities associated with supersonic flows can be measured with reasonable accuracy by using direct Doppler-shift measurements.

An optical beating technique for determining small Doppler shifts, which is equivalent to *heterodyning* in radio (signal mixing to obtain alternating constructive and destructive interference or beating), was first demonstrated by Yeh and Cummins[2] in 1964. Light scattered from particles carried in a water flow was mixed (heterodyned) with an unshifted reference beam of light from the laser to produce a beat frequency that is equal to the Doppler-shift frequency. The result of combining two signals with slightly different frequencies to obtain a beat frequency is illustrated schematically in Fig. 9.42.

A schematic illustration of a simple reference beam anemometer is shown in Fig. 9.43. As shown in the figure, light from the laser is divided with a beam splitter into an illuminating beam and a reference beam. Light from the illuminating beam is scattered in the direction of the reference beam by particles in the flow. When the light from the two beams is combined in the photodetector, an output signal is produced that contains a beat frequency that is equal to the Doppler-shift frequency produced by the movement (velocity) of the particles. This frequency can be determined by using spectrum analysis techniques. Optimum results are obtained when the intensity of the reference beam is approximately equal to that of the scattered beam. An attenuator (crossed polaroids) is often placed in the path of the reference beam to control the intensity. The reference beam anemometer is simple in principle; however, good signal-to-noise ratios are difficult to obtain in practice. The relationship between Doppler-shift frequency and particle velocity can be shown to be given by the following expression:

$$f_D = \frac{2V_o \cos \alpha}{\lambda} \sin \frac{\theta}{2} = \frac{2V}{\lambda} \sin \frac{\theta}{2} \qquad (9.57)$$

[2] "Localized Flow Measurements with a He-Ne Laser Spectrometer," by Y. Yeh and H. Z. Cummins, *Applied Physics Letters*, vol. 4, 1964, pp. 176–178.

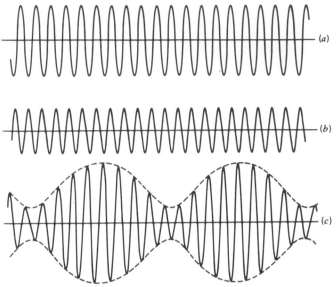

***Figure 9.42*** Signal mixing (heterodyning) yields a combined signal with a beat frequency equal to the difference in frequency of the original signals. (*a*) Signal No. 1. (*b*) Signal No. 2. (*c*) Combined Signal.

**where**    $f_D$ is the Doppler-shift frequency.

          $\lambda$   is the wavelength of the light.

          $V_o$ is the particle velocity.

          $V$   is the component of the particle velocity in a direction normal to the bisector of the angle between the illuminating and reference beams.

          $\theta$   is the angle between the illuminating and reference beams.

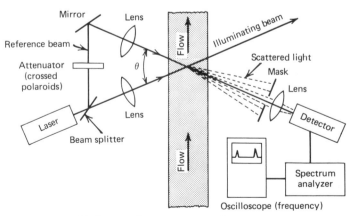

***Figure 9.43***   Reference beam anemometer.

α is the angle between the particle velocity vector and a normal to the bisector of the angle between the illuminating and reference beams.

A second type of velocity measuring instrument, known as a *dual-beam* or *differential–Doppler* anemometer, is shown schematically in Fig. 9.44. In this instrument, scattered light from two beams of equal intensity are combined to produce the beat signal. The frequency of the beat signal is equal to the difference between the Doppler shifts for the two angles of scattering. The primary advantage of this mode of operation is that the beat frequency is independent of the receiving direction; therefore, light can be collected over a large aperture and all light imposed on the photodetector contributes usefully to the output signal. When the particle concentration is low, this feature makes measurement possible when it would otherwise not be possible with the reference beam system. The differential–Doppler anemometer also has better signal-to-noise characteristics.

Operation of the differential–Doppler anemometer can also be explained in terms of the optical interference pattern (fringe pattern) formed in the crossover region of the two beams as shown in Fig. 9.45. The spacing of the interference fringes is given by the expression

$$s = \frac{\lambda}{2 \sin (\theta/2)} \tag{9.58}$$

**where** $s$ is the spacing between fringes.

$\lambda$ is the wavelength of the light that is producing the fringes.

$\theta$ is the angle between the two intersecting light beams.

A particle in the flow moving with a velocity $V_o$ in a direction that makes an angle $\alpha$ with respect to a normal to the fringe planes will experience a modulation

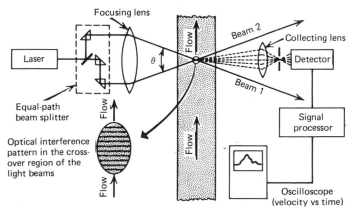

**Figure 9.44** Differential–Doppler anemometer.

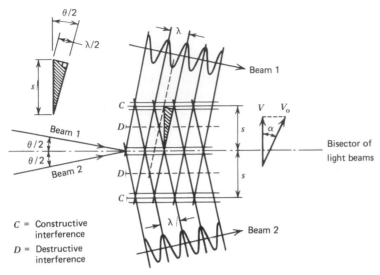

*Figure 9.45*  Optical interference pattern (fringe pattern) formed by constructive and destructive interference in the crossover region of the two light beams.

of light intensity as it moves through the fringes. Since the light scattered from the particle depends on this intensity variation associated with the fringes, it will also be modulated at the frequency $f_D$ and will be independent of the direction of observation. The frequency $f_D$ obtained from the particle velocity and fringe spacing is identical to that given by Eq. (9.57).

Output signals from the photodetectors can be processed in many ways to obtain the Doppler frequency $f_D$ required for velocity determinations. Included are spectrum analysis, frequency tracking, counter processing, filter bank processing, and photon correlation. Details of all of these procedures are beyond the purpose and scope of this book. The reader who is interested in application of laser–Doppler methods to flow measurements should consult one of a number of books that have been published on the subject.[3]

The advantages of laser–Doppler measurements include:

1. The probing is purely optical; therefore, the flow is not disturbed by the presence of a measuring instrument.
2. Velocity is measured in a direct manner; calibration is not required.
3. A component of velocity in a given direction can be measured.
4. System output is a linear function of the velocity component being measured.
5. Velocities can be measured in flows exhibiting high turbulence.
6. The method is suitable for a very wide range of velocities.

---

[3] *The Laser-Doppler Technique,* by L. E. Drain, Wiley, New York, 1980.

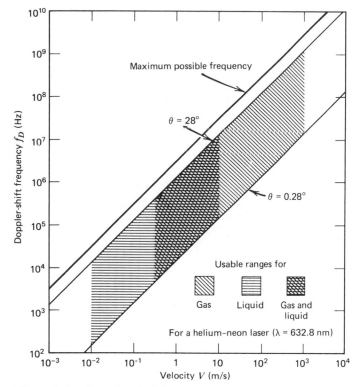

**Figure 9.46** Doppler-shift frequency versus velocity.

Typical frequency versus velocity curves for a reference beam anemometer are shown in Fig. 9.46. Since the wavelength of the helium–neon laser is known to an accuracy of 0.01 percent and since modern signal processing electronics can provide very accurate determinations of the Doppler-shift frequency $f_D$, the accuracy of velocity determinations is controlled essentially by the accuracy of the determination of the angle $\theta$ between the illuminating and reference beams. Useful ranges of beam angles for gas and liquid flow are shown crosshatched in Fig. 9.46.

## 9.8 SUMMARY

A wide variety of flow measurement methods and devices have been considered in this chapter. Other devices that have not been covered include swirl-meters, magnetic flowmeters, and a number of ultrasonic devices. The choice of an instrument depends on such factors as accuracy, range, reliability, cost, viscosity and density of the fluid, extremes of temperature and pressure, type of readout required, and nature of the fluid (clean, dirty, corrosive, etc.). Initial and subsequent calibration must also be given serious consideration.

## 9.9 TABLES OF PROPERTIES OF SOME COMMON LIQUIDS AND GASES

**TABLE 9.2** Physical Properties of Water

| Temperature<br>*English*<br>*units:* °F | Specific[a]<br>Weight<br>lb/ft³<br>γ | Density[a]<br>slug/ft³<br>ρ | Viscosity<br>lb-s/ft²<br>μ × 10⁵ | Kinematic<br>Viscosity<br>ft²/s<br>ν × 10⁵ |
|---|---|---|---|---|
| 32 | 62.42 | 1.940 | 3.746 | 1.931 |
| 40 | 62.43 | 1.940 | 3.229 | 1.664 |
| 50 | 62.41 | 1.940 | 2.735 | 1.410 |
| 60 | 62.37 | 1.938 | 2.359 | 1.217 |
| 70 | 62.30 | 1.936 | 2.050 | 1.059 |
| 80 | 62.22 | 1.934 | 1.799 | 0.930 |
| 90 | 62.11 | 1.931 | 1.595 | 0.826 |
| 100 | 62.00 | 1.927 | 1.424 | 0.739 |
| 110 | 61.86 | 1.923 | 1.284 | 0.667 |
| 120 | 61.71 | 1.918 | 1.168 | 0.609 |
| 130 | 61.55 | 1.913 | 1.069 | 0.558 |
| 140 | 61.38 | 1.908 | 0.981 | 0.514 |
| 150 | 61.20 | 1.902 | 0.905 | 0.476 |
| 160 | 61.00 | 1.896 | 0.838 | 0.442 |
| 170 | 60.80 | 1.890 | 0.780 | 0.413 |
| 180 | 60.58 | 1.883 | 0.726 | 0.385 |
| 190 | 60.36 | 1.876 | 0.678 | 0.362 |
| 200 | 60.12 | 1.868 | 0.637 | 0.341 |
| 212 | 59.83 | 1.860 | 0.593 | 0.319 |

| *SI units:* °C | kN/m³<br>γ | kg/m³<br>ρ | N · s/m²<br>μ × 10³ | m²/s<br>ν × 10⁶ |
|---|---|---|---|---|
| 0 | 9.805 | 999.8 | 1.781 | 1.785 |
| 5 | 9.807 | 1000.0 | 1.518 | 1.519 |
| 10 | 9.804 | 999.7 | 1.307 | 1.306 |
| 15 | 9.798 | 999.1 | 1.139 | 1.139 |
| 20 | 9.789 | 998.2 | 1.002 | 1.003 |
| 25 | 9.777 | 997.0 | 0.890 | 0.893 |
| 30 | 9.764 | 995.7 | 0.798 | 0.800 |
| 40 | 9.730 | 992.2 | 0.653 | 0.658 |
| 50 | 9.689 | 988.0 | 0.547 | 0.553 |
| 60 | 9.642 | 983.2 | 0.466 | 0.474 |
| 70 | 9.589 | 977.8 | 0.404 | 0.413 |
| 80 | 9.530 | 971.8 | 0.354 | 0.364 |
| 90 | 9.466 | 965.3 | 0.315 | 0.326 |
| 100 | 9.399 | 958.4 | 0.282 | 0.294 |

[a] At 14.7 psia (standard atmospheric pressure).

**TABLE 9.3** Approximate Properties of Some Common Liquids at Standard Atmospheric Pressure

| Liquid<br>*English Units:* | Temper-<br>ature<br>°F | Specific<br>Weight<br>lb/ft³<br>$\gamma$ | Density<br>slug/ft³<br>$\rho$ | Viscosity<br>lb-s/ft²<br>$\mu \times 10^5$ | Kinematic<br>Viscosity<br>ft²/s<br>$\nu \times 10^5$ |
|---|---|---|---|---|---|
| Benzene | 68 | 54.8 | 1.702 | 1.37 | 0.80 |
| Castor oil | 68 | 59.8 | 1.858 | 2060.0 | 1109.0 |
| Crude oil | 68 | 53.6 | 1.665 | 15.0 | 9.0 |
| Ethyl alcohol | 68 | 49.2 | 1.528 | 2.51 | 1.64 |
| Gasoline | 68 | 42.4 | 1.317 | 0.61 | 0.46 |
| Glycerin | 68 | 78.5 | 2.439 | 3120.0 | 1280.0 |
| Kerosene | 68 | 50.5 | 1.569 | 4.00 | 2.55 |
| Linseed oil | 68 | 58.6 | 1.820 | 92.0 | 50.0 |
| Mercury | 68 | 844.0 | 26.2 | 3.24 | 0.124 |
| Olive oil | 68 | 56.7 | 1.761 | 176.0 | 100.0 |
| Turpentine | 68 | 53.6 | 1.665 | 3.11 | 1.87 |
| Water | 68 | 62.32 | 1.936 | 2.10 | 1.085 |
| *SI units:* | °C | kN/m³<br>$\gamma$ | kg/m³<br>$\rho$ | N · s/m²<br>$\mu \times 10^3$ | m²/s<br>$\nu \times 10^6$ |
| Benzene | 20 | 8.77 | 895 | 0.65 | 0.73 |
| Castor oil | 20 | 9.39 | 958 | 979.0 | 1025.0 |
| Crude oil | 20 | 8.42 | 858 | 7.13 | 8.32 |
| Ethyl alcohol | 20 | 7.73 | 787 | 1.19 | 1.52 |
| Gasoline | 20 | 6.66 | 679 | 0.29 | 0.43 |
| Glycerin | 20 | 12.33 | 1257 | 1495.0 | 1189.0 |
| Kerosene | 20 | 7.93 | 809 | 1.90 | 2.36 |
| Linseed oil | 20 | 9.21 | 938 | 43.7 | 46.2 |
| Mercury | 20 | 132.6 | 13500 | 1.54 | 0.12 |
| Olive oil | 20 | 8.91 | 908 | 83.7 | 92.4 |
| Turpentine | 20 | 8.42 | 858 | 1.48 | 1.73 |
| Water | 20 | 9.79 | 998 | 1.00 | 1.00 |

**TABLE 9.4** Physical Properties of Air at Standard Atmospheric Pressure

| English units: | Temper-ature °F | Specific[a] Weight lb/ft³ $\gamma$ | Density[a] slug/ft³ $\rho$ | Viscosity lb-s/ft² $\mu \times 10^7$ | Kinematic Viscosity ft²/s $\nu \times 10^5$ |
|---|---|---|---|---|---|
| | 0 | 0.0866 | 0.00269 | 3.39 | 1.26 |
| | 10 | 0.0847 | 0.00263 | 3.45 | 1.31 |
| | 20 | 0.0828 | 0.00257 | 3.51 | 1.37 |
| | 30 | 0.0811 | 0.00252 | 3.57 | 1.42 |
| | 40 | 0.0794 | 0.00247 | 3.63 | 1.47 |
| | 50 | 0.0779 | 0.00242 | 3.68 | 1.52 |
| | 60 | 0.0764 | 0.00237 | 3.74 | 1.58 |
| | 70 | 0.0750 | 0.00233 | 3.79 | 1.63 |
| | 80 | 0.0735 | 0.00228 | 3.85 | 1.69 |
| | 90 | 0.0722 | 0.00224 | 3.90 | 1.74 |
| | 100 | 0.0709 | 0.00220 | 3.96 | 1.80 |
| | 150 | 0.0651 | 0.00202 | 4.23 | 2.09 |
| | 200 | 0.0601 | 0.00187 | 4.48 | 2.40 |
| | 300 | 0.0522 | 0.00162 | 4.96 | 3.05 |
| | 400 | 0.0462 | 0.00143 | 5.40 | 3.77 |
| SI units: | °C | N/m³ $\gamma$ | kg/m³ $\rho$ | N · s/m² $\mu \times 10^5$ | m²/s $\nu \times 10^5$ |
| | −20 | 13.7 | 1.40 | 1.61 | 1.16 |
| | −10 | 13.2 | 1.34 | 1.67 | 1.24 |
| | 0 | 12.7 | 1.29 | 1.72 | 1.33 |
| | 10 | 12.2 | 1.25 | 1.76 | 1.41 |
| | 20 | 11.8 | 1.20 | 1.81 | 1.51 |
| | 30 | 11.4 | 1.17 | 1.86 | 1.60 |
| | 40 | 11.1 | 1.13 | 1.91 | 1.69 |
| | 50 | 10.7 | 1.09 | 1.95 | 1.79 |
| | 60 | 10.4 | 1.06 | 2.00 | 1.89 |
| | 70 | 10.1 | 1.03 | 2.04 | 1.99 |
| | 80 | 9.81 | 1.00 | 2.09 | 2.09 |
| | 90 | 9.54 | 0.97 | 2.13 | 2.19 |
| | 100 | 9.28 | 0.95 | 2.17 | 2.29 |
| | 120 | 8.82 | 0.90 | 2.26 | 2.51 |
| | 140 | 8.38 | 0.85 | 2.34 | 2.74 |
| | 160 | 7.99 | 0.81 | 2.42 | 2.97 |
| | 180 | 7.65 | 0.78 | 2.50 | 3.20 |
| | 200 | 7.32 | 0.75 | 2.57 | 3.44 |

[a] At 14.7 psia (standard atmospheric pressure).

**TABLE 9.5**  Approximate Properties of Some Common Gases

| Gas<br>*English units:* | Density[a]<br>slug/ft³<br>$\rho$ | Engineering<br>Gas Constant<br>ft-lb/slug °R<br>$R$ | Specific<br>Heat<br>ft-lb/slug °R<br>$c_p$ | <br><br><br>$c_v$ | Adiabatic<br>Exponent<br><br>$k$ | Viscosity<br>lb-s/ft²<br>$\mu \times 10^5$ |
|---|---|---|---|---|---|---|
| Air | 0.00234 | 1715 | 6000 | 4285 | 1.40 | 0.0376 |
| Carbon dioxide | 0.00354 | 1123 | 5132 | 4009 | 1.28 | 0.0310 |
| Carbon monoxide | 0.00225 | 1778 | 6218 | 4440 | 1.40 | 0.0380 |
| Helium | 0.00032 | 12420 | 31230 | 18810 | 1.66 | 0.0411 |
| Hydrogen | 0.00016 | 24680 | 86390 | 61710 | 1.40 | 0.0189 |
| Methane | 0.00129 | 3100 | 13400 | 10300 | 1.30 | 0.0280 |
| Nitrogen | 0.00226 | 1773 | 6210 | 4437 | 1.40 | 0.0368 |
| Oxygen | 0.00258 | 1554 | 5437 | 3883 | 1.40 | 0.0418 |
| Water vapor | 0.00145 | 2760 | 11110 | 8350 | 1.33 | 0.0212 |
| *SI units:* | kg/m³<br>$\rho$ | N · m/kg K<br>$R$ | N · m/kg K<br>$c_p$ | <br>$c_v$ | <br>$k$ | N · s/m²<br>$\mu \times 10^5$ |
| Air | 1.205 | 287 | 1003 | 716 | 1.40 | 1.80 |
| Carbon dioxide | 1.84 | 188 | 858 | 670 | 1.28 | 1.48 |
| Carbon monoxide | 1.16 | 297 | 1040 | 743 | 1.40 | 1.82 |
| Helium | 0.166 | 2077 | 5220 | 3143 | 1.66 | 1.97 |
| Hydrogen | 0.0839 | 4120 | 14450 | 10330 | 1.40 | 0.90 |
| Methane | 0.668 | 520 | 2250 | 1730 | 1.30 | 1.34 |
| Nitrogen | 1.16 | 297 | 1040 | 743 | 1.40 | 1.76 |
| Oxygen | 1.33 | 260 | 909 | 649 | 1.40 | 2.00 |
| Water vapor | 0.747 | 462 | 1862 | 1400 | 1.33 | 1.01 |

[a] At 14.7 psia (standard atmospheric pressure).

# REFERENCES

1.  Benedict, R. P.: *Fundamentals of Temperature, Pressure, and Flow Measurement,* 2nd ed., Wiley, New York, 1977.

2.  Vennard, J. K., and R. L. Street: *Elementary Fluid Mechanics,* 5th ed., Wiley, New York, 1975.

3.  John, J. E. A., and W. L. Haberman: *Introduction to Fluid Mechanics,* 2nd ed., Prentice-Hall, Englewood Cliffs, New Jersey, 1980.

4.  Robertson, J. A., and C. T. Crowe: *Engineering Fluid Mechanics,* 2nd ed., Houghton Mifflin, Boston, Massachusetts, 1980.

5.  Vennard, J. K.: *Elementary Fluid Mechanics,* 4th ed., Wiley, New York, 1961.

6.  Streeter, V. L.: *Fluid Mechanics,* 5th ed., McGraw-Hill, New York, 1971.

7. Olson, R. M.: *Essentials of Engineering Fluid Mechanics,* 2nd ed., International Textbook, Scranton, Pennsylvania, 1966.

8. Daugherty, R. L., and J. B. Franzini: *Fluid Mechanics with Engineering Applications,* 6th ed., McGraw-Hill, New York, 1965.

9. Schlicting, H.: *Boundary Layer Theory,* 6th ed., McGraw-Hill, New York, 1968.

10. Tuve, G. L.: *Mechanical Engineering Experimentation,* McGraw-Hill, New York, 1961.

11. Rouse, H. (Ed.): *Engineering Hydraulics,* Wiley, New York, 1950.

12. Rayle, R. E.: Influence of Orifice Geometry on Static Pressure Measurements, ASME Paper 59-A-234, American Society of Mechanical Engineers, New York, December 1959.

13. Gracey, W., W. Letko, and W. R. Russel: Wind-Tunnel Investigation of a Number of Total-Pressure Tubes at High Angles of Attack, NACA TN2331, National Advisory Committee on Aeronautics, Washington, D.C., April 1951.

14. Hurd, C. W., K. P. Chesky, and A. H. Shapiro: Influence of Viscous Effects on Impact Tubes, *Transactions of the American Society of Mechanical Engineers, Journal of Applied Mechanics,* June 1953, p. 253.

15. Lighthill, M. J.: Contributions to the Theory of Pitot-tube Displacement Effects, *Journal of Fluid Mechanics,* vol. 2, part 2, 1957, p. 493.

16. Hall, I. M.: The Displacement Effect of a Sphere in a Two-dimensional Shear Flow, *Journal of Fluid Mechanics,* vol. 1, part 2, 1956, p. 142.

17. King, L. V.: On the Convection of Heat from Small Cylinders in a Stream of Fluid with Applications to Hot-wire Anemometry, *Philosophical Transactions of the Royal Society of London,* vol. 214, no. 14, 1914, p. 373.

18. Freymuth, P.: A Bibliography of Thermal Anemometry, *TSI Quarterly,* vol. 4, May/June, 1978. Available from Thermo Systems Incorporated, St. Paul, Minnesota.

19. Marris, A. W., and O. G. Brown: Hydrodynamically Excited Vibrations of Cantilever-Supported Rods, ASME Paper 62-HYD-7, American Society of Mechanical Engineers, New York, 1962.

20. White, D. F., A. E. Rodley, and C. L. McMurtrie: The Vortex Shedding Flowmeter, a paper in *Flow, Its Measurement and Control in Science and Industry,* vol. 1, part 2, Instrument Society of America, Pittsburgh, 1974, p. 967.

21. *Fluid Meters: Their Theory and Application,* 6th ed., American Society of Mechanical Engineers, New York, 1971.

22. Benedict, R. P.: Most Probable Discharge Coefficients for ASME Flow Nozzles, *Transactions of the American Society of Mechanical Engineers, Journal of Basic Engineering,* December 1966, p. 734.

23. Tuve, G. L., and R. E. Sprenkle: Orifice Discharge Coefficients for Viscous Liquids, *Instruments,* vol. 6, 1933, p. 201; also vol. 8, 1935, pp. 202, 225, 232.

24. Lansford, W. M.: The Use of an Elbow in a Pipe Line for Determining the Flow in a Pipe, Bulletin 289, Engineering Experiment Station, University of Illinois, Urbana, 1936.

25. Lenz, A. T.: Viscosity and Surface-Tension Effects on V-Notch Weir Coefficients, *Transactions of the American Society of Civil Engineers,* vol. 108, 1943, pp. 759–802.

26. Cheremisinoff, N. P.: *Applied Fluid Flow Measurement: Fundamentals and Technology,* Dekker, New York, 1979.

27. Yothers, M. T. (Ed.): *Standards and Practices for Instrumentation,* 5th ed., Instrument Society of America, Pittsburgh, 1977.

---

## EXERCISES

**9.1** For laminar flow in a circular pipe, show that the average velocity $V_{avg}$ is one-half of the centerline velocity $V_o$.

**9.2** The velocity profile for fully developed flow in circular pipes is given by Eq. (9.5). Show that the average velocity $V_{avg}$ is related to the centerline velocity $V_o$ by Eq. (9.6).

**9.3** Water at 60°F flows through a 2.0-ft-diameter pipe with an average velocity of 20 ft/s. Determine the weight flow rate, the mass flow rate, the energy per second being transmitted through the pipe in the form of kinetic energy ($\gamma Q V^2 / 2g$), the velocity profile exponent $n$, and the centerline velocity.

**9.4** For the rectangular duct shown in Fig. E9.4, the velocity profile can be approximated by the expression

$$V = V_C[1 - (x/a)^2][1 - (y/b)^2]$$

Show that $V_{avg} = 4/9 V_C$ for this case.

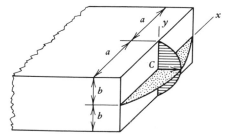

*Figure E9.4*

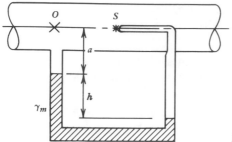

*Figure E9.5*

**9.5** A pitot tube is connected to a manometer filled with fluid of specific weight $\gamma_m$ as shown in Fig. E9.5. When the manometer connection pipes are filled with the flowing fluid of specific weight $\gamma$, the pressure differential is independent of dimension $a$. Show that the free-stream velocity is given by the expression

$$V_o = C_1 \sqrt{2g \frac{\gamma_m - \gamma}{\gamma} h} = K\sqrt{h}$$

Show that $K = 1.087$ when the manometer fluid is mercury, $C_1 = 0.98$, $V_o$ is expressed in meters per second, and $h$ is measured in centimeters. How would these results be altered if the pipe were inclined at an angle of 60 degrees from the horizontal?

**9.6** Compare the velocity measurement error associated with a 30-degree misalignment for the two types of pitot tubes shown in Fig. 9.7 when the true dynamic pressure is 4 in. of water for carbon dioxide flowing at 68°F.

**9.7** A 0.25-in.-diameter pitot tube is to be used to measure the velocity of two liquids (water and glycerin). What is the minimum velocity for each liquid below which viscous effects must be considered? Assume both liquids are at room temperature.

**9.8** A pitot-static tube is used to measure the speed of an aircraft. The air temperature and pressure are 30°F and 12.3 psia, respectively. What is the aircraft speed in miles per hour if the differential pressure is 30 in. of water? Solve by using both Eq. (9.10) and Eq. (9.16). Compare the results.

**9.9** Show that the fundamental form of Eq. (9.19) follows directly from King's law as expressed by Eq. (9.17) and that

$$C_0 = A^2/\rho B^2 \qquad \text{and} \qquad I_0 = \sqrt{A(T_a - T_f)/R_a}$$

**9.10** From the typical calibration curve for a hot-film anemometer operating in water as shown in Fig. 9.13, estimate the reference current $I_0$ and the calibration constant $C_0$. Assuming a nominal temperature difference of 80°F during calibration, how would you correct for a fluid temperature drop of 10°F occurring after start up of the experiment?

**9.11** A circular plate having a diameter of 3.81 cm is to be used as a drag-force velocity transducer to measure the velocity of water at 68°F. Determine the velocity sensitivity of the plate and the minimum velocity for which this sensitivity is valid.

**9.12** Current meter data were collected at 13 vertical locations similar to those shown in Fig. 9.19 on a river that is 72 ft wide. The calibration constant for the current meter can be expressed as $V = 2.45N$, where $V$ is the flow velocity in feet per second and $N$ is the speed of the rotating element in revolutions per second. The data collected at the 13 locations are as follows:

| | | Rotating Element (rev/min) | |
| :---: | :---: | :---: | :---: |
| Station | River Depth (ft) | 0.2h | 0.8h |
| 1 | 0.0 | — | — |
| 2 | 3.0 | 40.1 | 31.2 |
| 3 | 3.5 | 51.1 | 41.5 |
| 4 | 4.2 | 59.0 | 43.0 |
| 5 | 3.7 | 62.1 | 48.2 |
| 6 | 5.1 | 68.3 | 50.2 |
| 7 | 4.6 | 65.6 | 48.2 |
| 8 | 3.8 | 60.2 | 45.8 |
| 9 | 4.0 | 56.5 | 48.0 |
| 10 | 3.2 | 57.3 | 39.8 |
| 11 | 3.1 | 48.8 | 38.0 |
| 12 | 2.0 | 41.2 | 29.8 |
| 13 | 0.0 | — | — |

Show that the river flow rate is 482 ft³/s.

**9.13** A vortex shedding transducer is being designed to measure the velocity of water flowing in an open channel. The range of velocities to be measured is from 0.10 to 10.0 ft/s. The relationship between shedding frequency $f_s$ and velocity $V_o$ (subject to calibration) should be approximately $f_s = 10V_o$. Show that a cylinder having a diameter of 0.250 in. approximates these design requirements. What other factors must be considered?

**9.14** Design a bluff-body vortex shedding flowmeter similar to the one shown in Fig. 9.22 for use in an 8-in.-diameter pipe that is carrying water. Estimate the meter sensitivity (feet per second per hertz) and the range of velocities over which this sensitivity should be nearly constant. What is the equivalent volume flow-rate sensitivity?

**9.15** Show that the ideal and actual volume flow rates for the closed system shown in Fig. E9.15 are

$$Q_a = C_V Q_i$$

$$= \frac{C_V A_2}{\sqrt{1 - (A_2/A_1)^2}} \sqrt{\frac{2g(\gamma_m - \gamma)h}{\gamma}}$$

where $\gamma_m$ is the specific weight of the manometer fluid ($\gamma_m > \gamma$).

(a) Why do dimensions $a$ and $b$ not appear in the above equation?
(b) How would you change the manometer connection if $\gamma_m < \gamma$?
(c) What effect would $\gamma_m < \gamma$ have on the above equation?

**9.16** A 4.0-in. by 2.0-in.-diameter venturi meter is used to measure the volume flow rate of turpentine in a chemical processing plant. Pressure differences are measured with a manometer having water as the manometer fluid.

(a) Show that the ideal flow-rate sensitivity is 0.0210 ft³/s · in.¹/² when the pressure drop is measured in inches of water.
(b) How can the actual flow-rate sensitivity be estimated by using Fig. 9.25?
(c) For a manometer reading of 16 in. of water and a fluid temperature of 68°F, determine the actual flow rate.

**9.17** The flow of water at 150°F in an existing 8.0-in.-diameter pipe line is to be measured by using a 4.0-in.-diameter ASME long-radius flow nozzle. The flow rate will vary from 0.05 to 3.0 ft³/s.

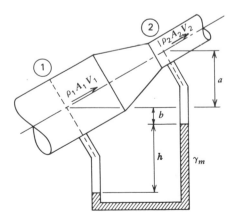

*Figure E9.15*

(a)  What range of pressure drops should the differential pressure transducer be able to measure?

(b)  If a manometer is to be used to measure these pressure drops, select a reasonable manometer fluid for use in this application if it is assumed that a manometer can be easily read to ±0.05 in.

**9.18**  A mercury manometer is connected to a standard orifice meter with a 40-mm-diameter hole that has been placed in a 100-mm-diameter pipe. For a manometer reading of 381 mm of mercury, show that the flow rate in the pipe is $8.05 \times 10^{-3}$ m³/s if the fluid is crude oil at 20°C.

**9.19**  An 8.0-in.-diameter elbow meter has a coefficient $C = 0.75$ when installed in a water line. The meter is connected to a mercury manometer having a 24-in. scale that is graduated in units of 0.05 in.

(a)  Determine the sensitivity of the instrument in terms of cubic feet per second and inches of mercury.

(b)  When $h = 20.0$ inches of mercury, what is the change in volume flow rate $\Delta Q$ corresponding to a scale reading error of 0.10 in.?

(c)  How significant is the 0.10-in. reading error when $h = 4.0$ in.?

**9.20**  Compare flow rates through 40-mm-diameter openings under 1.50-m static heads if the openings have been constructed as sharp-edged, rounded, short-tube, or Borda orifices. Ancient Rome's famous water system used sharp-edged orifices to meter water to Roman citizens. The clever citizens found that they could obtain 30 percent more water by inserting a short tube into the orifice and thus cheat Caesar out of significant water revenues.

**9.21**  A 10-in.-diameter opening is to be located in the side of a large tank. Water flows from the large tank into a large reservoir. Estimate the flow between the tank and the reservoir when the difference in free-surface elevations is 16 ft. Assume:

(a)  That the most efficient orifice construction from Table 9.1 is used

(b)  That the most inefficient orifice construction from Table 9.1 is used

(c)  What assumptions are made when the terms "large tank" and "large reservoir" are used?

**9.22**  A 4.0-m-wide by 0.50-m-deep sluice gate is used to control the overflow of water from a small reservoir with a surface area of 100,000 m². When the water surface in the reservoir is 2.5 m above the bottom of the spillway (see Fig. 9.32), estimate the flow rate through the sluice gate and the rate at which the reservoir surface is falling. Assume $C_C = 0.61$ and $C_V = 0.96$.

**9.23**  Derive Eq. (9.47) for the flow rate over a rectangular weir and carefully list any assumptions required for the derivation. If $C_D$ is assumed to be equal to 0.623 and if end contraction effects are neglected, show that

(a)  $Q_a = 3.32LH^{3/2}$ for the English System of units
(b)  $Q_a = 1.83LH^{3/2}$ for the SI System of units

**9.24**  T. Rehbock of the Karlsruhe Laboratory in Germany developed the following empirical expression for the weir discharge coefficient $C_D$, which yields good results for rectangular weirs with good ventilation, sharp edges, smooth weir faces, and adequate water stilling.

$$C_D = 0.605 + 0.08H/P + 1/305H$$

Plot values of $C_D$ as a function of $H$ (0.08 ft $< H <$ 2.0 ft) for $P$ equal to 0.33 ft, 1.00 ft, and 3.30 ft. *Note: H/P* must be less than 2 for the Rehbock equation to retain an accuracy of 1 percent.

**9.25**  A rectangular weir is to be placed in a 6-m-wide channel to measure a nominal flow rate of 5.0 m³/s while maintaining a minimal channel depth of 3.0 m. Determine a suitable rectangular weir (width $L$ and height $P$) if $H/P$ must be less than 0.4 to ensure that $V_1^2/2g$ is negligible. What would the height $P$ be for a 90-degree V-notch weir? If the flow rate doubles, which weir would experience the smaller change in weir head $H$?

**9.26**  The flow rate in a rectangular open channel of width $L$ must be measured while a nearly constant fluid depth $y$ is maintained. A floating sluice gate and a weir have been proposed as methods to achieve these goals. The two methods are shown in Fig. E9.26.

(a)  Show that the flow rate under the sluice gate as given by the following linear equation between $Q$ and $H$ is accurate within 5 percent, provided $C_{Ds}$ is constant and $H_s/P_s < 0.1$.

$$Q = C_{Ds} \sqrt{2gP_s}\, LH_s$$

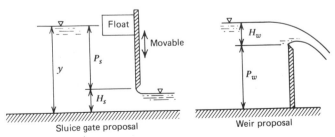

Sluice gate proposal          Weir proposal

*Figure E9.26*

    (b)  Compare the sluice gage and weir discharge equations and show that the sluice gate and weir readings are related by the expression

$$H_s = \frac{2}{3} \frac{C_{Dw}}{C_{Ds}} \frac{H_w^{3/2}}{\sqrt{P_s}}$$

    (c)  Based on the information of Part (b), which method will give the least variation of $H$ with flow rate?

    (d)  Which unit is least expensive to install and maintain?

    (e)  Which proposal would you select and why?

**9.27**  Air stored in a tank at 200 psia and 150°F flows into a second tank at 160 psia through a 1.70-in.-diameter flow nozzle. Show that 16.9 slugs of air per minute are moving from one tank to the other under these conditions. Why must the pressure in the second tank exceed 106 psia?

**9.28**  Oxygen at $-10°F$ and 150 psia is flowing in a 6-in.-diameter line at a rate of 25 lb/s. Estimate the pressure drop across a 3.0-in.-diameter venturi meter that would be available for measuring the flow rate.

**9.29**  A capillary-tube flowmeter is being constructed to measure the flow rate of water. The glass tubing has an inside diameter of 1.50 mm and the pressure taps are located 0.35 m apart.

    (a)  Estimate the flow rate if the manometer reading is 260 mm of mercury and the water temperature is 20°C.

    (b)  How much error results if the water temperature increases to 30°C? Neglect any effects due to expansion of the glass with temperature change.

**9.30**  Calibration of flowmeters is often performed with an experimental facility consisting of a pump, valve to regulate the flow from the pump, test meter and manometer, weigh tank and scale, stopwatch, and reservoir as shown in Fig. E9.30.

    (a)  What minimum length $L$ of pipe should be used on the inlet side of the test meter?

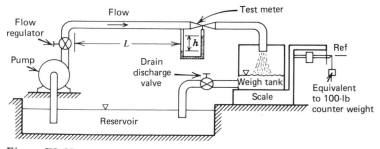

*Figure E9.30*

(b) What is the maximum flow rate that can be measured if errors are to be limited to ±1 percent, if water caught in the weigh tank is limited to 100 lb, and if the stop watch can be started and stopped within 0.05 s of the correct time?

(c) What additional information must be collected in order to properly calibrate the flowmeter?

(d) How should flow rate $Q$ versus manometer reading $h$ be plotted in order to obtain the meter calibration relationship?

(e) Find a standard relating to calibration of flowmeters in the library and study the calibration procedures recommended.

# TEN

# STATISTICAL METHODS IN EXPERIMENTAL MEASUREMENTS

## 10.1 INTRODUCTION

Experimental measurements of quantities such as pressure, temperature, force, or flow velocity will always exhibit some variation if the measurements are repeated a number of times with precise and accurate instruments. This variability is due to two different causes. First, the quantity being measured may exhibit significant variation due to changes in the process over the time interval required to make the measurement. For example, measurements of flow velocity in a section of pipe may exhibit large variations related to changes in the flow elsewhere in the system, to pump fluctuations, or to local changes in turbulence. Second, the instrumentation system, which includes the transducer, signal conditioning equipment, and recording instrument, or the operator may introduce error in the measurement. This error may be systematic or random, depending upon its source. An instrument operated out of calibration produces a systematic error, whereas, reading errors due to interpolation on a chart are random. The accumulation of random errors in a measuring system produces a variation that must be examined in relation to the magnitude of the quantity being measured.

The data obtained from repeated measurements are an assembly of readings, not an exact result. Maximum information can be extracted from such an assembly of readings by employing statistical methods. The first step in the statistical treatment of data is to establish the distribution. A graphical representation of the distribution is usually the most useful form for initial evaluation. Next, the statistical distribution is characterized with a measure of its central value, such as the mean, the median, or the mode. Finally, the spread of dispersion of the distribution is determined in terms of the variance or the standard deviation.

With elementary statistical methods, the experimentalist can reduce a large amount of data to a very compact and useful form by defining the type of distribution, establishing the single value that best represents the distribution (mean), and determining the variation from the mean value (standard deviation). Summarizing data in this manner is the most meaningful form for application to design problems or for communication to other engineers who may have need for the results of the experiments.

The treatment of statistical methods presented in this chapter is relatively brief; therefore, only the most commonly employed techniques for representing and interpreting data are presented. A formal course in statistics, which covers these techniques in much greater detail as well as many other useful techniques, should be included in the program of study of all engineering students.

## 10.2 CHARACTERIZING STATISTICAL DISTRIBUTIONS

For purposes of this discussion, consider that an experiment has been conducted $n$ times to determine the yield strength of a particular type of cold-drawn, mild-steel bar. The data obtained represent a sample of size $n$ from an infinite population of all possible measurements that could have been made. The simplest way to present these data is to list the measurements in order of increasing magnitude, as shown in Table 10.1.

**TABLE 10.1** Listing of Data in Order of Increasing Magnitude Yield Strength of Cold-Drawn Mild Steel

| Sample Number | Strength (ksi:MPa) | Sample Number | Strength (ksi:MPa) |
|---------------|--------------------|---------------|--------------------|
| 1 | 65.0 : 448 | 11 | 79.0 : 545 |
| 2 | 68.3 : 471 | 12 | 79.2 : 546 |
| 3 | 72.2 : 498 | 13 | 79.9 : 551 |
| 4 | 73.5 : 507 | 14 | 80.3 : 554 |
| 5 | 74.0 : 510 | 15 | 81.1 : 559 |
| 6 | 75.2 : 519 | 16 | 82.6 : 570 |
| 7 | 76.8 : 530 | 17 | 84.0 : 579 |
| 8 | 77.7 : 536 | 18 | 85.5 : 590 |
| 9 | 78.1 : 539 | 19 | 87.0 : 600 |
| 10 | 78.8 : 543 | 20 | 89.8 : 619 |

**TABLE 10.2**   Frequency Distribution of Yield Strength

| Group Intervals (ksi:MPa) | Observations in the Group | Relative Frequency | Cumulative Frequency |
|---|---|---|---|
| 65.0–69.9 : 448–482 | 2 | 0.10 | 0.10 |
| 70.0–74.9 : 483–516 | 3 | 0.15 | 0.25 |
| 75.0–79.9 : 517–551 | 8 | 0.40 | 0.65 |
| 80.0–84.9 : 552–585 | 4 | 0.20 | 0.85 |
| 85.0–89.9 : 586 620 | 3 | 0.15 | 1.00 |
| Total | 20 | | |

These data can be rearranged into five groups to give a frequency distribution, as shown in Table 10.2. The advantage of representing data in a frequency distribution is that the central tendency is more clearly illustrated.

## Graphical Representation of the Distribution

The shape of the distribution function representing the yield strength of the cold-drawn, mild-steel bars is indicated by the data groupings of Table 10.2. A graphical presentation of this group data, known as a histogram, is shown in Fig. 10.1. The histogram method of presentation shows the distribution with its central tendency and variability much more clearly than the tabular method of presentation of Table 10.2. Superimposed on the histogram is a plot showing the relative frequency of the occurrence of a group of measurements. Note that the points for the relative frequency plot are placed at the midpoint of the group interval.

A cumulative frequency diagram, shown in Fig. 10.2, is another way of representing the yield strength data from the experiments. The cumulative frequency is the number of readings having a value less than a specified value of

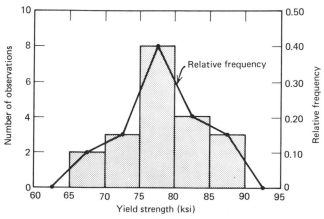

*Figure 10.1*   Yield strength distribution with superimposed relative frequency curve.

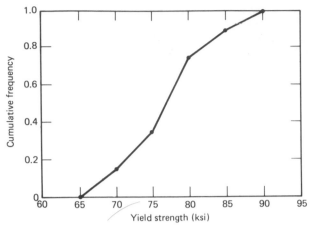

**Figure 10.2** Cumulative frequency curve for the yield strength data.

the quantity being measured (yield strength in this example) divided by the total number of measurements. As indicated in Table 10.2, the cumulative frequency is the running sum of the relative frequencies. When the graph of cumulative frequency versus the quantity being measured (yield strength) is prepared, the end value for the group interval is used to position the point along the abscissa.

## Analysis of Central Tendency

While histograms or frequency distributions are used to provide a visual representation of a distribution, numerical measures are used to define the characteristics of the distribution. One basic characteristic of a distribution is the central tendency of the data. The most commonly employed measure of the central tendency is the sample mean $\bar{x}$, which is defined as

$$\bar{x} = \sum_{i=1}^{n} \frac{x_i}{n} \tag{10.1}$$

**where** $x_i$ is the $i$th value of the quantity being measured.
$n$ is the total number of measurements.

Because of time and costs involved in conducting tests, the number of measurements is usually limited; therefore, the sample mean $\bar{x}$ is only an estimate of the true arithmetic mean $\mu$ of the population. It is shown later that $\bar{x}$ approaches $\mu$ as the number of measurements increases.

The median and mode are also measures of central tendency. The median is the middle value in a group of ordered data. For example, in an ordered set of 21 readings, the 11th reading represents the median value with 10 readings lower than the median and 10 readings higher than the median. In instances when an even number of readings are taken, the median is obtained by averaging the two middle values. For example, in an ordered set of 20 readings, the median is the average of the 10th and 11th readings.

The mode is the most frequent value of the data; therefore, it is the peak value on the relative frequency curve. In Fig. 10.1, the peak of the relative frequency curve occurs at a yield strength $S_y = 77.5$ ksi (535 MPa); therefore, this value is the mode of the data set presented in Table 10.1.

## Measures of Dispersion

It is possible for two different distributions of data to have the same mean but different dispersions, as shown in the relative frequency diagrams of Fig. 10.3. Different measures of dispersion are the range, the mean deviation, the variance, and the standard deviation. The standard deviation $S_x$ is the most popular and is defined as

$$ S_x = \left[ \sum_{i=1}^{n} \frac{(x_i - \bar{x})^2}{n - 1} \right]^{1/2} \tag{10.2} $$

When the sample size $n$ is small, the standard deviation $S_x$ of the sample represents an estimate of the true standard deviation $\sigma$ of the population. Determination of the mean and standard deviation for a distribution are easily and quickly performed today by utilizing the preprogrammed routines available in most small electronic calculators.

Expressions for the other measures of dispersion, namely, range $R$, mean deviation $d_x$, and variance $S_x^2$ are

$$ R = x_l - x_s \tag{10.3} $$

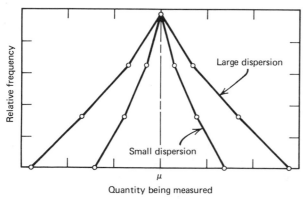

Figure 10.3    Relative frequency diagrams with large and small dispersion.

**where**     $x_l$ is the largest value of the quantity in the distribution.
$x_s$ is the smallest value of the quantity in the distribution.

$$d_x = \frac{\sum\limits_{i=1}^{n} |x_i - \bar{x}|}{n} \tag{10.4}$$

Equation (10.4) indicates that the deviation of each reading from the mean is determined and summed. The average of the $n$ deviations is the mean deviation. The absolute value of the difference $x_i - \bar{x}$ must be used in the summing process to avoid cancellation of positive and negative deviations.

$$S_x^2 = \frac{\sum\limits_{i=1}^{n} (x_i - \bar{x})^2}{n} \tag{10.5}$$

The variance is frequently used in statistics, since it is related to the standard deviation when the sample size is large. Also, it is useful in describing the dispersion characteristics of the most important statistical distribution function (the normal distribution).

Finally, a measure known as the coefficient of variation $C_v$ is used to express the standard deviation $S_x$ as a percentage of the mean $\bar{x}$. Thus

$$C_v = \frac{S_x}{\bar{x}} (100) \tag{10.6}$$

## 10.3   STATISTICAL DISTRIBUTION FUNCTIONS

As the sample size is increased, it is possible in tabulating the data to increase the number of group intervals and to decrease their width. The corresponding relative frequency plot, similar to the one illustrated in Fig. 10.1, will approach a smooth curve that can be represented by a frequency equation. This frequency equation defines the statistical properties of a distribution function that can be used to represent the population from which the sample was drawn.

Many distribution functions are used in statistical analyses. The best-known and most widely used distribution is the normal distribution, also known as the Gaussian distribution. Other useful distributions include binomial, $\chi^2$, $F$, Gumbel, Poisson, Student's $t$, and Weibull distributions. The normal distribution function is important in engineering, since it describes the statistical variation in many engineering processes. It also describes the variation resulting from random error. The normal distribution function, as represented by a normalized relative frequency diagram, is shown in Fig. 10.4.

The normal distribution is completely defined by two parameters; the mean

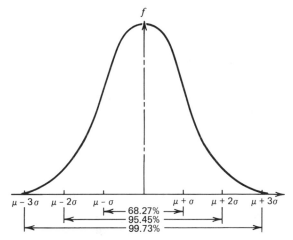

*Figure 10.4*  Normalized relative frequency diagram for the normal distribution function.

$\mu$ and the standard deviation $\sigma$. The equation for the relative frequency $f$ in terms of these two parameters is

$$f(z) = \frac{1}{\sqrt{2\pi}} e^{-z^2/2} \tag{10.7}$$

where

$$z = \frac{x - \mu}{\sigma} \tag{10.8}$$

Experimental data (with finite sample sizes) can be analyzed to obtain $\bar{x}$ as an estimate of $\mu$ and $S_x$ as an estimate of $\sigma$. Use of this procedure permits the experimentalist to utilize data drawn from small samples to represent the complete population.

Distribution functions are important since they provide a means for predicting population properties. This is an extremely valuable capability, since prediction of process behavior is vital in engineering applications. The method for predicting population properties from the normal distribution function utilizes the normalized relative frequency distribution shown in Fig. 10.4. The area $A$ under the entire normalized relative frequency curve is given by Eq. (10.7) as

$$A = \frac{1}{\sqrt{2\pi}} \int_{-\infty}^{\infty} e^{-z^2/2} \, dz = 1 \tag{10.9}$$

Equation (10.9) implies that the population has a value $z$ between $-\infty$ and $+\infty$ and that the probability of selecting a single sample from the population with a value $-\infty \leqslant z \leqslant +\infty$ is 100 percent. While the previous statement may appear trivial and obvious, it serves to illustrate the concept of using the area under the normalized relative frequency curve to determine the probability of drawing a single sample with a value $z$ between specified limits from a population. As a less obvious example, consider the probability of drawing a single sample with $z_1 \leqslant z \leqslant z_2$. This probability is equal to the area under the normalized relative frequency curve from $z = z_1$ to $z = z_2$. Thus, from Eq. (10.7),

$$p = \frac{1}{\sqrt{2\pi}} \int_{z_1}^{z_2} e^{-z^2/2} \, dz \qquad (10.10)$$

where $p$ is the probability of occurrence. Evaluation of Eq. (10.10) is most easily made by using tables that list the areas under the normalized relative frequency curve as a function of $z$. Table 10.3 lists one-side areas between limits of $z - z_1$ and $z = 0$ for the normal distribution function. Since the distribution function is symmetric about $z = 0$, this one-sided table is sufficient for all evaluations. For example,

$$A(-1,0) = A(0,+1)$$

therefore

$$A(-1,+1) = p(-1,+1) = 0.3413 + 0.3413 \quad = 0.6826$$
$$A(-2,+2) = p(-2,+2) = 0.4772 + 0.4772 \quad = 0.9544$$
$$A(-3,+3) = p(-3,+3) = 0.49865 + 0.49865 = 0.9973$$
$$A(-1,+2) = p(-1,+2) = 0.3413 + 0.4772 \quad = 0.8185$$

Since the normal distribution function has been well characterized, predictions can be made regarding the probability of a specific strength value or measurement error. For example, one may anticipate that 68.3 percent of the data will fall between limits of $\bar{x} \pm S_x$, 95.4 percent between limits of $\bar{x} \pm 2S_x$,

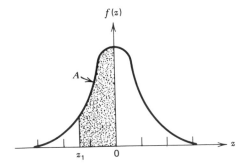

*Figure T10.3*  Definition of terms for Table 10.3.

**TABLE 10.3** Areas under the Normal Distribution Curve from $z_1$ to $z = 0$ (One Side)

| $z_1 = \dfrac{x_1 - \bar{x}}{S_x}$ | 0.00 | 0.01 | 0.02 | 0.03 | 0.04 | 0.05 | 0.06 | 0.07 | 0.08 | 0.09 |
|---|---|---|---|---|---|---|---|---|---|---|
| 0.0 | .0000 | .0040 | .0080 | .0120 | .0160 | .0199 | .0239 | .0279 | .0319 | .0359 |
| 0.1 | .0398 | .0438 | .0478 | .0517 | .0557 | .0596 | .0636 | .0675 | .0714 | .0753 |
| 0.2 | .0793 | .0832 | .0871 | .0910 | .0948 | .0987 | .1026 | .1064 | .1103 | .1141 |
| 0.3 | .1179 | .1217 | .1255 | .1293 | .1331 | .1368 | .1406 | .1443 | .1480 | .1517 |
| 0.4 | .1554 | .1591 | .1628 | .1664 | .1700 | .1736 | .1772 | .1808 | .1844 | .1879 |
| 0.5 | .1915 | .1950 | .1985 | .2019 | .2054 | .2088 | .2123 | .2157 | .2190 | .2224 |
| 0.6 | .2257 | .2291 | .2324 | .2357 | .2389 | .2422 | .2454 | .2486 | .2517 | .2549 |
| 0.7 | .2580 | .2611 | .2642 | .2673 | .2704 | .2734 | .2764 | .2794 | .2823 | .2852 |
| 0.8 | .2881 | .2910 | .2939 | .2967 | .2995 | .3023 | .3051 | .3078 | .3106 | .3233 |
| 0.9 | .3159 | .3186 | .3212 | .3238 | .3264 | .3289 | .3315 | .3340 | .3365 | .3389 |
| 1.0 | .3413 | .3438 | .3461 | .3485 | .3508 | .3531 | .3554 | .3577 | .3599 | .3621 |
| 1.1 | .3643 | .3665 | .3686 | .3708 | .3729 | .3749 | .3770 | .3790 | .3810 | .3830 |
| 1.2 | .3849 | .3869 | .3888 | .3907 | .3925 | .3944 | .3962 | .3980 | .3997 | .4015 |
| 1.3 | .4032 | .4049 | .4066 | .4082 | .4099 | .4115 | .4131 | .4147 | .4162 | .4177 |
| 1.4 | .4192 | .4207 | .4222 | .4236 | .4251 | .4265 | .4279 | .4292 | .4306 | .4319 |
| 1.5 | .4332 | .4345 | .4357 | .4370 | .4382 | .4394 | .4406 | .4418 | .4429 | .4441 |
| 1.6 | .4452 | .4463 | .4474 | .4484 | .4495 | .4505 | .4515 | .4525 | .4535 | .4545 |
| 1.7 | .4554 | .4564 | .4573 | .4582 | .4591 | .4599 | .4608 | .4616 | .4625 | .4633 |
| 1.8 | .4641 | .4649 | .4656 | .4664 | .4671 | .4678 | .4686 | .4693 | .4699 | .4706 |
| 1.9 | .4713 | .4719 | .4726 | .4732 | .4738 | .4744 | .4750 | .4758 | .4761 | .4767 |
| 2.0 | .4772 | .4778 | .4783 | .4788 | .4793 | .4799 | .4803 | .4808 | .4812 | .4817 |
| 2.1 | .4821 | .4826 | .4830 | .4834 | .4838 | .4842 | .4846 | .4850 | .4854 | .4857 |
| 2.2 | .4861 | .4864 | .4868 | .4871 | .4875 | .4878 | .4881 | .4884 | .4887 | .4890 |
| 2.3 | .4893 | .4896 | .4898 | .4901 | .4904 | .4906 | .4909 | .4911 | .4913 | .4916 |
| 2.4 | .4918 | .4920 | .4922 | .4925 | .4927 | .4929 | .4931 | .4932 | .4934 | .4936 |
| 2.5 | .4938 | .4940 | .4941 | .4943 | .4945 | .4946 | .4948 | .4949 | .4951 | .4952 |
| 2.6 | .4953 | .4955 | .4956 | .4957 | .4959 | .4960 | .4961 | .4962 | .4963 | .4964 |
| 2.7 | .4965 | .4966 | .4967 | .4968 | .4969 | .4970 | .4971 | .4972 | .4973 | .4974 |
| 2.8 | .4974 | .4975 | .4976 | .4977 | .4977 | .4978 | .4979 | .4979 | .4980 | .4981 |
| 2.9 | .4981 | .4982 | .4982 | .4983 | .4984 | .4984 | .4985 | .4985 | .4986 | .4986 |
| 3.0 | .49865 | .4987 | .4987 | .4988 | .4988 | .4988 | .4989 | .4989 | .4989 | .4990 |

and 99.7 percent between limits of $\bar{x} \pm 3S_x$. Also, 81.9 percent of the data should fall between limits of $\bar{x} - S_x$ and $\bar{x} + 2S_x$.

In many problems, the probability of a single sample exceeding a specified value $z_2$ must be determined. It is possible to determine this probability by using Table 10.3 together with the fact that the area under the entire curve is unity ($A = 1$); however, Table 10.4, which lists one-sided areas between limits of $z = z_2$ and $z \to \infty$, yields the results more directly.

Use of Tables 10.3 and 10.4 can be illustrated by considering the yield strength data presented in Table 10.1. By using Eqs. (10.1) and (10.2), it is easy to establish estimates for the mean and standard deviation as $\bar{x} = 78.4$ ksi (541 MPa) and $S_x = 6.04$ ksi (41.7 MPa). These values of $\bar{x}$ and $S_x$ characterize the population from which the data of Table 10.1 were drawn. It is possible to establish the probability that the yield strength of a single sample drawn randomly from the population will be between specified limits (by using Table 10.3), or that the strength of a single sample will not be above or below a specified value (by using Table 10.4). For example, one determines the probability that a single sample will exhibit a strength between 66 and 84 ksi by computing $z_1$ and $z_2$ and using Table 10.3. Thus,

$$z_1 = \frac{66 - 78.4}{6.04} = -2.05 \qquad z_2 = \frac{84 - 78.4}{6.04} = 0.93$$

$$p(-2.05, 0.93) = A(-2.05, 0) + A(0, 0.93)$$
$$= 0.4798 + 0.3238 = 0.8036$$

This simple calculation shows that the probability of obtaining a strength between 66 and 84 ksi from a single specimen is 80.4 percent. The probability of the strength of a single specimen being less than 65 ksi is determined by computing $z_1$ and using Table 10.4. Thus,

$$z_1 = \frac{65 - 78.4}{6.04} = -2.22$$

$$p(-\infty, -2.22) = A(-\infty, -2.22) = A(2.22, \infty) = 0.0132$$

Thus, the probability of drawing a single sample with a yield strength less than 65 ksi is 1.3 percent.

## 10.4  CONFIDENCE INTERVALS

Once experimental data are represented with a normal distribution by using estimates of the mean $\bar{x}$ and standard deviation $S_x$ and predictions are made about the occurrence of measurements, questions arise concerning the confi-

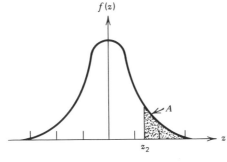

**Figure T10.4**  Definition of terms for Table 10.4

**TABLE 10.4**  Areas under the Normal-Distribution Curve from $z_2$ to $z \to \infty$ (One Side)

| $z_2 = \dfrac{x_2 - \bar{x}}{S_x}$ | 0.00 | 0.01 | 0.02 | 0.03 | 0.04 | 0.05 | 0.06 | 0.07 | 0.08 | 0.09 |
|---|---|---|---|---|---|---|---|---|---|---|
| 0.0 | .5000 | .4960 | .4920 | .4880 | .4840 | .4801 | .4761 | .4721 | .4681 | .4641 |
| 0.1 | .4602 | .4562 | .4522 | .4483 | .4443 | .4404 | .4364 | .4325 | .4286 | .4247 |
| 0.2 | .4207 | .4168 | .4129 | .4090 | .4052 | .4013 | .3974 | .3936 | .3897 | .3859 |
| 0.3 | .3821 | .3783 | .3745 | .3707 | .3669 | .3632 | .3594 | .3557 | .3520 | .3483 |
| 0.4 | .3446 | .3409 | .3372 | .3336 | .3300 | .3264 | .3228 | .3192 | .3156 | .3121 |
| 0.5 | .3085 | .3050 | .3015 | .2981 | .2946 | .2912 | .2877 | .2843 | .2810 | .2776 |
| 0.6 | .2743 | .2709 | .2676 | .2643 | .2611 | .2578 | .2546 | .2514 | .2483 | .2451 |
| 0.7 | .2420 | .2389 | .2358 | .2327 | .2296 | .2266 | .2236 | .2206 | .2177 | .2148 |
| 0.8 | .2119 | .2090 | .2061 | .2033 | .2005 | .1977 | .1949 | .1922 | .1894 | .1867 |
| 0.9 | .1841 | .1814 | .1788 | .1762 | .1736 | .1711 | .1685 | .1660 | .1635 | .1611 |
| 1.0 | .1587 | .1562 | .1539 | .1515 | .1492 | .1469 | .1446 | .1423 | .1401 | .1379 |
| 1.1 | .1357 | .1335 | .1314 | .1292 | .1271 | .1251 | .1230 | .1210 | .1190 | .1170 |
| 1.2 | .1151 | .1131 | .1112 | .1093 | .1075 | .1056 | .1038 | .1020 | .1003 | .0985 |
| 1.3 | .0968 | .0951 | .0934 | .0918 | .0901 | .0885 | .0869 | .0853 | .0838 | .0823 |
| 1.4 | .0808 | .0793 | .0778 | .0764 | .0749 | .0735 | .0721 | .0708 | .0694 | .0681 |
| 1.5 | .0668 | .0655 | .0643 | .0630 | .0618 | .0606 | .0594 | .0582 | .0571 | .0559 |
| 1.6 | .0548 | .0537 | .0526 | .0516 | .0505 | .0495 | .0485 | .0475 | .0465 | .0455 |
| 1.7 | .0446 | .0436 | .0427 | .0418 | .0409 | .0401 | .0392 | .0384 | .0375 | .0367 |
| 1.8 | .0359 | .0351 | .0344 | .0336 | .0329 | .0322 | .0314 | .0307 | .0301 | .0294 |
| 1.9 | .0287 | .0281 | .0274 | .0268 | .0262 | .0256 | .0250 | .0244 | .0239 | .0233 |
| 2.0 | .0228 | .0222 | .0217 | .0212 | .0207 | .0202 | .0197 | .0192 | .0188 | .0183 |
| 2.1 | .0179 | .0174 | .0170 | .0166 | .0162 | .0158 | .0154 | .0150 | .0146 | .0143 |
| 2.2 | .0139 | .0136 | .0132 | .0129 | .0125 | .0122 | .0119 | .0116 | .0113 | .0110 |
| 2.3 | .0107 | .0104 | .0102 | .00990 | .00964 | .00939 | .00914 | .00889 | .00866 | .00842 |
| 2.4 | .00820 | .00798 | .00776 | .00755 | .00734 | .00714 | .00695 | .00676 | .00657 | .00639 |
| 2.5 | .00621 | .00604 | .00587 | .00570 | .00554 | .00539 | .00523 | .00508 | .00494 | .00480 |
| 2.6 | .00466 | .00453 | .00440 | .00427 | .00415 | .00402 | .00391 | .00379 | .00368 | .00357 |
| 2.7 | .00347 | .00336 | .00326 | .00317 | .00307 | .00298 | .00288 | .00280 | .00272 | .00264 |
| 2.8 | .00256 | .00248 | .00240 | .00233 | .00226 | .00219 | .00212 | .00205 | .00199 | .00193 |
| 2.9 | .00187 | .00181 | .00175 | .00169 | .00164 | .00159 | .00154 | .00149 | .00144 | .00139 |

dence that can be placed on either the estimates or the predictions. One cannot be totally confident in the predictions or estimates because of the effects of sampling error.

Sampling error can be illustrated by drawing a series of samples (each containing $n$ measurements) from the same population and determining several estimates of the mean $\bar{x}_1, \bar{x}_2, \bar{x}_3, \ldots$. A variation in $\bar{x}$ will occur, but fortunately, this variation can also be characterized by a normal distribution function, as shown in Fig. 10.5. The mean of the $x$ and $\bar{x}$ distributions is the same; however,

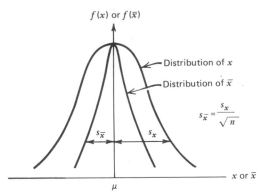

*Figure 10.5* Normal distribution of individual measurements of the quantity $x$ and of measurements of the mean $\bar{x}$ from samples of size $n$.

the standard deviation of the $\bar{x}$ distribution $S_{\bar{x}}$ (sometimes referred to as the standard error) is less than $S_x$ since

$$S_{\bar{x}} = \frac{S_x}{\sqrt{n}} \qquad (10.11)$$

Once the standard deviation of the population of $\bar{x}$'s is known, it is possible to place confidence limits on the determination of the true population mean $\mu$ from a sample of size $n$, provided $n$ is large ($n > 20$). The confidence interval within which the true population mean $\mu$ is located is given by the expression

$$(\bar{x} - zS_{\bar{x}}) < \mu < (\bar{x} + zS_{\bar{x}}) \qquad (10.12)$$

**where**   $\bar{x} - zS_{\bar{x}}$ is the lower confidence limit.
$\bar{x} + zS_{\bar{x}}$ is the upper confidence limit.

The width of the confidence interval depends upon the confidence level required. For instance, if $z = 3$ in Eq. (10.12), a relatively wide confidence interval exists; therefore, the probability that the population mean $\mu$ will be located within the confidence interval is high (99.7 percent). For confidence levels of 99.9, 99.0, and 95.0 percent, the corresponding values of $z$ in Eq. (10.12) are 3.30, 2.57, and 1.96, respectively. Thus, as the width of the confidence interval decreases, the level of confidence (probability that the population mean $\mu$ will fall within the interval) decreases. Commonly used confidence levels and their associated intervals are

99.9% Confidence level — interval $\bar{x} \pm 3.30 \, S_{\bar{x}}$

99.7% Confidence level — interval $\bar{x} \pm 3.00 \, S_{\bar{x}}$

99.0% Confidence level — interval $\bar{x} \pm 2.57 \, S_{\bar{x}}$

95.0% Confidence level — interval $\bar{x} \pm 1.96\ S_{\bar{x}}$

90.0% Confidence level — interval $\bar{x} \pm 1.65\ S_{\bar{x}}$

80.0% Confidence level — interval $\bar{x} \pm 1.28\ S_{\bar{x}}$

68.3% Confidence level — interval $\bar{x} \pm 1.00\ S_{\bar{x}}$

60.0% Confidence level — interval $\bar{x} \pm 0.84\ S_{\bar{x}}$

When the sample size is small ($n < 20$), the standard deviation $S_x$ does not provide a reliable estimate of the standard deviation $\sigma$ of the population; therefore, Eq. (10.12) should not be employed. The bias introduced by small sample size can be removed by modifying Eq. (10.12) as follows

$$[\bar{x} - t(\alpha)S_{\bar{x}}] < \mu < [\bar{x} + t(\alpha)S_{\bar{x}}] \tag{10.13}$$

where $t(\alpha)$ is the statistic known as *Student's t*. The Student $t$ distribution is defined by a relative frequency equation $f(t)$, which can be expressed as

$$f(t) = F_0\left(1 + \frac{t^2}{d}\right)^{(d+1)/2} \tag{10.14}$$

**where**   $F_0$ is the relative frequency at $t = 0$ required to make the total area under the $f(t)$ curve equal to unity.

$d$ is the number of degrees of freedom ($d = n - 1$ in this application).

The distribution function $f(t)$ is shown in Fig. 10.6 for several different degrees of freedom $d$. It is evident that as $d$ becomes large, Student's $t$ distribution approaches the normal distribution. The area under the Student $t$ distribution is an important quantity since it can be used to determine $t(\alpha)$ in Eq. (10.13). One-side areas for the $t$ distribution (i.e., between $t_1 \to -\infty$ and $t_2$) are listed in Table 10.5. Since $t(\alpha)$ in Eq. (10.13) is based on areas between $t_1$ and $t_2$ (two-

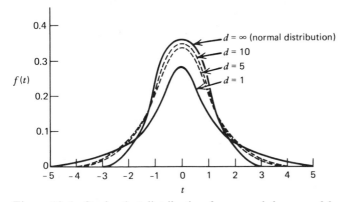

**Figure 10.6**   Student's $t$ distribution for several degrees of freedom $d$.

sided areas), care must be exercised in using these tabulated values, as illustrated in the example problem that follows. The term $t(\alpha)S_{\bar{x}}$ in Eq. (10.13) represents a measure of the interval between the estimated mean $\bar{x}$ and either of the confidence limits. Since $t(\alpha)$ depends on sample size $n$, the term $t(\alpha)S_{\bar{x}}$ can be used to estimate the sample size required to produce an estimate of the mean $\bar{x}$ with a specified reliability. By denoting the bandwidth of the confidence interval as $2\delta$ and using Eq. (10.11), sample size $n$ can be expressed as

$$n = [t(\alpha)S_x/\delta]^2 \qquad (10.15)$$

Use of Eq. (10.15) can be illustrated by considering the data in Table 10.1, where $S_x = 6.04$ ksi and $\bar{x} = 78.4$ ksi. If this estimate of $\mu$ is to be accurate to $\pm 5$ percent with a reliability of 95 percent then one-half the bandwidth $2S$ of the confidence interval is

$$\delta = (0.05)(78.4) = 3.92$$

Since $t(\alpha)$ depends on $n$, a trial-and-error solution is needed to establish sample size $n$ needed to satisfy the specifications. For the data of Table 10.1, $n = 20$; therefore, $d = 19$ and $t(\alpha) = t(0.975) = 2.09$ from Table 10.5. The value $t(\alpha) = t(0.975)$ is used since 2.5 percent of the distribution must be excluded on each end of the curve to give a two sided-area corresponding to a reliability of 95 percent. Substituting into Eq. (10.15) yields

$$n = [(2.09)(6.04)/(3.92)]^2 = 10.4$$

Now with $n = 11$, $d = 10$, and $t(\alpha) = 2.23$; therefore,

$$n = [(2.23)(6.04)/(3.92)]^2 - 11.8$$

Finally, with $n = 12$, $d = 11$, and $t(\alpha) = 2.20$; therefore,

$$n = [(2.20)(6.04)/(3.92)]^2 = 11.5$$

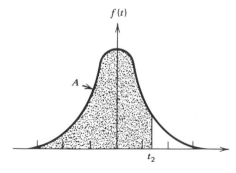

**Figure T10.5**   Definition of terms for Table 10.5.

**TABLE 10.5**  Student's *t* Distribution for *d* Degrees of Freedom Showing $t_2$ as a Function of Area *A* (One Side)

| $d$ | | | | | | $A$ | | | | |
|---|---|---|---|---|---|---|---|---|---|---|
| $\alpha^*$ | 0.995 | 0.99 | 0.975 | 0.95 | 0.90 | 0.80 | 0.75 | 0.70 | 0.60 | 0.55 |
| $(1-\alpha)$ | 0.005 | 0.01 | 0.025 | 0.05 | 0.10 | 0.20 | 0.25 | 0.30 | 0.40 | 0.45 |
| 1 | 63.66 | 31.82 | 12.71 | 6.31 | 3.08 | 1.376 | 1.000 | .727 | .325 | .158 |
| 2 | 9.92 | 6.96 | 4.30 | 2.92 | 1.89 | 1.061 | .816 | .617 | .289 | .142 |
| 3 | 5.84 | 4.54 | 3.18 | 2.35 | 1.64 | .978 | .765 | .584 | .277 | .137 |
| 4 | 4.60 | 3.75 | 2.78 | 2.13 | 1.53 | .941 | .741 | .569 | .271 | .134 |
| 5 | 4.03 | 3.36 | 2.57 | 2.02 | 1.48 | .920 | .727 | .559 | .267 | .132 |
| 6 | 3.71 | 3.14 | 2.45 | 1.94 | 1.44 | .906 | .718 | .553 | .265 | .131 |
| 7 | 3.50 | 3.00 | 2.36 | 1.90 | 1.42 | .896 | .711 | .549 | .263 | .130 |
| 8 | 3.36 | 2.90 | 2.31 | 1.86 | 1.40 | .889 | .706 | .546 | .262 | .130 |
| 9 | 3.25 | 2.82 | 2.26 | 1.83 | 1.38 | .883 | .703 | .543 | .261 | .129 |
| 10 | 3.17 | 2.76 | 2.23 | 1.81 | 1.37 | .879 | .700 | .542 | .260 | .129 |
| 11 | 3.11 | 2.72 | 2.20 | 1.80 | 1.36 | .876 | .697 | .540 | .260 | .129 |
| 12 | 3.06 | 2.68 | 2.18 | 1.78 | 1.36 | .873 | .695 | .539 | .259 | .128 |
| 13 | 3.01 | 2.65 | 2.16 | 1.77 | 1.35 | .870 | .694 | .538 | .259 | .128 |
| 14 | 2.98 | 2.62 | 2.14 | 1.76 | 1.34 | .868 | .692 | .537 | .258 | .128 |
| 15 | 2.95 | 2.60 | 2.13 | 1.75 | 1.34 | .866 | .691 | .536 | .258 | .128 |
| 16 | 2.92 | 2.58 | 2.12 | 1.75 | 1.34 | .865 | .690 | .535 | .258 | .128 |
| 17 | 2.90 | 2.57 | 2.11 | 1.74 | 1.33 | .863 | .689 | .534 | .257 | .128 |
| 18 | 2.88 | 2.55 | 2.10 | 1.73 | 1.33 | .862 | .688 | .534 | .257 | .127 |
| 19 | 2.86 | 2.54 | 2.09 | 1.73 | 1.33 | .861 | .688 | .533 | .257 | .127 |
| 20 | 2.84 | 2.53 | 2.09 | 1.72 | 1.32 | .860 | .687 | .533 | .257 | .127 |
| 21 | 2.83 | 2.52 | 2.08 | 1.72 | 1.32 | .859 | .686 | .532 | .257 | .127 |
| 22 | 2.82 | 2.51 | 2.07 | 1.72 | 1.32 | .858 | .686 | .532 | .256 | .127 |
| 23 | 2.81 | 2.50 | 2.07 | 1.71 | 1.32 | .858 | .685 | .532 | .256 | .127 |
| 24 | 2.80 | 2.49 | 2.06 | 1.71 | 1.32 | .857 | .685 | .531 | .256 | .127 |
| 25 | 2.79 | 2.48 | 2.06 | 1.71 | 1.32 | .856 | .684 | .531 | .256 | .127 |
| 26 | 2.78 | 2.48 | 2.06 | 1.71 | 1.32 | .856 | .684 | .531 | .256 | .127 |
| 27 | 2.77 | 2.47 | 2.05 | 1.70 | 1.31 | .855 | .684 | .531 | .256 | .127 |
| 28 | 2.76 | 2.47 | 2.05 | 1.70 | 1.31 | .855 | .683 | .530 | .256 | .127 |
| 29 | 2.76 | 2.46 | 2.04 | 1.70 | 1.31 | .854 | .683 | .530 | .256 | .127 |
| 30 | 2.75 | 2.46 | 2.04 | 1.70 | 1.31 | .854 | .683 | .530 | .256 | .127 |
| 40 | 2.70 | 2.42 | 2.02 | 1.68 | 1.30 | .851 | .681 | .529 | .255 | .126 |
| 60 | 2.66 | 2.39 | 2.00 | 1.67 | 1.30 | .848 | .679 | .527 | .254 | .126 |
| 120 | 2.62 | 2.36 | 1.98 | 1.66 | 1.29 | .845 | .677 | .526 | .254 | .126 |
| $\infty$ | 2.58 | 2.33 | 1.96 | 1.65 | 1.28 | .842 | .674 | .524 | .253 | .126 |

* The parameter $\alpha$ is the confidence level while the quantity $(1-\alpha)$ is the risk (chance that the limits based upon a sample do not include the population parameter being estimated).

Thus, a sample size of 12 would be sufficient to ensure an accuracy of $\pm 5$ percent with a confidence level of 95 percent. The sample size of 20 listed in Table 10.1 is too large for the degree of accuracy and confidence level specified. This simple example illustrates how sample size can be reduced and cost savings effected by using statistical methods to determine sample size.

## 10.5  COMPARISON OF MEANS

Since Student's $t$ distribution compensates for the effect of small-sample bias and converges to the normal distribution in large samples, it is a very useful statistic in engineering applications. A second important application utilizes the $t$ distribution as the basis for a test to determine if the difference between two means is significant or due to random variation. For example, consider the yield strength data of Table 10.1, where $n_1 = 20$, $\bar{x}_1 = 78.4$ ksi, and $S_{x1} = 6.04$ ksi. Suppose now that a second sample is tested to determine the yield strength of another shipment of steel and the results are $n_2 = 25$, $\bar{x}_2 = 81.6$ ksi, and $S_{x2} = 5.56$ ksi. Is the difference in means significant? The standard deviation of the difference in means $S_{(\bar{x}_2 - \bar{x}_1)}$ can be expressed as

$$S^2_{(\bar{x}_2 - \bar{x}_1)} = S^2_p \left( \frac{1}{n_1} + \frac{1}{n_2} \right) = S^2_p \frac{n_1 + n_2}{n_1 n_2} \tag{10.16}$$

where $S^2_p$ is the pooled variance that can be expressed as

$$S^2_p = \frac{(n_1 - 1)S^2_{x1} + (n_2 - 1)S^2_{x2}}{n_1 + n_2 - 2} \tag{10.17}$$

The statistic $t$ is then computed as

$$t = \frac{|\bar{x}_2 - \bar{x}_1|}{S_{(\bar{x}_2 - \bar{x}_1)}} \tag{10.18}$$

A comparison of the value of $t$ determined from Eq. (10.18) with a value of $t(\alpha)$ obtained from Table 10.5 provides a statistical basis for deciding whether the difference in means is significant. The value of $t(\alpha)$ to be used depends upon the degrees of freedom $d = n_1 + n_2 - 2$ and the level of significance required. Levels of significance commonly employed are 5 and 1 percent. The 5 percent level of significance means that the probability of a random variation being taken for a real difference is only 5 percent. Comparisons at the 1 percent level of significance are 99 percent certain; however, in such a strong test, real differences can often be attributed to random error.

In the example being considered, Eq. (10.17) yields $S^2_p = 33.37$ ksi, Eq.

(10.16) yields $S^2_{(\bar{x}_2 - \bar{x}_1)} = 3.00$ ksi, and Eq. (10.18) yields $t = 1.848$. For a 5 percent level of significancè test with $d = 43$ and $\alpha = 0.05$ (the comparison is one-sided, since $\bar{x}_2 > \bar{x}_1$), Table 10.5 indicates that $t(\alpha) = 1.68$. Since $t > t(\alpha)$, it can be concluded with a 95 percent level of confidence that the yield strength of the second shipment of steel was higher than the yield strength of the first shipment.

## 10.6  STATISTICAL CONDITIONING OF DATA

Previously it was indicated that measurement error can be characterized by a normal distribution function and that the standard deviation of the estimated mean $S_{\bar{x}}$ can be reduced by increasing the number of measurements. In most situations, cost places an upper limit on the number of measurements to be made. Also, it must be remembered that systematic error is not a random variable; therefore, statistical procedures cannot serve as a substitute for precise, accurately calibrated, and properly zeroed measuring instruments.

One area where statistical procedures can be used very effectively to condition experimental data is with the erroneous data point resulting from a measuring or recording mistake. Often, this data point appears questionable when compared with the other data collected, and the experimentalist must decide whether the deviation of the data point is due to a mistake (hence to be rejected) or due to some unusual but real condition (hence to be retained). A statistical procedure known as Chauvenet's criterion provides a consistent basis for making the decision to reject or retain such a point.

Application of Chauvenet's criterion requires computation of a deviation ratio $DR$ for each data point, followed by comparison with a standard deviation ratio $DR_o$. The standard deviation ratio $DR_o$ is a statistic that depends on the number of measurements, while the deviation ratio $DR$ for a point is defined as

$$DR = \frac{x_i - \bar{x}}{S_x} \tag{10.19}$$

The data point is rejected when

$$DR > DR_o \tag{10.20}$$

and retained when

$$DR \leqslant DR_o \tag{10.21}$$

Values for the standard deviation ratio $DR_o$ are listed in Table 10.6.

**TABLE 10.6** Deviation Ratio $DR_o$ Used for Statistical Conditioning of Data

| Number of Measurements $n$ | Deviation Ratio $DR_o$ | Number of Measurements $n$ | Deviation Ratio $DR_o$ |
|---|---|---|---|
| 2 | 1.15 | 15 | 2.13 |
| 3 | 1.38 | 25 | 2.33 |
| 4 | 1.54 | 50 | 2.57 |
| 5 | 1.65 | 100 | 2.81 |
| 7 | 1.80 | 300 | 3.14 |
| 10 | 1.96 | 500 | 3.29 |

If the statistical test of Eq. (10.20) indicates that a single data point in a sequence of $n$ data points should be rejected, then the data point should be removed from the sequence and the mean $\bar{x}$ and the standard deviation $S_x$ should be recalculated. Chauvenet's method can be applied only once to reject a data point that is questionable from a sequence of points. If several data points indicate that $DR > DR_o$, then it is likely that the instrumentation system is inadequate or that the process being investigated is extremely variable.

## 10.7 REGRESSION ANALYSIS

In experimental investigations, measurements are often made of two or more quantities to determine whether these quantities are related. In many engineering problems, the relationship between quantities is obvious (i.e., the stress in a typical beam is linearly proportional to the applied loads, with the constant of proportionality depending on the geometry of the beam). In other problems, however, the relationship is not known. Also, a process may be affected by more than one quantity and the process may exhibit variation. Regression analysis is an important statistical method that can be used to address this class of problems.

### Linear Regression Analysis

Suppose measurements are made of two quantities that are known to be important in describing the behavior of a certain process exhibiting variation. Let $y$ be the dependent variable and $x$ the independent variable. Since the process exhibits variation, there is not a unique relationship between $x$ and $y$ and the data, when plotted, exhibit scatter, as illustrated in Fig. 10.7. Frequently, the relationship between $x$ and $y$ that most closely represents the data, even with the scatter, is a linear function. Thus,

$$Y_i = mx_i + b \tag{10.22}$$

Where $Y_i$ is the predicted value of the dependent variable $y_i$ for a given value of the independent variable $x_i$. A statistical procedure used to fit a straight line

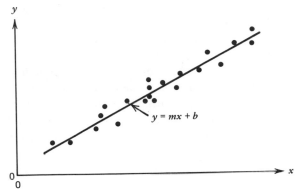

**Figure 10.7**   Least-squares line through scattered data points.

through scattered data points is called the *least-squares method.* With the least-squares method, the slope $m$ and the intercept $b$ in Eq. (10.22) are selected to minimize the sum of the squared deviations of the data points from the straight line as shown in Fig. 10.7. In other words, the quantity

$$\Delta^2 = \sum(y_i - Y_i)^2 \tag{10.23}$$

is minimized. After substituting Eq. (10.22) into Eq. (10.23) this implies that

$$\frac{\partial \Delta^2}{\partial b} = \frac{\partial}{\partial b} \sum(y_i - mx_i - b)^2 = 0$$

$$\frac{\partial \Delta^2}{\partial m} = \frac{\partial}{\partial m} \sum(y_i - mx_i - b)^2 = 0 \tag{10.24}$$

Differentiating yields

$$2\sum(y_i - mx_i - b)(-x_i) = 0$$
$$2\sum(y_i - mx_i - b)(-1) = 0 \tag{10.25}$$

Solving Eqs. (10.25) for $m$ and $b$ yields

$$m = \frac{\sum x_i \sum y_i - n \sum x_i y_i}{\left(\sum x_i\right)^2 - n \sum x_i^2}$$

$$b = \frac{\sum x_i \sum x_i y_i - \sum x_i^2 \sum y_i}{\left(\sum x_i\right)^2 - n \sum x_i^2} = \frac{\sum y_i - m \sum x_i}{n} \tag{10.26}$$

where $n$ is the number of data points. The slope $m$ and intercept $b$ define a straight line through the scattered data points such that Eq. (10.23) is minimized.

In any regression analysis it is important to establish the correlation between $x$ and $y$. Equation (10.22) does not predict the values that were measured exactly, because of the variation in the process. To illustrate, assume that the independent quantity $x$ is fixed at a value $x_1$ and that a sequence of measurements is made of the dependent quantity $y$. The data obtained would give a distribution of $y$, as illustrated in Fig. 10.8. The dispersion of the distribution of $y$ is a measure of the correlation. When the dispersion is small, the correlation is good and the regression analysis is effective in describing the variation in $y$. If the dispersion is large, the correlation is poor and the regression analysis may not be adequate to describe the variation in $y$.

The adequacy of a regression analysis can be evaluated by determining a correlation coefficient $r$ that is given by the following expression:

$$r = \frac{n\sum x_i y_i - \sum x_i \sum y_i}{\sqrt{n\sum x_i^2 - \left(\sum x_i\right)^2}\sqrt{n\sum y_i^2 - \left(\sum y_i\right)^2}} \qquad (10.27)$$

The sign of $r$ is always taken as positive. When the value of the correlation coefficient $r$ is relatively large (0.8 to 1.0), then most of the variation in the dependent variable $y$ has been accounted for in terms of the independent variable $x$, and the linear regression relationship provided by Eq. (10.22) is reliable. However, when the correlation coefficient $r$ is small, Eq. (10.22) is inadequate, possibly because the relationship between $x$ and $y$ is not linear or because other independent variables, not accounted for in Eq. (10.22), may be influencing $y$.

The method of linear regression can be extended to nonlinear cases when suitable nonlinear relationships are available to represent the process. For ex-

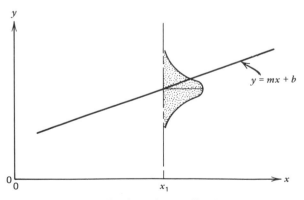

**Figure 10.8**   Distribution of $y$ at a fixed value of $x$ superimposed on the linear regression display.

ample, when the data suggest that is may be represented by an equation of the form

$$y = ax^m \tag{10.28}$$

it is often convenient to write the equation in the form

$$\log y = m \log x + \log a \tag{10.29}$$

In this way Eq. (10.29) can be transformed into

$$y' = mx' + b' \tag{10.30}$$

which is identical to Eq. (10.22) if the substitutions $y' = \log y$, $x' = \log x$, and $b' = \log a$ are made.

Similarly, when the data indicate that an exponential relationship of the form

$$y = ae^{mx} \tag{10.31}$$

may be appropriate, it is convenient to write the equation in the form

$$\ln y = mx + \ln a \tag{10.32}$$

Equation (10.32) can also be transformed into the form

$$y' = mx + b' \tag{10.33}$$

which is identical to Eq. (10.22) if the substitutions $y' = \ln y$ and $b' = \ln a$ are made.

These techniques for applying the least-squares method to nonlinear data minimize the sum of the squared deviations in logarithms of the data rather than in the data themselves. For at least some cases, which include the finite-life region of the stress-cycle diagram for ferrous metals, this procedure is useful, since the logarithms of fatigue life at a constant stress are normally distributed.

## Multiple Regression

Many experiments involve measurement of a dependent variable $y$, which depends upon several independent variables $x_1$, $x_2$, $x_3$, . . . , etc. It is possible to represent $y$ as a function of $x_1$, $x_2$, $x_3$, . . . by employing the multiple regression equation

$$Y_i = a + b_1 x_{1i} + b_2 x_{2i} + \cdots + b_k x_{ki} \tag{10.34}$$

where $a$, $b_1$, $b_2$, . . . , $b_k$ are regression coefficients.

The regression coefficients $a, b_1, b_2, \ldots, b_k$ are determined by using the method of least squares in a manner similar to that employed for linear regression analysis where the quantity $\Delta^2 = \Sigma(y_i - Y_i)^2$ is minimized. After substituting Eq. (10.34) into Eq. (10.23) to obtain

$$\Delta^2 = \sum (y_i - Y_i)^2 \tag{10.35}$$
$$= \sum (y_i - a - b_1 x_{1i} - b_2 x_{2i} - \cdots - b_k x_{ki})^2$$

differentiating yields

$$\frac{\partial \Delta^2}{\partial a} = 2\left[\sum (y_i - a - b_1 x_{1i} - b_2 x_{2i} - \cdots - b_k x_{ki})(-1)\right] = 0$$

$$\frac{\partial \Delta^2}{\partial b_1} = 2\left[\sum (y_i - a - b_1 x_{1i} - b_2 x_{2i} - \cdots - b_k x_{ki})(-x_{1i})\right] = 0$$

$$\frac{\partial \Delta^2}{\partial b_2} = 2\left[\sum (y_i - a - b_1 x_{1i} - b_2 x_{2i} - \cdots - b_k x_{ki})(-x_{2i})\right] = 0 \tag{10.36}$$

$$\cdots = \cdots \qquad\qquad\qquad\qquad = 0$$

$$\frac{\partial \Delta^2}{\partial b_k} = 2\left[\sum (y_i - a - b_1 x_{1i} - b_2 x_{2i} - \cdots - b_k x_{ki})(-x_{ki})\right] = 0$$

Equations (10.36) lead to the following set of $k+1$ equations, which can be solved for the unknown regression coefficients $a, b_1, b_2, \ldots, b_k$.

$$an + b_1 \sum x_{1i} + b_2 \sum x_{2i} + \cdots + b_k \sum x_{ki} = \sum y_i$$
$$a \sum x_{1i} + b_1 \sum x_{1i}^2 + b_2 \sum x_{1i}x_{2i} + \cdots + b_k \sum x_{1i}x_{ki} = \sum y_i x_{1i}$$
$$a \sum x_{2i} + b_1 \sum x_{1i}x_{2i} + b_2 \sum x_{2i}^2 + \cdots + b_k \sum x_{1i}x_{ki} = \sum y_i x_{2i} \tag{10.37}$$
$$\cdots = \cdots$$
$$a \sum x_{ki} + b_1 \sum x_{1i}x_{ki} + b_2 \sum x_{2i}x_{ki} + \cdots + b_k \sum x_{ki}^2 = \sum y_i x_{ki}$$

The correlation coefficient $r$ is again used to determine the degree of association between the dependent and independent variables. For multiple regression equations, the correlation coefficient $r$ is given as

$$r = 1 - \frac{n-1}{n-k}\left[\frac{\{y^2\} - b_1\{yx_1\} - b_2\{yx_2\} - \cdots - b_k\{yx_k\}}{y^2}\right] \tag{10.38}$$

where

$$\{yx_k\} = \sum y_i x_{ki} - \left(\sum y_i\right)\left(\sum x_{ki}\right)\Big/n$$

$$\{y^2\} = \sum y_i^2 - \left(\sum y_i\right)^2\Big/n$$

This analysis is for linear, noninteracting, independent variables; however, the analysis can be extended to include cases where the regression equations would have higher-order and cross-product terms. The nonlinear terms can enter the regression equation in an additive manner and are treated as an extra variable. With well-established computer routines for regression analysis, the set of $(k + 1)$ simultaneous equations given by Eqs. (10.37) can be solved quickly and inexpensively and no difficulties are encountered in adding extra terms to account for nonlinearities and interactions.

## 10.8 CHI-SQUARE TESTING

Once a series of measurements is made to obtain data to statistically characterize a population, it is possible to select a distribution function for the population and to predict population properties, such as the probability of occurrence of a certain event. This is an extremely important measurement procedure in engineering; therefore, to avoid error, it is essential that the correct distribution function be selected to represent the population. The data from a series of measurements can be subjected to a chi-square ($\chi^2$) test to check the validity of the assumed distribution function.

The $\chi^2$ statistic is defined as

$$\chi^2 = \sum \frac{(O - E)^2}{E} \tag{10.39}$$

**where**   $O$ is an observed number.
$E$ is an expected number, based on a specified statistical distribution function.

The value $\chi^2$ is used to determine how well the data fit the assumed statistical distribution. If $\chi^2 = 0$, the match is perfect. Values of $\chi^2 > 0$ indicate a possibility that the data are not represented by the specified distribution function. The probability $p$ of $\chi^2$ occurring as a result of random variation is listed in Table 10.7 and illustrated in Fig. 10.9. The number of degrees of freedom associated with any sequence of measurements is given by the expression

$$d = n - k \tag{10.40}$$

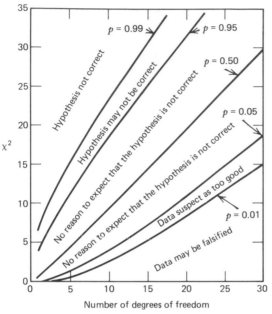

**Figure 10.9**   $\chi^2$ values as a function of the number of degrees of freedom for different probability levels.

**where**    $n$ is the number of observations.
        $k$ is the number of conditions imposed on the distribution.

As an example, consider the yield strength data presented in Table 10.1. A $\chi^2$ test can be used to determine how well these data are represented by a normal distribution having a mean $\bar{x} = 78.4$ ksi and a standard deviation $S_x = 6.04$ ksi. By using the properties of a normal distribution function, the number of specimens expected to fall in any strength group can be computed. The observed number of specimens in Table 10.1 exhibiting yield strengths within each of five group intervals, together with the computed number of specimens in a normal distribution in the same group intervals, are listed in Table 10.8. Also, the computation of the $\chi^2$ value ($\chi^2 = 1.129$) is illustrated in the table. Since the number of groups is 5 ($n = 5$) and since the two distribution parameters $\bar{x}$ and $S_x$ were determined by using these data (thus, $k = 2$), the number of degrees of freedom is $d = n - k = 5 - 2 = 3$. Thus, from Table 10.7, it can be concluded with a 77 percent probability that the data of Table 10.1 can be represented by a normal distribution function. This can be considered a very good fit, as shown in Fig. 10.9. One could conclude in other instances with higher $\chi^2$ values and much lower probability levels that the assumed distribution was representative.

**TABLE 10.7**   Chi-Squared ($\chi^2$) Values with Different Degrees of Freedom for Different Probability Levels

| Degrees of Freedom | Probability (%) | | | | | | | | | | | | | |
|---|---|---|---|---|---|---|---|---|---|---|---|---|---|---|
| | 1.0 | 2.5 | 5.0 | 10.0 | 20.0 | 30.0 | 40.0 | 50.0 | 60.0 | 70.0 | 80.0 | 90.0 | 95.0 | 99.0 |
| 1 | .00016 | .0098 | .00393 | .0158 | .0642 | .148 | .275 | .455 | .708 | 1.07 | 1.64 | 2.71 | 3.84 | 6.63 |
| 2 | .0201 | .0506 | .103 | .211 | .446 | .713 | 1.02 | 1.39 | 1.83 | 2.41 | 3.22 | 4.61 | 5.99 | 9.21 |
| 3 | .115 | .216 | .352 | .584 | 1.00 | 1.42 | 1.87 | 2.37 | 2.95 | 3.67 | 4.64 | 6.25 | 7.81 | 11.3 |
| 4 | .297 | .484 | .711 | 1.06 | 1.65 | 2.19 | 2.75 | 3.36 | 4.04 | 4.88 | 5.99 | 7.78 | 9.49 | 13.3 |
| 5 | .554 | .831 | 1.15 | 1.61 | 2.34 | 3.00 | 3.66 | 4.35 | 5.13 | 6.06 | 7.29 | 9.24 | 11.1 | 15.1 |
| 6 | .872 | 1.24 | 1.64 | 2.20 | 3.07 | 3.83 | 4.57 | 5.35 | 6.21 | 7.23 | 8.56 | 10.6 | 12.6 | 16.8 |
| 7 | 1.24 | 1.69 | 2.17 | 2.83 | 3.82 | 4.67 | 5.49 | 6.35 | 7.28 | 8.38 | 9.80 | 12.0 | 14.1 | 18.5 |
| 8 | 1.65 | 2.18 | 2.73 | 3.49 | 4.59 | 5.53 | 6.42 | 7.34 | 8.35 | 9.52 | 11.0 | 13.4 | 15.5 | 20.1 |
| 9 | 2.09 | 2.70 | 3.33 | 4.17 | 5.38 | 6.39 | 7.36 | 8.34 | 9.41 | 10.7 | 12.2 | 14.7 | 16.9 | 21.7 |
| 10 | 2.56 | 3.25 | 3.94 | 4.87 | 6.18 | 7.27 | 8.30 | 9.34 | 10.5 | 11.8 | 13.4 | 16.0 | 18.3 | 23.2 |
| 11 | 3.05 | 3.82 | 4.57 | 5.58 | 6.99 | 8.15 | 9.24 | 10.3 | 11.5 | 12.9 | 14.6 | 17.3 | 19.7 | 24.7 |
| 12 | 3.57 | 4.40 | 5.23 | 6.30 | 7.81 | 9.03 | 10.2 | 11.3 | 12.6 | 14.0 | 15.8 | 18.5 | 21.0 | 26.2 |
| 13 | 4.11 | 5.01 | 5.89 | 7.04 | 8.63 | 9.93 | 11.1 | 12.3 | 13.6 | 15.1 | 17.0 | 19.8 | 22.4 | 27.7 |
| 14 | 4.66 | 5.63 | 6.57 | 7.79 | 9.47 | 10.8 | 12.1 | 13.3 | 14.7 | 16.2 | 18.2 | 21.1 | 23.7 | 29.1 |
| 15 | 5.23 | 6.26 | 7.26 | 8.55 | 10.3 | 11.7 | 13.0 | 14.3 | 15.7 | 17.3 | 19.3 | 22.3 | 25.0 | 30.6 |
| 20 | 8.26 | 9.59 | 10.9 | 12.4 | 14.6 | 16.3 | 17.8 | 19.3 | 21.0 | 22.8 | 25.0 | 28.4 | 31.4 | 37.6 |
| 25 | 11.5 | 13.1 | 14.6 | 16.5 | 18.9 | 20.9 | 22.6 | 24.3 | 26.1 | 28.2 | 30.7 | 34.4 | 37.7 | 44.3 |
| 30 | 15.0 | 16.8 | 18.5 | 20.6 | 23.4 | 25.5 | 27.4 | 29.3 | 31.3 | 33.5 | 36.3 | 40.3 | 43.8 | 50.9 |
| 40 | 22.2 | 24.4 | 26.5 | 29.1 | 32.3 | 34.9 | 37.1 | 39.3 | 41.6 | 44.2 | 47.3 | 51.8 | 55.8 | 63.7 |
| 50 | 29.7 | 32.4 | 34.8 | 37.7 | 41.4 | 44.3 | 46.9 | 49.3 | 51.9 | 54.7 | 58.2 | 63.2 | 67.5 | 76.2 |
| 100 | 70.1 | 74.2 | 77.9 | 82.4 | 87.9 | 92.1 | 95.8 | 99.3 | 102.9 | 106.9 | 111.7 | 118.5 | 124.3 | 135.8 |

**TABLE 10.8** $\chi^2$ Computation for Grouped Yield Strength Data

| Group Interval | Number Observed | Number[a] Expected | $(O - E)^2/E$ |
|---|---|---|---|
| 0 –69.99 | 2 | 1.646 | 0.0761 |
| 70.0–74.99 | 3 | 4.108 | 0.2988 |
| 75.0–79.99 | 8 | 6.298 | 0.4600 |
| 80.0–84.99 | 4 | 5.190 | 0.2729 |
| 85.0–∞ | 3 | 2.758 | 0.0212 |
| Total | 20 | 20 | 1.129 $= \chi^2$ |

[a] Based on a normal (Gaussian) distribution of data.

The $\chi^2$ statistic can also be used in contingency testing where the sample is classified under one of two catagories—say pass or fail. Consider, for example, an inspection procedure with a particular type of strain gage where 10 percent of the gages are rejected due to etching imperfections in the grid. In an effort to reduce this rejection rate, the manufacturer has introduced new clean-room techniques that are expected to improve the quality of the grids. On the first lot of 1000 gages, the failure rate was reduced to 9 percent. Is this reduced failure rate due to chance variation, or have the new clean-room techniques improved the manufacturing process? A $\chi^2$ test can establish the probability of the improvement being the result of random variation. The computation of $\chi^2$ for this example is illustrated in Table 10.9.

The data in Table 10.7 with $d = 1$ give a probability $p$ of $\chi^2$ exceeding 1.111 of 29 percent; thus, the test provides a strong indication that the clean-room improvements have reduced the rejection rate. Stronger statistical statements could be made after testing of a second lot of 1000 gages is completed, if the trend continues. At this point 2000 gages would have been inspected with a value $\chi^2 = 2.222$. The probability of $\chi^2$ exceeding this value due to random variation is only 14 percent.

**TABLE 10.9** Observed and Expected Inspection Results

| | Observed Number | Expected Number | $(O - E)^2/E$ |
|---|---|---|---|
| Passed | 910 | 900 | 0.111 |
| Failed | 90 | 100 | 1.000 |
| | | | 1.111 $= \chi^2$ |

## 10.9 **ERROR PROPAGATION**

Previous discussions of error have been limited to error arising in the measurement of a single quantity; however, in many engineering applications, several quantities are measured (each with its associated error) and another quantity is predicted on the basis of these measurements. For example, the volume $V$ of a cylinder could be predicted on the basis of measurements of two quantities (diameter $D$ and length $L$). Thus errors in the measurements of diameter and length will propagate through the governing mathematical formula $V = \pi D^2 L/4$ to the quantity (volume, in this case) being predicted. Since the propagation of error depends upon the form of the mathematical expression being used to predict the reported quantity, standard deviations for several different mathematical operations are listed below.

For *addition* and/or *subtraction of quantities* $(y = x_1 \pm x_2 \pm \cdots \pm x_n)$, the standard deviation $S_{\bar{y}}$ of the mean $\bar{y}$ of the projected quantity $y$ is

$$S_{\bar{y}} = \sqrt{S_{\bar{x}_1}^2 + S_{\bar{x}_2}^2 + \cdots + S_{\bar{x}_n}^2} \tag{10.41}$$

For *multiplication of quantities* $(y = x_1 x_2 \ldots x_n)$, the standard deviation $S_{\bar{y}}$ is

$$S_{\bar{y}} = (\bar{x}_1 \bar{x}_2 \ldots \bar{x}_n) \sqrt{\frac{S_{\bar{x}_1}^2}{\bar{x}_1^2} + \frac{S_{\bar{x}_2}^2}{\bar{x}_2^2} + \cdots + \frac{S_{\bar{x}_n}^2}{\bar{x}_n^2}} \tag{10.42}$$

For *division of quantities* $(y = x_1/x_2)$, the standard deviation $S_{\bar{y}}$ is

$$S_{\bar{y}} = \frac{\bar{x}_1}{\bar{x}_2} \sqrt{\frac{S_{\bar{x}_1}^2}{\bar{x}_1^2} + \frac{S_{\bar{x}_2}^2}{\bar{x}_2^2}} \tag{10.43}$$

For *exponent calculations of the form* $(y = x_1^k)$, the standard deviation $S_{\bar{y}}$ is

$$S_{\bar{y}} = k\bar{x}_1^{k-1} S_{\bar{x}_1} \tag{10.44}$$

For *exponent calculations of the form* $(y = x_1^{1/k})$, the standard deviation $S_{\bar{y}}$ is

$$S_{\bar{y}} = \frac{\bar{x}_1^{1/k}}{k\bar{x}_1} S_{\bar{x}_1} \tag{10.45}$$

Consider, for example, the cylinder where measurements of the diameter have yielded $\bar{x}_1 = 1$ in., with a standard error $S_{\bar{x}_1} = 0.005$ in., and measurements of the length have yielded $\bar{x}_2 = 4$ in., with a standard error $S_{\bar{x}_2} = 0.040$ in. Since the volume $V$ of the cylinder is

$$V = \frac{\pi}{4}\bar{x}_1^2\bar{x}_2$$

the standard error of the volume $S_{\bar{y}}$ can be determined by using Eq. (10.42). Thus,

$$S_{\bar{y}} = \bar{x}_1^2\bar{x}_2\sqrt{\frac{2S_{\bar{x}_1}^2}{\bar{x}_1^2} + \frac{S_{\bar{x}_2}^2}{\bar{x}_2^2}}$$

$$= (1)^2(4)\sqrt{\frac{2(0.005)^2}{(1)^2} + \frac{(0.040)^2}{(4)^2}} \tag{10.46}$$

$$= 0.0490 \text{ in.}^3$$

This determination of $S_{\bar{y}}$ for the volume of the cylinder can be used (by utilizing the properties of a normal distribution) to predict the probability of producing cylinders having volumes within specified limits and to predict the deviation of the mean volume from a specified value.

The method of computing the standard error of a quantity $S_{\bar{y}}$ as given by Eqs. (10.41) to (10.45), which are based on the properties of the normal distribution function, should be used where possible. However, in many engineering applications, the number of measurements that can be made is small; therefore, the data $\bar{x}_1, \bar{x}_2, \ldots, \bar{x}_n$ and $S_{\bar{x}_1}, S_{\bar{x}_2}, \ldots, S_{\bar{x}_n}$ needed for statistical based estimtes of the error are not available. In these instances, error estimates can still be made but the results are less reliable.

A second method of estimating error when data are limited is based on the chain rule of differential calculus. For example, consider a quantity $y$ that is a function of several variables

$$y = f(x_1, x_2, \ldots, x_n) \tag{10.47}$$

Differentiating yields

$$dy = \frac{\partial y}{\partial x_1}dx_1 + \frac{\partial y}{\partial x_2}dx_2 + \cdots + \frac{\partial y}{\partial x_n}dx_n \tag{10.48}$$

In Eq. (10.48), $dy$ is the error in $y$ and $dx_1, dx_2, \ldots, dx_n$ are errors involved in the measurements of $x_1, x_2, \ldots, x_n$. The partial derivatives $\partial y/\partial x_1$,

$\partial y/\partial x_2, \ldots, \partial y/\partial x_n$ are determined exactly from Eq. (10.47). Frequently, the errors $dx_1, dx_2, \ldots, dx_n$ are estimates based on the experience and judgment of the experimentalist. An estimate of the maximum possible error can be obtained by summing the individual error terms in Eq. (10.48). Thus

$$dy \Big]_{max} = \left| \frac{\partial y}{\partial x_1} dx_1 \right| + \left| \frac{\partial y}{\partial x_2} dx_2 \right| + \cdots + \left| \frac{\partial y}{\partial x_n} dx_n \right| \qquad (10.49)$$

Use of Eq. (10.49) gives a worst case estimate of error, since the maximum errors $dx_1, dx_2, \ldots, dx_n$ are assumed to occur simultaneously and with the same sign. A more realistic method of estimating error under these conditions is given by the expression

$$dy = \sqrt{\left( \frac{\partial y}{\partial x_1} dx_1 \right)^2 + \left( \frac{\partial y}{\partial x_2} dx_2 \right)^2 + \cdots + \left( \frac{\partial y}{\partial x_n} dx_n \right)^2} \qquad (10.50)$$

Equation (10.50) cannot be derived; however, experience shows that it predicts error more accurately than Eq. (10.49); therefore, it is commonly employed.

## 10.10 SUMMARY

Statistical methods are extremely important in engineering, since they provide a means for representing large amounts of data in a concise form that is easily interpreted and understood. Usually, the data are represented with a statistical distribution function that can be characterized by a measure of central tendency (the mean $\bar{x}$) and a measure of dispersion (the standard deviation $S_x$). A normal or Gaussian distribution is by far the most commonly employed; however, in some special cases, other distribution functions may have to be employed to adequately represent the data.

The most significant advantage resulting from use of a statistical distribution function in engineering applications is the ability to predict the occurrence of an event based on a relatively small sample. The effects of sampling error are accounted for by placing confidence limits on the predictions and establishing the associated confidence levels. Sampling error can be controlled if the sample size is adequate. Use of Student's $t$ distribution function, which characterizes sampling error, provides a basis for determining sample size consistent with specified levels of confidence. The Student $t$ distribution also permits a comparison to be made of two means to determine whether the observed difference is significant or whether it is due to random variation.

Statistical methods can also be employed to condition data and to eliminate an erroneous data point (one) from a series of measurements. This is a useful technique that improves the data base by providing strong evidence when something unanticipated is affecting an experiment.

Regression analysis can be used effectively to interpret data when the behavior of one quantity (say, $y$) depends upon variations in one or more independent quantities (say, $x_1, x_2, \ldots, x_n$). Even though the functional relationship between quantities exhibiting variation remains unknown, it can be characterized statistically. Regression analysis provides a method to fit a straight line or a curve through a series of scattered data points on a graph. The adequacy of the regression analysis can be evaluated by determining a correlation coefficient. Methods for extending regression analysis to nonlinear functions and to multivariate functions exist. In principle, these methods are identical to linear regression analysis; however, the analysis becomes much more complex. The increase in complexity is not a concern, however, since computer subroutines are available that solve the tedious equations and provide the results in a convenient format.

Many distribution functions are used in statistical analyses to represent data and predict population properties. Once a distribution function has been selected to represent a population, any series of measurements can be subjected to a $\chi^2$ test to check the validity of the assumed distribution. Accurate predictions can be made only if the proper distribution function has been selected.

Finally, statistical methods for accessing error propagation were discussed. These methods provide a means for determining error in a quantity of interest $y$ based on measurements of related quantities $x_1, x_2, \ldots, x_n$ and the functional relationship $y = f(x_1, x_2, \ldots, x_n)$ between quantities.

## REFERENCES

1. Blackwell, D.: *Basic Statistics*, McGraw-Hill, New York, 1969.

2. Snedecor, G. W., and W. G. Cochran: *Statistical Methods*, 6th ed., Iowa State University Press, Ames, 1967.

3. Zehna, P. W.: *Introductory Statistics*, Prindle, Weber & Schmidt, Boston, 1974.

4. Chou, Y.: *Probability and Statistics for Decision Making*, Holt, Rinehart & Winston, New York, 1972.

5. Young, H. D.: *Statistical Treatment of Experimental Data*, McGraw-Hill, New York, 1962.

6. Bragg, G. M.: *Principles of Experimentation and Measurement*, Prentice-Hall, Englewood Cliffs, New Jersey, 1974.

7. Durelli, A. J., E. A. Phillips, and C. H. Tsao: *Theoretical and Experimental Analysis of Stress and Strain*, McGraw-Hill, New York, 1958.

8. Bethea, R. M., B. S. Duran, and T. L. Boullion: *Statistical Methods for Engineers and Scientists*, Dekker, New York, 1975.

9. McCall, C.H. Jr., *Sampling and Statistics Handbook for Research,* Iowa State University Press, Ames, Iowa, 1982.

**EXERCISES**

**10.1** The air pressure (in psi) at a point near the end of an air-supply line is monitored at 15-min intervals over an 8-h period. The readings are listed in the four columns below:

| Column 1 | Column 2 | Column 3 | Column 4 |
|----------|----------|----------|----------|
| 110 | 102 | 97 | 94 |
| 104 | 98 | 93 | 97 |
| 106 | 95 | 90 | 100 |
| 94 | 120 | 84 | 102 |
| 92 | 115 | 78 | 107 |
| 89 | 109 | 82 | 110 |
| 100 | 100 | 88 | 117 |
| 114 | 98 | 91 | 125 |

List the pressure readings in order of increasing magnitude. Rearrange the data in five group intervals to obtain a frequency distribution. Show both the relative frequency and the cumulative frequency in this tabular rearrangement. Select the median pressure reading from the data.

**10.2** Construct a histogram for the data listed in Exercise 10.1. Superimpose a plot of the relative frequency on the histogram.

**10.3** Prepare a cumulatve frequency curve for the data of Exercise 10.1.

**10.4** For the air-pressure data listed in Exercise 10.1:

(a) Compute the sample mean $\bar{x}$ of the individual columns.
(b) Compute the sample means of columns 1 + 2 and columns 3 + 4.
(c) Compute the sample mean of the complete set of data.
(d) Comment on the results of (a), (b), and (c).

**10.5** Determine the mode for the air-pressure data of Exercise 10.1 and compare it with the median and the mean of the data.

**10.6** A quality control laboratory monitors the tensile strength of paper by testing a small sample every 10 min. The data shown below were reported over an 8-h shift.

| Column 1 | Column 2 | Column 3 | Column 4 | Column 5 | Column 6 |
|----------|----------|----------|----------|----------|----------|
| 1030 | 937 | 1093 | 1018 | 970 | 983 |
| 1042 | 963 | 1081 | 1013 | 947 | 994 |
| 1019 | 1022 | 1049 | 1000 | 921 | 1030 |
| 979 | 1037 | 1040 | 992 | 897 | 1063 |
| 942 | 1071 | 1033 | 984 | 893 | 1093 |
| 910 | 1116 | 1032 | 986 | 918 | 1122 |
| 870 | 1129 | 1027 | 989 | 934 | 1141 |
| 912 | 1130 | 1021 | 972 | 957 | 1165 |

List the strength readings in order of increasing magnitude. Rearrange the data in seven group intervals to obtain a frequency distribution. Show both the relative frequency and the cumulative frequency in this tabular rearrangement. Select the median strength reading from the data.

**10.7** Construct a histogram for the data listed in Exercise 10.6. Superimpose a plot of relative frequency on the histogram.

**10.8** Prepare a cumulative frequency curve for the data of Exercise 10.6.

**10.9** For the tensile-strength data listed in Exercise 10.6,

(a) Compute the sample mean $\bar{x}$ of the individual columns.

(b) Compute the sample means of columns 1 + 2, 3 + 4 and 5 + 6.

(c) Compute the sample mean of the complete set of data.

(d) Comment on the results of (a), (b), and (c).

**10.10** Determine the mode for the tensile-strength data of Exercise 10.6 and compare it with the median and the mean of the data.

**10.11** The accuracy of a new flowmeter for diesel fuel is being checked by pumping 100 gal of the fuel into a tank and measuring the true volume with a calibrated sight glass. The readings from the sight glass were

| Column 1 | Column 2 | Column 3 | Column 4 | Column 5 | Column 6 |
|----------|----------|----------|----------|----------|----------|
| 100.84 | 100.12 | 100.42 | 100.04 | 100.32 | 99.97 |
| 100.62 | 100.07 | 99.70 | 100.17 | 100.11 | 99.73 |
| 99.88 | 100.33 | 99.81 | 99.73 | 99.94 | 100.02 |
| 99.64 | 100.26 | 99.63 | 99.77 | 99.82 | 100.14 |
| 99.72 | 99.73 | 100.13 | 99.89 | 99.97 | 99.90 |
| 99.81 | 99.47 | 100.32 | 100.03 | 99.86 | 100.24 |
| 99.99 | 99.90 | 100.24 | 100.09 | 100.01 | 99.76 |
| 100.00 | 99.86 | 99.74 | 99.92 | 99.80 | 100.00 |
| 100.63 | 99.40 | 99.92 | 99.88 | 100.12 | 99.95 |
| 100.22 | 99.73 | 100.16 | 99.62 | 100.23 | 100.13 |

List the volume readings in order of increasing magnitude. Rearrange the data in eight group intervals to obtain a frequency distribution. Show both the relative frequency and the cumulative frequency in this tabular rearrangement. Select the median volume delivered. Comment on the merits of the new flowmeter.

**10.12**  Construct a histogram for the data listed in Exercise 10.11. Superimpose a plot of relative frequency on the histogram.

**10.13**  Prepare a cumulative frequency curve for the data of Exercise 10.11.

**10.14**  For the volume-delivered data listed in Exercise 10.11,

    (a)  Compute the sample mean $\bar{x}$ of the individual columns.
    (b)  Compute the sample means of columns $1 + 2$, $3 + 4$, and $5 + 6$.
    (c)  Compute the sample means of columns $1 + 2 + 3$, and $4 + 5 + 6$.
    (d)  Compute the sample mean for the complete set of data.
    (e)  Comment on the results of (a), (b), (c), and (d).

**10.15**  Determine the mode for the volume-delivered data of Exercise 10.11 and compare it with the median and the mean of the data.

**10.16**  For the data presented in Exercise 10.1, determine:

    (a)  The standard deviation $S_x$
    (b)  The range $R$
    (c)  The mean deviation $d_x$
    (d)  The variance $S_x^2$
    (e)  The coefficient of variation $C_v$

**10.17**  For the data presented in Exercise 10.6, determine:

    (a)  The standard deviation $S_x$
    (b)  The range $R$
    (c)  The mean deviation $d_x$
    (d)  The variance $S_x^2$
    (e)  The coefficient of variation $C_v$

**10.18**  For the data presented in Exercise 10.11, determine:

    (a)  The standard deviation $S_x$
    (b)  The range $R$
    (c)  The mean deviation $d_x$
    (d)  The variance $S_x^2$
    (e)  The coefficient of variation $C_v$

**10.19**  A quality control laboratory associated with a manufacturing operation periodically makes a measurement that has been characterized as normal or Gaussian with a mean of 80 and a standard deviation of 4. Determine the probability of a single measurement:

(a) Falling between 76 and 84
(b) Falling between 72 and 88
(c) Falling between 68 and 92
(d) Falling between 80 and 82
(e) Falling between 79 and 80
(f) Being less than 77
(g) Being greater than 86

**10.20** Careful measurement of diameter was made after a large shipment of electric motor shafts was received. Measurements of the bearing journal indicate that the mean is 7 mm, with a standard deviation of 0.02 mm. Determine the probability of a journal measurement being

(a) Greater than 7.04 mm
(b) Greater than 7.03 mm
(c) Less than 7.00 mm
(d) Less than 6.94 mm
(e) Between 7.01 and 7.03 mm
(f) Between 6.96 and 6.99 mm
(g) Between 6.97 and 7.03 mm

**10.21** Use the data for the flowmeter in Exercise 10.11 to determine the probability of the pump delivering a volume that is

(a) In excess of the measured volume by 1.0 percent
(b) In excess of the measured volume by 0.5 percent
(c) In excess of the measured volume by 0.2 percent
(d) In excess of the measured volume by 0.1 percent
(e) Less than the measured volume by 1.0 percent
(f) Less than the measured volume by 0.5 percent
(g) Less than the measured volume by 0.2 percent
(h) Less than the measured volume by 0.1 percent

**10.22** Determine the standard deviation of the mean $S_{\bar{x}}$ (standard error) for the data of

(a) Exercise 10.1
(b) Exercise 10.6
(c) Exercise 10.11

**10.23** The data presented in Exercise 10.1 were drawn from a large population and provided an estimate $\bar{x}$ of the true mean $\mu$ of the population. Determine the confidence interval if the mean is to be stated with a confidence level of

(a) 99 percent          (c) 90 percent
(b) 95 percent          (d) 80 percent

**10.24** Repeat Exercise 10.23 with the data presented in Exercise 10.6.

**10.25** Repeat Exercise 10.23 with the data presented in Exercise 10.11.

**10.26** A small sample ($n < 20$) of measurements from a drag coefficient $C_D$ determination for an airfoil are listed below:

| | |
|---|---|
| 0.043 | 0.052 |
| 0.050 | 0.049 |
| 0.053 | 0.055 |
| 0.047 | 0.046 |
| 0.049 | 0.051 |

(a) Estimate the mean drag coefficient.
(b) Estimate the standard deviation.
(c) Determine the confidence level for the statement that $C_D$ = 0.0495.

**10.27** Determine the sample size needed in Exercise 10.26 to permit specification of the mean drag coefficient $C_D$ with a confidence level of

(a) 90 percent  (c) 99 percent
(b) 95 percent  (d) 99.9 percent

**10.28** Measurements of flue-gas temperature from a power-plant stack are listed below:

| | |
|---|---|
| 273°F | 281°F |
| 279°F | 287°F |
| 280°F | 286°F |
| 291°F | 287°F |
| 282°F | 276°F |

(a) Estimate the mean flue-gas temperature.
(b) Estimate the standard deviation.
(c) Determine the confidence level for the statement that the flue-gas temperature is 282.2°F.

**10.29** Determine the sample size required in Exercise 10.28 to permit specification of the flue-gas temperature with a confidence level of

(a) 90 percent  (c) 99 percent
(b) 95 percent  (d) 99.9 percent

**10.30** A manufacturing process yields aluminum rods with a mean yield strength of 35,000 psi and a standard deviation of 1000 psi. A customer places a very large order for rods with a minimum yield strength of 32,000 psi. Prepare a letter for submission to the customer that describes the yield strength to be expected and outline your firm's procedures for assuring that this quality level will be achieved and maintained.

**10.31**   An inspection laboratory samples two large shipments of dowel pins by measuring both length and diameter. For shipment A, the sample size was 40, the mean diameter was 6.12 mm, the mean length was 25.3 mm, the estimated standard deviation on diameter was 0.022 mm, and the estimated standard deviation on length was 0.140 mm. For shipment B, the sample size was 60, the mean diameter was 6.04 mm, the mean length was 25.05 mm, the estimated standard deviation on diameter was 0.034 mm, and the estimated standard deviation on length was 0.203 mm.

   (a)   Are the two shipments of dowel pins the same?
   (b)   What is the level of confidence in your prediction?
   (c)   Would it be safe to mix the two shipments of pins? Explain.

**10.32**   Repeat Exercise 10.31 for the following two shipments of dowel pins.

| | Shipment A | Shipment B |
|---|---|---|
| Number | 20 | 10 |
| Diameter | $\bar{x} = 6.05$ mm, $S_x = 0.03$ mm | $\bar{x} = 5.98$ mm, $S_x = 0.04$ mm |
| Length | $\bar{x} = 24.9$ mm, $S_x = 0.22$ mm | $\bar{x} = 25.4$ mm, $S_x = 0.18$ mm |

**10.33**   Employ Chauvenet's criterion to statistically condition the following sequence of measurements of atmospheric pressure (millimeters of mercury) obtained with a mercury barometer.

| | | |
|---|---|---|
| 764.3 | 764.6 | 764.4 |
| 765.2 | 764.5 | 764.5 |
| 765.7 | 765.4 | 764.8 |
| 765.3 | 765.2 | 764.9 |
| 764.6 | 765.1 | 764.6 |

**10.34**   After conditioning the data in Exercise 10.33, determine the mean and the standard deviation.

**10.35**   The weight of a shipment of gold has been measured $n$ times to obtain $\bar{x}$ and $S_x$. If the precision of the stated weight must be increased by a factor of 1.5, how many additional measurements of the weight should be made?

**10.36**   Repeat Exercise 10.36 if the required factor of improvement is

   (a)   1.10          (c)   2.00          (e)   5.0
   (b)   1.25          (d)   3.00          (f)   10.0

**10.37**   Prepare a graph showing relative precision as a function of the number of measurements for a population with $\mu = 1$ and $\sigma = 0.1$.

**10.38** Determine the slope $m$ and the intercept $b$ for a linear regression equation $y = mx + b$ representing the following set of data.

| $x$ | 1.1 | 1.9 | 2.4 | 3.0 | 3.3 | 3.8 | 4.2 | 4.5 | 4.9 | 5.2 |
|---|---|---|---|---|---|---|---|---|---|---|
| $y$ | 3.5 | 6.0 | 7.2 | 8.8 | 10.1 | 11.8 | 12.3 | 13.7 | 15.0 | 15.8 |

**10.39** Determine the slope $m$ and the intercept $b$ for a linear regression equation $y = mx + b$ representing the following set of data.

| $x$ | $y$ | $x$ | $y$ | $x$ | $y$ |
|---|---|---|---|---|---|
| 10.7 | 25.7 | 16.2 | 39.7 | 20.8 | 51.2 |
| 12.0 | 31.2 | 17.0 | 40.5 | 22.2 | 55.3 |
| 12.9 | 32.9 | 17.9 | 46.7 | 23.5 | 60.6 |
| 13.7 | 35.9 | 18.7 | 48.0 | 25.1 | 59.7 |
| 15.1 | 37.8 | 19.9 | 50.1 | 26.4 | 66.5 |

**10.40** Determine the slope $m$ and the intercept $b$ for a linear regression equation $y = mx + b$ representing the following set of data.

| $x$ | $y$ | $x$ | $y$ | $x$ | $y$ |
|---|---|---|---|---|---|
| 10.7 | 36.8 | 16.2 | 50.8 | 20.8 | 62.3 |
| 12.0 | 42.3 | 17.0 | 51.6 | 22.2 | 66.4 |
| 12.9 | 44.0 | 17.9 | 57.8 | 23.5 | 71.7 |
| 13.7 | 47.0 | 18.7 | 59.1 | 25.1 | 70.8 |
| 15.1 | 48.9 | 19.9 | 61.2 | 26.4 | 77.6 |

**10.41** Determine the correlation coefficient $r$ for the regression analysis of Exercise 10.38.

**10.42** Determine the correlation coefficient $r$ for the regression analysis of Exercise 10.39.

**10.43** Determine the correlation coefficient $r$ for the regression analysis of Exercise 10.40.

**10.44** The solvent concentration in a coating as a function of time is governed by an equation of the form

$$C = ae^{-mt}$$

**where** $C$ is the concentration.

$t$ is the time.

$a$ and $m$ are constants that depend upon the diffusion process.

| $t(s)$ | $C(\%)$ | $t(s)$ | $C(\%)$ |
|--------|---------|--------|---------|
| 10 | 1.98 | 50 | 1.88 |
| 20 | 2.01 | 100 | 1.81 |
| 30 | 1.95 | 200 | 1.60 |
| 40 | 1.93 | 500 | 1.21 |

For the given data, determine the constants $a$ and $m$ that provide the best fit.

**10.45** Determine the regression coefficients $a$, $b_1$, $b_2$, and $b_3$ for the following data set.

| $y$ | $x_1$ | $x_2$ | $x_3$ |
|-----|-------|-------|-------|
| 6.8 | 1.0 | 2.0 | 1.0 |
| 7.8 | 1.5 | 2.0 | 1.5 |
| 7.9 | 2.0 | 2.0 | 2.0 |
| 8.0 | 2.5 | 3.0 | 1.0 |
| 8.3 | 3.0 | 3.0 | 1.5 |
| 8.4 | 3.5 | 3.0 | 2.0 |
| 8.5 | 4.0 | 4.0 | 1.0 |
| 8.6 | 4.5 | 4.0 | 1.5 |
| 8.9 | 5.0 | 4.0 | 2.0 |
| 9.1 | 5.5 | 5.0 | 1.0 |
| 9.3 | 6.0 | 5.0 | 1.5 |
| 9.6 | 6.5 | 5.0 | 2.0 |

**10.46** Determine the correlation coefficient $r$ for the solution of Exercise 10.45.

**10.47** Batteries from a production process are weighed prior to packing as a routine quality check. The data from a typical 8-h run are listed below:

| Weight | Number |
|--------|--------|
| Less than 30.60 lb | 18 |
| 30.6 to 30.79 lb | 512 |
| 30.8 to 30.99 lb | 5,375 |
| 31.0 to 31.19 lb | 10,200 |
| 31.2 to 31.39 lb | 4,180 |
| 31.4 to 31.59 lb | 719 |
| Greater than 31.60 lb | 12 |

(a) Find the mean and the standard deviation of the weights.
(b) Determine whether the weights of the batteries can be expressed as a Gaussian distribution function.

   (c)   Would you expect quality control problems with batteries weighing less than 30.6 lb or more than 31.6 lb?

**10.48**  A die-casting operation produces bearing housings with a rejection
rate of 4 percent when the machine is operated over an 8-h shift to
produce a total output of 3200 housings. The method of die cooling
was changed in an attempt to reduce the rejection rate. After 2 h of
operation under the new cooling conditions, 775 acceptable castings
and 25 rejects had been produced.

   (a)   Did the change in the process improve the output?
   (b)   How certain are you of your answer?

**10.49**  A gear-shaft assembly consists of a shaft with a shoulder, a bearing,
a sleeve, a gear, a second sleeve, and a nut. Dimensional tolerances
for each of these components are listed below:

| Component | Tolerance (mm) |
|---|---|
| Shoulder | 0.050 |
| Bearing | 0.025 |
| First sleeve | 0.100 |
| Gear | 0.050 |
| Second sleeve | 0.100 |
| Nut | 0.100 |

   (a)   Determine the anticipated tolerance of the series assembly.
   (b)   What will be the frequency of occurrence of the tolerance of Part
   (a)?

**10.50**  Estimate the error in a determination of the volume of a sphere having
a diameter of 50 mm if the diameter is measured with a standard
deviation of the mean (standard error) of $S_{\bar{x}} = 0.05$ mm.

**10.51**  The stress $\sigma_x$ at a point on the free surface of a structure or machine
component can be expressed in terms of the normal strains $\varepsilon_x$ and $\varepsilon_y$
measured with electrical resistance strain gages as

$$\sigma_x = \frac{E}{1 - \nu^2} (\varepsilon_x + \nu\varepsilon_y)$$

If $\varepsilon_x$ and $\varepsilon_y$ are measured within $\pm 2$ percent and $E$ and $\nu$ are measured
within $\pm 5$ percent, estimate the error in $\sigma_x$.

# APPENDIX

**TABLE A.1** Temperature–Resistance Data for a Thermistor

| Temperature °C | Resistance | Temperature °C | Resistance | Temperature °C | Resistance |
|---|---|---|---|---|---|
| −80 | 2,210,400 | −67 | 731,700 | −54 | 268,560 |
| −79 | 2,022,100 | −66 | 675,060 | −53 | 249,600 |
| −78 | 1,851,100 | −65 | 623,160 | −52 | 232,110 |
| −77 | 1,695,800 | −64 | 575,610 | −51 | 215,970 |
| −76 | 1,554,500 | −63 | 531,990 | −50 | 201,030 |
| −75 | 1,425,900 | −62 | 491,970 | −49 | 187,230 |
| −74 | 1,308,900 | −61 | 455,220 | −48 | 174,450 |
| −73 | 1,202,200 | −60 | 421,470 | −47 | 162,660 |
| −72 | 1,105,000 | −59 | 390,420 | −46 | 151,710 |
| −71 | 1,016,300 | −58 | 361,890 | −45 | 141,570 |
| −70 | 935,250 | −57 | 335,610 | −44 | 132,180 |
| −69 | 861,240 | −56 | 311,400 | −43 | 123,480 |
| −68 | 793,590 | −55 | 289,110 | −42 | 115,410 |

**TABLE A.1** (Continued)

| Temper-ature °C | Resistance | Temper-ature °C | Resistance | Temper-ature °C | Resistance |
|---|---|---|---|---|---|
| −41 | 107,910 | 2 | 8,850.0 | 45 | 1,311.0 |
| −40 | 100,950 | 3 | 8,415.0 | 46 | 1,260.0 |
| −39 | 94,470 | 4 | 8,007.0 | 47 | 1,212.0 |
| −38 | 88,440 | 5 | 7,617.0 | 48 | 1,167.0 |
| −37 | 82,860 | 6 | 7,251.0 | 49 | 1,122.9 |
| −36 | 77,640 | 7 | 6,903.0 | 50 | 1,080.9 |
| −35 | 72,810 | 8 | 6,576.0 | 51 | 1,040.1 |
| −34 | 68,280 | 9 | 6,264.0 | 52 | 1,002.0 |
| −33 | 64,080 | 10 | 5,970.0 | 53 | 965.10 |
| −32 | 60,150 | 11 | 5,691.0 | 54 | 929.70 |
| −31 | 56,490 | 12 | 5,427.0 | 55 | 895.80 |
| −30 | 53,100 | 13 | 5,175.0 | 56 | 863.40 |
| −29 | 49,890 | 14 | 4,938.0 | 57 | 832.20 |
| −28 | 46,920 | 15 | 4,713.0 | 58 | 802.50 |
| −27 | 44,160 | 16 | 4,500.0 | 59 | 773.70 |
| −26 | 41,550 | 17 | 4,296.0 | 60 | 746.40 |
| −25 | 39,120 | 18 | 4,104.0 | 61 | 720.00 |
| −24 | 36,840 | 19 | 3,921.0 | 62 | 694.80 |
| −23 | 34,710 | 20 | 3,747.0 | 63 | 670.50 |
| −22 | 32,730 | 21 | 3,582.0 | 64 | 647.10 |
| −21 | 30,870 | 22 | 3,426.0 | 65 | 624.90 |
| −20 | 29,121 | 23 | 3,276.0 | 66 | 603.30 |
| −19 | 27,483 | 24 | 3,135.0 | 67 | 582.60 |
| −18 | 25,947 | 25 | 3,000.0 | 68 | 562.80 |
| −17 | 24,507 | 26 | 2,871.9 | 69 | 543.90 |
| −16 | 23,154 | 27 | 2,750.1 | 70 | 525.60 |
| −15 | 21,885 | 28 | 2,633.1 | 71 | 507.90 |
| −14 | 20,694 | 29 | 2,522.1 | 72 | 490.80 |
| −13 | 19,572 | 30 | 2,417.1 | 73 | 474.60 |
| −12 | 18,519 | 31 | 2,316.9 | 74 | 459.00 |
| −11 | 17,529 | 32 | 2,220.9 | 75 | 443.70 |
| −10 | 16,599 | 33 | 2,129.1 | 76 | 429.30 |
| −9 | 15,720 | 34 | 2,042.1 | 77 | 415.20 |
| −8 | 14,895 | 35 | 1,959.0 | 78 | 402.00 |
| −7 | 14,118 | 36 | 1,880.1 | 79 | 389.10 |
| −6 | 13,386 | 37 | 1,805.1 | 80 | 376.50 |
| −5 | 12,699 | 38 | 1,733.1 | 81 | 364.50 |
| −4 | 12,048 | 39 | 1,664.1 | 82 | 353.10 |
| −3 | 11,433 | 40 | 1,598.1 | 83 | 342.00 |
| −2 | 10,857 | 41 | 1,535.1 | 84 | 331.20 |
| −1 | 10,311 | 42 | 1,475.1 | 85 | 321.00 |
| 0 | 9,795.0 | 43 | 1,418.1 | 86 | 310.80 |
| 1 | 9,309.0 | 44 | 1,362.9 | 87 | 301.20 |

**TABLE A.1** (*Continued*)

| Temperature °C | Resistance | Temperature °C | Resistance | Temperature °C | Resistance |
|---|---|---|---|---|---|
| 88 | 292.11 | 109 | 157.50 | 130 | 90.279 |
| 89 | 283.20 | 110 | 153.09 | 131 | 88.041 |
| 90 | 274.59 | 111 | 149.01 | 132 | 85.869 |
| 91 | 266.31 | 112 | 144.90 | 133 | 83.751 |
| 92 | 258.30 | 113 | 141.00 | 134 | 81.609 |
| 93 | 250.59 | 114 | 137.19 | 135 | 79.710 |
| 94 | 243.09 | 115 | 133.50 | 136 | 77.790 |
| 95 | 236.01 | 116 | 129.99 | 137 | 75.900 |
| 96 | 228.99 | 117 | 126.51 | 138 | 74.079 |
| 97 | 222.30 | 118 | 123.21 | 139 | 72.309 |
| 98 | 215.79 | 119 | 120.00 | 140 | 70.581 |
| 99 | 209.61 | 120 | 116.79 | 141 | 68.910 |
| 100 | 203.49 | 121 | 113.79 | 142 | 67.290 |
| 101 | 197.70 | 122 | 110.91 | 143 | 65.700 |
| 102 | 192.09 | 123 | 108.00 | 144 | 64.170 |
| 103 | 186.60 | 124 | 105.18 | 145 | 62.661 |
| 104 | 181.29 | 125 | 102.51 | 146 | 61.209 |
| 105 | 176.19 | 126 | 99.930 | 147 | 59.799 |
| 106 | 171.30 | 127 | 97.410 | 148 | 58.431 |
| 107 | 166.50 | 128 | 94.950 | 149 | 57.099 |
| 108 | 161.91 | 129 | 92.580 | 150 | 55.791 |

**TABLE A.2** Thermoelectric Voltages for Iron–Constantan Thermocouples with the Reference Junction at 0°C (32°F)

| °C | 0 | 1 | 2 | 3 | 4 | 5 | 6 | 7 | 8 | 9 | 10 | °C |
|---|---|---|---|---|---|---|---|---|---|---|---|---|
| | | | | | Thermoelectric Voltage (absolute mV) | | | | | | | |
| −210 | −8.096 | | | | | | | | | | | −210 |
| −200 | −7.890 | −7.912 | −7.934 | −7.955 | −7.976 | −7.996 | −8.017 | −8.037 | −8.057 | −8.076 | −8.096 | −200 |
| −190 | −7.659 | −7.683 | −7.707 | −7.731 | −7.755 | −7.778 | −7.801 | −7.824 | −7.846 | −7.868 | −7.890 | −190 |
| −180 | −7.402 | −7.429 | −7.455 | −7.482 | −7.508 | −7.533 | −7.559 | −7.584 | −7.609 | −7.634 | −7.659 | −180 |
| −170 | −7.122 | −7.151 | −7.180 | −7.209 | −7.237 | −7.265 | −7.293 | −7.321 | −7.348 | −7.375 | −7.402 | −170 |
| −160 | −6.821 | −6.852 | −6.883 | −6.914 | −6.944 | −6.974 | −7.004 | −7.034 | −7.064 | −7.093 | −7.122 | −160 |
| −150 | −6.499 | −6.532 | −6.565 | −6.598 | −6.630 | 6.663 | −6.695 | −6.727 | −6.758 | −6.790 | −6.821 | −150 |
| −140 | −6.159 | −6.194 | −6.228 | −6.263 | −6.297 | −6.331 | −6.365 | −6.399 | −6.433 | −6.466 | −6.499 | −140 |
| −130 | −5.801 | −5.837 | −5.874 | −5.910 | −5.946 | −5.982 | −6.018 | −6.053 | −6.089 | −6.124 | −6.159 | −130 |
| −120 | −5.426 | −5.464 | −5.502 | −5.540 | −5.578 | −5.615 | −5.653 | −5.690 | −5.727 | −5.764 | −5.801 | −120 |
| −110 | −5.036 | −5.076 | −5.115 | −5.155 | −5.194 | −5.233 | −5.272 | −5.311 | −5.349 | −5.388 | −5.426 | −110 |
| −100 | −4.632 | −4.673 | −4.714 | −4.755 | −4.795 | −4.836 | −4.876 | −4.916 | −4.956 | −4.996 | −5.036 | −100 |
| −90 | −4.215 | −4.257 | −4.299 | −4.341 | −4.383 | −4.425 | −4.467 | −4.508 | −4.550 | −4.591 | −4.632 | −90 |
| −80 | −3.785 | −3.829 | −3.872 | −3.915 | −3.958 | −4.001 | −4.044 | −4.087 | −4.130 | −4.172 | −4.215 | −80 |
| −70 | −3.344 | −3.389 | −3.433 | −3.478 | −3.522 | −3.566 | −3.610 | −3.654 | −3.698 | −3.742 | −3.785 | −70 |
| −60 | −2.892 | −2.938 | −2.984 | −3.029 | −3.074 | −3.120 | −3.165 | −3.210 | −3.255 | −3.299 | −3.344 | −60 |
| −50 | −2.431 | −2.478 | −2.524 | −2.570 | −2.617 | −2.663 | −2.709 | −2.755 | −2.801 | −2.847 | −2.892 | −50 |
| −40 | −1.960 | −2.008 | −2.055 | −2.102 | −2.150 | −2.197 | −2.244 | −2.291 | −2.338 | −2.384 | −2.431 | −40 |
| −30 | −1.481 | −1.530 | −1.578 | −1.626 | −1.674 | −1.722 | −1.770 | −1.818 | −1.865 | −1.913 | −1.960 | −30 |
| −20 | −0.995 | −1.044 | −1.093 | −1.141 | −1.190 | −1.239 | −1.288 | −1.336 | −1.385 | −1.433 | −1.481 | −20 |
| −10 | −0.501 | −0.550 | −0.600 | −0.650 | −0.699 | −0.748 | −0.798 | −0.847 | −0.896 | −0.945 | −0.995 | −10 |
| 0 | 0.000 | −0.050 | −0.101 | −0.151 | −0.201 | −0.251 | −0.301 | −0.351 | −0.401 | −0.451 | −0.501 | 0 |
| 0 | 0.000 | 0.050 | 0.101 | 0.151 | 0.202 | 0.253 | 0.303 | 0.354 | 0.405 | 0.456 | 0.507 | 0 |
| 10 | 0.507 | 0.558 | 0.609 | 0.660 | 0.711 | 0.762 | 0.813 | 0.865 | 0.916 | 0.967 | 1.019 | 10 |
| 20 | 1.019 | 1.070 | 1.122 | 1.174 | 1.225 | 1.277 | 1.329 | 1.381 | 1.432 | 1.484 | 1.536 | 20 |
| 30 | 1.536 | 1.588 | 1.640 | 1.693 | 1.745 | 1.797 | 1.849 | 1.901 | 1.954 | 2.006 | 2.058 | 30 |
| 40 | 2.058 | 2.111 | 2.163 | 2.216 | 2.268 | 2.321 | 2.374 | 2.426 | 2.479 | 2.532 | 2.585 | 40 |
| 50 | 2.585 | 2.638 | 2.691 | 2.743 | 2.796 | 2.849 | 2.902 | 2.956 | 3.009 | 3.062 | 3.115 | 50 |
| 60 | 3.115 | 3.168 | 3.221 | 3.275 | 3.328 | 3.381 | 3.435 | 3.488 | 3.542 | 3.595 | 3.649 | 60 |
| 70 | 3.649 | 3.702 | 3.756 | 3.809 | 3.863 | 3.917 | 3.971 | 4.024 | 4.078 | 4.132 | 4.186 | 70 |
| 80 | 4.186 | 4.239 | 4.293 | 4.347 | 4.401 | 4.455 | 4.509 | 4.563 | 4.617 | 4.671 | 4.725 | 80 |
| 90 | 4.725 | 4.780 | 4.834 | 4.888 | 4.942 | 4.996 | 5.050 | 5.105 | 5.159 | 5.213 | 5.268 | 90 |
| 100 | 5.268 | 5.322 | 5.376 | 5.431 | 5.485 | 5.540 | 5.594 | 5.649 | 5.703 | 5.758 | 5.812 | 100 |
| 110 | 5.812 | 5.867 | 5.921 | 5.976 | 6.031 | 6.085 | 6.140 | 6.195 | 6.249 | 6.304 | 6.359 | 110 |
| 120 | 6.359 | 6.414 | 6.468 | 6.523 | 6.578 | 6.633 | 6.688 | 6.742 | 6.797 | 6.852 | 6.907 | 120 |
| 130 | 6.907 | 6.962 | 7.017 | 7.072 | 7.127 | 7.182 | 7.237 | 7.292 | 7.347 | 7.402 | 7.457 | 130 |
| 140 | 7.457 | 7.512 | 7.567 | 7.622 | 7.677 | 7.732 | 7.787 | 7.843 | 7.898 | 7.953 | 8.008 | 140 |
| 150 | 8.008 | 8.063 | 8.118 | 8.174 | 8.229 | 8.284 | 8.339 | 8.394 | 8.450 | 8.505 | 8.560 | 150 |
| 160 | 8.560 | 8.616 | 8.671 | 8.726 | 8.781 | 8.837 | 8.892 | 8.947 | 9.003 | 9.058 | 9.113 | 160 |
| 170 | 9.113 | 9.169 | 9.224 | 9.279 | 9.335 | 9.390 | 9.446 | 9.501 | 9.556 | 9.612 | 9.667 | 170 |
| 180 | 9.667 | 9.723 | 9.778 | 9.834 | 9.889 | 9.944 | 10.000 | 10.055 | 10.111 | 10.166 | 10.222 | 180 |
| 190 | 10.222 | 10.277 | 10.333 | 10.388 | 10.444 | 10.499 | 10.555 | 10.610 | 10.666 | 10.721 | 10.777 | 190 |
| 200 | 10.777 | 10.832 | 10.888 | 10.943 | 10.999 | 11.054 | 11.110 | 11.165 | 11.221 | 11.276 | 11.332 | 200 |
| 210 | 11.332 | 11.387 | 11.443 | 11.498 | 11.554 | 11.609 | 11.665 | 11.720 | 11.776 | 11.831 | 11.887 | 210 |
| 220 | 11.887 | 11.943 | 11.998 | 12.054 | 12.109 | 12.165 | 12.220 | 12.276 | 12.331 | 12.387 | 12.442 | 220 |
| 230 | 12.442 | 12.498 | 12.553 | 12.609 | 12.664 | 12.720 | 12.776 | 12.831 | 12.887 | 12.942 | 12.998 | 230 |
| 240 | 12.998 | 13.053 | 13.109 | 13.164 | 13.220 | 13.275 | 13.331 | 13.386 | 13.442 | 13.497 | 13.553 | 240 |
| 250 | 13.553 | 13.608 | 13.664 | 13.719 | 13.775 | 13.830 | 13.886 | 13.941 | 13.997 | 14.052 | 14.108 | 250 |
| 260 | 14.108 | 14.163 | 14.219 | 14.274 | 14.330 | 14.385 | 14.441 | 14.496 | 14.552 | 14.607 | 14.663 | 260 |
| 270 | 14.663 | 14.718 | 14.774 | 14.829 | 14.885 | 14.940 | 14.995 | 15.051 | 15.106 | 15.162 | 15.217 | 270 |

## TABLE A.2 (Continued)

| °C | 0 | 1 | 2 | 3 | 4 | 5 | 6 | 7 | 8 | 9 | 10 | °C |
|----|---|---|---|---|---|---|---|---|---|---|----|----|
| | | | | | Thermoelectric Voltage (absolute mV) | | | | | | | |
| 280 | 15.217 | 15.273 | 15.328 | 15.383 | 15.439 | 15.494 | 15.550 | 15.605 | 15.661 | 15.716 | 15.771 | 280 |
| 290 | 15.771 | 15.827 | 15.882 | 15.938 | 15.993 | 16.048 | 16.104 | 16.159 | 16.214 | 16.270 | 16.325 | 290 |
| 300 | 16.325 | 16.380 | 16.436 | 16.491 | 16.547 | 16.602 | 16.657 | 16.713 | 16.768 | 16.823 | 16.879 | 300 |
| 310 | 16.879 | 16.934 | 16.989 | 17.044 | 17.100 | 17.155 | 17.210 | 17.266 | 17.321 | 17.376 | 17.432 | 310 |
| 320 | 17.432 | 17.487 | 17.542 | 17.597 | 17.653 | 17.708 | 17.763 | 17.818 | 17.874 | 17.929 | 17.984 | 320 |
| 330 | 17.984 | 18.039 | 18.095 | 18.150 | 18.205 | 18.260 | 18.316 | 18.371 | 18.426 | 18.481 | 18.537 | 330 |
| 340 | 18.537 | 18.592 | 18.647 | 18.702 | 18.757 | 18.813 | 18.868 | 18.923 | 18.978 | 19.033 | 19.089 | 340 |
| 350 | 19.089 | 19.144 | 19.199 | 19.254 | 19.309 | 19.364 | 19.420 | 19.475 | 19.530 | 19.585 | 19.640 | 350 |
| 360 | 19.640 | 19.695 | 19.751 | 19.806 | 19.861 | 19.916 | 19.971 | 20.026 | 20.081 | 20.137 | 20.192 | 360 |
| 370 | 20.192 | 20.247 | 20.302 | 20.357 | 20.412 | 20.467 | 20.523 | 20.578 | 20.633 | 20.688 | 20.743 | 370 |
| 380 | 20.743 | 20.798 | 20.853 | 20.909 | 20.964 | 21.019 | 21.074 | 21.129 | 21.184 | 21.239 | 21.295 | 380 |
| 390 | 21.295 | 21.350 | 21.405 | 21.460 | 21.515 | 21.570 | 21.625 | 21.680 | 21.736 | 21.791 | 21.846 | 390 |
| 400 | 21.846 | 21.901 | 21.956 | 22.011 | 22.066 | 22.122 | 22.177 | 22.232 | 22.287 | 22.342 | 22.397 | 400 |
| 410 | 22.397 | 22.453 | 22.508 | 22.563 | 22.618 | 22.673 | 22.728 | 22.784 | 22.839 | 22.894 | 22.949 | 410 |
| 420 | 22.949 | 23.004 | 23.060 | 23.115 | 23.170 | 23.225 | 23.280 | 23.336 | 23.391 | 23.446 | 23.501 | 420 |
| 430 | 23.501 | 23.556 | 23.612 | 23.667 | 23.722 | 23.777 | 23.833 | 23.888 | 23.943 | 23.999 | 24.054 | 430 |
| 440 | 24.054 | 24.109 | 24.164 | 24.220 | 24.275 | 24.330 | 24.386 | 24.441 | 24.496 | 24.552 | 24.607 | 440 |
| 450 | 24.607 | 24.662 | 24.718 | 24.773 | 24.829 | 24.884 | 24.939 | 24.995 | 25.050 | 25.106 | 25.161 | 450 |
| 460 | 25.161 | 25.217 | 25.272 | 25.327 | 25.383 | 25.438 | 25.494 | 25.549 | 25.605 | 25.661 | 25.716 | 460 |
| 470 | 25.716 | 25.772 | 25.827 | 25.883 | 25.938 | 25.994 | 26.050 | 26.105 | 26.161 | 26.216 | 26.272 | 470 |
| 480 | 26.272 | 26.328 | 26.383 | 26.439 | 26.495 | 26.551 | 26.606 | 26.662 | 26.718 | 26.774 | 26.829 | 480 |
| 490 | 26.829 | 26.885 | 26.941 | 26.997 | 27.053 | 27.109 | 27.165 | 27.220 | 27.276 | 27.332 | 27.388 | 490 |
| 500 | 27.388 | 27.444 | 27.500 | 27.556 | 27.612 | 27.668 | 27.724 | 27.780 | 27.836 | 27.893 | 27.949 | 500 |
| 510 | 27.949 | 28.005 | 28.061 | 28.117 | 28.173 | 28.230 | 28.286 | 28.342 | 28.398 | 28.455 | 28.511 | 510 |
| 520 | 28.511 | 28.567 | 28.624 | 28.680 | 28.736 | 28.793 | 28.849 | 28.906 | 28.962 | 29.019 | 29.075 | 520 |
| 530 | 29.075 | 29.132 | 29.188 | 29.245 | 29.301 | 29.358 | 29.415 | 29.471 | 29.528 | 29.585 | 29.642 | 530 |
| 540 | 29.642 | 29.698 | 29.755 | 29.812 | 29.869 | 29.926 | 29.983 | 30.039 | 30.096 | 30.153 | 30.210 | 540 |
| 550 | 30.210 | 30.267 | 30.324 | 30.381 | 30.439 | 30.496 | 30.553 | 30.610 | 30.667 | 30.724 | 30.782 | 550 |
| 560 | 30.782 | 30.839 | 30.896 | 30.954 | 31.011 | 31.068 | 31.126 | 31.183 | 31.241 | 31.298 | 31.356 | 560 |
| 570 | 31.356 | 31.413 | 31.471 | 31.528 | 31.586 | 31.644 | 31.702 | 31.759 | 31.817 | 31.875 | 31.933 | 570 |
| 580 | 31.933 | 31.991 | 32.048 | 32.106 | 32.164 | 32.222 | 32.280 | 32.338 | 32.396 | 32.455 | 32.513 | 580 |
| 590 | 32.513 | 32.571 | 32.629 | 32.687 | 32.746 | 32.804 | 32.862 | 32.921 | 32.979 | 33.038 | 33.096 | 590 |
| 600 | 33.096 | 33.155 | 33.213 | 33.272 | 33.330 | 33.389 | 33.448 | 33.506 | 33.565 | 33.624 | 33.683 | 600 |
| 610 | 33.683 | 33.742 | 33.800 | 33.859 | 33.918 | 33.977 | 34.036 | 34.095 | 34.155 | 34.214 | 34.273 | 610 |
| 620 | 34.273 | 34.332 | 34.391 | 34.451 | 34.510 | 34.569 | 34.629 | 34.688 | 34.748 | 34.807 | 34.867 | 620 |
| 630 | 34.867 | 34.926 | 34.986 | 35.046 | 35.105 | 35.165 | 35.225 | 35.285 | 35.344 | 35.404 | 35.464 | 630 |
| 640 | 35.464 | 35.524 | 35.584 | 35.644 | 35.704 | 35.764 | 35.825 | 35.885 | 35.945 | 36.005 | 36.066 | 640 |
| 650 | 36.066 | 36.126 | 36.186 | 36.247 | 36.307 | 36.368 | 36.428 | 36.489 | 36.549 | 36.610 | 36.671 | 650 |
| 660 | 36.671 | 36.732 | 36.792 | 36.853 | 36.914 | 36.975 | 37.036 | 37.097 | 37.158 | 37.219 | 37.280 | 660 |
| 670 | 37.280 | 37.341 | 37.402 | 37.463 | 37.525 | 37.586 | 37.647 | 37.709 | 37.770 | 37.831 | 37.893 | 670 |
| 680 | 37.893 | 37.954 | 38.016 | 38.078 | 38.139 | 38.201 | 38.262 | 38.324 | 38.386 | 38.448 | 38.510 | 680 |
| 690 | 38.510 | 38.572 | 38.633 | 38.695 | 38.757 | 38.819 | 38.882 | 38.944 | 39.006 | 39.068 | 39.130 | 690 |
| 700 | 39.130 | 39.192 | 39.255 | 39.317 | 39.379 | 39.442 | 39.504 | 39.567 | 39.629 | 39.692 | 39.754 | 700 |
| 710 | 39.754 | 39.817 | 39.880 | 39.942 | 40.005 | 40.068 | 40.131 | 40.193 | 40.256 | 40.319 | 40.382 | 710 |
| 720 | 40.382 | 40.445 | 40.508 | 40.571 | 40.634 | 40.697 | 40.760 | 40.823 | 40.886 | 40.950 | 41.013 | 720 |
| 730 | 41.013 | 41.076 | 41.139 | 41.203 | 41.266 | 41.329 | 41.393 | 41.456 | 41.520 | 41.583 | 41.647 | 730 |
| 740 | 41.647 | 41.710 | 41.774 | 41.837 | 41.901 | 41.965 | 42.028 | 42.092 | 42.156 | 42.219 | 42.283 | 740 |
| 750 | 42.283 | 42.347 | 42.411 | 42.475 | 42.538 | 42.602 | 42.666 | 42.730 | 42.794 | 42.858 | 42.922 | 750 |
| 760 | 42.922 | | | | | | | | | | | 760 |

**TABLE A.3** Thermoelectric Voltages for Chromel–Alumel Thermocouples with the Reference Junction at 0°C (32°F)

| °C | 0 | 1 | 2 | 3 | 4 | 5 | 6 | 7 | 8 | 9 | 10 | °C |
|---|---|---|---|---|---|---|---|---|---|---|---|---|
| | | | | | Thermoelectric Voltage (absolute mV) | | | | | | | |
| −270 | −6.458 | | | | | | | | | | | −270 |
| −260 | −6.441 | −6.444 | −6.446 | −6.448 | −6.450 | −6.452 | −6.453 | −6.455 | −6.456 | −6.457 | −6.458 | −260 |
| −250 | −6.404 | −6.408 | −6.413 | −6.417 | −6.421 | −6.425 | −6.429 | −6.432 | −6.435 | −6.438 | −6.441 | −250 |
| −240 | −6.344 | −6.351 | −6.358 | −6.364 | −6.371 | −6.377 | −6.382 | −6.388 | −6.394 | −o.399 | −6.404 | −240 |
| −230 | −6.262 | −6.271 | −6.280 | −6.289 | −6.297 | −6.306 | −6.314 | −6.322 | −6.329 | −6.337 | −6.344 | −230 |
| −220 | −6.158 | −6.170 | −6.181 | −6.192 | −6.202 | −6.213 | −6.223 | −6.233 | −6.243 | −6.253 | −6.262 | −220 |
| −210 | −6.035 | −6.048 | −6.061 | −6.074 | −6.087 | −6.099 | −6.111 | −6.123 | −6.135 | 6.147 | 6.158 | 210 |
| −200 | −5.891 | −5.907 | −5.922 | −5.936 | −5.951 | −5.965 | −5.980 | −5.994 | −6.007 | −6.021 | −6.035 | −200 |
| −190 | −5.730 | −5.747 | −5.763 | −5.780 | −5.796 | −5.813 | −5.829 | −5.845 | −5.860 | −5.876 | −5.891 | −190 |
| −180 | −5.550 | −5.569 | −5.587 | −5.606 | −5.624 | −5.642 | −5.660 | −5.678 | −5.695 | −5.712 | −5.730 | −180 |
| −170 | −5.354 | −5.374 | −5.394 | −5.414 | −5.434 | −5.454 | −5.474 | −5.493 | −5.512 | −5.531 | −5.550 | −170 |
| −160 | −5.141 | −5.163 | −5.185 | −5.207 | −5.228 | −5.249 | −5.271 | −5.292 | −5.313 | −5.333 | −5.354 | −160 |
| −150 | −4.912 | −4.936 | −4.959 | −4.983 | −5.006 | −5.029 | −5.051 | −5.074 | −5.097 | −5.119 | −5.141 | −150 |
| −140 | −4.669 | −4.694 | −4.719 | −4.743 | −4.768 | −4.792 | −4.817 | −4.841 | −4.865 | −4.889 | −4.912 | −140 |
| −130 | −4.410 | −4.437 | −4.463 | −4.489 | −4.515 | −4.541 | −4.567 | −4.593 | −4.618 | −4.644 | −4.669 | −130 |
| −120 | −4.138 | −4.166 | −4.193 | −4.221 | −4.248 | −4.276 | −4.303 | −4.330 | −4.357 | −4.384 | −4.410 | −120 |
| −110 | −3.852 | −3.881 | −3.910 | −3.939 | −3.968 | −3.997 | −4.025 | −4.053 | −4.082 | −4.110 | −4.138 | −110 |
| −100 | −3.553 | −3.584 | −3.614 | −3.644 | −3.674 | −3.704 | −3.734 | −3.764 | −3.793 | −3.823 | −3.852 | −100 |
| −90 | −3.242 | −3.274 | −3.305 | −3.337 | −3.368 | −3.399 | −3.430 | −3.461 | −3.492 | −3.523 | −3.553 | −90 |
| −80 | −2.920 | −2.953 | −2.985 | −3.018 | −3.050 | −3.082 | −3.115 | −3.147 | −3.179 | −3.211 | −3.242 | −80 |
| −70 | −2.586 | −2.620 | −2.654 | −2.687 | −2.721 | −2.754 | −2.788 | −2.821 | −2.854 | −2.887 | −2.920 | −70 |
| −60 | −2.243 | −2.277 | −2.312 | −2.347 | −2.381 | −2.416 | −2.450 | −2.484 | −2.518 | −2.552 | −2.586 | −60 |
| −50 | −1.889 | −1.925 | −1.961 | −1.996 | −2.032 | −2.067 | −2.102 | −2.137 | −2.173 | −2.208 | −2.243 | −50 |
| −40 | −1.527 | −1.563 | −1.600 | −1.636 | −1.673 | −1.709 | −1.745 | −1.781 | −1.817 | −1.853 | −1.889 | −40 |
| −30 | −1.156 | −1.193 | −1.231 | −1.268 | −1.305 | −1.342 | −1.379 | −1.416 | −1.453 | −1.490 | −1.527 | −30 |
| −20 | −0.777 | −0.816 | −0.854 | −0.892 | −0.930 | −0.968 | −1.005 | −1.043 | −1.081 | −1.118 | −1.156 | −20 |
| −10 | −0.392 | −0.431 | −0.469 | −0.508 | −0.547 | −0.585 | −0.624 | −0.662 | −0.701 | −0.739 | −0.777 | −10 |
| 0 | 0.000 | −0.039 | −0.079 | −0.118 | −0.157 | −0.197 | −0.236 | −0.275 | −0.314 | −0.353 | −0.392 | 0 |
| 0 | 0.000 | 0.039 | 0.079 | 0.119 | 0.158 | 0.198 | 0.238 | 0.277 | 0.317 | 0.357 | 0.397 | 0 |
| 10 | 0.397 | 0.437 | 0.477 | 0.517 | 0.557 | 0.597 | 0.637 | 0.677 | 0.718 | 0.758 | 0.798 | 10 |
| 20 | 0.798 | 0.838 | 0.879 | 0.919 | 0.960 | 1.000 | 1.041 | 1.081 | 1.122 | 1.162 | 1.203 | 20 |
| 30 | 1.203 | 1.244 | 1.285 | 1.325 | 1.366 | 1.407 | 1.448 | 1.489 | 1.529 | 1.570 | 1.611 | 30 |
| 40 | 1.611 | 1.652 | 1.693 | 1.734 | 1.776 | 1.817 | 1.858 | 1.899 | 1.940 | 1.981 | 2.022 | 40 |
| 50 | 2.022 | 2.064 | 2.105 | 2.146 | 2.188 | 2.229 | 2.270 | 2.312 | 2.353 | 2.394 | 2.436 | 50 |
| 60 | 2.436 | 2.477 | 2.519 | 2.560 | 2.601 | 2.643 | 2.684 | 2.726 | 2.767 | 2.809 | 2.850 | 60 |
| 70 | 2.850 | 2.892 | 2.933 | 2.975 | 3.016 | 3.058 | 3.100 | 3.141 | 3.183 | 3.224 | 3.266 | 70 |
| 80 | 3.266 | 3.307 | 3.349 | 3.390 | 3.432 | 3.473 | 3.515 | 3.556 | 3.598 | 3.639 | 3.681 | 80 |
| 90 | 3.681 | 3.722 | 3.764 | 3.805 | 3.847 | 3.888 | 3.930 | 3.971 | 4.012 | 4.054 | 4.095 | 90 |
| 100 | 4.095 | 4.137 | 4.178 | 4.219 | 4.261 | 4.302 | 4.343 | 4.384 | 4.426 | 4.467 | 4.508 | 100 |
| 110 | 4.508 | 4.549 | 4.590 | 4.632 | 4.673 | 4.714 | 4.755 | 4.796 | 4.837 | 4.878 | 4.919 | 110 |
| 120 | 4.919 | 4.960 | 5.001 | 5.042 | 5.083 | 5.124 | 5.164 | 5.205 | 5.246 | 5.287 | 5.327 | 120 |
| 130 | 5.327 | 5.368 | 5.409 | 5.450 | 5.490 | 5.531 | 5.571 | 5.612 | 5.652 | 5.693 | 5.733 | 130 |
| 140 | 5.733 | 5.774 | 5.814 | 5.855 | 5.895 | 5.936 | 5.976 | 6.016 | 6.057 | 6.097 | 6.137 | 140 |
| 150 | 6.137 | 6.177 | 6.218 | 6.258 | 6.298 | 6.338 | 6.378 | 6.419 | 6.459 | 6.499 | 6.539 | 150 |
| 160 | 6.539 | 6.579 | 6.619 | 6.659 | 6.699 | 6.739 | 6.779 | 6.819 | 6.859 | 6.899 | 6.939 | 160 |
| 170 | 6.939 | 6.979 | 7.019 | 7.059 | 7.099 | 7.139 | 7.179 | 7.219 | 7.259 | 7.299 | 7.338 | 170 |
| 180 | 7.338 | 7.378 | 7.418 | 7.458 | 7.498 | 7.538 | 7.578 | 7.618 | 7.658 | 7.697 | 7.737 | 180 |
| 190 | 7.737 | 7.777 | 7.817 | 7.857 | 7.897 | 7.937 | 7.977 | 8.017 | 8.057 | 8.097 | 8.137 | 190 |
| 200 | 8.137 | 8.177 | 8.216 | 8.256 | 8.296 | 8.336 | 8.376 | 8.416 | 8.456 | 8.497 | 8.537 | 200 |
| 210 | 8.537 | 8.577 | 8.617 | 8.657 | 8.697 | 8.737 | 8.777 | 8.817 | 8.857 | 8.898 | 8.938 | 210 |
| 220 | 8.938 | 8.978 | 9.018 | 9.058 | 9.099 | 9.139 | 9.179 | 9.220 | 9.260 | 9.300 | 9.341 | 220 |
| 230 | 9.341 | 9.381 | 9.421 | 9.462 | 9.502 | 9.543 | 9.583 | 9.624 | 9.664 | 9.705 | 9.745 | 230 |
| 240 | 9.745 | 9.786 | 9.826 | 9.867 | 9.907 | 9.948 | 9.989 | 10.029 | 10.070 | 10.111 | 10.151 | 240 |
| 250 | 10.151 | 10.192 | 10.233 | 10.274 | 10.315 | 10.355 | 10.396 | 10.437 | 10.478 | 10.519 | 10.560 | 250 |
| 260 | 10.560 | 10.600 | 10.641 | 10.682 | 10.723 | 10.764 | 10.805 | 10.846 | 10.887 | 10.928 | 10.969 | 260 |
| 270 | 10.969 | 11.010 | 11.051 | 11.093 | 11.134 | 11.175 | 11.216 | 11.257 | 11.298 | 11.339 | 11.381 | 270 |
| 280 | 11.381 | 11.422 | 11.463 | 11.504 | 11.546 | 11.587 | 11.628 | 11.669 | 11.711 | 11.752 | 11.793 | 280 |
| 290 | 11.793 | 11.835 | 11.876 | 11.918 | 11.959 | 12.000 | 12.042 | 12.083 | 12.125 | 12.166 | 12.207 | 290 |

## TABLE A.3 (*Continued*)

| °C | 0 | 1 | 2 | 3 | 4 | 5 | 6 | 7 | 8 | 9 | 10 | °C |
|---|---|---|---|---|---|---|---|---|---|---|---|---|
| | | | | | Thermoelectric Voltage (absolute mV) | | | | | | | |
| 300 | 12.207 | 12.249 | 12.290 | 12.332 | 12.373 | 12.415 | 12.456 | 12.498 | 12.539 | 12.581 | 12.623 | 300 |
| 310 | 12.623 | 12.664 | 12.706 | 12.747 | 12.789 | 12.831 | 12.872 | 12.914 | 12.955 | 12.997 | 13.039 | 310 |
| 320 | 13.039 | 13.080 | 13.122 | 13.164 | 13.205 | 13.247 | 13.289 | 13.331 | 13.372 | 13.414 | 13.456 | 320 |
| 330 | 13.456 | 13.497 | 13.539 | 13.581 | 13.623 | 13.665 | 13.706 | 13.748 | 13.790 | 13.832 | 13.874 | 330 |
| 340 | 13.874 | 13.915 | 13.957 | 13.999 | 14.041 | 14.083 | 14.125 | 14.167 | 14.208 | 14.250 | 14.292 | 340 |
| 350 | 14.292 | 14.334 | 14.376 | 14.418 | 14.460 | 14.502 | 14.544 | 14.586 | 14.628 | 14.670 | 14.712 | 350 |
| 360 | 14.712 | 14.754 | 14.796 | 14.838 | 14.880 | 14.922 | 14.964 | 15.006 | 15.048 | 15.090 | 15.132 | 360 |
| 370 | 15.132 | 15.174 | 15.216 | 15.258 | 15.300 | 15.342 | 15.384 | 15.426 | 15.468 | 15.510 | 15.552 | 370 |
| 380 | 15.552 | 15.594 | 15.636 | 15.679 | 15.721 | 15.763 | 15.805 | 15.847 | 15.889 | 15.931 | 15.974 | 380 |
| 390 | 15.974 | 16.016 | 16.058 | 16.100 | 16.142 | 16.184 | 16.227 | 16.269 | 16.311 | 16.353 | 16.395 | 390 |
| 400 | 16.395 | 16.438 | 16.480 | 16.522 | 16.564 | 16.607 | 16.649 | 16.691 | 16.733 | 16.776 | 16.818 | 400 |
| 410 | 16.818 | 16.860 | 16.902 | 16.945 | 16.987 | 17.029 | 17.072 | 17.114 | 17.156 | 17.199 | 17.241 | 410 |
| 420 | 17.241 | 17.283 | 17.326 | 17.368 | 17.410 | 17.453 | 17.495 | 17.537 | 17.580 | 17.622 | 17.664 | 420 |
| 430 | 17.664 | 17.707 | 17.749 | 17.792 | 17.834 | 17.876 | 17.919 | 17.961 | 18.004 | 18.046 | 18.088 | 430 |
| 440 | 18.088 | 18.131 | 18.173 | 18.216 | 18.258 | 18.301 | 18.343 | 18.385 | 18.428 | 18.470 | 18.513 | 440 |
| 450 | 18.513 | 18.555 | 18.598 | 18.640 | 18.683 | 18.725 | 18.768 | 18.810 | 18.853 | 18.895 | 18.938 | 450 |
| 460 | 18.938 | 18.980 | 19.023 | 19.065 | 19.108 | 19.150 | 19.193 | 19.235 | 19.278 | 19.320 | 19.363 | 460 |
| 470 | 19.363 | 19.405 | 19.448 | 19.490 | 19.533 | 19.576 | 19.618 | 19.661 | 19.703 | 19.746 | 19.788 | 470 |
| 480 | 19.788 | 19.831 | 19.873 | 19.916 | 19.959 | 20.001 | 20.044 | 20.086 | 20.129 | 20.172 | 20.214 | 480 |
| 490 | 20.214 | 20.257 | 20.299 | 20.342 | 20.385 | 20.427 | 20.470 | 20.512 | 20.555 | 20.598 | 20.640 | 490 |
| 500 | 20.640 | 20.683 | 20.725 | 20.768 | 20.811 | 20.853 | 20.896 | 20.938 | 20.981 | 21.024 | 21.066 | 500 |
| 510 | 21.066 | 21.109 | 21.152 | 21.194 | 21.237 | 21.280 | 21.322 | 21.365 | 21.407 | 21.450 | 21.493 | 510 |
| 520 | 21.493 | 21.535 | 21.578 | 21.621 | 21.663 | 21.706 | 21.749 | 21.791 | 21.834 | 21.876 | 21.919 | 520 |
| 530 | 21.919 | 21.962 | 22.004 | 22.047 | 22.090 | 22.132 | 22.175 | 22.218 | 22.260 | 22.303 | 22.346 | 530 |
| 540 | 22.346 | 22.388 | 22.431 | 22.473 | 22.516 | 22.559 | 22.601 | 22.644 | 22.687 | 22.729 | 22.772 | 540 |
| 550 | 22.772 | 22.815 | 22.857 | 22.900 | 22.942 | 22.985 | 23.028 | 23.070 | 23.113 | 23.156 | 23.198 | 550 |
| 560 | 23.198 | 23.241 | 23.284 | 23.326 | 23.369 | 23.411 | 23.454 | 23.497 | 23.539 | 23.582 | 23.624 | 560 |
| 570 | 23.624 | 23.667 | 23.710 | 23.752 | 23.795 | 23.837 | 23.880 | 23.923 | 23.965 | 24.008 | 24.050 | 570 |
| 580 | 24.050 | 24.093 | 24.136 | 24.178 | 24.221 | 24.263 | 24.306 | 24.348 | 24.391 | 24.434 | 24.476 | 580 |
| 590 | 24.476 | 24.519 | 24.561 | 24.604 | 24.646 | 24.689 | 24.731 | 24.774 | 24.817 | 24.859 | 24.902 | 590 |
| 600 | 24.902 | 24.944 | 24.987 | 25.029 | 25.072 | 25.114 | 25.157 | 25.199 | 25.242 | 25.284 | 25.327 | 600 |
| 610 | 25.327 | 25.369 | 25.412 | 25.454 | 25.497 | 25.539 | 25.582 | 25.624 | 25.666 | 25.709 | 25.751 | 610 |
| 620 | 25.751 | 25.794 | 25.836 | 25.879 | 25.921 | 25.964 | 26.006 | 26.048 | 26.091 | 26.133 | 26.176 | 620 |
| 630 | 26.176 | 26.218 | 26.260 | 26.303 | 26.345 | 26.387 | 26.430 | 26.472 | 26.515 | 26.557 | 26.599 | 630 |
| 640 | 26.599 | 26.642 | 26.684 | 26.726 | 26.769 | 26.811 | 26.853 | 26.896 | 26.938 | 26.980 | 27.022 | 640 |
| 650 | 27.022 | 27.065 | 27.107 | 27.149 | 27.192 | 27.234 | 27.276 | 27.318 | 27.361 | 27.403 | 27.445 | 650 |
| 660 | 27.445 | 27.487 | 27.529 | 27.572 | 27.614 | 27.656 | 27.698 | 27.740 | 27.783 | 27.825 | 27.867 | 660 |
| 670 | 27.867 | 27.909 | 27.951 | 27.993 | 28.035 | 28.078 | 28.120 | 28.162 | 28.204 | 28.246 | 28.288 | 670 |
| 680 | 28.288 | 28.330 | 28.372 | 28.414 | 28.456 | 28.498 | 28.540 | 28.583 | 28.625 | 28.667 | 28.709 | 680 |
| 690 | 28.709 | 28.751 | 28.793 | 28.835 | 28.877 | 28.919 | 28.961 | 29.002 | 29.044 | 29.086 | 29.128 | 690 |
| 700 | 29.128 | 29.170 | 29.212 | 29.254 | 29.296 | 29.338 | 29.380 | 29.422 | 29.464 | 29.505 | 29.547 | 700 |
| 710 | 29.547 | 29.589 | 29.631 | 29.673 | 29.715 | 29.756 | 29.798 | 29.840 | 29.882 | 29.924 | 29.965 | 710 |
| 720 | 29.965 | 30.007 | 30.049 | 30.091 | 30.132 | 30.174 | 30.216 | 30.257 | 30.299 | 30.341 | 30.383 | 720 |
| 730 | 30.383 | 30.424 | 30.466 | 30.508 | 30.549 | 30.591 | 30.632 | 30.674 | 30.716 | 30.757 | 30.799 | 730 |
| 740 | 30.799 | 30.840 | 30.882 | 30.924 | 30.965 | 31.007 | 31.048 | 31.090 | 31.131 | 31.173 | 31.214 | 740 |
| 750 | 31.214 | 31.256 | 31.297 | 31.339 | 31.380 | 31.422 | 31.463 | 31.504 | 31.546 | 31.587 | 31.629 | 750 |
| 760 | 31.629 | 31.670 | 31.712 | 31.753 | 31.794 | 31.836 | 31.877 | 31.918 | 31.960 | 32.001 | 32.042 | 760 |
| 770 | 32.042 | 32.084 | 32.125 | 32.166 | 32.207 | 32.249 | 32.290 | 32.331 | 32.372 | 32.414 | 32.455 | 770 |
| 780 | 32.455 | 32.496 | 32.537 | 32.578 | 32.619 | 32.661 | 32.702 | 32.743 | 32.784 | 32.825 | 32.866 | 780 |
| 790 | 32.866 | 32.907 | 32.948 | 32.990 | 33.031 | 33.072 | 33.113 | 33.154 | 33.195 | 33.236 | 33.277 | 790 |
| 800 | 33.277 | 33.318 | 33.359 | 33.400 | 33.441 | 33.482 | 33.523 | 33.564 | 33.604 | 33.645 | 33.686 | 800 |
| 810 | 33.686 | 33.727 | 33.768 | 33.809 | 33.850 | 33.891 | 33.931 | 33.972 | 34.013 | 34.054 | 34.095 | 810 |
| 820 | 34.095 | 34.136 | 34.176 | 34.217 | 34.258 | 34.299 | 34.339 | 34.380 | 34.421 | 34.461 | 34.502 | 820 |
| 830 | 34.502 | 34.543 | 34.583 | 34.624 | 34.665 | 34.705 | 34.746 | 34.787 | 34.827 | 34.868 | 34.909 | 830 |
| 840 | 34.909 | 34.949 | 34.990 | 35.030 | 35.071 | 35.111 | 35.152 | 35.192 | 35.233 | 35.273 | 35.314 | 840 |
| 850 | 35.314 | 35.354 | 35.395 | 35.435 | 35.476 | 35.516 | 35.557 | 35.597 | 35.637 | 35.678 | 35.718 | 850 |
| 860 | 35.718 | 35.758 | 35.799 | 35.839 | 35.880 | 35.920 | 35.960 | 36.000 | 36.041 | 36.081 | 36.121 | 860 |
| 870 | 36.121 | 36.162 | 36.202 | 36.242 | 36.282 | 36.323 | 36.363 | 36.403 | 36.443 | 36.483 | 36.524 | 870 |
| 880 | 36.524 | 36.564 | 36.604 | 36.644 | 36.684 | 36.724 | 36.764 | 36.804 | 36.844 | 36.885 | 36.925 | 880 |
| 890 | 36.925 | 36.965 | 37.005 | 37.045 | 37.085 | 37.125 | 37.165 | 37.205 | 37.245 | 37.285 | 37.325 | 890 |

**TABLE A.3** (*Continued*)

| °C | 0 | 1 | 2 | 3 | 4 | 5 | 6 | 7 | 8 | 9 | 10 | °C |
|----|---|---|---|---|---|---|---|---|---|---|----|----|
| | | | | | Thermoelectric Voltage (absolute mV) | | | | | | | |
| 900 | 37.325 | 37.365 | 37.405 | 37.445 | 37.484 | 37.524 | 37.564 | 37.604 | 37.644 | 37.684 | 37.724 | 900 |
| 910 | 37.724 | 37.764 | 37.803 | 37.843 | 37.883 | 37.923 | 37.963 | 38.002 | 38.042 | 38.082 | 38.122 | 910 |
| 920 | 38.122 | 38.162 | 38.201 | 38.241 | 38.281 | 38.320 | 38.360 | 38.400 | 38.439 | 38.479 | 38.519 | 920 |
| 930 | 38.519 | 38.558 | 38.598 | 38.638 | 38.677 | 38.717 | 38.756 | 38.796 | 38.836 | 38.875 | 38.915 | 930 |
| 940 | 38.915 | 38.954 | 38.994 | 39.033 | 39.073 | 39.112 | 39.152 | 39.191 | 39.231 | 39.270 | 39.310 | 940 |
| 950 | 39.310 | 39.349 | 39.388 | 39.428 | 39.467 | 39.507 | 39.546 | 39.585 | 39.625 | 39.664 | 39.703 | 950 |
| 960 | 39.703 | 39.743 | 39.782 | 39.821 | 39.861 | 39.900 | 39.939 | 39.979 | 40.018 | 40.057 | 40.096 | 960 |
| 970 | 40.096 | 40.136 | 40.175 | 40.214 | 40.253 | 40.292 | 40.332 | 40.371 | 40.410 | 40.449 | 40.488 | 970 |
| 980 | 40.488 | 40.527 | 40.566 | 40.605 | 40.645 | 40.684 | 40.723 | 40.762 | 40.801 | 40.840 | 40.879 | 980 |
| 990 | 40.879 | 40.918 | 40.957 | 40.996 | 41.035 | 41.074 | 41.113 | 41.152 | 41.191 | 41.230 | 41.269 | 990 |
| 1,000 | 41.269 | 41.308 | 41.347 | 41.385 | 41.424 | 41.463 | 41.502 | 41.541 | 41.580 | 41.619 | 41.657 | 1,000 |
| 1,010 | 41.657 | 41.696 | 41.735 | 41.774 | 41.813 | 41.851 | 41.890 | 41.929 | 41.968 | 42.006 | 42.045 | 1,010 |
| 1,020 | 42.045 | 42.084 | 42.123 | 42.161 | 42.200 | 42.239 | 42.277 | 42.316 | 42.355 | 42.393 | 42.432 | 1,020 |
| 1,030 | 42.432 | 42.470 | 42.509 | 42.548 | 42.586 | 42.625 | 42.663 | 42.702 | 42.740 | 42.779 | 42.817 | 1,030 |
| 1,040 | 42.817 | 42.856 | 42.894 | 42.933 | 42.971 | 43.010 | 43.048 | 43.087 | 43.125 | 43.164 | 43.202 | 1,040 |
| 1,050 | 43.202 | 43.240 | 43.279 | 43.317 | 43.356 | 43.394 | 43.432 | 43.471 | 43.509 | 43.547 | 43.585 | 1,050 |
| 1,060 | 43.585 | 43.624 | 43.662 | 43.700 | 43.739 | 43.777 | 43.815 | 43.853 | 43.891 | 43.930 | 43.968 | 1,060 |
| 1,070 | 43.968 | 44.006 | 44.044 | 44.082 | 44.121 | 44.159 | 44.197 | 44.235 | 44.273 | 44.311 | 44.349 | 1,070 |
| 1,080 | 44.349 | 44.387 | 44.425 | 44.463 | 44.501 | 44.539 | 44.577 | 44.615 | 44.653 | 44.691 | 44.729 | 1,080 |
| 1,090 | 44.729 | 44.767 | 44.805 | 44.843 | 44.881 | 44.919 | 44.957 | 44.995 | 45.033 | 45.070 | 45.108 | 1,090 |
| 1,100 | 45.108 | 45.146 | 45.184 | 45.222 | 45.260 | 45.297 | 45.335 | 45.373 | 45.411 | 45.448 | 45.486 | 1,100 |
| 1,110 | 45.486 | 45.524 | 45.561 | 45.599 | 45.637 | 45.675 | 45.712 | 45.750 | 45.787 | 45.825 | 45.863 | 1,110 |
| 1,120 | 45.863 | 45.900 | 45.938 | 45.975 | 46.013 | 46.051 | 46.088 | 46.126 | 46.163 | 46.201 | 46.238 | 1,120 |
| 1,130 | 46.238 | 46.275 | 46.313 | 46.350 | 46.388 | 46.425 | 46.463 | 46.500 | 46.537 | 46.575 | 46.612 | 1,130 |
| 1,140 | 46.612 | 46.649 | 46.687 | 46.724 | 46.761 | 46.799 | 46.836 | 46.873 | 46.910 | 46.948 | 46.985 | 1,140 |
| 1,150 | 46.985 | 47.022 | 47.059 | 47.096 | 47.134 | 47.171 | 47.208 | 47.245 | 47.282 | 47.319 | 47.356 | 1,150 |
| 1,160 | 47.356 | 47.393 | 47.430 | 47.468 | 47.505 | 47.542 | 47.579 | 47.616 | 47.653 | 47.689 | 47.726 | 1,160 |
| 1,170 | 47.726 | 47.763 | 47.800 | 47.837 | 47.874 | 47.911 | 47.948 | 47.985 | 48.021 | 48.058 | 48.095 | 1,170 |
| 1,180 | 48.095 | 48.132 | 48.169 | 48.205 | 48.242 | 48.279 | 48.316 | 48.352 | 48.389 | 48.426 | 48.462 | 1,180 |
| 1,190 | 48.462 | 48.499 | 48.536 | 48.572 | 48.609 | 48.645 | 48.682 | 48.718 | 48.755 | 48.792 | 48.828 | 1,190 |
| 1,200 | 48.828 | 48.865 | 48.901 | 48.937 | 48.974 | 49.010 | 49.047 | 49.083 | 49.120 | 49.156 | 49.192 | 1,200 |
| 1,210 | 49.192 | 49.229 | 49.265 | 49.301 | 49.338 | 49.374 | 49.410 | 49.446 | 49.483 | 49.519 | 49.555 | 1,210 |
| 1,220 | 49.555 | 49.591 | 49.627 | 49.663 | 49.700 | 49.736 | 49.772 | 49.808 | 49.844 | 49.880 | 49.916 | 1,220 |
| 1,230 | 49.916 | 49.952 | 49.988 | 50.024 | 50.060 | 50.096 | 50.132 | 50.168 | 50.204 | 50.240 | 50.276 | 1,230 |
| 1,240 | 50.276 | 50.311 | 50.347 | 50.383 | 50.419 | 50.455 | 50.491 | 50.526 | 50.562 | 50.598 | 50.633 | 1,240 |
| 1,250 | 50.633 | 50.669 | 50.705 | 50.741 | 50.776 | 50.812 | 50.847 | 50.883 | 50.919 | 50.954 | 50.990 | 1,250 |
| 1,260 | 50.990 | 51.025 | 51.061 | 51.096 | 51.132 | 51.167 | 51.203 | 51.238 | 51.274 | 51.309 | 51.344 | 1,260 |
| 1,270 | 51.344 | 51.380 | 51.415 | 51.450 | 51.486 | 51.521 | 51.556 | 51.592 | 51.627 | 51.662 | 51.697 | 1,270 |
| 1,280 | 51.697 | 51.733 | 51.768 | 51.803 | 51.838 | 51.873 | 51.908 | 51.943 | 51.979 | 52.014 | 52.049 | 1,280 |
| 1,290 | 52.049 | 52.084 | 52.119 | 52.154 | 52.189 | 52.224 | 52.259 | 52.294 | 52.329 | 52.364 | 52.398 | 1,290 |
| 1,300 | 52.398 | 52.433 | 52.468 | 52.503 | 52.538 | 52.573 | 52.608 | 52.642 | 52.677 | 52.712 | 52.747 | 1,300 |
| 1,310 | 52.747 | 52.781 | 52.816 | 52.851 | 52.886 | 52.920 | 52.955 | 52.989 | 53.024 | 53.059 | 53.093 | 1,310 |
| 1,320 | 53.093 | 53.128 | 53.162 | 53.197 | 53.232 | 53.266 | 53.301 | 53.335 | 53.370 | 53.404 | 53.439 | 1,320 |
| 1,330 | 53.439 | 53.473 | 53.507 | 53.542 | 53.576 | 53.611 | 53.645 | 53.679 | 53.714 | 53.748 | 53.782 | 1,330 |
| 1,340 | 53.782 | 53.817 | 53.851 | 53.885 | 53.920 | 53.954 | 53.988 | 54.022 | 54.057 | 54.091 | 54.125 | 1,340 |
| 1,350 | 54.125 | 54.159 | 54.193 | 54.228 | 54.262 | 54.296 | 54.330 | 54.364 | 54.398 | 54.432 | 54.466 | 1,350 |
| 1,360 | 54.466 | 54.501 | 54.535 | 54.569 | 54.603 | 54.637 | 54.671 | 54.705 | 54.739 | 54.773 | 54.807 | 1,360 |
| 1,370 | 54.807 | 54.841 | 54.875 | | | | | | | | | 1,370 |

**TABLE A.4** Thermoelectric Voltages for Chromel–Constantan Thermocouples with the Reference Junction at 0°C (32°F)

| °C | 0 | 1 | 2 | 3 | 4 | 5 | 6 | 7 | 8 | 9 | 10 | °C |
|---|---|---|---|---|---|---|---|---|---|---|---|---|
| | | | | | Thermoelectric Voltage (absolute mV) | | | | | | | |
| −270 | −9.835 | | | | | | | | | | | −270 |
| −260 | −9.797 | −9.802 | −9.808 | −9.813 | −9.817 | −9.821 | −9.825 | −9.828 | −9.831 | −9.833 | −9.835 | −260 |
| −250 | −9.719 | −9.728 | −9.737 | −9.746 | −9.754 | −9.762 | −9.770 | −9.777 | −9.784 | −9.791 | −9.797 | −250 |
| −240 | −9.604 | −9.617 | −9.630 | −9.642 | −9.654 | −9.666 | −9.677 | −9.688 | −9.699 | −9.709 | −9.719 | −240 |
| −230 | −9.455 | −9.472 | −9.488 | −9.503 | −9.519 | −9.534 | −9.549 | −9.563 | −9.577 | −9.591 | −9.604 | −230 |
| −220 | −9.274 | −9.293 | −9.313 | −9.332 | −9.350 | −9.368 | −9.386 | −9.404 | −9.421 | −9.438 | −9.455 | −220 |
| −210 | −9.063 | −9.085 | −9.107 | −9.129 | −9.151 | −9.172 | −9.193 | −9.214 | −9.234 | −9.254 | −9.274 | −210 |
| −200 | −8.824 | −8.850 | −8.874 | −8.899 | −8.923 | −8.947 | −8.971 | −8.994 | −9.017 | −9.040 | −9.063 | −200 |
| −190 | −8.561 | −8.588 | −8.615 | −8.642 | −8.669 | −8.696 | −8.722 | −8.748 | −8.774 | −8.799 | −8.824 | −190 |
| −180 | −8.273 | −8.303 | −8.333 | −8.362 | −8.391 | −8.420 | −8.449 | −8.477 | −8.505 | −8.533 | −8.561 | −180 |
| −170 | −7.963 | −7.995 | −8.027 | −8.058 | −8.090 | −8.121 | −8.152 | −8.183 | −8.213 | −8.243 | −8.273 | −170 |
| −160 | −7.631 | −7.665 | −7.699 | −7.733 | −7.767 | −7.800 | −7.833 | −7.866 | −7.898 | −7.931 | −7.963 | −160 |
| −150 | −7.279 | −7.315 | −7.351 | −7.387 | −7.422 | −7.458 | −7.493 | −7.528 | −7.562 | −7.597 | −7.631 | −150 |
| −140 | −6.907 | −6.945 | −6.983 | −7.020 | −7.058 | −7.095 | −7.132 | −7.169 | −7.206 | −7.243 | −7.279 | −140 |
| −130 | −6.516 | −6.556 | −6.596 | −6.635 | −6.675 | −6.714 | −6.753 | −6.792 | −6.830 | −6.869 | −6.907 | −130 |
| −120 | −6.107 | −6.149 | −6.190 | −6.231 | −6.273 | −6.314 | −6.354 | −6.395 | −6.436 | −6.476 | −6.516 | −120 |
| −110 | −5.680 | −5.724 | −5.767 | −5.810 | −5.853 | −5.896 | −5.938 | −5.981 | −6.023 | −6.065 | −6.107 | −110 |
| −100 | −5.237 | −5.282 | −5.327 | −5.371 | −5.416 | −5.460 | −5.505 | −5.549 | −5.593 | −5.637 | −5.680 | −100 |
| −90 | −4.777 | −4.824 | −4.870 | −4.916 | −4.963 | −5.009 | −5.055 | −5.100 | −5.146 | −5.191 | −5.237 | −90 |
| −80 | −4.301 | −4.350 | −4.398 | −4.446 | −4.493 | −4.541 | −4.588 | −4.636 | −4.683 | −4.730 | −4.777 | −80 |
| −70 | −3.811 | −3.860 | −3.910 | −3.959 | −4.009 | −4.058 | −4.107 | −4.156 | −4.204 | −4.253 | −4.301 | −70 |
| −60 | −3.306 | −3.357 | −3.408 | −3.459 | −3.509 | −3.560 | −3.610 | −3.661 | −3.711 | −3.761 | −3.811 | −60 |
| −50 | −2.787 | −2.839 | −2.892 | −2.944 | −2.996 | −3.048 | −3.100 | −3.152 | −3.203 | −3.254 | −3.306 | −50 |
| −40 | −2.254 | −2.308 | −2.362 | −2.416 | −2.469 | −2.522 | −2.575 | −2.628 | −2.681 | −2.734 | −2.787 | −40 |
| −30 | −1.709 | −1.764 | −1.819 | −1.874 | −1.929 | −1.983 | −2.038 | −2.092 | −2.146 | −2.200 | −2.254 | −30 |
| −20 | −1.151 | −1.208 | −1.264 | −1.320 | −1.376 | −1.432 | −1.487 | −1.543 | −1.599 | −1.654 | −1.709 | −20 |
| −10 | −0.581 | −0.639 | −0.696 | −0.754 | −0.811 | −0.868 | −0.925 | −0.982 | −1.038 | −1.095 | −1.151 | −10 |
| 0 | 0.000 | −0.059 | −0.117 | −0.176 | −0.234 | −0.292 | −0.350 | −0.408 | −0.466 | −0.524 | −0.581 | 0 |
| 0 | 0.000 | 0.059 | 0.118 | 0.176 | 0.235 | 0.295 | 0.354 | 0.413 | 0.472 | 0.532 | 0.591 | 0 |
| 10 | 0.591 | 0.651 | 0.711 | 0.770 | 0.830 | 0.890 | 0.950 | 1.011 | 1.071 | 1.131 | 1.192 | 10 |
| 20 | 1.192 | 1.252 | 1.313 | 1.373 | 1.434 | 1.495 | 1.556 | 1.617 | 1.678 | 1.739 | 1.801 | 20 |
| 30 | 1.801 | 1.862 | 1.924 | 1.985 | 2.047 | 2.109 | 2.171 | 2.233 | 2.295 | 2.357 | 2.419 | 30 |
| 40 | 2.419 | 2.482 | 2.544 | 2.607 | 2.669 | 2.732 | 2.795 | 2.858 | 2.921 | 2.984 | 3.047 | 40 |
| 50 | 3.047 | 3.110 | 3.173 | 3.237 | 3.300 | 3.364 | 3.428 | 3.491 | 3.555 | 3.619 | 3.683 | 50 |
| 60 | 3.683 | 3.748 | 3.812 | 3.876 | 3.941 | 4.005 | 4.070 | 4.134 | 4.199 | 4.264 | 4.329 | 60 |
| 70 | 4.329 | 4.394 | 4.459 | 4.524 | 4.590 | 4.655 | 4.720 | 4.786 | 4.852 | 4.917 | 4.983 | 70 |
| 80 | 4.983 | 5.049 | 5.115 | 5.181 | 5.247 | 5.314 | 5.380 | 5.446 | 5.513 | 5.579 | 5.646 | 80 |
| 90 | 5.646 | 5.713 | 5.780 | 5.846 | 5.913 | 5.981 | 6.048 | 6.115 | 6.182 | 6.250 | 6.317 | 90 |
| 100 | 6.317 | 6.385 | 6.452 | 6.520 | 6.588 | 6.656 | 6.724 | 6.792 | 6.860 | 6.928 | 6.996 | 100 |
| 110 | 6.996 | 7.064 | 7.133 | 7.201 | 7.270 | 7.339 | 7.407 | 7.476 | 7.545 | 7.614 | 7.683 | 110 |
| 120 | 7.683 | 7.752 | 7.821 | 7.890 | 7.960 | 8.029 | 8.099 | 8.168 | 8.238 | 8.307 | 8.377 | 120 |
| 130 | 8.377 | 8.447 | 8.517 | 8.587 | 8.657 | 8.727 | 8.797 | 8.867 | 8.938 | 9.008 | 9.078 | 130 |
| 140 | 9.078 | 9.149 | 9.220 | 9.290 | 9.361 | 9.432 | 9.503 | 9.573 | 9.644 | 9.715 | 9.787 | 140 |
| 150 | 9.787 | 9.858 | 9.929 | 10.000 | 10.072 | 10.143 | 10.215 | 10.286 | 10.358 | 10.429 | 10.501 | 150 |
| 160 | 10.501 | 10.573 | 10.645 | 10.717 | 10.789 | 10.861 | 10.933 | 11.005 | 11.077 | 11.150 | 11.222 | 160 |
| 170 | 11.222 | 11.294 | 11.367 | 11.439 | 11.512 | 11.585 | 11.657 | 11.730 | 11.803 | 11.876 | 11.949 | 170 |
| 180 | 11.949 | 12.022 | 12.095 | 12.168 | 12.241 | 12.314 | 12.387 | 12.461 | 12.534 | 12.608 | 12.681 | 180 |
| 190 | 12.681 | 12.755 | 12.828 | 12.902 | 12.975 | 13.049 | 13.123 | 13.197 | 13.271 | 13.345 | 13.419 | 190 |
| 200 | 13.419 | 13.493 | 13.567 | 13.641 | 13.715 | 13.789 | 13.864 | 13.938 | 14.012 | 14.087 | 14.161 | 200 |
| 210 | 14.161 | 14.236 | 14.310 | 14.385 | 14.460 | 14.534 | 14.609 | 14.684 | 14.759 | 14.834 | 14.909 | 210 |
| 220 | 14.909 | 14.984 | 15.059 | 15.134 | 15.209 | 15.284 | 15.359 | 15.435 | 15.510 | 15.585 | 15.661 | 220 |
| 230 | 15.661 | 15.736 | 15.812 | 15.887 | 15.963 | 16.038 | 16.114 | 16.190 | 16.266 | 16.341 | 16.417 | 230 |
| 240 | 16.417 | 16.493 | 16.569 | 16.645 | 16.721 | 16.797 | 16.873 | 16.949 | 17.025 | 17.101 | 17.178 | 240 |
| 250 | 17.178 | 17.254 | 17.330 | 17.406 | 17.483 | 17.559 | 17.636 | 17.712 | 17.789 | 17.865 | 17.942 | 250 |
| 260 | 17.942 | 18.018 | 18.095 | 18.172 | 18.248 | 18.325 | 18.402 | 18.479 | 18.556 | 18.633 | 18.710 | 260 |
| 270 | 18.710 | 18.787 | 18.864 | 18.941 | 19.018 | 19.095 | 19.172 | 19.249 | 19.326 | 19.404 | 19.481 | 270 |
| 280 | 19.481 | 19.558 | 19.636 | 19.713 | 19.790 | 19.868 | 19.945 | 20.023 | 20.100 | 20.178 | 20.256 | 280 |
| 290 | 20.256 | 20.333 | 20.411 | 20.488 | 20.566 | 20.644 | 20.722 | 20.800 | 20.877 | 20.955 | 21.033 | 290 |

**TABLE A.4** *(Continued)*

| °C | 0 | 1 | 2 | 3 | 4 | 5 | 6 | 7 | 8 | 9 | 10 | °C |
|----|---|---|---|---|---|---|---|---|---|---|----|----|
| | | | | | Thermoelectric Voltage (absolute mV) | | | | | | | |
| 300 | 21.033 | 21.111 | 21.189 | 21.267 | 21.345 | 21.423 | 21.501 | 21.579 | 21.657 | 21.735 | 21.814 | 300 |
| 310 | 21.814 | 21.892 | 21.970 | 22.048 | 22.127 | 22.205 | 22.283 | 22.362 | 22.440 | 22.518 | 22.597 | 310 |
| 320 | 22.597 | 22.675 | 22.754 | 22.832 | 22.911 | 22.989 | 23.068 | 23.147 | 23.225 | 23.304 | 23.383 | 320 |
| 330 | 23.383 | 23.461 | 23.540 | 23.619 | 23.698 | 23.777 | 23.855 | 23.934 | 24.013 | 24.092 | 24.171 | 330 |
| 340 | 24.171 | 24.250 | 24.329 | 24.408 | 24.487 | 24.566 | 24.645 | 24.724 | 24.803 | 24.882 | 24.961 | 340 |
| 350 | 24.961 | 25.041 | 25.120 | 25.199 | 25.278 | 25.357 | 25.437 | 25.516 | 25.595 | 25.675 | 25.754 | 350 |
| 360 | 25.754 | 25.833 | 25.913 | 25.992 | 26.072 | 26.151 | 26.230 | 26.310 | 26.389 | 26.469 | 26.549 | 360 |
| 370 | 26.549 | 26.628 | 26.708 | 26.787 | 26.867 | 26.947 | 27.026 | 27.106 | 27.186 | 27.265 | 27.345 | 370 |
| 380 | 27.345 | 27.425 | 27.504 | 27.584 | 27.664 | 27.744 | 27.824 | 27.903 | 27.983 | 28.063 | 28.143 | 380 |
| 390 | 28.143 | 28.223 | 28.303 | 28.383 | 28.463 | 28.543 | 28.623 | 28.703 | 28.783 | 28.863 | 28.943 | 390 |
| 400 | 28.943 | 29.023 | 29.103 | 29.183 | 29.263 | 29.343 | 29.423 | 29.503 | 29.584 | 29.664 | 29.744 | 400 |
| 410 | 29.744 | 29.824 | 29.904 | 29.984 | 30.065 | 30.145 | 30.225 | 30.305 | 30.386 | 30.466 | 30.546 | 410 |
| 420 | 30.546 | 30.627 | 30.707 | 30.787 | 30.868 | 30.948 | 31.028 | 31.109 | 31.189 | 31.270 | 31.350 | 420 |
| 430 | 31.350 | 31.430 | 31.511 | 31.591 | 31.672 | 31.752 | 31.833 | 31.913 | 31.994 | 32.074 | 32.155 | 430 |
| 440 | 32.155 | 32.235 | 32.316 | 32.396 | 32.477 | 32.557 | 32.638 | 32.719 | 32.799 | 32.880 | 32.960 | 440 |
| 450 | 32.960 | 33.041 | 33.122 | 33.202 | 33.283 | 33.364 | 33.444 | 33.525 | 33.605 | 33.686 | 33.767 | 450 |
| 460 | 33.767 | 33.848 | 33.928 | 34.009 | 34.090 | 34.170 | 34.251 | 34.332 | 34.413 | 34.493 | 34.574 | 460 |
| 470 | 34.574 | 34.655 | 34.736 | 34.816 | 34.897 | 34.978 | 35.059 | 35.140 | 35.220 | 35.301 | 35.382 | 470 |
| 480 | 35.382 | 35.463 | 35.544 | 35.624 | 35.705 | 35.786 | 35.867 | 35.948 | 36.029 | 36.109 | 36.190 | 480 |
| 490 | 36.190 | 36.271 | 36.352 | 36.433 | 36.514 | 36.595 | 36.675 | 36.756 | 36.837 | 36.918 | 36.999 | 490 |
| 500 | 36.999 | 37.080 | 37.161 | 37.242 | 37.323 | 37.403 | 37.484 | 37.565 | 37.646 | 37.727 | 37.808 | 500 |
| 510 | 37.808 | 37.889 | 37.970 | 38.051 | 38.132 | 38.213 | 38.293 | 38.374 | 38.455 | 38.536 | 38.617 | 510 |
| 520 | 38.617 | 38.698 | 38.779 | 38.860 | 38.941 | 39.022 | 39.103 | 39.184 | 39.264 | 39.345 | 39.426 | 520 |
| 530 | 39.426 | 39.507 | 39.588 | 39.669 | 39.750 | 39.831 | 39.912 | 39.993 | 40.074 | 40.155 | 40.236 | 530 |
| 540 | 40.236 | 40.316 | 40.397 | 40.478 | 40.559 | 40.640 | 40.721 | 40.802 | 40.883 | 40.964 | 41.045 | 540 |
| 550 | 41.045 | 41.125 | 41.206 | 41.287 | 41.368 | 41.449 | 41.530 | 41.611 | 41.692 | 41.773 | 41.853 | 550 |
| 560 | 41.853 | 41.934 | 42.015 | 42.096 | 42.177 | 42.258 | 42.339 | 42.419 | 42.500 | 42.581 | 42.662 | 560 |
| 570 | 42.662 | 42.743 | 42.824 | 42.904 | 42.985 | 43.066 | 43.147 | 43.228 | 43.308 | 43.389 | 43.470 | 570 |
| 580 | 43.470 | 43.551 | 43.632 | 43.712 | 43.793 | 43.874 | 43.955 | 44.035 | 44.116 | 44.197 | 44.278 | 580 |
| 590 | 44.278 | 44.358 | 44.439 | 44.520 | 44.601 | 44.681 | 44.762 | 44.843 | 44.923 | 45.004 | 45.085 | 590 |
| 600 | 45.085 | 45.165 | 45.246 | 45.327 | 45.407 | 45.488 | 45.569 | 45.649 | 45.730 | 45.811 | 45.891 | 600 |
| 610 | 45.891 | 45.972 | 46.052 | 46.133 | 46.213 | 46.294 | 46.375 | 46.455 | 46.536 | 46.616 | 46.697 | 610 |
| 620 | 46.697 | 46.777 | 46.858 | 46.938 | 47.019 | 47.099 | 47.180 | 47.260 | 47.341 | 47.421 | 47.502 | 620 |
| 630 | 47.502 | 47.582 | 47.663 | 47.743 | 47.824 | 47.904 | 47.984 | 48.065 | 48.145 | 48.226 | 48.306 | 630 |
| 640 | 48.306 | 48.386 | 48.467 | 48.547 | 48.627 | 48.708 | 48.788 | 48.868 | 48.949 | 49.029 | 49.109 | 640 |
| 650 | 49.109 | 49.189 | 49.270 | 49.350 | 49.430 | 49.510 | 49.591 | 49.671 | 49.751 | 49.831 | 49.911 | 650 |
| 660 | 49.911 | 49.992 | 50.072 | 50.152 | 50.232 | 50.312 | 50.392 | 50.472 | 50.553 | 50.633 | 50.713 | 660 |
| 670 | 50.713 | 50.793 | 50.873 | 50.953 | 51.033 | 51.113 | 51.193 | 51.273 | 51.353 | 51.433 | 51.513 | 670 |
| 680 | 51.513 | 51.593 | 51.673 | 51.753 | 51.833 | 51.913 | 51.993 | 52.073 | 52.152 | 52.232 | 52.312 | 680 |
| 690 | 52.312 | 52.392 | 52.472 | 52.552 | 52.632 | 52.711 | 52.791 | 52.871 | 52.951 | 53.031 | 53.110 | 690 |
| 700 | 53.110 | 53.190 | 53.270 | 53.350 | 53.429 | 53.509 | 53.589 | 53.668 | 53.748 | 53.828 | 53.907 | 700 |
| 710 | 53.907 | 53.987 | 54.066 | 54.146 | 54.226 | 54.305 | 54.385 | 54.464 | 54.544 | 54.623 | 54.703 | 710 |
| 720 | 54.703 | 54.782 | 54.862 | 54.941 | 55.021 | 55.100 | 55.180 | 55.259 | 55.339 | 55.418 | 55.498 | 720 |
| 730 | 55.498 | 55.577 | 55.656 | 55.736 | 55.815 | 55.894 | 55.974 | 56.053 | 56.132 | 56.212 | 56.291 | 730 |
| 740 | 56.291 | 56.370 | 56.449 | 56.529 | 56.608 | 56.687 | 56.766 | 56.845 | 56.924 | 57.004 | 57.083 | 740 |
| 750 | 57.083 | 57.162 | 57.241 | 57.320 | 57.399 | 57.478 | 57.557 | 57.636 | 57.715 | 57.794 | 57.873 | 750 |
| 760 | 57.873 | 57.952 | 58.031 | 58.110 | 58.189 | 58.268 | 58.347 | 58.426 | 58.505 | 58.584 | 58.663 | 760 |
| 770 | 58.663 | 58.742 | 58.820 | 58.899 | 58.978 | 59.057 | 59.136 | 59.214 | 59.293 | 59.372 | 59.451 | 770 |
| 780 | 59.451 | 59.529 | 59.608 | 59.687 | 59.765 | 59.844 | 59.923 | 60.001 | 60.080 | 60.159 | 60.237 | 780 |
| 790 | 60.237 | 60.316 | 60.394 | 60.473 | 60.551 | 60.630 | 60.708 | 60.787 | 60.865 | 60.944 | 61.022 | 790 |
| 800 | 61.022 | 61.101 | 61.179 | 61.258 | 61.336 | 61.414 | 61.493 | 61.571 | 61.649 | 61.728 | 61.806 | 800 |
| 810 | 61.806 | 61.884 | 61.962 | 62.041 | 62.119 | 62.197 | 62.275 | 62.353 | 62.432 | 62.510 | 62.588 | 810 |
| 820 | 62.588 | 62.666 | 62.744 | 62.822 | 62.900 | 62.978 | 63.056 | 63.134 | 63.212 | 63.290 | 63.368 | 820 |
| 830 | 63.368 | 63.446 | 63.524 | 63.602 | 63.680 | 63.758 | 63.836 | 63.914 | 63.992 | 64.069 | 64.147 | 830 |
| 840 | 64.147 | 64.225 | 64.303 | 64.380 | 64.458 | 64.536 | 64.614 | 64.691 | 64.769 | 64.847 | 64.924 | 840 |
| 850 | 64.924 | 65.002 | 65.080 | 65.157 | 65.235 | 65.312 | 65.390 | 65.467 | 65.545 | 65.622 | 65.700 | 850 |
| 860 | 65.700 | 65.777 | 65.855 | 65.932 | 66.009 | 66.087 | 66.164 | 66.241 | 66.319 | 66.396 | 66.473 | 860 |
| 870 | 66.473 | 66.551 | 66.628 | 66.705 | 66.782 | 66.859 | 66.937 | 67.014 | 67.091 | 67.168 | 67.245 | 870 |
| 880 | 67.245 | 67.322 | 67.399 | 67.476 | 67.553 | 67.630 | 67.707 | 67.784 | 67.861 | 67.938 | 68.015 | 880 |
| 890 | 68.015 | 68.092 | 68.169 | 68.246 | 68.323 | 68.399 | 68.476 | 68.553 | 68.630 | 68.706 | 68.783 | 890 |

## TABLE A.4 (Continued)

| °C | 0 | 1 | 2 | 3 | 4 | 5 | 6 | 7 | 8 | 9 | 10 | °C |
|---|---|---|---|---|---|---|---|---|---|---|---|---|
| | | | | | Thermoelectric Voltage (absolute mV) | | | | | | | |
| 900 | 68.783 | 68.860 | 68.936 | 69.013 | 69.090 | 69.166 | 69.243 | 69.320 | 69.396 | 69.473 | 69.549 | 900 |
| 910 | 69.549 | 69.626 | 69.702 | 69.779 | 69.855 | 69.931 | 70.008 | 70.084 | 70.161 | 70.237 | 70.313 | 910 |
| 920 | 70.313 | 70.390 | 70.466 | 70.542 | 70.618 | 70.694 | 70.771 | 70.847 | 70.923 | 70.999 | 71.075 | 920 |
| 930 | 71.075 | 71.151 | 71.227 | 71.304 | 71.380 | 71.456 | 71.532 | 71.608 | 71.683 | 71.759 | 71.835 | 930 |
| 940 | 71.835 | 71.911 | 71.987 | 72.063 | 72.139 | 72.215 | 72.290 | 72.366 | 72.442 | 72.518 | 72.593 | 940 |
| 950 | 72.593 | 72.669 | 72.745 | 72.820 | 72.896 | 72.972 | 73.047 | 73.123 | 73.199 | 73.274 | 73.350 | 950 |
| 960 | 73.350 | 73.425 | 73.501 | 73.576 | 73.652 | 73.727 | 73.802 | 73.878 | 73.953 | 74.029 | 74.104 | 960 |
| 970 | 74.104 | 74.179 | 74.255 | 74.330 | 74.405 | 74.480 | 74.556 | 74.631 | 74.706 | 74.781 | 74.857 | 970 |
| 980 | 74.857 | 74.932 | 75.007 | 75.082 | 75.157 | 75.232 | 75.307 | 75.382 | 75.458 | 75.533 | 75.608 | 980 |
| 990 | 75.608 | 75.683 | 75.758 | 75.833 | 75.908 | 75.983 | 76.058 | 76.133 | 76.208 | 76.283 | 76.358 | 990 |
| 1,000 | 76.358 | | | | | | | | | | | 1,000 |

## TABLE A.5 Thermoelectric Voltages for Copper–Constantan Thermocouples with the Reference Junction at 0°C (32°F)

| °C | 0 | 1 | 2 | 3 | 4 | 5 | 6 | 7 | 8 | 9 | 10 | °C |
|---|---|---|---|---|---|---|---|---|---|---|---|---|
| | | | | | Thermoelectric Voltage (absolute mV) | | | | | | | |
| −270 | −6.258 | | | | | | | | | | | −270 |
| −260 | −6.232 | −6.236 | −6.239 | −6.242 | −6.245 | −6.248 | −6.251 | −6.253 | −6.255 | −6.256 | −6.258 | −260 |
| −250 | −6.181 | −6.187 | −6.193 | −6.198 | −6.204 | −6.209 | −6.214 | −6.219 | −6.224 | −6.228 | −6.232 | −250 |
| −240 | −6.105 | −6.114 | −6.122 | −6.130 | −6.138 | −6.146 | −6.153 | −6.160 | −6.167 | −6.174 | −6.181 | −240 |
| −230 | −6.007 | −6.018 | −6.028 | −6.039 | −6.049 | −6.059 | −6.068 | −6.078 | −6.087 | −6.096 | −6.105 | −230 |
| −220 | −5.889 | −5.901 | −5.914 | −5.926 | −5.938 | −5.950 | −5.962 | −5.973 | −5.985 | −5.996 | −6.007 | −220 |
| −210 | −5.753 | −5.767 | −5.782 | −5.795 | −5.809 | −5.823 | −5.836 | −5.850 | −5.863 | −5.876 | −5.889 | −210 |
| −200 | −5.603 | −5.619 | −5.634 | −5.650 | −5.665 | −5.680 | −5.695 | −5.710 | −5.724 | −5.739 | −5.753 | −200 |
| −190 | −5.439 | −5.456 | −5.473 | −5.489 | −5.506 | −5.522 | −5.539 | −5.555 | −5.571 | −5.587 | −5.603 | −190 |
| −180 | −5.261 | −5.279 | −5.297 | −5.315 | −5.333 | −5.351 | −5.369 | −5.387 | −5.404 | −5.421 | −5.439 | −180 |
| −170 | −5.069 | −5.089 | −5.109 | −5.128 | −5.147 | −5.167 | −5.186 | −5.205 | −5.223 | −5.242 | −5.261 | −170 |
| −160 | −4.865 | −4.886 | −4.907 | −4.928 | −4.948 | −4.969 | −4.989 | −5.010 | −5.030 | −5.050 | −5.069 | −160 |
| −150 | −4.648 | −4.670 | −4.693 | −4.715 | −4.737 | −4.758 | −4.780 | −4.801 | −4.823 | −4.844 | −4.865 | −150 |
| −140 | −4.419 | −4.442 | −4.466 | −4.489 | −4.512 | −4.535 | −4.558 | −4.581 | −4.603 | −4.626 | −4.648 | −140 |
| −130 | −4.177 | −4.202 | −4.226 | −4.251 | −4.275 | −4.299 | −4.323 | −4.347 | −4.371 | −4.395 | −4.419 | −130 |
| −120 | −3.923 | −3.949 | −3.974 | −4.000 | −4.026 | −4.051 | −4.077 | −4.102 | −4.127 | −4.152 | −4.177 | −120 |
| −110 | −3.656 | −3.684 | −3.711 | −3.737 | −3.764 | −3.791 | −3.818 | −3.844 | −3.870 | −3.897 | −3.923 | −110 |
| −100 | −3.378 | 3.407 | −3.435 | −3.463 | −3.491 | −3.519 | −3.547 | −3.574 | −3.602 | −3.629 | −3.656 | −100 |
| −90 | −3.089 | −3.118 | −3.147 | −3.177 | −3.206 | −3.235 | −3.264 | −3.293 | −3.321 | −3.350 | −3.378 | −90 |
| −80 | −2.788 | −2.818 | −2.849 | −2.879 | −2.909 | −2.939 | −2.970 | −2.999 | −3.029 | −3.059 | −3.089 | −80 |
| −70 | −2.475 | −2.507 | −2.539 | −2.570 | −2.602 | −2.633 | −2.664 | −2.695 | −2.726 | −2.757 | −2.788 | −70 |
| −60 | −2.152 | −2.185 | −2.218 | −2.250 | −2.283 | −2.315 | −2.348 | −2.380 | −2.412 | −2.444 | −2.475 | −60 |
| −50 | −1.819 | −1.853 | −1.886 | −1.920 | −1.953 | −1.987 | −2.020 | −2.053 | −2.087 | −2.120 | −2.152 | −50 |
| −40 | −1.475 | −1.510 | −1.544 | −1.579 | −1.614 | −1.648 | −1.682 | −1.717 | −1.751 | −1.785 | −1.819 | −40 |
| −30 | −1.121 | −1.157 | −1.192 | −1.228 | −1.263 | −1.299 | −1.334 | −1.370 | −1.405 | −1.440 | −1.475 | −30 |
| −20 | −0.757 | −0.794 | −0.830 | −0.867 | −0.903 | −0.940 | −0.976 | −1.013 | −1.049 | −1.085. | −1.121 | −20 |
| −10 | −0.383 | −0.421 | −0.458 | −0.496 | −0.534 | −0.571 | −0.608 | −0.646 | −0.683 | −0.720 | −0.757 | −10 |
| 0 | 0.000 | −0.039 | −0.077 | −0.116 | −0.154 | −0.193 | −0.231 | −0.269 | −0.307 | −0.345 | −0.383 | 0 |
| 0 | 0.000 | 0.039 | 0.078 | 0.117 | 0.156 | 0.195 | 0.234 | 0.273 | 0.312 | 0.351 | 0.391 | 0 |
| 10 | 0.391 | 0.430 | 0.470 | 0.510 | 0.549 | 0.589 | 0.629 | 0.669 | 0.709 | 0.749 | 0.789 | 10 |
| 20 | 0.789 | 0.830 | 0.870 | 0.911 | 0.951 | 0.992 | 1.032 | 1.073 | 1.114 | 1.155 | 1.196 | 20 |
| 30 | 1.196 | 1.237 | 1.279 | 1.320 | 1.361 | 1.403 | 1.444 | 1.486 | 1.528 | 1.569 | 1.611 | 30 |
| 40 | 1.611 | 1.653 | 1.695 | 1.738 | 1.780 | 1.822 | 1.865 | 1.907 | 1.950 | 1.992 | 2.035 | 40 |
| 50 | 2.035 | 2.078 | 2.121 | 2.164 | 2.207 | 2.250 | 2.294 | 2.337 | 2.380 | 2.424 | 2.467 | 50 |
| 60 | 2.467 | 2.511 | 2.555 | 2.599 | 2.643 | 2.687 | 2.731 | 2.775 | 2.819 | 2.864 | 2.908 | 60 |
| 70 | 2.908 | 2.953 | 2.997 | 3.042 | 3.087 | 3.131 | 3.176 | 3.221 | 3.266 | 3.312 | 3.357 | 70 |
| 80 | 3.357 | 3.402 | 3.447 | 3.493 | 3.538 | 3.584 | 3.630 | 3.676 | 3.721 | 3.767 | 3.813 | 80 |
| 90 | 3.813 | 3.859 | 3.906 | 3.952 | 3.998 | 4.044 | 4.091 | 4.137 | 4.184 | 4.231 | 4.277 | 90 |

## TABLE A.5 (*Continued*)

| °C | 0 | 1 | 2 | 3 | 4 | 5 | 6 | 7 | 8 | 9 | 10 | °C |
|----|---|---|---|---|---|---|---|---|---|---|----|----|
| | | | | | Thermoelectric Voltage (absolute mV) | | | | | | | |
| 100 | 4.277 | 4.324 | 4.371 | 4.418 | 4.465 | 4.512 | 4.559 | 4.607 | 4.654 | 4.701 | 4.749 | 100 |
| 110 | 4.749 | 4.796 | 4.844 | 4.891 | 4.939 | 4.987 | 5.035 | 5.083 | 5.131 | 5.179 | 5.227 | 110 |
| 120 | 5.227 | 5.275 | 5.324 | 5.372 | 5.420 | 5.469 | 5.517 | 5.566 | 5.615 | 5.663 | 5.712 | 120 |
| 130 | 5.712 | 5.761 | 5.810 | 5.859 | 5.908 | 5.957 | 6.007 | 6.056 | 6.105 | 6.155 | 6.204 | 130 |
| 140 | 6.204 | 6.254 | 6.303 | 6.353 | 6.403 | 6.452 | 6.502 | 6.552 | 6.602 | 6.652 | 6.702 | 140 |
| 150 | 6.702 | 6.753 | 6.803 | 6.853 | 6.903 | 6.954 | 7.004 | 7.055 | 7.106 | 7.156 | 7.207 | 150 |
| 160 | 7.207 | 7.258 | 7.309 | 7.360 | 7.411 | 7.462 | 7.513 | 7.564 | 7.615 | 7.666 | 7.718 | 160 |
| 170 | 7.718 | 7.769 | 7.821 | 7.872 | 7.924 | 7.975 | 8.027 | 8.079 | 8.131 | 8.183 | 8.235 | 170 |
| 180 | 8.235 | 8.287 | 8.339 | 8.391 | 8.443 | 8.495 | 8.548 | 8.600 | 8.652 | 8.705 | 8.757 | 180 |
| 190 | 8.757 | 8.810 | 8.863 | 8.915 | 8.968 | 9.021 | 9.074 | 9.127 | 9.180 | 9.233 | 9.286 | 190 |
| 200 | 9.286 | 9.339 | 9.392 | 9.446 | 9.499 | 9.553 | 9.606 | 9.659 | 9.713 | 9.767 | 9.820 | 200 |
| 210 | 9.820 | 9.874 | 9.928 | 9.982 | 10.036 | 10.090 | 10.144 | 10.198 | 10.252 | 10.306 | 10.360 | 210 |
| 220 | 10.360 | 10.414 | 10.469 | 10.523 | 10.578 | 10.632 | 10.687 | 10.741 | 10.796 | 10.851 | 10.905 | 220 |
| 230 | 10.905 | 10.960 | 11.015 | 11.070 | 11.125 | 11.180 | 11.235 | 11.290 | 11.345 | 11.401 | 11.456 | 230 |
| 240 | 11.456 | 11.511 | 11.566 | 11.622 | 11.677 | 11.733 | 11.788 | 11.844 | 11.900 | 11.956 | 12.011 | 240 |
| 250 | 12.011 | 12.067 | 12.123 | 12.179 | 12.235 | 12.291 | 12.347 | 12.403 | 12.459 | 12.515 | 12.572 | 250 |
| 260 | 12.572 | 12.628 | 12.684 | 12.741 | 12.797 | 12.854 | 12.910 | 12.967 | 13.024 | 13.080 | 13.137 | 260 |
| 270 | 13.137 | 13.194 | 13.251 | 13.307 | 13.364 | 13.421 | 13.478 | 13.535 | 13.592 | 13.650 | 13.707 | 270 |
| 280 | 13.707 | 13.764 | 13.821 | 13.879 | 13.936 | 13.993 | 14.051 | 14.108 | 14.166 | 14.223 | 14.281 | 280 |
| 290 | 14.281 | 14.339 | 14.396 | 14.454 | 14.512 | 14.570 | 14.628 | 14.686 | 14.744 | 14.802 | 14.860 | 290 |
| 300 | 14.860 | 14.918 | 14.976 | 15.034 | 15.092 | 15.151 | 15.209 | 15.267 | 15.326 | 15.384 | 15.443 | 300 |
| 310 | 15.443 | 15.501 | 15.560 | 15.619 | 15.677 | 15.736 | 15.795 | 15.853 | 15.912 | 15.971 | 16.030 | 310 |
| 320 | 16.030 | 16.089 | 16.148 | 16.207 | 16.266 | 16.325 | 16.384 | 16.444 | 16.503 | 16.562 | 16.621 | 320 |
| 330 | 16.621 | 16.681 | 16.740 | 16.800 | 16.859 | 16.919 | 16.978 | 17.038 | 17.097 | 17.157 | 17.217 | 330 |
| 340 | 17.217 | 17.277 | 17.336 | 17.396 | 17.456 | 17.516 | 17.576 | 17.636 | 17.696 | 17.756 | 17.816 | 340 |
| 350 | 17.816 | 17.877 | 17.937 | 17.997 | 18.057 | 18.118 | 18.178 | 18.238 | 18.299 | 18.359 | 18.420 | 350 |
| 360 | 18.420 | 18.480 | 18.541 | 18.602 | 18.662 | 18.723 | 18.784 | 18.845 | 18.905 | 18.966 | 19.027 | 360 |
| 370 | 19.027 | 19.088 | 19.149 | 19.210 | 19.271 | 19.332 | 19.393 | 19.455 | 19.516 | 19.577 | 19.638 | 370 |
| 380 | 19.638 | 19.699 | 19.761 | 19.822 | 19.883 | 19.945 | 20.006 | 20.068 | 20.129 | 20.191 | 20.252 | 380 |
| 390 | 20.252 | 20.314 | 20.376 | 20.437 | 20.499 | 20.560 | 20.622 | 20.684 | 20.746 | 20.807 | 20.869 | 390 |
| 400 | 20.869 | | | | | | | | | | | 400 |

# INDEX